Intelligent Processing on Image and Optical Information

Intelligent Processing on Image and Optical Information

Editor

Seokwon Yeom

MDPI • Basel • Beijing • Wuhan • Barcelona • Belgrade • Manchester • Tokyo • Cluj • Tianjin

Editor
Seokwon Yeom
Daegu University
Korea

Editorial Office
MDPI
St. Alban-Anlage 66
4052 Basel, Switzerland

This is a reprint of articles from the Special Issue published online in the open access journal *Applied Sciences* (ISSN 2076-3417) (available at: https://www.mdpi.com/journal/applsci/special_issues/image_and_optical_information).

For citation purposes, cite each article independently as indicated on the article page online and as indicated below:

LastName, A.A.; LastName, B.B.; LastName, C.C. Article Title. *Journal Name* **Year**, *Article Number*, Page Range.

ISBN 978-3-03936-944-7 (Hbk)
ISBN 978-3-03936-945-4 (PDF)

Contents

About the Editor . . . . **vii**

Seokwon Yeom
Special Issue on Intelligent Processing on Image and Optical Information
Reprinted from: *Appl. Sci.* **2020**, *10*, 3911, doi:10.3390/app10113911 . . . . **1**

Wang Xin, Tang Can, Wang Wei and Li Ji
Change Detection of Water Resources via Remote Sensing: An L-V-NSCT Approach
Reprinted from: *Appl. Sci.* **2019**, *9*, 1223, doi:10.3390/app9061223 . . . . **5**

Mingwei Wang and Lang Gao, Xiaohui Huang and Ying Jiang and Xianjun Gao
A Texture Classification Approach Based on the Integrated Optimization for Parameters and
Features of Gabor Filter via Hybrid Ant Lion Optimizer
Reprinted from: *Appl. Sci.* **2019**, *9*, 2173, doi:10.3390/app9112173 . . . . **17**

Wahyu Rahmaniar and Wen-June Wang
Real-Time Automated Segmentation and Classification of Calcaneal Fractures in CT Images
Reprinted from: *Appl. Sci.* **2019**, *9*, 3011, doi:10.3390/app9153011 . . . . **31**

Shang Shang, Ling Long, Sijie Lin and Fengyu Cong
Automatic Zebrafish Egg Phenotype Recognition from Bright-Field Microscopic Images Using
Deep Convolutional Neural Network
Reprinted from: *Appl. Sci.* **2019**, *9*, 3362, doi:10.3390/app9163362 - **49**

Shang Shang, Sijie Lin and Fengyu Cong
Zebrafish Larvae Phenotype Classification from Bright-field Microscopic Images Using a
Two-Tier Deep-Learning Pipeline
Reprinted from: *Appl. Sci.* **2020**, *10*, 1247, doi:10.3390/app10041247 . . . . **61**

Jeong gi Kwak and Hanseok Ko
Unsupervised Generation and Synthesis of Facial Images via an Auto-Encoder-Based Deep
Generative Adversarial Network
Reprinted from: *Appl. Sci.* **2020**, *10*, 1995, doi:10.3390/app10061995 . . . . **73**

Yuriy Vashpanov, Gwanghee Heo, Yongsuk Kim, Tetiana Venkel and Jung-Young Son
Detecting Green Mold Pathogens on Lemons Using Hyperspectral Images
Reprinted from: *Appl. Sci.* **2020**, *10*, 1209, doi:10.3390/app10041209 . . . . **85**

Wenhui Hou, Dashan Zhang, Ye Wei, Jie Guo and Xiaolong Zhang
Review on Computer Aided Weld Defect Detection from Radiography Images
Reprinted from: *Appl. Sci.* **2020**, *10*, 1878, doi:10.3390/app10051878 . . . . **101**

Xiaoming Liu, Ke Xu, Peng Zhou and Huajie Liu
Feature Extraction with Discrete Non-Separable Shearlet Transform and Its Application to
Surface Inspection of Continuous Casting Slabs
Reprinted from: *Appl. Sci.* **2019**, *9*, 4668, doi:10.3390/app9214668 . . . . **117**

Wending Liu, Hanxing Liu, Yuan Wang, Xiaorui Zheng and Junguo Zhang
A Novel Extraction Method for Wildlife Monitoring Images with Wireless Multimedia Sensor
Networks (WMSNs)
Reprinted from: *Appl. Sci.* **2019**, *9*, 2276, doi:10.3390/app9112276 . . . . **131**

Hanzhang Xue, Hao Fu and Bin Dai
IMU-Aided High-Frequency Lidar Odometry for Autonomous Driving
Reprinted from: *Appl. Sci.* **2019**, *9*, 1506, doi:10.3390/app9071506 . . . . **147**

Lino Antoni Giefer, Michael Lütjen, Ann-Kathrin Rohde and Michael Freitag
Determination of the Optimal State of Dough Fermentation in Bread Production by Using
Optical Sensors and Deep Learning
Reprinted from: *Appl. Sci.* **2019**, *9*, 4266, doi:10.3390/app9204266 . . . . **167**

Wei Li, Mingli Dong, Naiguang Lu, Xiaoping Lou and Wanyong Zhou
Multi-Sensor Face Registration Based on Global and Local Structures
Reprinted from: *Appl. Sci.* **2019**, *9*, 4623, doi:10.3390/app9214623 . . . . **183**

Bo-Lin Jian, Wen-Lin Chu, Yu-Chung Li and Her-Terng Yau
Multifocus Image Fusion Using a Sparse and Low-Rank Matrix Decomposition for Aviator's
Night Vision Goggle
Reprinted from: *Appl. Sci.* **2020**, *10*, 2178, doi:10.3390/app10062178 . . . . **199**

Hsiang-Cheh Huang, Po-Liang Chen and Feng-Cheng Chang
Error Resilience for Block Compressed Sensing with Multiple-Channel Transmission
Reprinted from: *Appl. Sci.* **2020**, *10*, 161, doi:10.3390/app10010161 . . . . **219**

Rafał Zdunek and Tomasz Sadowski
Image Completion with Hybrid Interpolation in Tensor Representation
Reprinted from: *Appl. Sci.* **2020**, *10*, 797, doi:10.3390/app10030797 . . . . **237**

Chang Ma, Zhoumo Zeng, Hui Zhang and Xiaobo Rui
A Correction Method for Heat Wave Distortion in Digital Image Correlation Measurements
Based on Background-Oriented Schlieren
Reprinted from: *Appl. Sci.* **2019**, *9*, 3851, doi:10.3390/app9183851 . . . . **253**

Dokkyun Yi, Jaehyun Ahn and Sangmin Ji
An Effective Optimization Method for Machine Learning Based on ADAM
Reprinted from: *Appl. Sci.* **2020**, *10*, 1073, doi:10.3390/app10031073 . . . . **271**

Qi Li, Shihong Yue, Yaru Wang, Mingliang Ding, Jia Li and Zeying Wang
Boundary Matching and Interior Connectivity-Based Cluster Validity Anlysis
Reprinted from: *Appl. Sci.* **2020**, *10*, 1337, doi:10.3390/app10041337 . . . . **291**

About the Editor

Seokwon Yeom (Prof.) has been a faculty member of Daegu University since 2007. He is now a full Professor of the same university, the School of ICT Convergence. He has a Ph.D. in Electrical and Computer Engineering from the University of Connecticut in 2006. His research interests are intelligent processing of image and optical information, machine learning, and target tracking. He has researched multiple target tracking for airborne early warnings; three-dimensional image processing with digital holography and integral imaging; photon-counting linear discriminant analysis and nonlinear matched filter; millimeter wave and infrared image analysis; low-resolution object recognition; and aerial surveillance with small unmanned vehicle systems. He has been a Guest Editor of *Applied Sciences* Special Issue on Intelligent Processing on Image and Optical Information in 2019, and Volume II in 2020. He has been a member of the Editorial Board of *Applied Sciences* since 2019, a board member of the Korean Institute of Intelligent Systems since 2016, and a member of the board of directors of the Korean Institute of Convergence Signal Processing since 2014. He has served as program chair of the ICCCS2015, ISIS2017, iFUZZY2018, and ICCCS2019. He was a visiting scholar at the University of Maryland in 2014 and a director of the Gyeongbuk techno-park specialization center in 2013. He is currently working on projects related with smart surveillance using drones.

 *applied sciences*

Editorial

Special Issue on Intelligent Processing on Image and Optical Information

Seokwon Yeom

School of Information and Convergence Technology, Daegu University, Gyeongsan 38453, Korea; yeom@daegu.ac.kr

Received: 14 May 2020; Accepted: 25 May 2020; Published: 5 June 2020

1. Introduction

Intelligent image and optical information processing have paved the way for the recent epoch of new intelligence and information era. Certainly, information acquired by various imaging techniques is of tremendous value, thus, an intelligent analysis of them is necessary to make the best use of it.

The objectives of intelligent processing range from the refinement of raw data to the symbolic representation and visualization of the real world. The extraction and manipulation of the descriptive features are essential for such a task [1]. It comes through unsupervised or supervised learning based on statistical and mathematical models or computational algorithms. With recent advances in computing power and learning algorithms, many applications have become more practical and further development is expected.

This Special Issue focuses on the intelligent processing of images and optical information acquired by various imaging methods. Images are commonly formed via visible light; there are other imaging sources that represent scenes and objects in a multi-dimensional frame [2]. For example, radiography can be used to see internal objects for medical diagnosis and non-destructive inspection. Infrared imaging detects thermal variations in a non-visible environment. Spectral imaging is useful for collecting information at high spectral resolution or over a wide spectrum. Microscopic imaging reproduces the microscopic world, while satellite imaging probes the Earth from space. Since the statistical distribution depends on the nature of the source, suitable intelligent processing should be contrived accordingly. A broad range of research fields is included in the Special Issue. Many studies focus on object classification and detection. Registration, segmentation, and fusion are performed between a series of images. Many valuable and up-to-most recent technologies are provided to solve the real problems in selected papers.

A total of 61 manuscripts were submitted and only 18 research, and one review, papers were verified through a thorough review process. The first volume of the Special Issue on the topic is closed; more in-depth research of the same topic is expected in the second volume of the Special Issue. It is anticipated that the scope of the intelligent processing would be even broader in the future.

2. Intelligent Processing on Image and Optical Information

This Special Issue was introduced to collect the latest research on relevant topics, and more importantly, to address the current practical and theoretical challenges. In the following, the papers are categorized into several subtopics; classification and detection, feature extraction and segmentation; estimation and localization; registration and fusion; compression, completion, and correction; optimization and clustering.

2.1. Classification and Detection

In the first paper of this category, entitled 'Change Detection of Water Resources via Remote Sensing: An L-V-NSCT Approach', Wang Xin, Tang Can, Wang Wei, and Li Ji [3] presented landscape monitoring via satellite remote sensing. The authors analyzed the texture features of the image to address the characteristics of changes in certain types of surface, such as rivers and lakes.

Texture classification is an important topic for many applications in image analysis. A paper, entitled 'A Texture Classification Approach Based on the Integrated Optimization for Parameters and Features of Gabor Filter via Hybrid Ant Lion Optimizer' by Mingwei Wang, Lang Gao, Xiaohui Huang, Ying Jiang, and Xianjun Gao [4], proposed a new texture classification based on the integrated optimization of the parameters and features of the Gabor filter.

Calcaneal fractures are often caused by accidents during exercise or activity. In general, the detection of calcaneal fractures is still performed manually through computer tomography (CT) image observation. Wahyu Rahmaniar and Wen-June Wang [5], in their paper, 'Real-Time Automated Segmentation and Classification of Calcaneal Fractures in CT Images' proposed a method for detecting calcaneal fractures through localization and color segmentation of calcaneus.

Two papers on biological research, entitled 'Automatic Zebrafish Egg Phenotype Recognition from Bright-Field Microscopic Images Using Deep Convolutional Neural Network' [6] and 'Zebrafish Larvae Phenotype Classification from Bright-field Microscopic Images Using a Two-Tier Deep-Learning Pipeline' [7] are presented for zebrafish egg recognition. In the former paper, Shang, Ling Long, Sijie Lin, and Fengyu Cong addressed an automated zebrafish egg microscopic image analysis based on deep convolution neural networks (CNN). This study applies deep learning technique to classify fertilized and unfertilized zebrafish eggs from bright-field microscopic images. Transfer learning and data augmentation schemes were used to overcome the problem of small unbalanced training datasets. In the latter paper, Sijie Lin and his coworkers proposed a deep-learning pipeline for zebrafish larvae phenotype classification from brightfield microscopic images. Facing the difficulties of scarce training data, they used a two-tier classification pipeline.

As the deep-learning model develops, a huge number of images is required for training. Jeong gi Kwak and Hanseok Ko [8] described in their article: 'Unsupervised Generation and Synthesis of Facial Images via an Auto-Encoder-Based Deep Generative Adversarial Network (GAN)' an auto-encoder-based GAN with an enhanced network structure and training scheme for database augmentation and image synthesis.

As fruit production worldwide increases, the use of spectroscopy to monitor the surface condition of food continues to increase. A paper entitled 'Detecting Green Mold Pathogens on Lemons Using Hyperspectral Images' by Yuriy Vashpanov, Gwanghee Heo, Yongsuk Kim, Tetiana Venkel, and Jung-Young Son [9] investigates spectral images to detect green mold pathogens that parasitize on the lemon surface.

In their review paper entitled 'Computer Aided Weld Defect Detection from Radiography Images', Wenhui Hou, Dashan Zhang, Ye Wei, Jie Guo, and Xiaolong Zhang [10] reviewed the automatic defect inspection in three aspects: pre-processing, defect segmentation and defect classification. The achievement and limitations of classification based on feature extraction, selection and classifier are summarized and the application of new models based on deep learning were introduced.

2.2. Feature Extraction and Segmentation

The paper entitled 'Feature Extraction with Discrete Non-Separable Shearlet Transform and Its Application to Surface Inspection of Continuous Casting Slabs' by Xiaoming Liu, Ke Xu, Peng Zhou, and Huajie Liu [11] proposes a new feature extraction technique for the direction and texture information of surface defects of continuous casting slabs with complex backgrounds.

Wending Liu, Hanxing Liu, Yuan Wang, Xiaorui Zheng, and Junguo Zhang [12], in their paper on 'A Novel Extraction Method for Wildlife Monitoring Images with Wireless Multimedia Sensor Networks (WMSNs)', proposed an segmentation method with a low computational complexity, which can extract the target area from the background based on the image texture and color information. This proposed method can realize more accurate extraction of wildlife monitoring images and effectively support image transmission in wireless sensor networks.

2.3. Estimation and Localization

For autonomous driving, it is important to obtain precise and high-frequency localization of the vehicle. The inertial measurement unit (IMU), wheel encoder, and lidar odometry are utilized together to estimate the ego-motion of the unmanned ground vehicle in the paper entitled 'IMU-Aided High-Frequency Lidar Odometry for Autonomous Driving' by Hanzhang Xue, Hao Fu, and Bin Dai [13].

Dough fermentation plays an essential role in the bread production process, and its success is critical to producing high-quality products. In the paper entitled 'Determination of the Optimal State of Dough Fermentation in Bread Production by Using Optical Sensors and Deep Learning', Lino Antoni Giefer, Michael Lütjen, Ann-Kathrin Rohde, and Michael Freitag [14] proposed a novel method for the continuous monitoring of the volumetric parameters of dough piece during the fermentation process.

2.4. Registration and Fusion

In the paper entitled 'Multi-Sensor Face Registration Based on Global and Local Structures', Wei Li 1, Mingli Dong, Naiguang Lu, Xiaoping Lou, and Wanyong Zhou [15] introduced a novel multi-sensor face image registration method. This work utilizes global geometrical relationships and local shape features to register visible and infrared (IR) facial images for fusion-based recognition.

The paper entitled 'Multifocus Image Fusion Using a Sparse and Low-Rank Matrix Decomposition for Aviator's Night Vision Goggle' by Bo-Lin Jian, Wen-Lin Chu, Yu-Chung Li, and Her-Terng Yau [16] proposed an autofocusing and image fusion algorithm with sparse and low-rank matrix decomposition, to inspect the night vision goggles (NVG) of the aircraft. Their method can solve the multi-focusing problem caused by the mechanism error of the NVG.

2.5. Compression, Completion, and Correction

The paper entitled 'Error Resilience for Block Compressed Sensing with Multiple-Channel Transmission' by Hsiang-Cheh Huang, Po-Liang Chen, and Feng-Cheng Chang [17] proposed an error resilient transmission scheme for block compressed sensing. In their paper, the compressed information is transmitted over the multiple independent and lossy channels. By introducing correlation to multiple description coding, errors induced in the lossy channel are effectively alleviated.

The incomplete images can contain many missing pixels distributed over the entire image. In the paper entitled 'Image Completion with Hybrid Interpolation in Tensor Representation', Rafał Zdunek and Tomasz Sadowski [18] proposes an interpolation algorithm for a wide spectrum of image-completion problems. Their algorithm outperforms other conventional methods with a considerably shorter computational runtime.

In the paper entitled 'A Correction Method for HeatWave Distortion in Digital Image Correlation Measurements Based on Background-Oriented Schlieren', Chang Ma, Zhoumo Zeng, Hui Zhang, and Xiaobo Rui [19] proposed a correction method based on the background-oriented schlieren technique. The method can correct the distortion caused by heat waves in measuring image correlation.

2.6. Optimization and Clustering

The theoretical aspects of optimization and clustering are emphasized in the last two papers. On the paper entitled 'An Effective Optimization Method for Machine Learning Based on ADAM', Dokkyun Yi, Jaehyun Ahn, and Sangmin Ji [20] proposed a novel optimization method for non-convex cost functions to make machine learning more efficient. The proposed method solves the problem of falling to a local minimum by adding a cost function in the parameter update rule of the adaptive moment estimation (ADAM) method.

The paper entitled 'Boundary Matching and Interior Connectivity-Based Cluster Validity Analysis' by Qi Li, Shihong Yue, Yaru Wang, Mingliang Ding, Jia Li, and Zeying Wang [21] proposed a novel method to evaluate clustering results by the validity index. Their analysis can be applied to select the optimal clustering parameters including the number of clusters.

Acknowledgments: This issue would not be possible without the contributions of many outstanding authors. I would like to thanks to all authors who submitted their papers to this issue regardless of the final decision. I sincerely hope that comments and suggestions from the reviewers and the editors helped the authors improve their papers. I also appreciate the dedicated reviewers for their efforts to expertise. Finally, I place on record my gratitude to the professional editorial team of Applied Sciences, and special thanks to Stella Zhang, Assistant Editor from MDPI Branch Office, Beijing.

Conflicts of Interest: The author declares no conflict of interest.

References

1. Gonzalez, R.C.; Woods, R.E. *Digital Image Processing*, 4th ed.; Pearson: London, UK, 2018.
2. Hecht, E. *Optics*, 5th ed.; Pearson: London, UK, 2016.
3. Xin, W.; Can, T.; Wei, W.; Ji, L. Change Detection of Water Resources via Remote Sensing: An L-V-NSCT approach. *Appl. Sci.* **2019**, *9*, 1223. [CrossRef]
4. Wang, M.; Gao, L.; Huang, X.; Jiang, Y.; Gao, X. A Texture Classification Approach Based on the Integrated Optimization for Parameters and Features of Gabor Filter via Hybrid Ant Lion Optimizer. *Appl. Sci.* **2019**, *9*, 2173. [CrossRef]
5. Rahmaniar, W.; Wang, W. Real-Time Automated Segmentation and Classification of Calcaneal Fractures in CT Images. *Appl. Sci.* **2019**, *9*, 3011. [CrossRef]
6. Shang, S.; Long, L.; Lin, S.; Cong, F. Automatic Zebrafish Egg Phenotype Recognition from Bright-Field Microscopic Images Using Deep Convolutional Neural Network. *Appl. Sci.* **2019**, *9*, 3362. [CrossRef]
7. Shang, S.; Lin, S.; Cong, F. Zebrafish Larvae Phenotype Classification from Bright-field Microscopic Images Using a Two-Tier Deep-Learning Pipeline. *Appl. Sci.* **2020**, *10*, 1247. [CrossRef]
8. Kwak, J.; Ko, H. Unsupervised Generation and Synthesis of Facial Images via an Auto-Encoder-Based Deep Generative Adversarial Network. *Appl. Sci.* **2020**, *10*, 1995. [CrossRef]
9. Vashpanov, Y.; Heo, G.; Kim, Y.; Venkel, T.; Son, J.Y. Detecting Green Mold Pathogens on Lemons Using Hyperspectral Images. *Appl. Sci.* **2020**, *10*, 1209. [CrossRef]
10. Hou, W.; Zhang, D.; Wei, Y.; Guo, J.; Zhang, X. Review on Computer Aided Weld Defect Detection from Radiography Images. *Appl. Sci.* **2020**, *10*, 1878. [CrossRef]
11. Liu, X.; Xu, K.; Zhou, P.; Liu, H. Feature Extraction with Discrete Non-Separable Shearlet Transform and Its Application to Surface Inspection of Continuous Casting Slabs. *Appl. Sci.* **2019**, *9*, 4668. [CrossRef]
12. Liu, W.; Liu, H.; Wang, Y.; Zheng, X.; Zhang, J. A Novel Extraction Method for Wildlife Monitoring Images with Wireless Multimedia Sensor Networks (WMSNs). *Appl. Sci.* **2019**, *9*, 2276. [CrossRef]
13. Xue, H.; Fu, H.; Dai, B. IMU-Aided High-Frequency Lidar Odometry for Autonomous Driving. *Appl. Sci.* **2019**, *9*, 1506. [CrossRef]
14. Giefer, L.A.; Lütjen, M.; Rohde, A.K.; Freitag, M. Determination of the Optimal State of Dough Fermentation in Bread Production by Using Optical Sensors and Deep Learning. *Appl. Sci.* **2019**, *9*, 4266. [CrossRef]
15. Li, W.; Dong, M.; Lu, N.; Lou, X.; Zhou, W. Multi-Sensor Face Registration Based on Global and Local Structures. *Appl. Sci.* **2019**, *9*, 4623. [CrossRef]
16. Jian, B.L.; Chu, W.L.; Li, Y.C.; Yau, H.T. Multifocus Image Fusion Using a Sparse and Low-Rank Matrix Decomposition for Aviator's Night Vision Goggle. *Appl. Sci.* **2020**, *10*, 2178. [CrossRef]
17. Huang, H.C.; Chen, P.L.; Chang, F.C. Error Resilience for Block Compressed Sensing with Multiple-Channel Transmission. *Appl. Sci.* **2020**, *10*, 161. [CrossRef]
18. Zdunek, R.; Sadowski, T. Image Completion with Hybrid Interpolation in Tensor Representation. *Appl. Sci.* **2020**, *10*, 797. [CrossRef]
19. Ma, C.; Zeng, Z.; Zhang, H.; Rui, X. A Correction Method for Heat Wave Distortion in Digital Image Correlation Measurements Based on Background-Oriented Schlieren. *Appl. Sci.* **2019**, *9*, 3851. [CrossRef]
20. Yi, D.; Ahn, J.; Ji, S. An Effective Optimization Method for Machine Learning Based on ADAM. *Appl. Sci.* **2020**, *10*, 1073. [CrossRef]
21. Li, Q.; Yue, S.; Wang, Y.; Ding, M.; Li, J.; Wang, Z. Boundary Matching and Interior Connectivity-Based Cluster Validity Analysis. *Appl. Sci.* **2020**, *10*, 1337. [CrossRef]

 *applied
sciences*

Article

Change Detection of Water Resources via Remote Sensing: An L-V-NSCT Approach

Wang Xin, Tang Can, Wang Wei * and Li Ji

School of Computer and Communication Engineering, Changsha University of Science and Technology,
Changsha 410114, Hunan, China; wangxin@csust.edu.cn (W.X.); tangcan1237@163.com (T.C.);
hangliji@163.com (L.J.)
* Correspondence: wangwei@csust.edu.cn

Received: 19 February 2019; Accepted: 19 March 2019; Published: 22 March 2019

Abstract: Aiming at the change detection of water resources via remote sensing, the non-subsampling contour transformation method combining a log-vari model and the Stractural Similarity of Variogram (VSSIM) model, namely log-vari and VSSIM based non-subsampled contourlet transform (L-V-NSCT) approach, is proposed. Firstly, a differential image construction method based on non-subsampled contourlet transform (NSCT) texture analysis is designed to extract the low-frequency and high-frequency texture features of the objects in the images. Secondly, the texture features of rivers, lakes and other objects in the images are accurately classified. Finally, the change detection results of regions of interest are extracted and evaluated. In this experiment, the L-V-NSCT approach is compared with other methods with the results showing the effectiveness of this method. The change in Dongting Lake is also analyzed, which can be used as a reference for relevant administrative departments.

Keywords: change detection; NSCT; variogram function; structure similarity; Dongting Lake

---

1. Introduction

With improvement in the level of software and hardware for remote sensing acquisition along with the increase in the quantity of data, the need for change detection is also increasing, and the requirements for precision are getting higher and higher. Remote sensing plays an important role in water environment monitoring. In disaster prevention and mitigation, water conservation involves flood and waterlogging disasters, drought and wading geological disaster monitoring and assessment, flood forecasting and early warning. In water resource monitoring and protection, it involves satellite remote sensing precipitation forecasts, soil moisture and evapotranspiration remote sensing estimation, surface water monitoring, groundwater monitoring, water environment monitoring and surface water body monitoring. With respect to ecological protection, water conservation involves the monitoring of the environment, the investigation of soil and water conservation and irrigation area, the monitoring and control of river course and estuary change.

In lake water area change detection, remote sensing technology is used to monitor lake area change. On the one hand, the law along with the trend of annual and interannual changes in the lake area should be analyzed from a natural point of view to better protect lakes through watershed management. On the other hand, we must supervise the illegal occupation of water in real time to provide clues for law enforcement. In the Hubei Province, China, from January to July 2015, nearly 48 suspected illegal spots were found in lake monitoring by remote sensing. Law enforcement departments checked 15 suspected illegal spots in the field and confirmed seven real illegal events. This work will be further extended to the monitoring of reservoirs and rivers. In addition, surface water monitoring is also an important part of drought monitoring.

In the remote sensing of water resources, Xu et al. [1] used the NDWI model to extract the water bodies of the Tangjiashan barrier lake from the multisource satellite images to detect the changes in the barrier lake, pre- and post-earthquake. Qiao et al. [2] compared paleo lakes with modern lakes, showing that lakes on the Tibetan Plateau have shrunk significantly since the great lake period, which provides fundamental information to support research on both global paleo-climatology and paleo-hydrology change. Zhao et al. [3], by combining ecological quantity analysis with GIS technology based on land use data and remote sensing imagery, analyzed the changes in land use and land cover as well as the driving force in the mainstream of the Tarim River from 1973 to 2005. Markogianni et al. [4] used Landsat and Systeme Probatoire d'Observation de la Terre (SPOT) images to estimate the Normalized Difference Vegetation Index (NDVI) and land-use changes at the Plastira artificial lake catchment for the period 1984–2009. Adesina et al. [5] focused on change detection on the Jebba Lake Basin between 1978 (five years before the dam was established) and 1995 (twelve years after the dam impoundment). Refice et al. [6] applied high-resolution, X-band, and stripmap Cosmo-SkyMed data to the monitoring of flood events in the Basilicata region (Southern Italy). By using the mapping results with post-earthquake high-resolution images from Google Earth, Zhao et al. [7] showed that the pixel-based landslide mapping method was able to identify landslides with relatively high accuracy. Sun et al. [8] used Cosmo-Skymed ScanSAR mode (HH) data for one-year monitoring of seasonal changes in the water surface areas of Poyang Lake from January 2014 to December 2014. Li et al. [9] proposed a change detection method based on Gabor wavelet features for very high resolution (VHR) remote sensing images. The fuzzy c-means cluster algorithm was employed to obtain the final change map. Setiawan et al. [10] implemented the canny edge detection method, combining with Otsu thresholding to detect the edges, achieved good edge detection results.

Gao et al. [11] presented a change detection method for multitemporal synthetic aperture radar images based on PCANet and exploited representative neighborhood features from each pixel using PCA filters as convolutional filters. Zhao et al. [12] proposed a difference image analysis approach based on deep neural networks for use on image change detection problems. Gong et al. [13] put forward a novel approach for change detection in SAR images. This approach classified changed and unchanged regions by fuzzy c-means (FCM) clustering with a Markov random field (MRF) energy function.

At present, commonly used change detection methods are not accurate enough to detect the edge contours of rivers, lakes, urban roads, and forest vegetation in severely damaged areas and there are too many small areas of debris in the disaster areas, which are judged as image changes. At the same time, the impact of this noise will lead to high error rates and false alarm rates.

To dynamically monitor the changes of inland surface resources such as Ocean Lake wetlands and lake swamps, this paper proposes an L-V-NSCT approach based on the NSCT and the logarithmic variation function. Firstly, NSCT transform was used to decompose the detected image and the VSSIM algorithm was used to extract the texture information difference at different scales after multi-scale decomposition. Secondly, the texture difference images of rivers, lakes and other objects were fused by utilizing the NSCT inverse transform. Finally, the spectral information, texture information, and space of the objects were fully considered. The fuzzy c-means clustering model (CFCM) was designed to classify the difference features of the above images.

The experiments in subsequent chapters can prove the effectiveness of the method. At the same time, the greatest innovation of this method was that it has good practicability and has been applied in the protection of Dongting Lake. On the basis of the GF-2 image, the method regularly provided the relevant departments with test results and reference.

2. Directional Logarithmic Variogram and the VSSIM Model

At present, variogram analysis [14,15] is typically used to study the texture structure and spatial correlation of an entire image. This paper mainly uses a variation function to consider the structural characteristics of surface features in different period's images and highlights the edge contour structure of target features such as rivers and lakes in images, as well as the variation characteristics of the main edges in the detection area. Based on the classic mathematical model of the variogram, a logarithmic variogram model with horizontal (0°), vertical (90°), and diagonal (45°, 135°) directions was proposed. The model can quickly locate and analyze the main edge contour information of image objects.

Set $f(x,y)$ as the size of the collected image M × N; the sliding window with the designed width $L = 2d + 1$ was designed to analyze and process different types of ground object data in the image source. The center pixel of the ground object in the window can be set as (i_0, j_0), the coordinates of the ground object pixels in other positions can be set as (i, j), and the edge analysis model (log-vari) of the directional logarithmic variation function can be set as follows:

$$r_h^{0}(i_0, j_0) = \frac{1}{2N_0(h)} \sum_{i-i_0-d}^{i_0+d} \sum_{j-j_0-d}^{j_0+d-h} |\ln[f(i,j)] - \ln[f(i,j+h)]| \tag{1}$$

$$r_h^{45}(i_0, j_0) = \frac{1}{2N_{45}(h)} \sum_{i-i_0-d}^{i_0+d-h} \sum_{j-j_0-d}^{j_0+d-h} |\ln[f(i,j+h)] - \ln[f(i+h,j)]| \tag{2}$$

$$r_h^{90}(i_0, j_0) = \frac{1}{2N_{90}(h)} \sum_{i-i_0-d}^{i_0+d-h} \sum_{j-j_0-d}^{j_0+d} |\ln[f(i,j)] - \ln[f(i+h,j)]| \tag{3}$$

$$r_h^{135}(i_0, j_0) = \frac{1}{2N_{135}(h)} \sum_{i-i_0-d}^{i_0+d-h} \sum_{j-j_0-d}^{j_0+d-h} |\ln[f(i,j)] - \ln[f(i+h,j+h)]| \tag{4}$$

where $f(i, j+h)$ and $f(i, j)$ are two coordinate values with distance h in the area. The values of the directional value θ are {0°, 45°, 90°, 135°}. $N_\theta(h)$ is the number of all two-coordinate points in the region with a distance of h, and the final output mathematical matrix sequence $r^\theta(x,y)$ is the texture coefficient matrix in the direction of θ.

In the process of feature similarity analysis, the structure similarity model (SSIM) can be constructed from the brightness, contrast, and structure factor of the image [16]. The expression is as follows:

$$SSIM(x,y) = [l(x,y)]^\alpha \cdot [c(x,y)]^\beta \cdot [s(x,y)]^\gamma \tag{5}$$

where $l(x,y)$ is the luminance coefficient matrix, $c(x,y)$ is the contrast parameter, and $s(x,y)$ is the structural parameter. The parameters α, β, γ are all less than 1.

To improve the detection accuracy of texture details of different types of objects in seismic images and the clearer edge structure of lakes and rivers, we used the edge texture coefficient matrix obtained in the upper section to improve and optimize the SSIM model and propose the structure of directional logarithmic variogram model similarity (VSSIM).

We selected two images of size M × N, which are represented by X and Y. First, the average value of the edge texture $r^\theta(x,y)$ of the two images was calculated by using the log-vari model in the direction of θ. The extracted edge texture feature coefficients are $r_X(x,y)$ and $r_Y(x,y)$. At the same time, the structural similarity algorithm $v(x,y)$ of the logarithmic variation function was constructed to judge the change degree of edge features in the texture feature matrix.

$$v(x,y) = \frac{k \sum_{j=1}^{M} \sum_{i=1}^{N} [r_X(i,j) \times r_Y(i,j)] + c_1}{\sum_{j=1}^{M} \sum_{i=1}^{N} [r_X(i,j)]^2 + \sum_{j=1}^{M} \sum_{i=1}^{N} [r_Y(i,j)]^2 + c_2} \tag{6}$$

Through these experiments, we found that the optimal values of the parameters in the above expressions were generally $k = 2$, $c_1 = c_2 = 0.001$.

In the SSIM expression, the structure factor $s(x,y)$ can be replaced by the edge structure factor $v(x,y)$. The constructed an edge structure similarity model (VSSIM) based on the logarithmic variation function is expressed as follows:

$$VSSIM(x,y) = [l(x,y)]^\alpha \cdot [c(x,y)]^\beta \cdot [v(x,y)]^\gamma \tag{7}$$

3. Differential Image Construction

By using edge texture enhancement, the goal of change detection with the logarithmic variation function was to construct more accurate image difference features in the process of NSCT multiscale transformation. The main steps were as follows:

(1) NSCT multiscale transform of the original image; the multiscale NSCT transform was used to obtain low-frequency texture coefficients and high-frequency texture coefficients in different scales. The expressions are as follows:

$$NSCT(I_1) = \{L_1, H_1^{1,1}, H_1^{1,2}, \ldots, H_1^{1,k}\} \tag{8}$$

$$NSCT(I_2) = \{L_2, H_2^{1,1}, H_2^{1,2}, \ldots, H_2^{1,k}\} \tag{9}$$

where $L_1(x,y)$ and $L_2(x,y)$ are the low-frequency coefficient matrices of the two-period images, and $H_1^{l,k}(x,y)$ and $H_2^{l,k}(x,y)$ are the high-frequency coefficient matrices in the kth direction of the first level in the multiscale decomposition process.

(2) Construction of the low-frequency texture difference coefficient $L(x,y)$ in the NSCT domain; after I_1 and I_2 decomposition, the low-frequency features $L_1(x,y)$ and $L_2(x,y)$ were extracted. VSSIM was used to measure the difference in the low-frequency texture features, and the coefficient matrix $vssim(x,y)$ of low-frequency texture similarity was extracted. Then, the coefficient matrix was used to weigh the low-frequency texture difference to obtain the feature difference. Finally, threshold and FCM clustering were used to extract the low-frequency texture feature difference image $L(x,y)$.

$$L(x,y)= \text{FCM}[|L_1(x,y) - L_2(x,y)| \times vssim(x,y)] \tag{10}$$

(3) Construction of the high-frequency texture difference coefficient $H^{l,k}(x,y)$ in the NSCT domain; the log-variogram was used to analyze high-frequency texture features and to extract the high-frequency edge texture coefficient matrix $r_1^{l,k}(x,y)$ and $r_2^{l,k}(x,y)$. These texture coefficients were used to calculate the similarity coefficient matrix $s^{l,k}(x,y)$ of high-frequency texture and to judge the degree of difference of high-frequency texture features in these scale-space images. Finally, FCM clustering was used to extract the high-frequency edge difference texture features.

$$s^{l,k}(x,y) = \sum_{i=1}^{M} \sum_{j=1}^{N} \left\{1 - \frac{\left|r_1^{l,k}(i,j) - r_2^{l,k}(i,j)\right|}{\max[r_1^{l,k}(i,j), r_2^{l,k}(i,j)]}\right\} \tag{11}$$

The difference coefficient expression of high-frequency edge texture features in NSCT domain is as follows:

$$H^{l,k}(x,y) = \text{FCM}[\left|H_1^{l,k}(x,y) - H_2^{l,k}(x,y)\right| \times s^{l,k}(x,y)] \tag{12}$$

(4) Construction of image texture difference features based on the NSCT transform; the low-frequency texture difference coefficients $L(x,y)$ and high-frequency texture difference coefficients $H^{l,k}(x,y)$ on different scales were extracted. The NSCT inverse transform was used to construct texture difference images $D(x,y)$ of multi-temporal remote sensing images.

$$D(x,y) = NSCTRE\{L(x,y), H^{l,k}(x,y)\} \tag{13}$$

4. Difference Image Clustering

To improve the classification accuracy of remote sensing images, an improved fuzzy C-means clustering method (CFCM) was used based on the degree of correlation. Firstly, the correlation between image data were accurately described by the similarity between their key feature vectors. The improved weighted coefficient similarity measure takes the similarity degree between the central vertex O_i and the neighboring vertex O_j as the weight(e_{ij}) of any edge e_{ij} in a 3×3 sliding window.

$$\text{weight}(e_{ij}) = \sum_{k=1}^{3} \frac{1}{1 + m_i \times \left| d_{ik} - d_{jk} \right|} \tag{14}$$

where m_i is a coefficient, and the values of k and i belong to {1,2,3}. Secondly, the weight of an edge e_{ij} of the center vertex O_i of the texture information in the window was calculated.

$$k_{ij} = \frac{\text{weight}(e_{ij})}{\sum\limits_{e_{ij} \in E} weight(e_{ij})} \tag{15}$$

Next, the correlation degree of data objects based on pixel $D(x,y)$ in difference images was calculated.

$$\text{Cor}(D_i) = \sum_{e_{ij} \in E} \{ k_{ij} \times weight(e_{ij}) \} \tag{16}$$

FCM clustering was then used to accurately classify the correlation degree of all data objects in the extracted differential images.

$$D(p,q) = \begin{cases} w_c, \text{Cor}(D_i) > T_1 \\ w_n, T_2 \le \text{Cor}(D_i) \le T_1 \\ w_u, \text{Cor}(D_i) < T_2 \end{cases} \tag{17}$$

where w_c, w_n, w_u denote change class, fuzzy class and invariant class, respectively. T_1, T_2 are thresholds. Lastly, a region growing method was used to optimize the classification results of CFCM based on the classified difference features and the appropriate threshold processing was used to extract more accurate change detection results.

5. Experimental Results and Analysis

To verify the effectiveness of the L-V-NSCT method proposed in this paper, we used real datasets to conduct experiments and analyze the results. Before the change detection, we first carried out the image gray normalization, texture enhancement, and other preprocessing. The flow chart of L-V-NSCT change detection is shown in Figure 1.

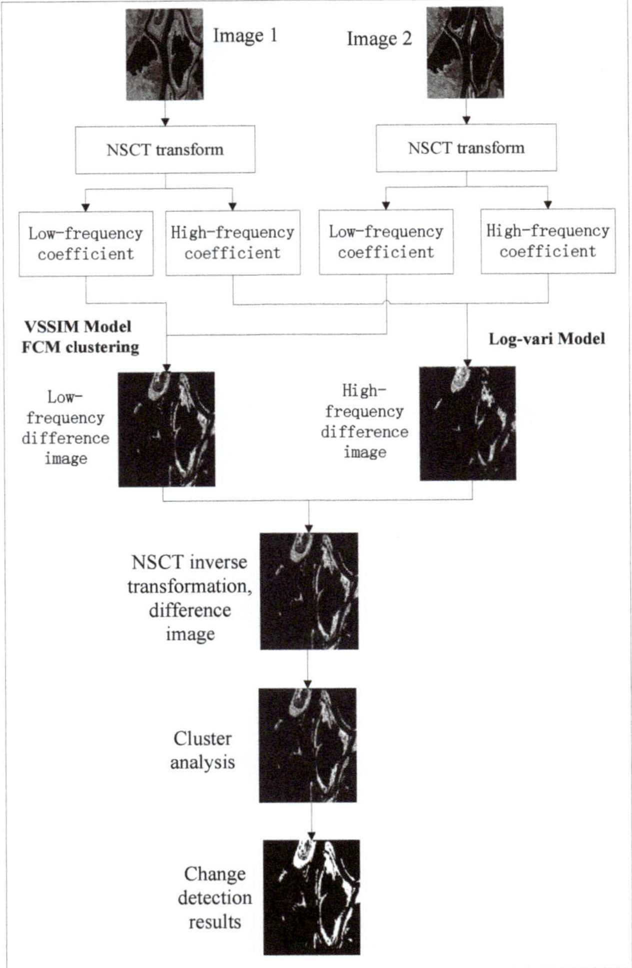

Figure 1. The flow chart of L-V-NSCT change detection.

We first detected the changes of lakes and water bodies in Sardinia area [17], as shown in Figure 2. The size of the image was 412 × 300. The comparison algorithms were NSCTKFCM [18] and NSCTFCM [19], both of which were improvements of the NSCT algorithm. Figure 2a is the image of the Sardinia region in April 1999, Figure 2b is the image of the Sardinia region in May 1999, Figure 2c is the result of the NSCTKFCM method, Figure 2d is the result of the NSCTFCM method, Figure 2e is the result of L-V-NSCT and Figure 2f is the reference image of the result, where the number of changed pixels was 7626.

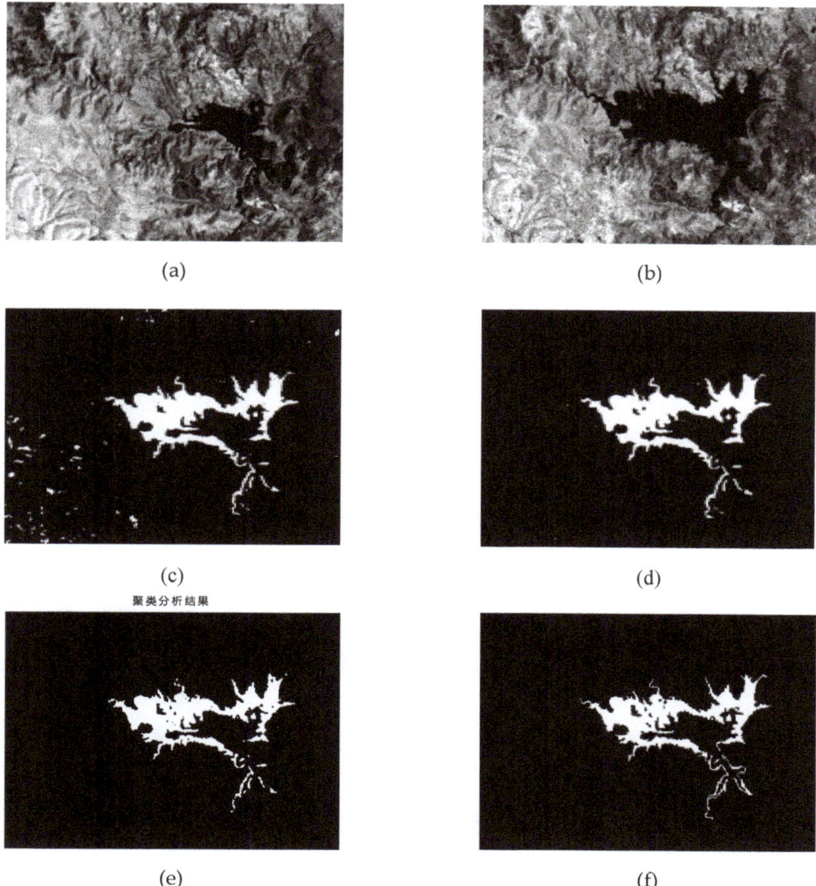

Figure 2. Detection results in Sardinia area. (**a**) the image of the Sardinia region in April 1999, (**b**) the image of the Sardinia region in May 1999, (**c**) the result of the NSCTKFCM method, (**d**) the result of the NSCTFCM method, (**e**) the result of L-V-NSCT, and (**f**) the reference image of the result.

From the results, we can see that three detection algorithms achieved good results for the change detection of rivers and lakes in the image. However, the other two algorithms were greatly affected by noise, the edges of the river change areas were blurred, and the fragments change areas (areas of no concern) caused by false alarm were relatively increased. Nonetheless, the L-V-NSCT approach can effectively preserve the details of the change area and show that the edge of the main change area is clear. The above experiments showed that L-V-NSCT approach has some advantages in detecting changes in remote sensing images.

In the experiment, we found that the indexes of the L-V-NSCT approach were different when different types of images were analyzed. For low-brightness images, the bit error rate of change detection increased, and the accuracy decreased.

Radarsat images of Ottawa, Canada, from May 1997 and August 1997 were the original images of the dataset in the second experiments [11–13], as shown in Figure 3. The image size is 290 × 350, and the number of changed pixels is 16049.

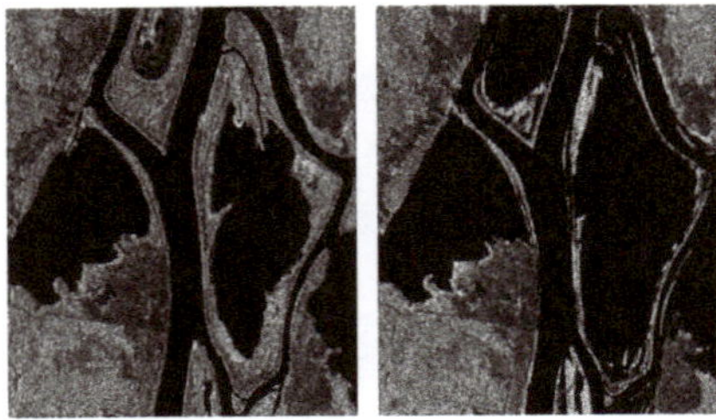

Figure 3. Original image and change detection map in Ottawa area.

Figure 4 shows the low-frequency difference image, the high-frequency difference image, and the final detection result. Figure 4a,b are the low-frequency difference image based on the logarithmic variation function and the low-frequency difference coefficient image after further clustering, respectively. Figure 4c is the high-frequency difference coefficient image after similarity analysis. Figure 4d is the NSCT inverse transform fusion image. Figure 4e is the clustering result image, and Figure 4f is the last change detection binary image.

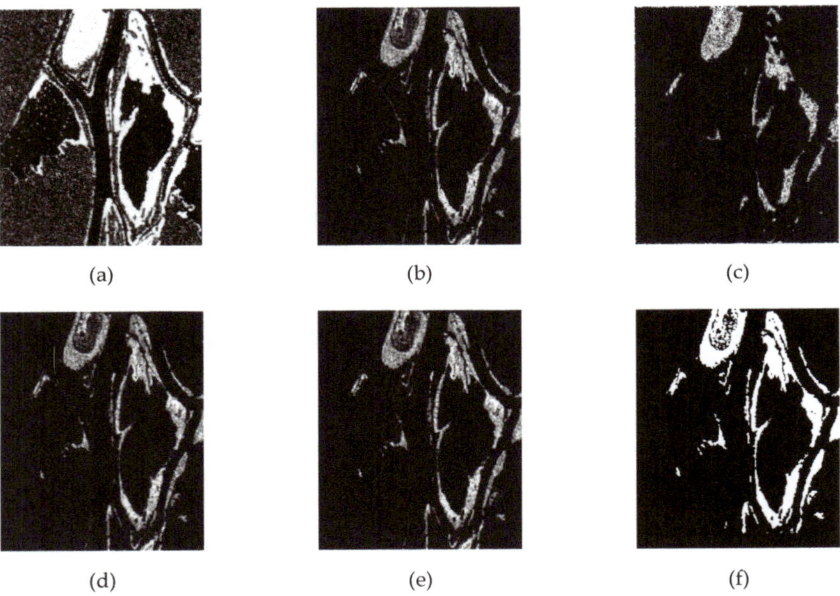

Figure 4. Test results in Ottawa area. (**a**) the low-frequency difference image based on the logarithmic variation function, (**b**) the low-frequency difference coefficient image after further clustering, (**c**) the high-frequency difference coefficient image after similarity analysis, (**d**) the NSCT inverse transform fusion image, (**e**) the clustering result image, and (**f**) the last change detection binary image.

As seen from the images, L-V-NSCT can detect the changes in the registered images very well, and the details are relatively complete.

Figure 5 further reflects the detail effect of L-V-NSCT change detection. The left image is the manual labeled change detection reference image, and the right image is the detection result of this method. The labeling parts in the right figures show that the proposed method can obtain results consistent with the manual judgment in most local areas. The left parts with circles are marked with false alarm or missing alarm areas, which need further analysis and judgment. Therefore, we will further compare the indicators with other methods.

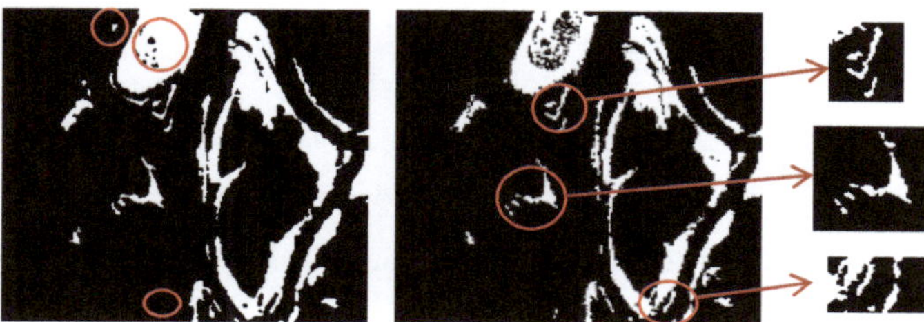

Figure 5. Detection results with details.

We used statistical accuracy (*PCC*) and Kappa index (*KC*) as experimental indicators to evaluate the performance of the algorithm [12]. If *FP* is the number of the pixels belonging to the unchanged class but falsely classified as the changed class, and *FN* is the number of the pixels belonging to the changed class but falsely classified as the unchanged class. *TP* and *TN* represent the number of changed pixels and unchanged pixels that are correctly detected, respectively. Then the correct classification accuracy (*PCC*) is:

$$PCC = \frac{TP + TN}{N} \tag{18}$$

where N is the total number of pixels in the image. And the Kappa index KC is calculated as:

$$KC = \frac{PCC - PRE}{1 - PRE} \tag{19}$$

where $PRE = \frac{(TP+FP)\times(TP+FN)+(FN+TN)\times(FP+TN)}{N^2}$.

In the experimental comparison, besides NSCTKFCM and NSCTFCM, the L-V-NSCT method was also compared with other methods in recent years, namely PCANet [11], DNN [12], and MRFFCM [13]. The experimental results are shown in Table 1.

Table 1. Change detection results under evaluation by different algorithms.

Method	PCC (%)	FN	FP	KC
NSCTKFCM	95.24	2085	2749	0.8241
NSCTFCM	97.51	236	2286	0.9113
PCANet	98.22	944	863	0.9300
DNN	98.09	1059	883	0.9278
MRFFCM	97.69	712	1636	0.9151
L-V-NSCT	98.36	948	681	0.9379

From the data analysis in Table 1, we can see that both the *PCC* and *KC* of L-V-NSCT were higher than the other five approaches. Although the *FN* of L-V-NSCT was higher than NSCTFCM, MRFFCM and PCANet, the *FP* was comparatively lower. So, the final *PCC* and *KC* of L-V-NSCT were also better than that of the 3 algorithms.

We also conducted experiments with 20 of our remote sensing image databases, with the average accuracy and Kappa index obtained are shown in Table 2. As can be seen from the table, among all the six methods, NSCTKFCM performed the worst and the experimental results were unstable. The experimental results of PCANet, DNN and L-V-NSCT were better. DNN, in particular, performed very well due to the use of deep neural networks. The performance of L-V-NSCT was relatively stable. Although *KC* of L-V-NSCT was slightly lower than that of DNN, the *PCC* was still higher than that of DNN.

Table 2. Average results of change detection by different methods.

Method	PCC (%)	KC
NSCTKFCM	82.63	0.5527
NSCTFCM	84.51	0.7816
PCANet	90.78	0.8569
DNN	91.57	0.8932
MRFFCM	88.34	0.8544
L-V-NSCT	92.96	0.8845

This method can be used to detect and analyze changes in the water environment and provide help for water environment management. We examined the changes in the Dongting Lake in China. The remote sensing images of Dongting Lake from January 1973 and February 2014 are shown in Figure 6a,b. Figure 6a was derived from landsat1 with a resolution of 79 m, while Figure 6b was derived from landsat8 with a resolution of 15 m.

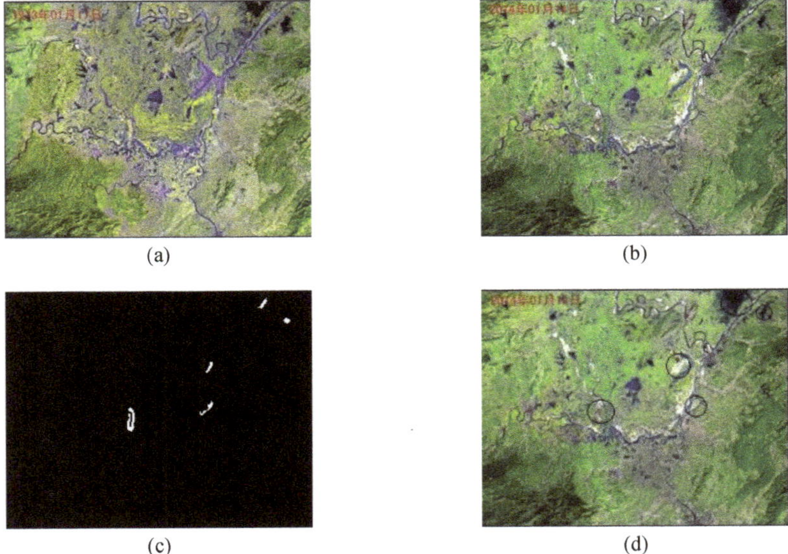

Figure 6. Overall changes under large scale conditions. (a) remote sensing images of Dongting Lake from January 1973, (b) remote sensing images of Dongting Lake from February 2014, (c) the changes in Dongting Lake on a large scale, and (d) the region with the greatest change in the map.

Images collected in the same season can better reflect the real changes in the lake. Figure 6c shows the changes in Dongting Lake on a large scale. The region with the greatest change in the map showed that, after 40 years, the area of some local watersheds had changed, and some local watersheds have changed their original route. Although the results of large-scale detection do not provide specific details, the results can provide location reference for further research.

Upon further analysis, changes in the rainy season and dry season in Dongting Lake are shown in Figure 7. The original image size is 4400 × 4000, and the resolution is 10 kilometers [20]. Because of the cloud interference in Figure 7b, the detection result is affected. After removing the cloud and text interference information by color detection [21], the detection result is shown in Figure 7d, and the number of pixels is 8786. The results of change area detection provide a basis for water storage, flood prevention and landform change analysis of Dongting Lake.

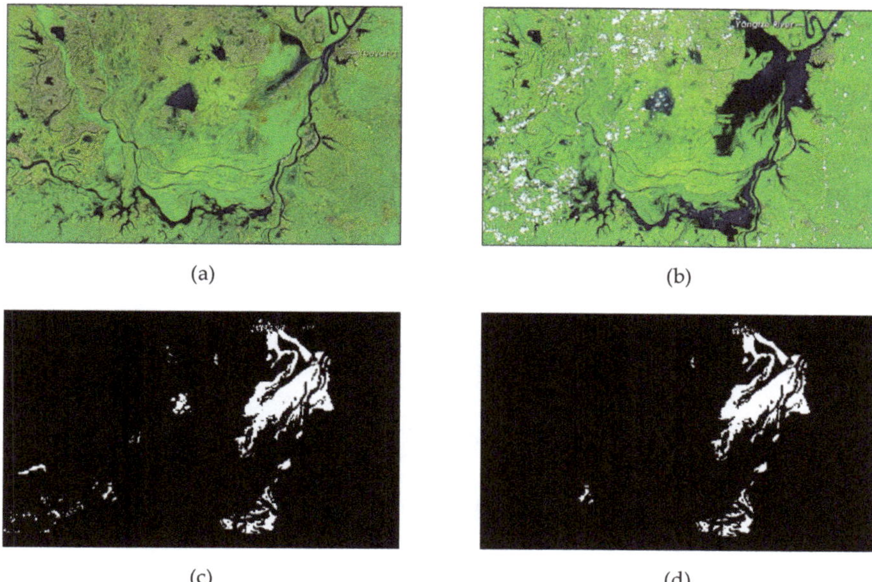

(a)   (b)

(c)   (d)

Figure 7. Changes in the different seasons of Dongting Lake. (**a**) image in dry season, (**b**) rainy season image, (**c**) change detection results, and (**d**) Results after cloud elimination.

6. Conclusions

Addressing the change characteristics of surface types with complex texture information such as rivers and lakes, this paper used an improved directional logarithmic variation function model to analyze the texture features of images and proposes an L-V-NSCT approach. This approach can not only effectively reflect the characteristics of land surface change in different periods but also has good anti-noise performance.

Through our experiments, we can also see that the detection accuracy and other indicators of the proposed method are not stable for different types of images. The stability and parameter settings of the method merit further study.

Author Contributions: W.X. contributed to the conception of the study and wrote the manuscript; T.C. performed the experiments and data analyses; W.W. contributed significantly to algorithm design and manuscript preparation; L.J. helped perform the analysis with constructive discussions.

Funding: National Defense Pre-Research Foundation of China (7301506); National Natural Science Foundation of China (61070040); Education Department of Hunan Province (17C0043).

Conflicts of Interest: The authors declare no conflict of interest.

References

1. Xu, M.; Cao, C.; Zhang, H.; Xue, Y.; Li, Y.; Guo, J.; Chang, C.; He, Q.; Gao, M.; Li, X. Change detection of the Tangjiashan barrier lake based on multi-source remote sensing data. In Proceedings of the IEEE International Geoscience and Remote Sensing Symposium, Cape Town, South Africa, 12–17 July 2009; pp. 303–306.
2. Qiao, C.; Luo, J.; Sheng, Y.; Shen, Z.; Li, J. Lake shrinkage analysis using spectral-spatial coupled remote sensing on Tibetan Plateau. In Proceedings of the 2010 IEEE Geoscience and Remote Sensing Symposium, Honolulu, HI, USA, 25–30 July 2010; pp. 926–929.
3. Zhao, R.; Chen, Y.; Shi, P.; Zhang, L.; Pan, J.; Zhao, H. Land use and land cover change and driving mechanism in the arid inland river basin: A case study of Tarim River, Xinjiang, China. *Environ. Earth Sci.* **2013**, *68*, 591–604. [CrossRef]

4. Markogianni, V.; Dimitriou, E.; Kalivas, D.P. Land-use and vegetation change detection in Plastira artificial lake catchment (Greece) by using remote-sensing and GIS techniques. *Int. J. Remote Sens.* **2013**, *34*, 1265–1281. [CrossRef]

5. Adesina, G.O.; Mavomi, I. Landuse and Landcover Change Detection of Jebba Lake Basin Nigeria: Remote Sensing and GIS Approach. *J. Environ. Earth Sci.* **2014**, *4*, 119–127.

6. Refice, A.; Capolongo, D.; Pasquariello, G.; D'Addabbo, A.; Bovenga, F.; Nutricato, R.; Lovergine, F.P.; Pietranera, L.; et al. SAR and InSAR for Flood Monitoring: Examples With COSMO-SkyMed Data. *IEEE J. Sel. Top. Appl. Earth Obs. Remote Sens.* **2014**, *7*, 2711–2722. [CrossRef]

7. Zhao, W.; Li, A.; Nan, X.; Zhang, Z.; Lei, G. Postearthquake Landslides Mapping from Landsat-8 Data for the 2015 Nepal Earthquake Using a Pixel-Based Change Detection Method. *IEEE J. Sel. Top. Appl. Earth Obs. Remote Sens.* **2017**, *10*, 1758–1768. [CrossRef]

8. Sun, Y.; Huang, S.; Li, J.; Xiaotao, L.; Jianwei, M.; Hui, W. Monitoring seasonal changes in the water surface areas of Poyang Lake using Cosmo-Skymed time series data in PR China. In Proceedings of the 2016 IEEE Geoscience and Remote Sensing Symposium, Beijing, China, 10–15 July 2016; pp. 7180–7183.

9. Li, Z.; Shi, W.; Zhang, H.; Hao, M. Change Detection Based on Gabor Wavelet Features for Very High Resolution Remote Sensing Images. *IEEE Geosci. Remote Sens. Lett.* **2017**, *14*, 783–787. [CrossRef]

10. Setiawan, B.D.; Rusydi, A.N.; Pradityo, K. Lake edge detection using Canny algorithm and Otsu thresholding. In Proceedings of the 2018 IEEE International Symposium on Geoinformatics, Malang, Indonesia, 24–25 November 2018; pp. 72–76.

11. Gao, F.; Dong, J.; Li, B.; Xu, Q. Automatic Change Detection in Synthetic Aperture Radar Images Based on PCANet. *IEEE Geosci. Remote Sens. Lett.* **2016**, *13*, 1792–1796. [CrossRef]

12. Zhao, J.; Gong, M.; Liu, J.; Jiao, L. Deep learning to classify difference image for image change detection. In Proceedings of the 2014 International Joint Conference on Neural Networks (IJCNN), Beijing, China, 6–11 July 2014.

13. Gong, M.; Su, L.; Jia, M.; Chen, W. Fuzzy Clustering with a Modified MRF Energy Function for Change Detection in Synthetic Aperture Radar Images. *IEEE Trans. Fuzzy Syst.* **2014**, *22*, 98–109. [CrossRef]

14. Woodcock, C.E.; Strahler, A.H. The use of variograms in remote sensing: I. scene models and simulated images. *Remote Sens. Environ.* **1988**, *25*, 323–348. [CrossRef]

15. Woodcock, C.E.; Strahler, A.H.; Jupp, D.L.B. The use of variograms in remote sensing: II. Real digital images. *Remote Sens. Environ.* **1988**, *25*, 349–379. [CrossRef]

16. Wang, Z.; Bovik, A.C.; Sheikh, H.R.; Simoncelli, E.P. Image quality assessment: From error visibility to structural similarity. *IEEE Trans Image Process* **2004**, *13*, 600–612. [CrossRef] [PubMed]

17. Gou, S.; Yu, T. Graph based SAR images change detection. In Proceedings of the 2012 IEEE Geoscience & Remote Sensing Symposium, Munich, Germany, 22–27 July 2012.

18. Wu, C.; Wu, Y. Multitemporal Images Change Detection Using Nonsubsampled Contourlet Transform and Kernel Fuzzy C-Means Clustering. In Proceedings of the IEEE International Symposium on Intelligence Information Processing and Trusted Computing, Wuhan, China, 22–23 October 2011; pp. 96–99.

19. Li, Q.; Qin, X.; Jia, Z. An unsupervised change detection of SAR images based on NSCT and FCM Clustering. *Laser J.* **2013**, *34*, 20–22. (In Chinese)

20. Floods in Central China. Available online: https://earthobservatory.nasa.gov/images/18664/floods-in-central-china (accessed on 22 March 2019).

21. Ishida, H.; Oishi, Y.; Morita, K.; Moriwaki, K.; Nakajima, T.Y. Development of a support vector machine based cloud detection method for MODIS with the adjustability to various conditions. *Remote Sens. Environ.* **2018**, *205*, 390–407. [CrossRef]

Article

A Texture Classification Approach Based on the Integrated Optimization for Parameters and Features of Gabor Filter via Hybrid Ant Lion Optimizer

Mingwei Wang [1,*], Lang Gao [2], Xiaohui Huang [2], Ying Jiang [3] and Xianjun Gao [4]

[1]   Institute of Geological Survey, China University of Geosciences, Wuhan 430074, China
[2]   School of Computer Science, China University of Geosciences, Wuhan 430074, China; langgao@cug.edu.cn (L.G.); xhhuang@cug.edu.cn (X.H.)
[3]   Changjiang River Scientific Research Institute, Changjiang Water Resources Commission, Wuhan 430010, China; jiangying@mail.crsri.cn
[4]   School of Geoscience, Yangtze University, Wuhan 430100, China; junxgao@whu.edu.cn
*   Correspondence: wangmingwei@cug.edu.cn; Tel.: +86-137-2012-1823

Received: 15 March 2019; Accepted: 16 May 2019; Published: 28 May 2019

Abstract: Texture classification is an important topic for many applications in machine vision and image analysis, and Gabor filter is considered one of the most efficient tools for analyzing texture features at multiple orientations and scales. However, the parameter settings of each filter are crucial for obtaining accurate results, and they may not be adaptable to different kinds of texture features. Moreover, there is redundant information included in the process of texture feature extraction that contributes little to the classification. In this paper, a new texture classification technique is detailed. The approach is based on the integrated optimization of the parameters and features of Gabor filter, and obtaining satisfactory parameters and the best feature subset is viewed as a combinatorial optimization problem that can be solved by maximizing the objective function using hybrid ant lion optimizer (HALO). Experimental results, particularly fitness values, demonstrate that HALO is more effective than the other algorithms discussed in this paper, and the optimal parameters and features of Gabor filter are balanced between efficiency and accuracy. The method is feasible, reasonable, and can be utilized for practical applications of texture classification.

Keywords: texture classification; Gabor filter; parameter optimization; feature selection; hybrid ant lion optimizer

---

1. Introduction

Texture [1,2] is a core property of object appearance in natural scenes, ranging from large-scale samples to microscopic ones. It is also active visual information that is used to describe and recognize objects in the real environment. Texture classification is one of the key problems in texture analysis, and it has been a long-standing research topic because of its importance in understanding the process of texture classification by humans and its extensive applications in computer vision and image analysis [3]. The main applications of texture classification include understanding medical images, extracting visible objects, retrieving content-based images, inspecting industrial faults [4–7], and so on.

In general, the primary focus of studies on texture classification has been the determination of methods to extract texture features. It is generally believed that the extraction of powerful texture features is more important than that of weak texture features since they do not lead to good classification results, even when using excellent classifiers [8]. However, texture features can be found in various orientations and at different scales, and these cannot be characterized effectively by commonly used methods [9–12]. Gabor filter has been used for this purpose and performs better in the discrimination of individual texture features, especially those with similar descriptions. However, a single filter is difficult to apply to multi-orientation and multiscale texture features. Thus, a bank of Gabor filters with the ability to extract multi-orientation and multiscale texture features was proposed to address the issue. Gabor filter has been extensively used for texture feature extraction and texture classification [13]. Li [14] captured the dependence of Gabor filter on different channels, and had better performance results than that of state-of-the-art approaches. Younesi [15] proposed a palm print recognition method that used a bank of Gabor filters to extract texture features from images, and the method achieved higher accuracy than texture classification approaches based on single Gabor filter. Huang [16] proposed a new technique of identifying group-housed pigs on the basis of Gabor filter, and the experimental results demonstrate that the accuracy outperformed the compared approach by 91.86%, and PCA parameter drifted in the range of 0.85–0.9. Lu [17] presented a texture classification method for fracture risk estimation based on Gabor filter combined with the total T-score, and the shape texture features could be measured in a wider area compared with other methods. Kim [18] proposed a novel texture classification technique with directional statistical (DS)-Gabor filter and had satisfactory rotation invariance owing to the combination of a number of directional statistics. However, the parameters of the above method need to be set by experience, thus it may not be adaptable to different types of texture features. Furthermore, the parameter settings of each Gabor filter are crucial to classification accuracy.

On the other hand, the process of obtaining the optimal parameters of Gabor filter can be seen as a combinatorial optimization problem that can be tackled by swarm intelligence algorithms. For example, Khan [19,20] presented a technique for the optimization of each filter individually via particle swarm optimization (PSO) and cuckoo search (CS) algorithm, and the approach successfully represents local texture changes at multiple scales and orientations from mammograms with an effective improvement in classification accuracy. Tong [21] proposed a defect detection technique via Gabor filter to inspect flaws in woven fabrics in the fashion industry, and the differential evolution (DE) algorithm was utilized to obtain the optimal parameters of the filter bank. The method achieved high successful detection and low false alarm rates. However, the optimization ability of PSO, DE, and CS algorithms is mainly based on a random search, which cannot ensure convergence toward the optimal solution. Moreover, because of the multi-orientation and multiscale nature of the filter bank, there is redundant information in the process of texture feature extraction, and it contributes little to the classification [22]. Thus, numerous feature selection approaches based on Gabor filter have been proposed to reduce the data dimension [23,24], but the previous works have ignored the relevance of parameter optimization and feature selection. In essence, parameter optimization and feature selection are both considered combinatorial optimization problems, and the integrated optimization of the parameters and features of Gabor filter could be obtained by swarm intelligence algorithms at the same time.

Ant lion optimizer (ALO) [25] is a novel swarm intelligence algorithm. Nowadays, ALO has been applied in various fields, such as power system design, fault detection, schedule planning, path searching and so on [26–29]. The optimization ability of ALO does not rely on any parameters, and it is more likely to obtain the satisfactory solution. However, feature selection is considered a discrete optimization problem and is difficult to solve using ALO with decimal coding. Mafarja [30] presented a binary coded ALO (BALO) to select the optimal feature subset for some well-known datasets from the UCI repository. BALO obtained superior results by searching for the best feature subset, and the performance was independent of the step-length and classifier. Hence, in this paper, a new texture classification method is proposed that blends the use of Gabor filter and a hybrid binary-decimal coded ALO (HALO) to, respectively, solve the problems of parameter optimization and feature selection.

The rest of this paper is structured as follow. The basic principle of HALO is illustrated in Section 2. In Section 3, the proposed method to obtain the integrated optimization of the parameters and features of Gabor filter is detailed. Section 4 displays the experimental results and discussion. Finally, the paper is concluded in Section 5.

2. Overview of HALO

2.1. ALO for Parameter Optimization

In 2015, Mirjalili introduced a swarm intelligence algorithm called ALO, which imitates the hunting behavior of antlions and has no parameters to be set [25]. In nature, an antlion larva digs a cone-shaped pit in the sand by moving along a circular path and throwing out sand with its massive jaw, and the larva then hides underneath the bottom of the cone and waits for ants to be trapped in the pit. The exploratory behavior of ALO is similar to an antlion's digging in a circular path. Additionally, the exploitation behavior of ALO is similar to the boundary adjustment of antlions' traps. As each ant in ALO moves randomly in the solution space, its strategy is similar to that of spinning a roulette wheel and is represented below:

$$X^t = [0, cumsum(2r(t_1) - 1), cumsum(2r(t_2) - 1), ...,$$
$$cumsum(2r(t_n) - 1)] \tag{1}$$

where X^t is the position of ants after a random walk, n is the number of ants in the population, *cumsum* represents the cumulative sum, t is the iteration number, and $r(t)$ is a stochastic distribution function as defined below:

$$r(t) = \begin{cases} 1 & if\ rand > 0.5 \\ 0 & if\ rand \leq 0.5 \end{cases} \tag{2}$$

where *rand* is a random number taken from a uniform distribution in the interval of [0,1].

To maintain the random walk in the scope of the search space, the positions of each ant are normalized by min-max normalization as defined by the following:

$$X_i^t = \frac{(X_i^t - a_i)(b_i - c_i^t)}{d_t^t - a_i} + c_i^t \tag{3}$$

where a_i is the minimum of the random walk of ith variable, b_i is the maximum of the random walk of ith variable, c_i^t is the minimum of ith variable at iteration t, and d_i^t indicates the maximum of ith variable at iteration t.

Antlions build traps that are proportional to the size of the solution space, and the ants move stochastically. Moreover, antlions shoot sand outward from the center of the pit once they realize that an ant is in the trap to keep the ant trapped as it tries to escape. To mathematically define this behavior, the radius of an ant's random walk on a hypersphere is reduced adaptively, and the process of trapping can be described as follows:

$$c^t = \frac{c^t}{I} \tag{4}$$

$$d^t = \frac{d^t}{I} \tag{5}$$

where $I = 10^\omega \frac{t}{T}$, T is the maximum number of iterations, and ω is a constant variable that is defined based on the current iteration ($\omega = 2$ when $t > 0.1\ T$, $\omega = 3$ when $t > 0.5\ T$, $\omega = 4$ when $t > 0.75\ T$, $\omega = 5$ when $t > 0.9\ T$, and $\omega = 6$ when $t > 0.95\ T$). Basically, ω is utilized to adjust the optimization level of exploration.

2.2. Binary Coding for Feature Selection

In general, feature selection is regarded as a discrete combinatorial optimization problem and is difficult to solve directly by decimal coding. In the binary coding environment, moving through each band means that the position of each individual changes from 0 to 1 or vice versa [31]. To introduce a binary coding form of ALO, the coding process of each ant can be seen as continuous optimization of similarity. The major difference between the decimal and binary coding form is that ants are characterized by a realistic value or by switching to "0" or "1" in the transform function. That is to say, the position of each ant calculated by Equation (3) is set to either 0 or 1 with a certain probability and calculated with a conditional function, as shown in Equation (6):

$$R_i^t = \begin{cases} 1 & if \quad rand \leq |\tanh(X_i^t)| \\ 0 & otherwise \end{cases} \tag{6}$$

where $tanh(\cdot)$ is the hyperbolic tangent function and R_i^t is the form of binary coding for the position of each ant that is utilized to solve the problem of feature selection.

To summarize, in this paper, HALO is proposed with the combination of binary and decimal coding to, respectively, solve the problems of parameter optimization and feature selection using an ant, and the aim of this approach is to obtain the optimal classification accuracy for different texture features. The overall operating process of the proposed texture classification method is expressed in the following section.

3. The Proposed Method

An efficient texture classification method that applies the integrated optimization of the parameters and features of Gabor filter using HALO is presented in this section. The main goal of texture classification is to precisely classify several texture features in each sample.

3.1. Fundamental of Gabor Filter

Gabor filter is calculated by a convolutional kernel function, which has been utilized in image processing applications, especially for texture classification. To represent multiscale and multi-orientation texture features, a filter bank is structured with a number of parameters in different orientations and scales. The essential variable $g(x,y)$ of a 2-D Gabor filter is expressed by a Gaussian kernel and adjusted by a complex sinusoidal wave: [32]:

$$g(x,y) = \left(\frac{1}{2\pi\sigma_x\sigma_y}\right) exp\left(-\frac{1}{2}\left(\frac{\tilde{x}^2 + \tilde{y}^2}{\sigma_x^2 + \sigma_y^2}\right) + 2\pi jW\tilde{x}\right) \tag{7}$$

$$\tilde{x} = xcos\theta + ysin\theta \quad \tilde{y} = -xsin\theta + ycos\theta \tag{8}$$

where σ_x and σ_y are the parameters that describe the spread of the current pixel in the neighborhood in which weighted summation occurs, W is the central frequency of the complex sinusoid, and $\theta \in [0, \pi)$ is the orientation of the horizontal to vertical stripes in the equation above.

A bank of Gabor filters includes multiple individuals and can be adjusted for different orientations and frequencies. Gabor filter considered here is GS5O8 (including 40 filters: 5 scales $S \times 8$ orientations O with an initial max frequency equal to 0.2 and an initial orientation set to 0. The parameter settings of the orientations and frequencies for a bank are tuned by the following formulas [33]:

$$Orientation(i) = \frac{(i-1)\pi}{O}, where\ i = 1, 2, ..., O \tag{9}$$

$$Frequency(i) = \frac{f_{max}}{\sqrt{(2)^{i-1}}}, where\ i = 1, 2, ..., S \tag{10}$$

where O and S are, respectively, the orientations and scales in total, and f_{max} is the maximum value of the frequency and is set to $f_{max} = 0.2$.

3.2. Encoding Schema of Parameter Optimization and Feature Selection

The key objective of the proposed texture classification method is to achieve satisfactory parameters and features of Gabor filter and create a reasonable mapping between the solution and HALO. The parameter optimization process needs $O + S$ bits to apply coding to the parameters, which are the orientations θ and frequencies W in Equations (9) and (10). The former O bits represent each orientation, the latter S is each scale, and the ranges of the parameters are, respectively, set to $(0, 1)$ and $(0, 0.2]$ with decimal coding. For feature selection, the length of coding is equal to the total number of features in the database, and here it is set to $O \times S$. For binary coding, every bit is expressed by the fixed value "0" or "1" by Equation (6), where "1" illustrates that the current feature will be chosen for classification, and "0" illustrates that the current feature will be abandoned. Thus, the total coding length of an ant is equal to $O \times S + O + S$. The encoding form of an ant for Gabor filter with O orientations and S scales is shown in Figure 1. The entire code simultaneously indicates the solution of the parameter settings and the feature subset of Gabor filter.

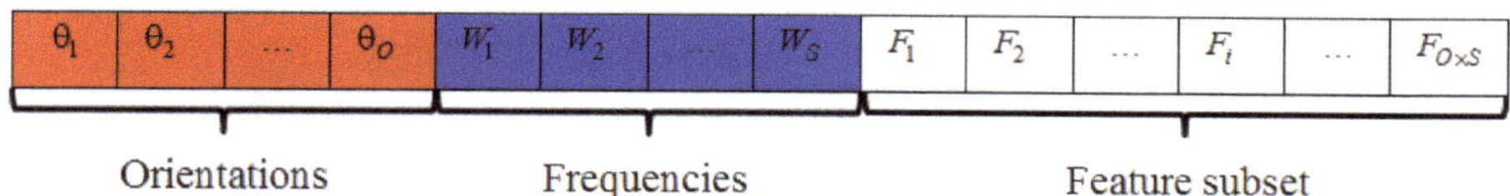

Figure 1. The encoding form of an ant for Gabor filter.

3.3. Definition of Objective Function

To evaluate the optimization ability of HALO and other evolutionary algorithms, it is necessary to choose a suitable objective function. Because the process of training can be considered prior knowledge, each texture feature is a category. The Fisher criterion has good performance in the calculation of heuristic knowledge. The Fisher criterion aims to maximize the inter-class difference and minimize the intra-class difference, and it is usually applied to assess binary classification problems [34]. For multi-classification problems, the objective function should be improved to synthetically weight the differences between each category. Therefore, it is defined as shown in Equation (11):

$$fit = \frac{\sum_{i=1}^{l-1} \sum_{j=i+1}^{l} |\mu_i - \mu_j|}{\sum_{i=1}^{l} \sigma_i^2} \tag{11}$$

where μ_i and σ_i are, respectively, the average and the standard deviation of the eigenvalues for ith category of texture features, and l is the number of classes. A larger value resulting from the above equation demonstrates better-quality parameters and features for Gabor filter.

3.4. Implementation of the Proposed Technique

The overall scheme of the proposed texture classification technique is shown below. First, it performs encoding for each ant of HALO, and the process of parameter optimization and feature selection is then constructed by binary and decimal encoding. The result constitutes a feature vector that is input into a classifier and yields a final label for each sample.

The value of the feature vector is composed of different orientations and frequencies of Gabor filter, and each feature vector is normalized to a uniform magnitude in the range of [0,1]. In general, the process of texture classification has two stages: training and testing. The former stage provides a certain extent of prior knowledge for classification, and the latter stage verifies the availability of the current model [35].

Training samples with different scales and rotations are utilized to avoid overfitting for fair performance evaluation. However, at the same time, the volume of the training dataset is increased by the 40 filters. As a result, interval sampling [36] with a length of 8 pixels and max-pooling with a 4×4 window is utilized for each sample to decrease the data volume in the process of convolution.

- Step 1: Extract the training and testing samples from each database.
- Step 2: For each ant, generate the decimal coding for parameter optimization, and the binary coding for feature selection.
- Step 3: Perform convolution between each training sample and the filter bank.
- Step 4: Reduce the data dimension via interval sampling and max-polling, and compute the fitness value by using Equation (11).
- Step 5: Operation of HALO:

  Step 5-1: Trap ants in antlion's pits by using Equations (4) and (5).
  Step 5-2: Random walk ants by using Equation (3).
  Step 5-3: Change the encoding of feature selection to the binary form by Equation (6);

- Step 6: Perform convolution between each training sample and the filter bank and build the feature subset according to the binary encoding.
- Step 7: Reduce the data dimension via interval sampling and max-polling, and compute the fitness value by using Equation (11).
- Step 8: If the solution is better, replace the position of the current ant; otherwise, do not change the position and find the current global best solution.
- Step 9: Judge whether the terminal criterion is satisfied: if it is, go to Step 10; otherwise, go to Step 5.
- Step 10: Output the optimal feature subset, and conduct classification for testing samples.

4. Experimental Results and Discussion

The proposed method was implemented in the language of MATLAB 2014b on a personal computer with a 2.30 GHz CPU and 8.00 GB RAM on a Windows 8 operating system. To assess the quality of the proposed technique, public databases were utilized to extract features based on Gabor filter with several orientations and frequencies, as described in this section.

4.1. Databases Description

To evaluate the performance of the proposed texture classification method optimized by HALO, three public texture databases were used in the experiment. The first database CGT [37] offers digital pictures of all sorts of materials with the pictures of fabric, wood, metal, bricks, plastic, and these texture images can be used for graphic design and visual effects. In the experiments, 18 homogeneous texture images from the database were as shown in Figure 2, all chosen texture images without any rotation and 10 images for each class were utilized as training samples, while the other 50 images were used as testing samples.

Figure 2. Samples of the 18 categories randomly selected from the CGT database.

The second database Kylberg [38] was imaged under only one light setting from one direction on the same distance. Textured surfaces are arranged, such as oatmeals, linseeds, lentils, the texture samples with the same category have 12 different angles of rotation with 30 degrees increment. In the experiments, 20 homogeneous texture images from the database were as shown in Figure 3, and images without any rotation were utilized as training samples with the number per category set as 15, and 60 images with other angles of rotation per category were used as testing samples.

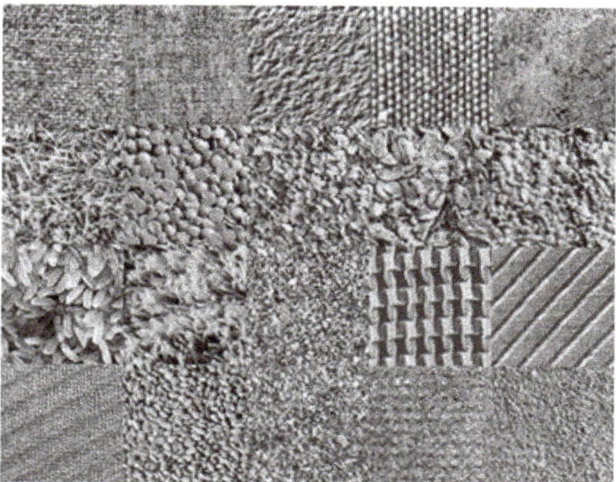

Figure 3. Samples of the 20 categories randomly selected from the Kylberg database.

The third database is Brodatz [39] with different background intensities, and the figure below gives an example of 40 different texture features organized into 5 columns. For example, D6 has a black background, whereas D10 has gray and white backgrounds, and D101 is a regular texture, whereas the texture type of D111 is irregular. In the experiments, 30 homogeneous texture images from the database were as shown in Figure 4, all of the images were rotated with a 20° step, images with 20° rotation were utilized as training samples with the number set as 6 for each class. The rest of the 48 images were utilized as testing samples.

Figure 4. Samples of the 30 categories randomly selected from the Brodatz database.

4.2. Parameters Setting for Different Algorithms

As detailed in Section 2, the optimization ability does not rely on any parameter settings in HALO, thus it is prevented from becoming trapped in the local optimal solution to a great extent. In addition, some commonly used swarm intelligence algorithms, such as PSO algorithm [40], DE algorithm [41], CS algorithm [42] and gray wolf optimizer (GWO) [43], were used for an intuition comparison between the optimization abilities. All of these algorithms are based on hybrid decimal and binary coding. Moreover, other types of Gabor filter were used for further comparison. For a relatively fair comparison, the number of function evaluations was used as the terminal criterion; that is, all algorithms stopped when the iteration number reached 20 combined with part of experimental results as the fitness value did not improve, and all algorithms performed 30 independent operations. Our primary interest was the integrated optimization of parameters and features of Gabor filter, and this was shown by the fitness value of the objective function and the classification accuracy for the testing samples in each database. Table 1 shows the parameter settings of the above algorithms.

Table 1. Parameter settings for different algorithms.

Parameters	Value
Population size	8
Dimension	40 + 13
Number of runs for each technique	30
c_1, c_2 Acceleration coefficient in PSO algorithm	2.0
f_m Mutation factor in DE algorithm	0.6
C_R Crossover rate in DE algorithm	0.9
G_0 Initial of gravitational variable in GSA	100
α User specified constant in GSA	10
p_a Detecting probability in CS algorithm	0.25
β Parameter in CS algorithm	1.5
a Correlation coefficient in GWO	[2,0]
r_1, r_2 Random vectors in GWO	[0,1]

4.3. Experiments for Different Swarm Intelligence Algorithms

A preliminary test of the proposed texture classification approach on three public texture databases, namely CGT, Kylberg, and Brodatz, was conducted, as described here. The size of the training samples was extracted as 128 × 128 with different angles of rotation. In Table 2, *Fiv* is the average fitness value calculated by Equation (11) using the integrated optimization for each filter bank handled by different swarm intelligence algorithms, and *Fn* and Time, respectively, indicate the selected number of features and CPU time, on average, for each training process.

Table 2. Result of different algorithms for public texture databases.

Database	Meas.	PSO	DE	CS	GWO	HALO
	Fiv	15.3503	16.1756	16.9323	17.5238	18.3323
CGT	*Fn*	20.7667	20.3333	19.8667	19.5000	19.0333
	Time	11.2552	10.9532	10.8711	11.0332	10.6732
	Fiv	15.9803	16.8198	18.1516	19.2445	19.7469
Kylberg	*Fn*	20.8667	20.4333	20.2000	19.6667	19.2000
	Time	15.3124	14.9997	14.9056	15.1336	14.6008
	Fiv	13.2545	14.2234	14.9204	16.0043	16.9607
Brodatz	*Fn*	18.7333	18.4000	17.8333	17.2667	16.9000
	Time	25.2872	24.9488	24.8150	25.0316	24.5248

As shown in Table 2, the average fitness value using the proposed method was the highest for all databases, proving that the optimization ability of HALO was superior to that of PSO, DE, CS, and GWO. The discrimination ability of each category was enhanced since the fitness value exceeded 16. More importantly, the process of ALO is based on the hunting behavior of antlions and has no parameters to be set. Hybrid decimal and binary encoding qA utilized to conduct integrated optimization of the parameters and features of Gabor filter; this strategy improved the exploitation ability compared with the use of only one ant encoding. Moreover, the selected number of features was the lowest among all of the algorithms involved: the number was higher than 20 for the CGT and Kylberg databases using PSO and DE algorithms, and HALO abandoned more than 50% of the redundant features from Gabor filter. From the aspect of operating efficiency, HALO had a faster convergence rate because it had fewer multiplications compared with the other algorithms. The difference in CPU time reached 0.7 s for the Kylberg and Brodatz databases, and HALO only needed

10.6732 s to select the best feature subset by obvious distinction for the CGT database. Overall, it can be deduced that the optimization ability and operating efficiency are improved by using HALO, which has the desired adaptability to obtain suitable parameters and features of Gabor filter.

4.4. Application for Texture Classification

Next, each texture sample was classified by using the optimal parameters and features of Gabor filter. Figures 5–7 indicate the difference in the average eigenvalues between the training and testing samples for the categories with the highest classification accuracy using the selected parameters and features of Gabor filter. Tables 3–5 show the classification accuracy using the newly proposed Log-Gabor filter [44], DS-Gabor filter [18], the only parameter optimization-based Gabor (OP-Gabor) filter, and the proposed method. In the tables, OA and Kappa are the overall classification accuracy and Kappa coefficient, respectively, obtained by using different texture classification methods. The Kappa coefficient is defined below:

$$k = \frac{P_o - P_e}{1 - P_e} \tag{12}$$

where P_o is the relative observed agreement, and P_e is the hypothetical probability of chance agreement.

Table 3. Overall classification accuracy and Kappa coefficient for CGT database.

Category	Log-Gabor	DS-Gabor	OP-Gabor	Proposed	Category	Log-Gabor	DS-Gabor	OP-Gabor	Proposed
Fabric	96.00	96.00	96.00	100.00	Brick	48.00	54.00	86.00	86.00
Leather	92.00	90.00	94.00	92.00	Vegetation	86.00	80.00	88.00	90.00
Wicker	64.00	90.00	96.00	96.00	Window	94.00	94.00	94.00	96.00
Metal	94.00	94.00	96.00	96.00	Tree	70.00	60.00	70.00	90.00
Plastic	98.00	98.00	98.00	100.00	Wood	96.00	94.00	98.00	96.00
Wheel	46.00	66.00	78.00	86.00	Soil	100.00	100.00	100.00	100.00
Road	88.00	74.00	92.00	90.00	Wool	96.00	96.00	96.00	100.00
Paper	100.00	100.00	100.00	100.00	Water	98.00	98.00	100.00	100.00
Decoration	60.00	98.00	98.00	100.00	Tile	94.00	76.00	96.00	98.00
					OA(%)	84.44	86.56	93.11	95.33
					Kappa	0.8353	0.8576	0.9271	0.9506

Table 4. Overall classification accuracy and Kappa coefficient for Kylberg database.

Category	Log-Gabor	DS-Gabor	OP-Gabor	Proposed	Category	Log-Gabor	DS-Gabor	OP-Gabor	Proposed
Blanket	96.67	83.00	96.67	96.67	Rice	75.00	81.67	86.67	90.00
Canvas	91.67	91.67	91.67	93.33	Rug	66.67	75.00	95.00	96.67
Ceiling	93.33	88.33	95.00	96.67	Sand	55.00	71.67	86.67	93.33
Cushion	86.67	81.67	88.33	91.67	Scarf	96.67	96.67	98.33	98.33
Floor	83.33	86.67	90.00	91.67	Screen	93.33	93.33	95.00	100.00
Grass	70.00	58.33	76.67	90.00	Seat	96.67	96.67	96.67	98.33
Lentils	73.33	76.67	88.33	95.00	Sesameseed	81.67	93.33	100.00	100.00
Linseed	73.33	90.00	95.00	96.67	Stone	78.33	86.67	90.00	91.67
Oatmeal	65.00	80.00	91.67	95.00	Stoneslab	35.00	38.33	56.67	73.33
Perlsugar	83.33	90.00	91.67	91.67	Wall	90.00	91.67	95.00	96.67
					OA(%)	79.25	82.58	90.25	93.83
					Kappa	0.7816	0.8167	0.8974	0.9351

Table 5. Overall classification accuracy and Kappa coefficient for Brodatz database.

Category	Log-Gabor	DS-Gabor	OP-Gabor	Proposed	Category	Log-Gabor	DS-Gabor	OP-Gabor	Proposed
D1	87.50	97.92	97.92	97.92	D49	100.00	100.00	100.00	100.00
D3	75.00	97.92	97.92	97.92	D52	85.42	93.75	97.92	97.92
D6	97.92	95.83	97.92	97.92	D56	95.83	89.58	95.83	95.83
D10	100.00	100.00	100.00	100.00	D64	85.42	87.50	97.92	97.92
D11	97.92	97.92	100.00	100.00	D66	89.58	97.92	97.92	97.92
D14	97.92	93.75	97.92	97.92	D74	95.83	95.83	95.83	95.83
D18	81.25	91.67	93.75	93.75	D83	95.83	97.92	97.92	97.92
D20	91.67	95.83	95.83	95.83	D87	89.58	87.50	95.83	95.83
D25	95.83	95.83	95.83	95.83	D88	83.33	89.58	95.83	95.83
D31	70.83	85.42	91.67	91.67	D93	87.50	91.67	93.75	93.75
D37	85.42	95.83	95.83	95.83	D94	62.50	81.25	89.58	89.58
D41	97.92	97.92	97.92	97.92	D99	83.33	87.50	91.67	91.67
D46	93.75	95.83	95.83	95.83	D101	91.67	97.92	100.00	100.00
D47	93.75	95.83	95.83	95.83	D104	93.75	89.58	95.83	95.83
D48	85.42	95.83	95.83	95.83	D111	85.42	87.50	93.75	93.75
					OA(%)	89.24	93.61	96.32	96.32
					Kappa	0.8886	0.9339	0.9619	0.9619

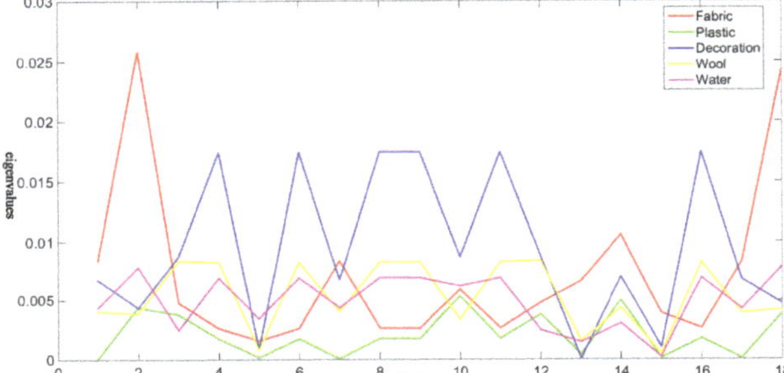

Figure 5. Difference of average eigenvalues between training and testing samples for CGT database.

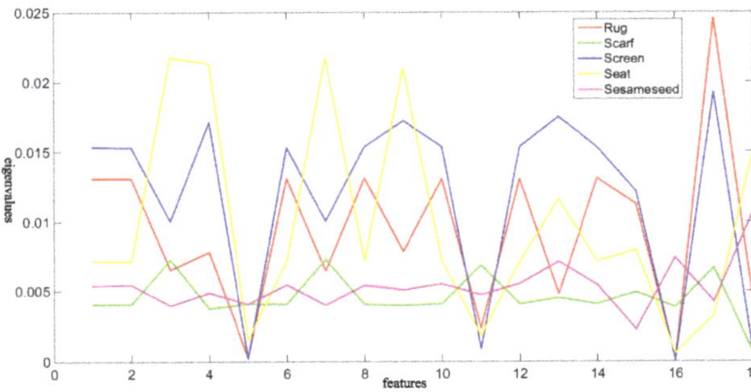

Figure 6. Difference of average eigenvalues between training and testing samples for Kylberg database.

Tables 3–5 reveal that the classification accuracy was increased, with a difference of more than 2%, by removing some redundant features of Gabor filter. In addition, more than 40 samples were misclassified for the Kylberg database. Although the classification accuracy for the Leather, Road, and Wood classes in the CGT database was relatively high, misclassification still had a certain influence on the overall process. Moreover, the Kappa coefficient was more than 0.93 for all databases, illustrating that the precision could adapt to a number of application demands. The difference in the average eigenvalues between the training and testing samples was lower than 0.03, and the change trend was similar to that shown in Figures 5–7, thus proving the discriminability and identity of each Gabor filter. The overall classification accuracy of the Log-Gabor filter and DS-Gabor filter was unsatisfactory: it was lower than 85% for the CGT database and only 86.56% and 82.58%, respectively, for the Kylberg database. With these two methods, very few samples were correctly classified for the Wheel, Brick, Sand, and Stone slab categories, thus it was difficult for them to extract texture features from the whole database. In brief, the proposed approach is a reliable, efficient, and reasonable method for texture classification.

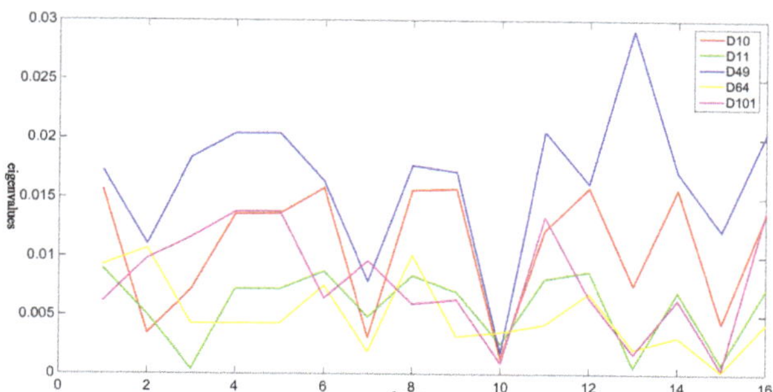

Figure 7. Difference of average eigenvalues between training and testing samples for Brodatz database.

5. Conclusions

This paper details a texture classification method based on the integrated optimization of the parameters and features of Gabor filter using HALO. Three public texture databases with different types of texture features were utilized for its evaluation. The experimental results were firstly compared with those of some commonly used swarm intelligence algorithms, such as PSO, DE, CS, and GWO. In general, it was demonstrated that swarm intelligence algorithms are well coded to solve parameter optimization and feature selection problems at the same time. Among them, HALO with hybrid binary-decimal coding has great optimization ability, and its fitness value is distinctly higher than that of the other algorithms. Thus, the proposed method is more appropriate for texture classification, and it is fast enough to meet real-time application needs. Moreover, for a more comprehensive comparison, Log-Gabor, DS-Gabor filter, and the only parameter optimization-based Gabor filter were also utilized. It was observed that the proposed texture classification method is robust, and the classification accuracy is satisfactory for multi-classification problems, especially those based on texture features. In sum, the multi-orientation and multiscale nature of Gabor filter can enhance the discrimination of texture features. Further, the disadvantage of high time complexity can be overcome to a great extent with HALO. The proposed texture classification method has an excellent balance between efficiency and accuracy, making it a good candidate to deal with practical applications. In the future, we will collect some geological texture samples and use them for the classification of more texture features.

Author Contributions: M.W. methodology; L.G. validation; X.H. formal analysis; Y.J. investigation; X.G. writing.

Funding: This work was funded by the National Key Research & Development Program of China under Grant No. 2017YFC1502406-03; the National Natural Science Foundation of China under Grant No. U1711266; and the Open Fund of State Laboratory of Information Engineering in Surveying, Mapping and Remote Sensing, Wuhan University under Grant No. 18R04.

Conflicts of Interest: The authors declare no conflict of interest. The funders had no role in the design of the study; in the collection, analyses, or interpretation of data; in the writing of the manuscript, or in the decision to publish the results.

References

1. Haralick, R.M. Statistical and structural approaches to texture. *Proc. IEEE* **2005**, *67*, 786–804. [CrossRef]
2. Chen, W.; Li, X.; He, H.; Wang, L. Assessing different feature sets' effects on land cover classification in complex surface-mined landscapes by ZiYuan-3 satellite imagery. *Remote Sens.* **2017**, *10*, 23. [CrossRef]

3. Dong, Y.; Feng, J.; Liang, L.; Zheng, L.; Wu, Q. Multiscale sampling based texture image classification. *IEEE Signal Process. Lett.* **2017**, *24*, 614–618. [CrossRef]

4. Zhou, W.; Zhang, L.; Wang, K.; Chen, S.; Wang, G.; Liu, Z.; Liang, C. Malignancy characterization of hepatocellular carcinomas based on texture analysis of contrast-enhanced MR images. *J. Magn. Reson. Imaging* **2017**, *45*, 1476–1484. [CrossRef]

5. Liang, Y.; Li, Y.; Zhao, K; Meng, L. Object Tracking Algorithm based on Multi-channel Extraction of AHLBP Texture Features. In Proceedings of the International Conference on Advanced Mechatronic Systems (ICAMechS), Zhengzhou, China, 30 August–2 September 2018; pp. 332–336.

6. Banerjee, P.; Bhunia, A.K.; Bhattacharyya, A.; Roy, P.P.; Murala, S. Local Neighborhood Intensity Pattern—A new texture feature descriptor for image retrieval. *Expert Syst. Appl.* **2018**, *113*, 100–115. [CrossRef]

7. Islam, R.; Uddin, J.; Kim, J.M. Texture analysis based feature extraction using Gabor filter and SVD for reliable fault diagnosis of an induction motor. *Int. J. Inf. Technol. Manag.* **2018**, *17*, 20–32. [CrossRef]

8. Yuan, B.; Xia, B.; Zhang, D. Polarization Image Texture Feature Extraction Algorithm Based on CS-LBP Operator. *Procedia Comput. Sci.* **2018**, *131*, 295–301. [CrossRef]

9. Lloyd, K.; Rosin, P.L.; Marshall, D.; Moore, S.C. Detecting violent and abnormal crowd activity using temporal analysis of grey level co-occurrence matrix (GLCM)-based texture measures. *Mach. Vis. Appl.* **2017**, *28*, 361–371. [CrossRef]

10. Omar, M.; Khelifi, F.; Tahir, M.A. Detection and classification of retinal fundus images exudates using region based multiscale LBP texture approach. In Proceedings of the International Conference on Control, Decision and Information Technologies (CoDIT), St. Julian's, Malta, 6–8 April 2016; pp. 227–232.

11. Zuniga, A.G.; Florindo, J.B.; Bruno, O.M. Gabor wavelets combined with volumetric fractal dimension applied to texture analysis. *Pattern Recognit. Lett.* **2014**, *36*, 135–143. [CrossRef]

12. Wang, S.H.; Lin, C.Y.; Fu, J.T. Gender classification based on multi-scale and run-length features. *J. Electron. Sci. Technol.* **2017**, *15*, 251–257.

13. LLee, G.G.C.; Huang, C.H.; Chen, C.F.R.; Wang, T.P. Complexity-Aware Gabor Filter Bank Architecture Using Principal Component Analysis. *J. Signal Process. Syst.* **2017**, *89*, 431–444.

14. Li, C.; Huang, Y.; Zhu, L. Color texture image retrieval based on Gaussian copula models of Gabor wavelets. *Pattern Recognit.* **2017**, *64*, 118–129. [CrossRef]

15. Younesi, A.; Amirani, M.C. Gabor filter and texture based features for palmprint recognition. *Procedia Comput. Sci.* **2017**, *108*, 2488–2495. [CrossRef]

16. Huang, W.; Zhu, W.; Ma, C.; Guo, Y.; Chen, C. Identification of group-housed pigs based on Gabor and Local Binary Pattern features. *Biosyst. Eng.* **2018**, *166*, 90–100. [CrossRef]

17. Lu, R.S.; Dennison, E.; Denison, H.; Cooper, C.; Taylor, M.; Bottema, M.J. Texture analysis based on Gabor filters improves the estimate of bone fracture risk from DXA images. *Comput. Methods Biomech. Biomed. Eng. Imaging Vis.* **2018**, *6*, 453–464. [CrossRef]

18. Kim, N.C.; So, H.J. Directional statistical Gabor features for texture classification. *Pattern Recognit. Lett.* **2018**, *112*, 18–26. [CrossRef]

19. Khan, S.; Hussain, M.; Aboalsamh, H.; Mathkour, H.; Bebis, G.; Zakariah, M. Optimized Gabor features for mass classification in mammography. *Appl. Soft Comput.* **2016**, *44*, 267–280. [CrossRef]

20. Khan, S.; Khan, A.; Maqsood, M.; Aadil, F.; Ghazanfar, M.A. Optimized gabor feature extraction for mass classification using cuckoo search for big data e-healthcare. *J. Grid Comput.* **2018**, *8*, 1–16. [CrossRef]

21. Tong, L.; Wong, W.K.; Kwong, C.K. Differential evolution-based optimal Gabor filter model for fabric inspection. *Neurocomputing* **2016**, *173*, 1386–1401. [CrossRef]

22. Murugappan, V.; Sabeenian R.S. Texture based medical image classification by using multi-scale Gabor rotation-invariant local binary pattern (MGRLBP). *Cluster Comput.* **2017**, *12*, 1–14. [CrossRef]

23. Shen, L.; Zhu, Z.; Jia, S.; Zhu, J.; Sun, Y. Discriminative Gabor feature selection for hyperspectral image classification. *IEEE Geosci. Remote Sens. Lett.* **2013**, *10*, 29–33. [CrossRef]

24. Hussain, M. False positive reduction using Gabor feature subset selection. In Proceedings of the International Conference on Information Science and Applications (ICISA), Suwon, South Korea, 24–26 June 2013; pp. 1–5.

25. Mirjalili, S. The ant lion optimizer. *Adv. Eng. Softw.* **2015**, *83*, 80–98. [CrossRef]

26. Saikia, L.C.; Sinha, N. Automatic generation control of a multi-area system using ant lion optimizer algorithm based PID plus second order derivative controller. *Int. J. Electr. Power Energy Syst.* **2016**, *80*, 52–63.

27. Mouassa, S.; Bouktir, T.; Salhi, A. Ant lion optimizer for solving optimal reactive power dispatch problem in power systems. *Eng. Sci. Technol. Int. J.* **2017**, *20*, 885–895. [CrossRef]

28. Petrović, M.; Petronijević, J.; Mitić, M.; Vuković, N.; Miljković, Z.; Babić, B. The ant lion optimization algorithm for integrated process planning and scheduling. *Appl. Mech. Mater. Trans Tech Publ.* **2016**, *834*, 187–192. [CrossRef]

29. Yao, P.; Wang, H. Dynamic Adaptive Ant Lion Optimizer applied to route planning for unmanned aerial vehicle. *Soft Comput.* **2017**, *21*, 5475–5488. [CrossRef]

30. Mafarja, M.M.; Mirjalili, S. Hybrid binary ant lion optimizer with rough set and approximate entropy reducts for feature selection. *Soft Comput.* **2018**, *6*, 1–17. [CrossRef]

31. Shen, F.; Zhou, X.; Yang, Y.; Song, J.; Shen, H.T.; Tao, D. A fast optimization method for general binary code learning. *IEEE Trans. Image Process.* **2016**, *25*, 5610–5621. [CrossRef]

32. Song, X.; Liu, F.; Zhang, Z.; Yang, C.; Luo, X.; Chen, L. 2D Gabor filters-based steganalysis of content-adaptive JPEG steganography. *Multimedia Tools Appl.* **2017**, *76*, 26391–26419. [CrossRef]

33. Kong. R.; Zhang, B. Design of Gabor filters' parameter. *Control Decis.* **2012**, *27*, 1277–1280.

34. Lu, J.; Liu, F.; Luo, X. Selection of image features for steganalysis based on the Fisher criterion. *Digit. Investig.* **2014**, *11*, 57–66. [CrossRef]

35. Sheppard, JM.; Young, W.B. Agility literature review: Classifications, training and testing. *J. Sports Sci.* **2006**, *24*, 919–932. [CrossRef]

36. Shen, B.; Wang, Z.; Huang, T. Stabilization for sampled-data systems under noisy sampling interval. *Automatica* **2016**, *63*, 162–166. [CrossRef]

37. Textures.com, CGT Database. 26 September 2014. Available online: https://www.textures.com/ (accessed on 25 February 2019).

38. Kylberg, G. Kylberg Database. 1 September 2011. Available online: http://www.cb.uu.se/gustaf (accessed on 25 February 2019).

39. Picard, R.W.; Kabir, T.; Liu, F. Real-time recognition with the entire Brodatz texture database. In Proceedings of the IEEE Conference on Computer Vision and Pattern Recognition, New York, NY, USA, 15–17 June 1993; pp. 638–639.

40. Kenney, J. Particle swarm optimization. In Proceedings of the IEEE International Conference on Neural Networks, Perth, Australia, 27 November–1 December 1995; pp.1942–1948.

41. Storn, R; Price, K. Differential evolution—A simple and efficient heuristic for global optimization over continuous spaces. *J. Glob. Optim.* **1997**, *11*, 341–359. [CrossRef]

42. Yang, X.S.; Deb, S. Cuckoo search via Lévy flights. In Proceedings of the World Congress on Nature & Biologically Inspired Computing (NaBIC), Coimbatore, India, 9–11 December 2009; pp. 210–214.

43. Mirjalili, S.; Mirjalili, S.M.; Lewis, A. Grey wolf optimizer. *Adv. Eng. Softw.* **2014**, *69*, 46–61. [CrossRef]

44. Nunes, F.G.; Pádua, L.C. A local feature descriptor based on Log-Gabor filters for key point matching in multispectral images. *IEEE Geosci. Remote Sens. Lett.* **2017**, *14*, 1850–1854. [CrossRef]

Article

Real-Time Automated Segmentation and Classification of Calcaneal Fractures in CT Images

Wahyu Rahmaniar [1,2,*] **and Wen-June Wang** [1]

[1]   Department of Electrical Engineering, National Central University, Zhongli 32001, Taiwan
[2]   Department of Computer Science and Electronics, Universitas Gadjah Mada, Yogyakarta 55281, Indonesia
*   Correspondence: wahyurahmaniarncu@g.ncu.edu.tw or wahyurahmaniar@yahoo.com

Received: 14 June 2019; Accepted: 23 July 2019; Published: 26 July 2019

Featured Application: Segmentation and classification of calcaneal fractures.

Abstract: Calcaneal fractures often occur because of accidents during exercise or activities. In general, the detection of the calcaneal fracture is still carried out manually through CT image observation, and as a result, there is a lack of precision in the analysis. This paper proposes a computer-aid method for the calcaneal fracture detection to acquire a faster and more detailed observation. First, the anatomical plane orientation of the tarsal bone in the input image is selected to determine the location of the calcaneus. Then, several fragments of the calcaneus image are detected and marked by color segmentation. The Sanders system is used to classify fractures in transverse and coronal images into four types, based on the number of fragments. In sagittal image, fractures are classified into three types based on the involvement of the fracture area. The experimental results show that the proposed method achieves a high precision rate of 86%, with a fast computational performance of 133 frames per second (fps), used to analyze the severity of injury to the calcaneus. The results in the test image are validated based on the assessment and evaluation carried out by the physician on the reference datasets.

Keywords: biomedical imaging; bone fracture; calcaneus; CT image; segmentation

---

1. Introduction

The calcaneus is the largest tarsal bone that has the responsibility of supporting the axial load of the body's weight [1]. A calcaneal fracture is the most common in the tarsal bone fractures, most of which are intra-articular fractures, which usually occur as a result of falling from a height, sports injuries, and vehicle accident [2]. The severity of fracture displacement and the extent of soft tissue injury are directly related to the amount of energy absorbed by the limbs in producing injury [3]. Injuries with higher energies produces soft tissue disorders that are more severe and may cause open fractures [4]. Bleeding fractures into the plane of the fascial, which surrounds the heel, produce severe pain above the fracture and may result in a leg compartment syndrome [5]. Furthermore, calcaneal fractures have presented a significant challenge for orthopedic surgeons in patients' treatment [6,7]. So, detection of the calcaneal fracture is an important subject for patient diagnostic decisions and treatment planning [8,9].

Modern calcaneal fracture classification systems rely on three-dimensional computed tomography (CT) rather than two-dimensional conventional radiography [10]. Although a CT image contains a significant amount of medical information, it is not accurate enough to examine fractures through manual visual inspection due to its low resolution [11]. Thus, details, such as the skeletal structures, the boundary between internal organs and bone, and soft tissue in the bone may not be accurately seen and assessed depending on the experience of the physician [12]. Therefore, a computer-aided method for fracture detection is needed which can significantly assist physicians to examine CT images, and it is crucial for physicians to make diagnostic decisions, as well as plan treatments, based on this information [13]. In addition, with a computer-aid, faster and more detailed fracture detection can be achieved.

Recently, several approaches have been proposed to detect bone fractures in CT images of various areas of the human body. Wu et al. [14] proposed a method for automatic fracture detection of CT images of traumatic pelvic injuries. These fractures are detected using a segmentation technique, which consists of four main parts: Pre-processing, edge detection, shape matching, and Registered Active Shape Model (RASM) with an automatic initialization. However, this segmentation method only applies to the reference frame, which was generated based on previously known pelvic bone anatomy information. Roth et al. [15] implemented a method for the automatic detection of posterior element fractures on spinal CT images. This method used the multi-atlas label fusion to segment the spinal vertebral body and computed the edge map of its posterior elements. But, this segmentation region is predicted on the set of probabilities for fractures along the edges of the image, which are limited to the spine. These aforementioned works [14,15] show that a computer-aided method provides more accurate results for fracture detection and has the potential to accelerate the assessment of trauma cases, reduce the possibility of misclassification of bone fractures, and reduce variability between observers. But, these methods cannot be applied to the calcaneus bones, which have different shapes, features, and types of fracture than their study. In the case of the calcaneus, segmentation and detection of the calcaneal fracture are very challenging, due to the low resolution of CT images, the complex anatomy, and soft tissue structures of the calcaneus [16,17]. At present, there are deficiencies in the operative or conservative management planning of the calcaneus, which are caused by the lack of a standard system in the classification of calcaneal fractures [18].

A study to apply image processing for the detection and classification of calcaneal fractures has not been widely performed. Pranata et al. [19] has proposed automated classification and detection of calcaneus bone fracture locations in CT images. This method classifies the calcaneal fractures into two classes: Fractured and non-fractured. However, the classification results do not provide more detailed conclusions about the fractures type of each anatomical plane orientation, where the identification of fracture alone is not sufficient to assess injury severity. Moreover, the orientation of the calcaneus anatomical plane is manually registered and the computation time is 5 min and 10 s, which are not fast enough for real-time applications.

In this paper, a new method for the automated segmentation and detection of calcaneal fractures for real-time applications is proposed. The main objective is to provide results for calcaneus segmentation and detection of fractures for each anatomical plane orientation, so that physicians can better, and further assess fractures in the calcaneus region with a shorter processing time. One of the challenges in detecting calcaneal fractures is the different shape of the calcaneus in each patient's CT image [10–12]. As such, the shape approach cannot be used to find the calcaneus based on its shape in the tarsal bone. Furthermore, due to limited features and colors in the CT images, it is a difficult task to identify calcaneus in the tarsal bone. To solve this problem, the region of interest (ROI) of the calcaneus is determined based on the anatomical plane orientation of the tarsal bones, i.e., transverse, coronal, and sagittal. In addition, due to quality limitations in CT images, it is a challenging task to classify the type of fracture for calcaneus. In particular, mild and small fractures, that only appear slightly in CT images, cannot be detected a physician at first inspection because the physician may not be able to reliably call them to rely on these fractures, due to the quality of CT and the amount of data to be processed [14,15]. Thus, for fractures that appear slightly in CT images, repeated inspections are required to identify the existence and details of the fracture. Computer-based analyses can be used to process detailed information from several neighboring slices to give instructions to the physician on whether a particular slice contains a fracture, then details of the separation between pieces can be identified [12,14] can be extracted. However, due to the anatomical variability between individuals, the accuracy in segmentation and detection of fractures in the calcaneus structure remain a challenging task. This proposed method provides the calcaneus segmentation to show the details of the fractures in the calcaneus region, which can be fragments or lines. This method has performed the segmentation by detecting the calcaneus bone structure, based on the white bone area on the CT images. Then, the fracture type is classified based on the amount of the fragments and the location of the lines fractures.

The remainder of this paper is organized as follows. Section 2 introduces a literature review of the classification of calcaneal fractures and describes the main algorithm. Section 3 illustrates the performance of fracture detection results in CT images. Finally, conclusions are drawn in Section 4.

2. Materials and Methods

2.1. Materials

We obtained 2210 CT Digital Imaging and Communications in Medicine (DICOM) images from the datasets with assessment and evaluation of the fracture type information [1–13], radiopedia (https://radiopaedia.org/), anesthesia key (https://aneskey.com/fractures-and-dislocations-of-the-tarsal-bones/), and CT images from two patients with three anatomical plane orientations. The dataset consists of 815 coronal images, 777 transverse images, and 618 sagittal images. The experiment was conducted using Visual Studio C++ 2017 in the 3.40 GHz CPU with 8 GB RAM.

2.2. Fractures Classification in Calcaneus

On the basis of subtalar joint involvement, calcaneal fractures observed in CT images have been divided into two categories: Intra-articular and extra-articular. Intra-articular is a fracture involving the joints, including damage to the connective tissue between two bones. Extra-articular is a fracture that does not involve the joints, rather, it includes pieces of bone drawn from the calcaneus by the ligament.

The intra-articular fractures of the calcaneus represent about 75% of all calcaneal fractures in adults [1,2], where the prospect of recovery depends on how severely the calcaneus was damaged during injury. Several classification systems for intra-articular fractures of the calcaneus have been developed, in which the Sanders system [20,21] is the most commonly used due to its correlation with clinical outcomes and lower inter-observer variability [22]. This classification divides the intra-articular fractures into four types, based on the number of fragments in the calcaneus, i.e., Type I, Type II, Type III, and Type IV.

Figure 1 shows examples of calcaneal fractures on the transverse and coronal planes, the red circle shows the location of the calcaneus in the tarsal structure. Type I is a non-displaced extra-articular fracture with a displacement of less than 2 mm; this is also knowns as a line fracture. Types II–IV are displaced intra-articular fractures corresponding to an increase in the number of fragments, in which Type IV has more than three fragments.

Extra-articular fractures represent about 25% of calcaneal fractures and include all fractures that do not involve the posterior aspect of the subtalar joint [1,8]. Extra-articular fractures are normally caused by trauma, such as a crush or mild injury. Based on the location of the fracture, several methods [1,10,12] classify extra-articular fractures of the calcaneus into three types, i.e., Type A, Type B, and Type C. Figure 2 shows an example of fractures on the sagittal plane. The red circle shows the location of the calcaneal fractures in sagittal images. The Type A fracture involves the anterior process of the calcaneus (Figure 2b). The Type B fracture extends through the calcaneus or middle calcaneus, including the lateral processes (Figure 2c); and the Type C fracture is a calcaneal fracture involving the posterior (Figure 2d).

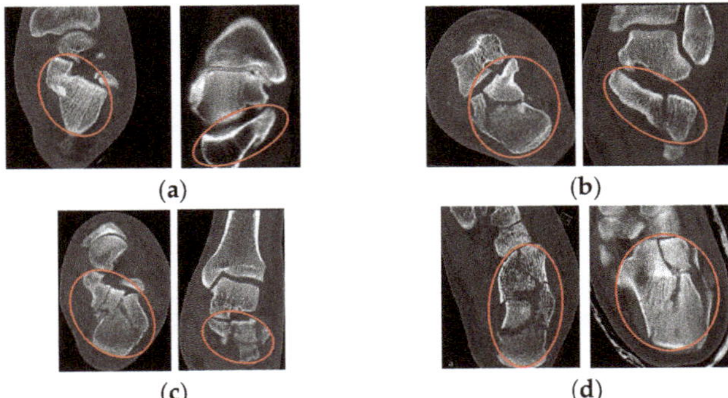

Figure 1. The Sanders system of fracture classification on the calcaneus [1,2,20]: (**a**) Normal calcaneus in transverse and coronal images; (**b**) Type II; (**c**) Type III; and (**d**) Type IV.

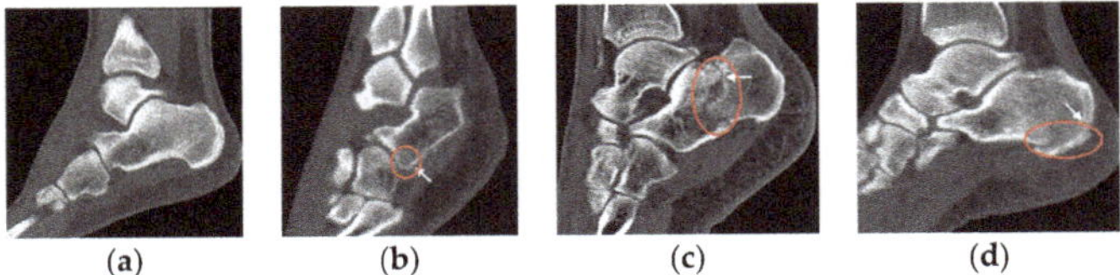

Figure 2. Classification of calcaneus fractures in sagittal images [1,8,10]: (**a**) Normal calcaneus; (**b**) Type A; (**c**) Type B; and (**d**) Type C.

2.3 System Overview

Figure 3 shows a general overview of the system proposed in this paper. The algorithm contains two steps to complete the main task: Step 1 involves the detection of the calcaneus location in the input image using a machine learning approach; and Step 2 involves the segmentation of the calcaneus ROI, based on several morphology methods, and contour detection of the fragmented region so as to determine the type of calcaneal fracture.

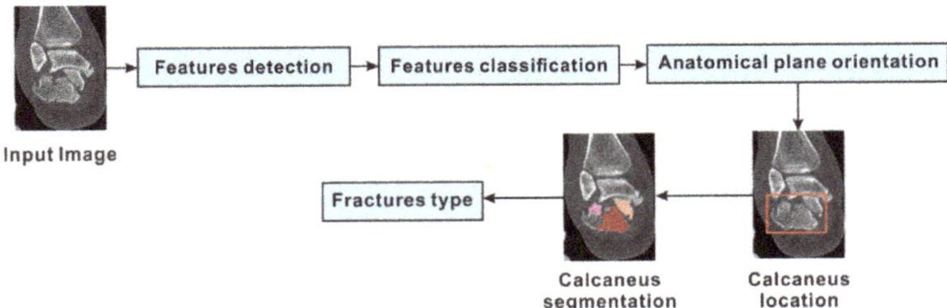

Figure 3. System overview of the proposed method.

2.4. Step 1: Detection of Calcaneus Location

In the first step, the extended local binary pattern (LBP) [23] and cascade classifier [24] are used to determine the anatomical plane orientation in the input image based on the shape of the tarsal bone. The basic idea behind LBP is that the image is composed of a micro-pattern. LBP is the first-order circular derivative pattern generated by concatenating the direction of the binary gradient. CT scans are generally available as DICOM files, where each image contains a 2D array with pixel intensity. The CT DICOM image is a grayscale image in which the bone area is represented by white pixels surrounded by gray pixels. Therefore, LBP is suitable for defining the information in CT DICOM images, based on texture descriptors.

The LBP operator is the sum of the gray-level intensities label computed at each pixel location. The LBP code labels can be expressed as:

$$LBP(S,R) = \sum_{S=0}^{S-1} b(g)2^S \tag{1}$$

where $g = g_s - g_c$, S is the number of pixels in a small circular neighborhood with radius R (R can be the value within 1–3, in this study we set $R = 1$), g_s is the grey-level intensities label of S, g_c is the grey-level intensity of the center pixel, and the function $b(g)$ is defined as:

$$b(g) = \begin{cases} 1, & \text{if } g \geq 0 \\ 0, & \text{otherwise} \end{cases} \tag{2}$$

The LBP code label histogram contains information about edge distribution and other local features in the image, so it can be used to describe the image texture in the CT DICOM image. The LBP feature is extracted from the input image at the pixel location (i,j). Then, the image is divided into several small non-overlapping blocks to get the feature histogram. The region blocks A_k are the same size, where $k = 0,\ldots,K$ and K is the number of blocks in the image. The LBP histogram of the labeled image is defined as:

$$H(L) = \sum_{(i,j)\in A_k} h_{(i,j)}(l) \tag{3}$$

where $h_{(i,j)}(l)$ is the value of the bin l which consists of a look-up table of $2^9-1 = 511$ bins, l is the LBP feature computed at location (i,j) and L is the number of different labels produced by the LBP operator. Figure 4 shows the process that is used to provide three types of information: Code labels for local textures, histograms in local regions, and feature histograms whereby the tarsal bone structure in each anatomical plane orientation is described. The feature histogram is the concatenation of each region histogram into a single LBP histogram.

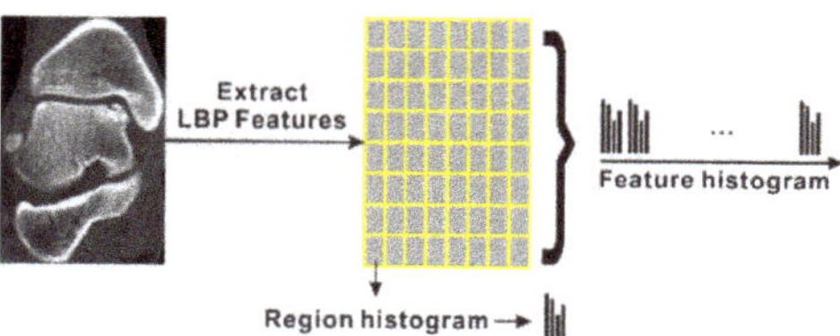

Figure 4. Input image region.

In this step, the AdaBoost algorithm [25,26] is used to select the features of the classifier in the training stage. The classifier is selected to evaluate a single LBP histogram which maximizes the margin between positive and negative samples. The classifier determines the best threshold classification function for each histogram so that the number of misclassified instances can be minimized. The negative image input of each anatomical plane is the orientation of the other plane. In the training stages, the features in positive and negative images will be learned as positive, and negative labels, respectively. Figure 5 shows the example images for training, which are the tarsal bone shape in each anatomical plane of various sizes of input images. In this study, the images used have minimum and maximum sizes of 126 × 147, and 980 × 1024, respectively.

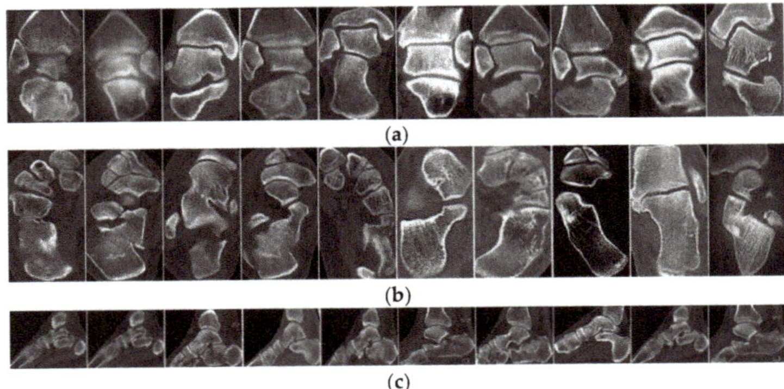

Figure 5. Example images of the tarsal bone: (**a**) Coronal; (**b**) Transverse; and (**c**) Sagittal.

A training sample is set as (x_m, y_m), $m = 1, 2, \ldots, M$, where $y_m = 0$, or 1 for negative, or positive labels, respectively, is the class label for the sample x_m. Firstly, the initial weight vector is set as $\omega_1(m) = 1/M$. The classifier is defined as $\lambda_n(x_m) = 0$, or 1 where $n = 1, 2, \ldots, N$ is the iteration number. The error associated with the classifier is evaluated as:

$$\varepsilon_n = \sum_{m=1}^{M} \omega_n(m), \text{ if } |\lambda_n(x_m) \neq y_m|. \tag{4}$$

The selected classifier is used to update the weight vector as:

$$\omega_{n+1}(m) = \omega_n(s)\beta_n^{1-\tau_m}$$
$$\text{where } \tau_m = \begin{cases} 0, \text{ if } x_m \text{ classified correctly} \\ 1, \text{ otherwise} \end{cases} \tag{5}$$

and β_n is the weighting parameter computed from:

$$\beta_n = \frac{\varepsilon_n}{1 - \varepsilon_n} \tag{6}$$

The result of the training stage is the labeled result of each region which is represented as:

$$W(x) = \begin{cases} 1, \text{ if } \sum_{n=1}^{N}\left[\lambda_n(x) \times \log\left(\frac{1}{\beta_n}\right)\right] \geq \frac{1}{2} \sum_{n=1}^{N} \log\left(\frac{1}{\beta_n}\right) \\ 0, \text{ otherwise} \end{cases} . \tag{7}$$

In the testing stages, the shape of the tarsal bones is detected by the sliding window method [24], in which each sub-window contains labels for each anatomical plane orientation from the training stage. Each area, which passes by each sub-window, is labeled at each classification stage either as, positive (1) or negative (0). If the label detected is positive, the region is recognized as the object and the classifier passes to the next stage. Otherwise, the region is rejected immediately. Then, the last stage will show the result of the detector in the current window, as shown in Figure 6. If there is more than one anatomical plane orientation detected in the input image, then the largest area of the current window is selected. The location of the calcaneus, based on the selected anatomical plane orientation, is determined according to Figure 7. Then, the calcaneus ROI is used as the location of the input image for the next step.

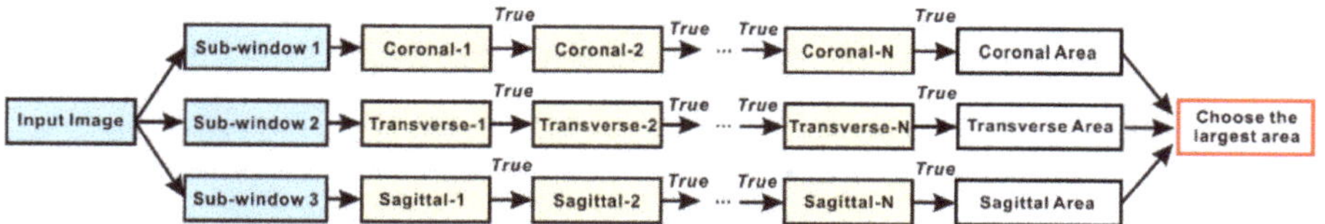

Figure 6. Cascade classifier stage to select anatomical plane orientation.

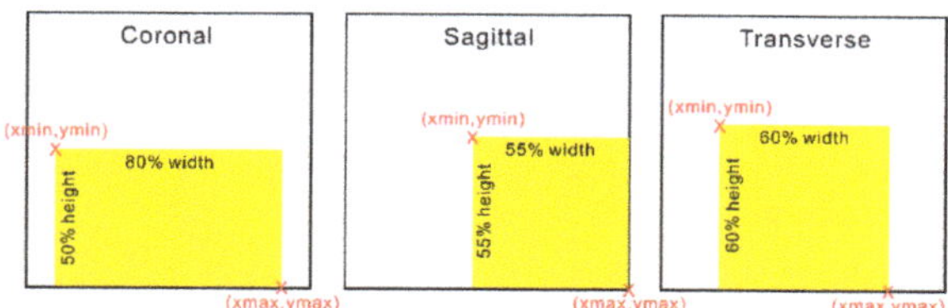

Figure 7. Region of interest (ROI) of the calcaneus based on the anatomical plane orientation.

2.5. Step 2: Segmentation of the Calcaneus Fragments

2.5.1. Classification of Calcaneal Fractures in Coronal and Transverse Images

In the next step, the regional segmentation method, based on the contour detection [27], is applied to determine the type of fracture. Figure 8 shows the steps for determining the type of calcaneal fracture in coronal and transverse images.

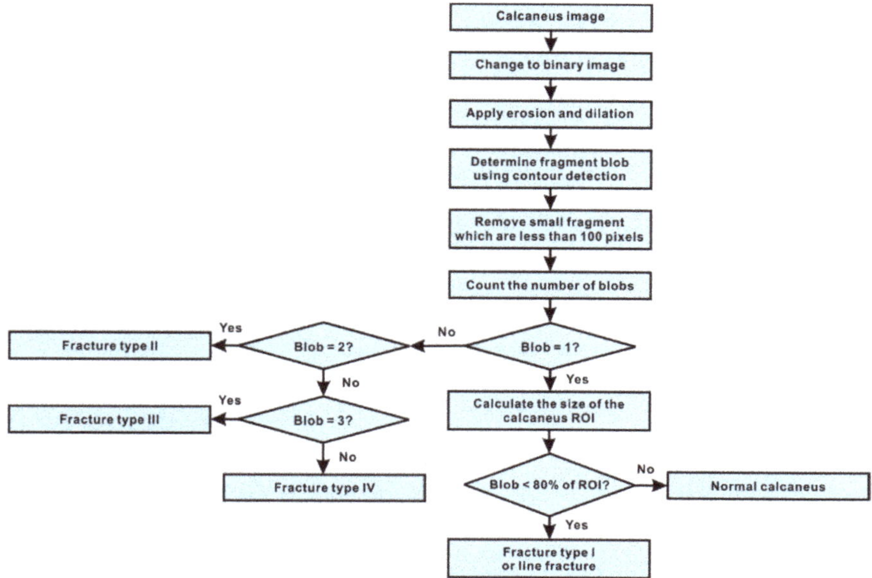

Figure 8. Algorithm for determining the type of calcaneal fracture in coronal and transverse image.

Figure 9a shows the selected tarsal bone region using LBP features and cascade classifier. Figure 9b shows the location of the calcaneus determined in accordance with Figure 7 (see Section 2.4). Several morphological operations are applied in the calcaneus region to separate the fractional area. Figure 9c shows the calcaneus region having been converted into a binary image using Otsu segmentation [28].

This step selects a threshold to minimize the intra-class variance in black (background) and white (foreground) pixels. Then, the automatic threshold selection is used to segment the image. The optimal threshold selection is based on thresholds that minimize within-class weighted variance [28] which is equivalent to maximizing the intra-class variance.

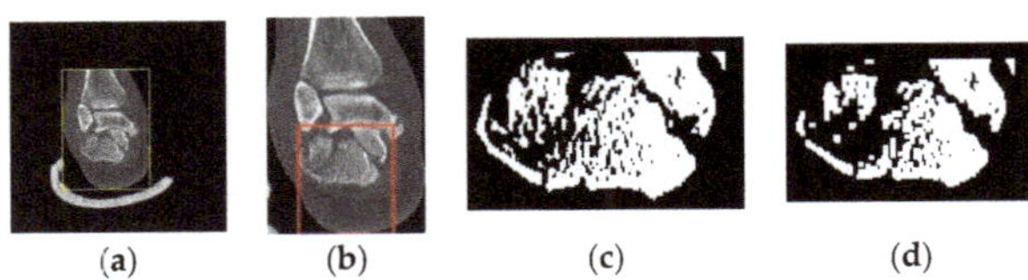

(a)   (b)   (c)   (d)

Figure 9. (**a**) Tarsal bone detection; (**b**) Calcaneus region; (**c**) Binary image of calcaneus region; and (**d**) Calcaneus region after morphological operations.

For each given threshold intra-class variance can be computed by

$$\sigma_B^2(z) \ = \ q_b(z)(1 - q_b(z))\big(\mu_b(z) - \mu_f(z)\big)^2 \tag{8}$$

where $z \ = \ 1, 2, \ldots, 256$ is the gray level of the input calcaneus image, μ_b and μ_f are the means of background and foreground, respectively, which are defined as:

$$\mu_b(z) \ = \ \sum_{i=0}^{z-1} \frac{i \times P(i)}{q_b(z)} \tag{9}$$

$$\mu_f(z) \ = \ \sum_{i=z}^{E-1} \frac{i \times P(i)}{q_f(z)} \tag{10}$$

where E is the bins of the histogram, $q_b(z) \ = \ \sum_{i=0}^{z-1} P(i)$ and $q_f(z) \ = \ \sum_{i=z}^{E-1} P(i)$ are the gray level probability distributions $P(i)$ for the background, and foreground, respectively. Then, the optimum threshold is obtained by maximizing the between-class variances.

$$Th \ = \ \arg\left(\max_{1 \le z < 256} \sigma_B^2(z) \right). \tag{11}$$

Figure 9d shows the erosion and dilation morphological operations [29], which are implemented on binary images. The erosion filter removes white pixels along the foreground boundaries in the form of fragments of the calcaneus, so that the value of neighboring pixels in the foreground becomes minimum. Dilation adds white pixels to the foreground boundaries in the image so that the neighboring pixels in the foreground can have a maximum value.

2.5.2. Contour Detection

The final step is to use the contour detection algorithm to extract regional boundaries in a calcaneal fracture. The initial value is set to 1 to determine the affiliation of the new contours and other contours. The outer border is the boundary with white pixels (1-component) which are enclosed by the black pixels (0-component). The hole border is the 1-component boundary which is enclosed by the 0-component. The frame is the boundary of the image that is wrapped by 1-component. NBD is the number of contours in the current calculation. LNBD is the last border number encountered when scanning a starting point. Figure 10 shows the connected components and borders in the image.

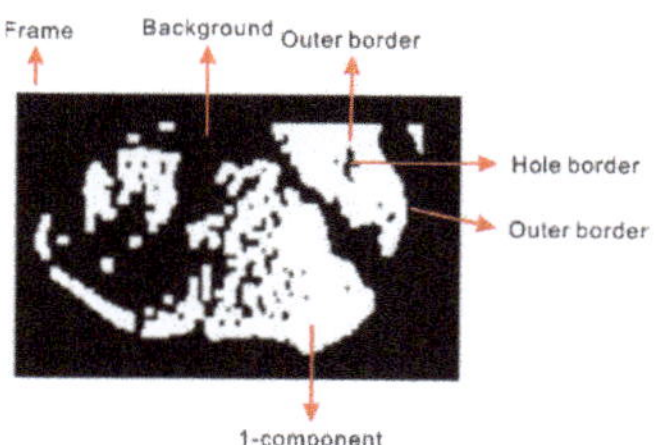

Figure 10. Connected components and borders.

The sub-steps for detection of contour of the calcaneus ROI are as follows:

1. Find the starting point of the contour

Start from the upper-left pixel at location (0,0) of the calcaneus ROI, then the image is scanned row-by-row. This condition indicates that the outer contour that has not been passed as the pixel value is changed in the next contour. Values greater than 1 indicates an 8-connected case that can represent the hole contours that have not been passed.

2. Contour Following

Next, start from the starting point that obtained from the previous step. Then the image is scanned row by row and the edges of 1-component is recorded. The square-tracing algorithm [30] is used since it is suitable for 4-connected cases as follows:

a. If the current pixel value is 1, change the scan direction to the left and move 1 pixel. Conversely, when the pixel value is 0, change the scan direction to the right and move 1 pixel.
b. Continue sub-steps 2a until the current contour point returns to the starting point.

The outer contour is the contour that passes through the entire contour in different directions. Then, change the pixel value in the following contour procedure. If the neighboring point $(p + 1, q)$ of the current contour point (p, q) has a pixel value of 0, set the current point to $-NBD$. In addition, if the current point value is more than 1 NBD, it remains unchanged. Otherwise, it is set to NBD.

Algorithm 1: Build Contour Hierarchies.

S = *corresponding LNDB*
S' = *new contour found*
If S = *OUTER* and S' = *OUTER*
  If S = *HOLE* and S' = *HOLE*
    Build the same hierarchy parent for S and S'
    Add S to the last of children linked list of S'
  Else
    Let S' be the parent of S
    If $SS'S$ child = 0
      Let S to be the first child of S'
    Else
      Add S to the last of children linked list of S'
    End
  End
End

---

3. Build Contour Hierarchies

The next sub-step is to find out the relationship between contours. NBD that hits the contour during step 1 will be scanned and recorded in LNBD. Each boundary extracted is stored as a point vector which is retrieved and reconstructed with a full hierarchy of nested contours. Then, each contour is encoded with four points. The result is a closed two-dimensional contour for each remaining region as shown in Figure 11a. The contour boundary is a dense set of sequential neighbor points called a blob. A blob larger than 100 pixels is selected and the minimum closing circle is found so that each blob represents one fragment as shown in Figure 11b. The algorithm is shown as follows:

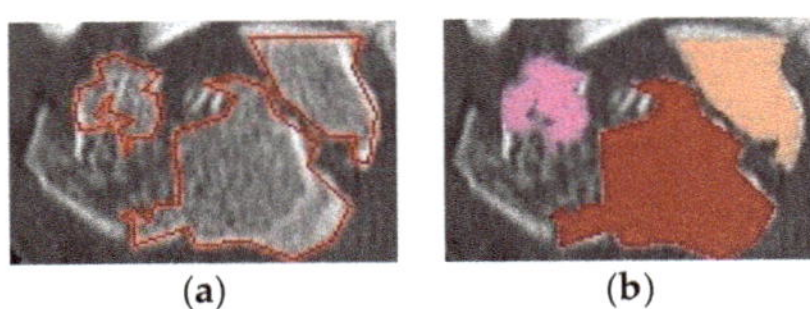

(a)   (b)

Figure 11. Fragment segmentation: (**a**) Contour detection and (**b**) Color segmentation.

In transverse and coronal images, if the calcaneus is segmented into a single full blob then it is classified as a normal calcaneus. If the calcaneus is detected as one blob but the area is not full, it is classified as type I (line fracture). Otherwise, the type of fracture is classified by a number of blobs.

2.5.3. Classification of Calcaneal Fractures in Sagittal Images

Figure 12 defines the steps for determining the type of calcaneal fracture in sagittal images. The calcaneus region is divided into three equal areas where each area represents an extra-articular region: Type A (left), Type B (middle), and Type C (right). The fracture classification in sagittal image is based on the area of the fracture line. However, it is slightly difficult to separate the lines from the bone structure. Thus, the median filter [31] is applied to reduce the ambiguity between a bone structure and fracture, as shown in Figure 13a. Then, the automatic contrast enhancement [32] is implemented to obtain a clear fragment separation in bone structure, as shown in Figure 13b. The image histogram, which shows the relationship between the gray level and the corresponding frequency, can be expressed as:

$$I_{hist}(z) = \frac{\Delta_z}{\Delta} \tag{12}$$

where Δ_z is the number of pixels in z, and Δ is the total number of pixels in the input image. The histogram equalization (HE) accumulates a histogram of the pixel values in the input image, then displaces all pixel values to enhance the contrast. HE with interval $[0, D-1]$ can be computed by:

$$Map_z = F(j) = (D-1)\sum_{z=1}^{j-1} I_{hist}(z) \tag{13}$$

where Map_z indicates an $F(j)$ mapping function that maps every pixel value j from the input image to Map_z, and D is the dynamic range of HE in the output image.

Figure 13c shows the Otsu segmentation, which is implemented to convert Figure 13b into a binary image. Figure 13d shows the erosion and dilation which makes the fracture line more obvious. Figure 13e shows the fracture lines and fragmented areas using a contour detection algorithm (see Section 2.5.2). The largest blob is selected as the calcaneus region and other blobs can be recognized as fractures. Figure 13e indicates that there is more than one fracture line detected in the calcaneus region. Then, the largest line is selected to determine the type of fracture, based on the line position in the calcaneus body, as shown in Figure 14a.

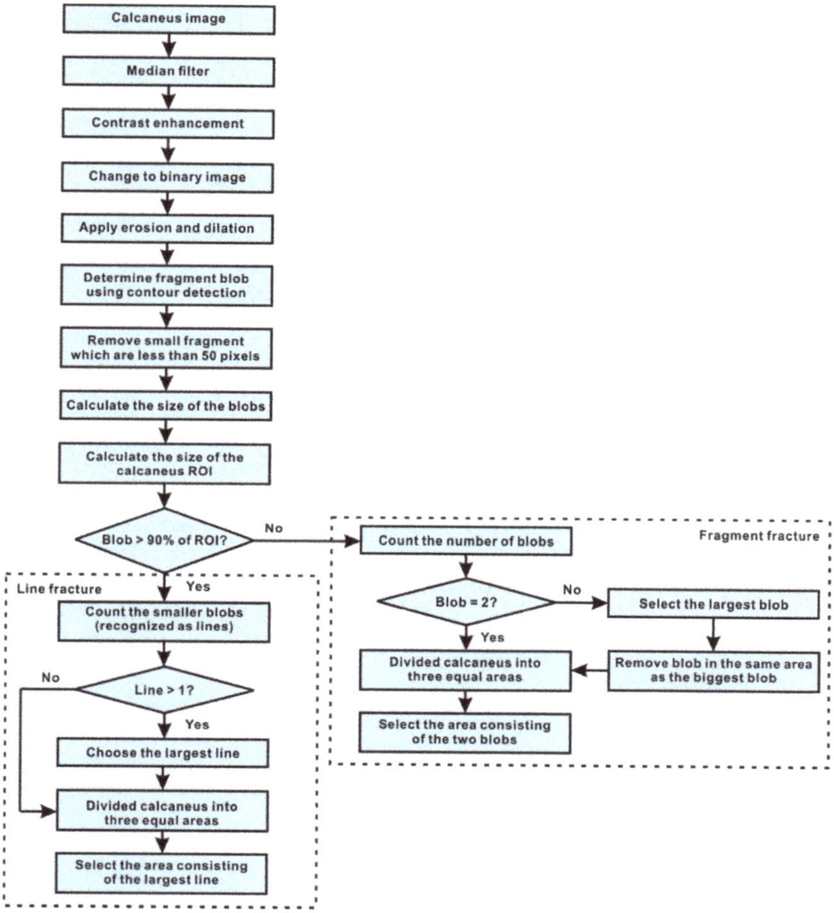

Figure 12. Algorithm for determining the type of calcaneal fracture in sagittal images.

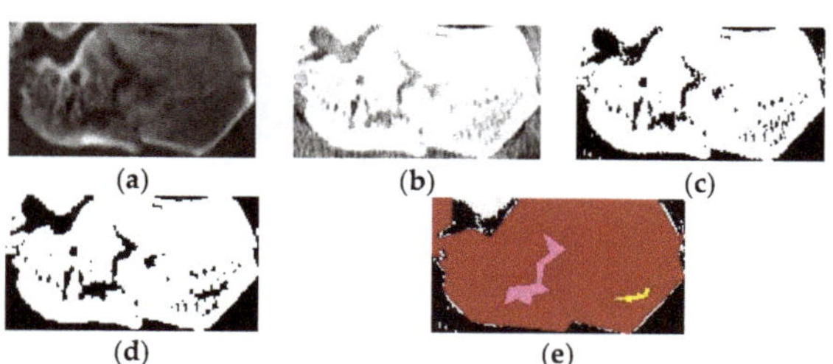

Figure 13. (**a**) Apply median filter; (**b**) Contrast enhancement; (**c**) Otsu segmentation; (**d**) Morphological operations; and (**e**) Fracture detection.

Figure 14a indicates the area of the selected fracture line. In Figure 14b, the calcaneal fracture is classified as Type B, since most of the area of the fracture is in the middle region. In the case of fragments, the region of each fragment is selected, as shown in Figure 15a. Figure 15b shows the fracture type (called Type B in the middle region), which is determined by the area between the two fragments.

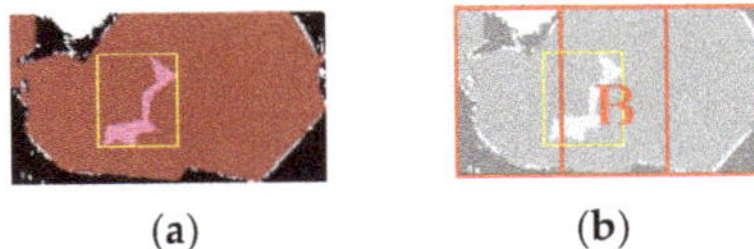

(**a**)   (**b**)

Figure 14. Steps to determine the type of fracture in case of a line fracture: (**a**) The region of the fracture line is represented by a yellow box and (**b**) The fracture type is based on the line region.

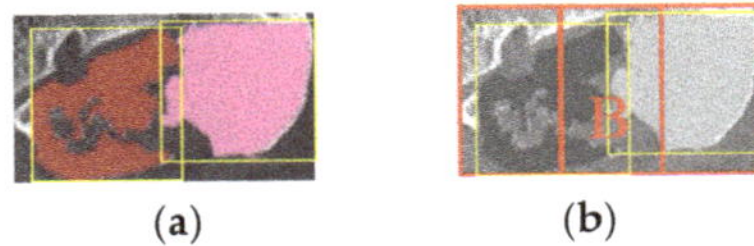

(**a**)   (**b**)

Figure 15. Steps to determine the type of fracture in case of fragment fracture: (**a**) The region of the fragment is represented by a yellow box and (**b**) The fracture type is based on the area between the two fragments.

Figure 16 shows another case of fragment detection. Figure 16a shows the area of two fragments detected in the same region. Since the smaller fragment is inside the larger fragmented region, only the largest fragment will be selected as shown in Figure 16. Figure 16c indicates this calcaneal fracture is classified as Type A, since the left area consists of two fragmented regions.

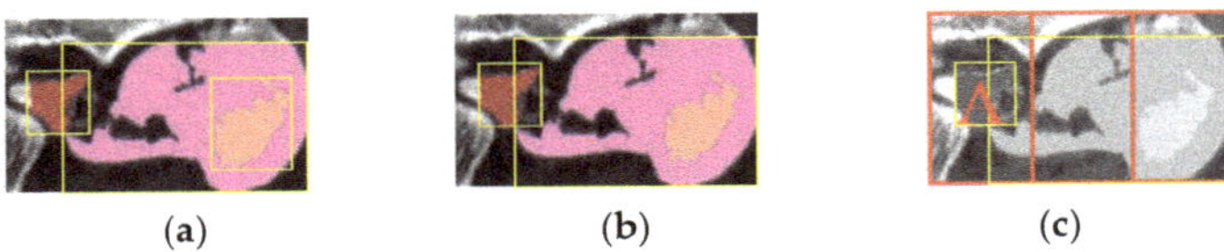

(**a**)   (**b**)   (**c**)

Figure 16. (**a**) Two fragments are in the same region; (**b**) The largest fragment is selected; and (**c**) The fracture type is based on the area between these two fragments.

3. Experimental Results and Discussion

Figures 17 and 18a show the calcaneus classified as non-fracture and Type I, in which both have one segmented region, represented by only one color of the calcaneus segmentation. In non-fractured calcaneus images, the calcaneus region may be fully segmented. In Type I fracture images, the calcaneus region cannot be fully segmented, as there are several fracture lines in the calcaneus body.

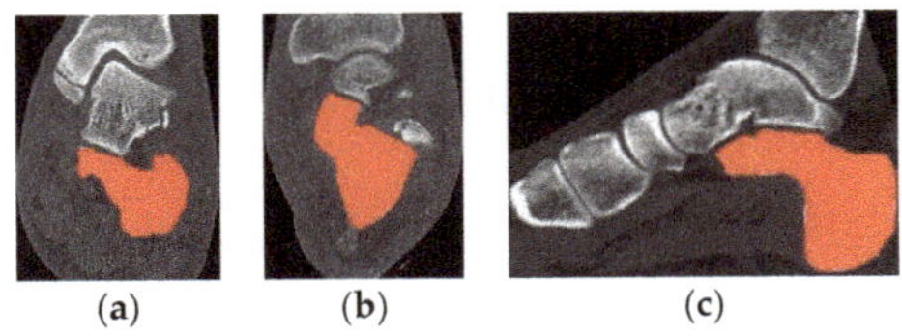

(**a**)   (**b**)   (**c**)

Figure 17. Normal calcaneus segmentation: (**a**) Coronal; (**b**) Transverse; and (**c**) Sagittal.

Figure 18 shows each color on the calcaneus segmentation representing one fragment. Although the type of fracture can be determined based on the number of fragments, the location of the fragments in the calcaneus is different with different shapes. In images of Type II fractures, failure of segmentation results will cause errors in the detection of fracture types. Thus, the calcaneus will be detected as a Type III fracture. This is due to the ambiguity in the bone structure that causes errors in separating a single blob into two blobs. In images of Type III fractures, failure of segmentation results will cause errors in detecting two blobs as a single blob. This is due to the ambiguity of the CT image, which makes the algorithm unable to separate the white pixels between the two fragments.

Due to several factors mentioned, the most accurate detection results relate to Type IV fracture images. Since Type IV fractures have more than three fragments, some errors in the segmentation results do not significantly affect the detection of fracture types, which are determined by the number of fragments.

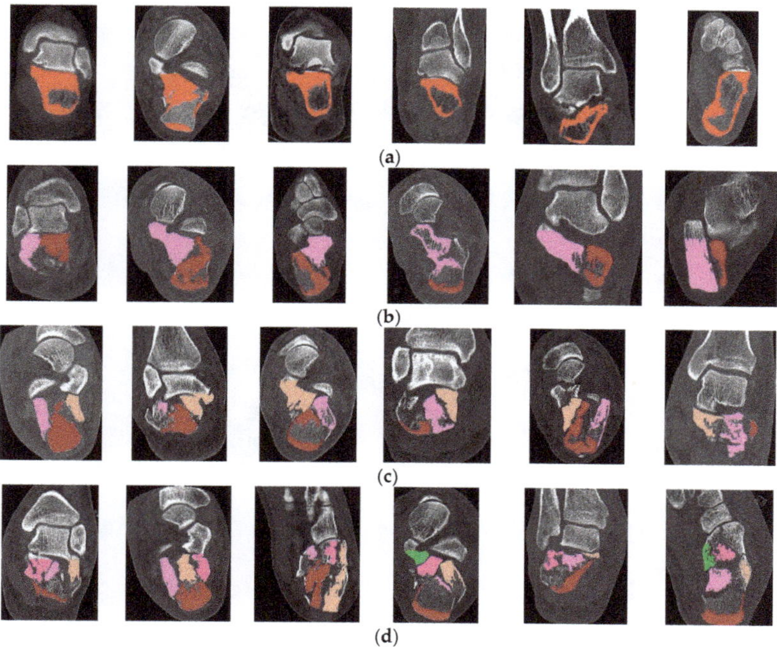

Figure 18. Segmentation result of the calcaneal fractures based on Sanders classification: (**a**) Type I; (**b**) Type II; (**c**) Type III; and (**d**) Type IV.

Table 1 shows the accuracy of fracture detection in coronal and transverse images. In coronal images, fragment separation is shown more clearly than in transverse images. In transverse images, the bone structure is slightly ambiguous, thus making segmentation fail to separate the foreground (bone structure) and background. If the segmentation fails, the fragment region cannot be filled as a single blob and it will be considered as an additional fragment. The proposed method achieves an average accuracy of 87.25% for coronal images and 83.25% for transverse images.

Table 1. Accuracy of the fractures type detection based on Sanders classification.

Fractures Type	Coronal (%)	Transverse (%)
Type I	88	84
Type II	86	81
Type III	83	80
Type IV	92	88
Average	87.25	83.25

Figure 19 shows the fracture segmentation results in sagittal images. Table 2 shows the accuracy of the fracture type detection in sagittal images. In sagittal images, fractures in the calcaneus region are too ambiguous and almost similar to the bone structure. Common mistakes in fracture detection in sagittal image are as follows:

- Bone structure is detected as a fracture,
- The fracture in the image is too similar to the bone structure, so it is not recognized as a fracture.

The most accurate results in sagittal image is for Type B fractures, where the area of the fracture is located in the middle of the calcaneus. In images with type A (left) and type C (right) fractures, the fracture line is too vague within the bone structure, so that some errors occur in the segmentation result. The proposed method achieves an average accuracy of 82% for sagittal images. Thus, the average accuracy performance for fracture type detection in the calcaneus is 84.17%.

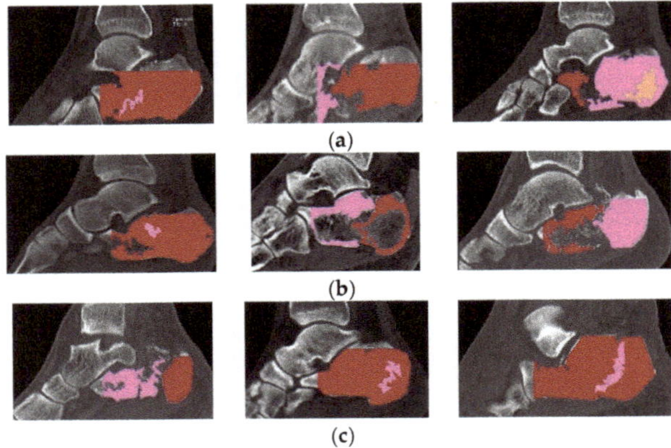

Figure 19. Segmentation result of the calcaneal fractures in sagittal images: (**a**) Type A; (**b**) Type B; and (**c**) Type C.

Table 2. Accuracy of the fractures type detection in sagittal image.

Fractures Type	Sagittal (%)
Type A	81
Type B	85
Type C	80
Average	82

The results of computation performance are summarized in Table 3 in respect of frames per second (fps). The average time cost is about 7.5 ms per frame or 133 fps. The fastest processing time is for sagittal images that have the smallest calcaneus region, compared to other anatomical planes. The slowest processing time is for coronal images, which have the largest calcaneus region. The part that uses the greatest cost in computational time is segmentation so that the size of the calcaneus region determines the processing speed of a CT image.

Table 3. Computational performance in fps.

Anatomical Plane	Average fps
Coronal	92.33
Transverse	146.67
Sagittal	160.33

Table 4 shows the performance accuracy in terms of True Positive (TP), False Positive (FP), False Negative (FN), Precision Rate (PR), recall, and F-measure, based on the classification results, which are indicated in the segmented area. This performance result is checked for each image manually. TP is the detected area corresponds to the associated fracture. FP is the detected area not related to the fracture. FN is the area associated with a fracture which is not detected. The accuracy of the performance can be computed by:

$$PR = \frac{TP}{TP + FP} \tag{14}$$

$$Recall = \frac{TP}{TP + FN} \tag{15}$$

$$F\text{-measure} = 2 \times \frac{PR \times Recall}{PR + Recall} \tag{16}$$

The proposed method achieves an average precision rate of 0.86 and recall of 0.89. The highest recall is 0.92 for coronal images that have a clearer separation between the bone structure on the calcaneus, and the background compared to transverse and sagittal images. The sagittal image has the lowest recall caused by ambiguity on the fracture line with the bone structure.

Table 4. Detection result performance.

Anatomical Plane	TP	FP	FN	PR	Recall	F-Measure
Coronal	92	13	7	0.87	0.92	0.89
Transverse	87	11	10	0.88	0.89	0.88
Sagittal	82	15	12	0.85	0.87	0.86
Average Accuracy	87	13	9.6	0.86	0.89	0.8

4. Conclusions

This paper presents a new method in automatically segmenting and detecting calcaneal fractures in CT images in real time applications. As shown in the result, the proposed algorithm in detecting the calcaneal fractures is relatively accurate. Using the proposed algorithm, bone fractures can be further highlighted in the processed images. This can help physicians to analyze the CT images better and increase the possibility of getting the real fracture condition. In addition, the method designed by us can estimate the distance of fracture separation and the angle between broken bone pieces, as well as other quantitative fracture assessments, which may not be easily accessed and measured through visual inspection. The proposed algorithm provides an analysis guide to fracture detection automatically with fast processing of more than one hundred images in one second. Thus, this method can help physicians to reduce decision-making and diagnostic time, which is very important for traumatic calcaneus injury.

Author Contributions: W.R. contributed to the conception of the study and wrote the manuscript; performed the experiment and data analyses; and contributed significantly to algorithm design and manuscript preparation; W.-J.W. helped perform the analysis and contribute to constructive discussions.

Funding: This research received no external funding.

References

1. Badillo, K.; Pacheco, J.A.; Padua, S.O.; Gomez, A.A.; Colon, E.; Vidal, J.A. Multidetector CT evaluation of calcaneal fractures. *Radiographics* **2011**, *31*, 81–92. [CrossRef] [PubMed]
2. Swanson, S.A.; Clare, M.P.; Sanders, R.W. Management of Intra-Articular Fractures of the Calcaneus. *Foot Ankle Clin.* **2008**, *13*, 659–678. [CrossRef] [PubMed]
3. Rammelt, S.; Zwipp, H. Calcaneus fractures: Facts, controversies and recent developments. *Injury* **2004**, *35*, 443–461. [CrossRef] [PubMed]

4. Park, C.H.; Lee, D.Y. Surgical treatment of sanders type 2 calcaneal fractures using a sinus tarsi approach. *Indian J. Orthop.* **2017**, *51*, 461–467. [CrossRef] [PubMed]

5. Feng, Y.; Shui, X.; Ying, X.; Cai, L.; Yu, Y.; Hong, J. Closed Reduction and Percutaneous Fixation of Calcaneal Fractures in Children. *Orthopedics* **2016**, *39*, e744–e748. [CrossRef] [PubMed]

6. Hoelscher-Doht, S.; Frey, S.P.; Kiesel, S.; Meffert, R.H.; Jansen, H. Subtalar Dislocation: Long-Term Follow-Up and CT-Morphology. *Open J. Orthop.* **2015**, *5*, 53–59. [CrossRef]

7. Mostafa, M.F.; El-Adl, G.; Hassanin, E.Y.; Abdellatif, M.S. Surgical treatment of displaced intra-articular calcaneal fracture using a single small lateral approach. *Strat. Trauma Limb Reconstr.* **2010**, *5*, 87–95. [CrossRef]

8. Rammelt, S.; Godoy-Santos, A.L.; Schneiders, W.; Fitze, G.; Zwipp, H. Foot and ankle fractures during childhood: Review of the literature and scientific evidence for appropriate treatment. *Rev. Bras. Ortop.* **2016**, *51*, 630–639. [CrossRef]

9. Dhillon, M.S.; Prabhakar, S. Treatment of displaced intra-articular calcaneus fractures: A current concepts review. *SICOT-J* **2017**, *3*, 59. [CrossRef]

10. Moussa, K.M.; Hassaan, M.A.E.; Moharram, A.N.; Elmahdi, M.D. The role of multidetector CT in evaluation of calcaneal fractures. *Egypt. J. Radiol. Nucl. Med.* **2015**, *46*, 413–421. [CrossRef]

11. Daftary, A.; Haims, A.H.; Baumgaertner, M.R. Fractures of the Calcaneus: A Review with Emphasis on CT. *Radiographics* **2005**, *25*, 1215–1226. [CrossRef] [PubMed]

12. Galluzzo, M.; Greco, F.; Pietragalla, M.; De Renzis, A.; Carbone, M.; Zappia, M.; Maggialetti, N.; D'andrea, A.; Caracchini, G.; Miele, V. Calcaneal fractures: Radiological and CT evaluation and classification systems. *Acta Biomed.* **2018**, *89*, 138–150. [PubMed]

13. Harnroongroj, T.; Suntharapa, T.; Arunakul, M. The new intra-articular calcaneal fracture classification system in term of sustentacular fragment configurations and incorporation of posterior calcaneal facet fractures with fracture components of the calcaneal body. *Acta Orthop. Traumatol. Turc.* **2016**, *50*, 519–526. [CrossRef] [PubMed]

14. Wu, J.; Davuluri, P.; Ward, K.R.; Cockrell, C.; Hobson, R.; Najarian, K. Fracture Detection in Traumatic Pelvic CT Images. *Int. J. Biomed. Imaging* **2012**, *2012*, 1–10. [CrossRef] [PubMed]

15. Roth, H.R.; Wang, Y.; Yao, J.; Lu, L.; Burns, J.E.; Summers, R.M. Deep convolutional networks for automated detection of posterior-element fractures on spine CT. *Med. Imaging 2016 Comput. Aided Diagn.* **2016**, *9785*, 97850.

16. Long, C.; Fang, Y.; Huang, F.G.; Zhang, H.; Wang, G.L.; Yang, T.F.; Liu, L. Sanders II–III calcaneal fractures fixed with locking plate in elderly patients. *Chin. J. Traumatol.* **2016**, *19*, 164–167. [CrossRef] [PubMed]

17. Biz, C.; Barison, E.; Ruggieri, P.; Iacobellis, C. Radiographic and functional outcomes after displaced intra-articular calcaneal fractures: A comparative cohort study among the traditional open technique (ORIF) and percutaneous surgical procedures (PS). *J. Orthop. Surg. Res.* **2016**, *11*, 92. [CrossRef] [PubMed]

18. Dhillon, M.S.; Bali, K.; Prabhakar, S. Controversies in calcaneus fracture management: A systematic review of the literature. *Musculoskelet. Surg.* **2011**, *95*, 171–181. [CrossRef]

19. Pranata, Y.D.; Wang, K.C.; Wang, J.C.; Idram, I.; Lai, J.Y.; Liu, J.W.; Hsieh, I.H. Deep learning and SURF for automated classification and detection of calcaneus fractures in CT images. *Comput. Methods Programs Biomed.* **2019**, *171*, 27–37. [CrossRef]

20. Sanders, R.; Fortin, P.; DiPasquale, T.; Walling, A. Operative treatment in 120 displaced intraarticular calcaneal fractures. Results using a prognostic computed tomography scan classification. *Clin. Orthop. Relat. Res.* **1993**, *290*, 87–95.

21 Sanders, R. Displaced Intra-Articular Fractures of the Calcaneus. *J. Bone Jt. Surg. Am.* **2000**, *82*, 225–250. [CrossRef] [PubMed]

22. Janzen, D.L.; Connell, D.G.; Munk, P.L.; E Buckley, R.; Meek, R.N.; Schechter, M.T. Intraarticular fractures of the calcaneus: Value of CT findings in determining prognosis. *Am. J. Roentgenol.* **1992**, *158*, 1271–1274. [CrossRef] [PubMed]

23. Huang, D.; Shan, C.; Ardabilian, M.; Wang, Y.; Chen, L. Local Binary Patterns and its application to facial image analysis: A Survey. *IEEE Trans. Syst. Man Cybern. Part C* **2011**, *41*, 765–781. [CrossRef]

24. Ludwig, O.; Nunes, U.J.C.; Ribeiro, B.; Premebida, C. Improving the generalization capacity of cascade classifiers. *IEEE Trans. Cybern.* **2013**, *43*, 2135–2146. [CrossRef] [PubMed]

25. Yu, H.; Moulin, P. Regularized Adaboost learning for identification of time-varying content. *IEEE Trans. Inf. Forensics Secur.* **2014**, *9*, 1606–1616. [CrossRef]

Appl. Sci. **2019**, *9*, 3011

26. Viola, P.; Jones, M.J. Robust real-time object detection. *Int. J. Comput. Vis.* **2004**, *57*, 137–154. [CrossRef]
27. Suzuki, S.; Be, K. Topological structural analysis of digitized binary images by border following. *Comput. Vis. Graph. Image Process.* **1985**, *30*, 32–46. [CrossRef]
28. Yang, X.; Shen, X.; Long, J.; Chen, H. An improved median-based otsu image thresholding algorithm. *AASRI Procedia* **2012**, *3*, 468–473. [CrossRef]
29. Gonzalez, R.C.; Woods, R.E. *Digital Image Processing*; Prentice Hall: Upper Saddle River, NJ, USA, 2002; ISBN 0201180758.
30. Seo, J.; Chae, S.; Shim, J.; Kim, D.; Cheong, C.; Han, T.D. Fast Contour-Tracing Algorithm Based on a Pixel-Following Method for Image Sensors. *Sensors* **2016**, *16*, 353. [CrossRef]
31. Zhu, Y.; Huang, C. An Improved Median Filtering Algorithm for Image Noise Reduction. *Phys. Procedia* **2012**, *25*, 609–616. [CrossRef]
32. Chen, C.M.; Chen, C.C.; Wu, M.C.; Horng, G.; Wu, H.C.; Hsueh, S.H.; Ho, H.Y. Automatic Contrast Enhancement of Brain MR Images Using Hierarchical Correlation Histogram Analysis. *J. Med. Biol. Eng.* **2015**, *35*, 724–734. [CrossRef] [PubMed]

Article

Automatic Zebrafish Egg Phenotype Recognition from Bright-Field Microscopic Images Using Deep Convolutional Neural Network

Shang Shang [1], Ling Long [2], Sijie Lin [2,*] and Fengyu Cong [1,*]

[1] School of Biomedical Engineering, Faculty of Electronic Information and Electrical Engineering, Dalian University of Technology, No.2 Linggong Street, Dalian 116024, China
[2] College of Environmental Science and Engineering, Biomedical Multidisciplinary Innovation Research Institute, Shanghai Institute of Pollution Control and Ecological Security, Key Laboratory of Yangtze River Water Environment, Ministry of Education, Tongji University, 1239 Siping Road, Shanghai 200092, China
* Correspondence: lin.sijie@tongji.edu.cn (S.L.); cong@dlut.edu.cn (F.C.);
    Tel.: +86-021-6598-1217 (S.L.); +86-0411-8470-8704 (F.C.)

Received: 9 June 2019; Accepted: 8 July 2019; Published: 15 August 2019

Featured Application: Automatic analysis of high throughput zebrafish egg microscopic images.

Abstract: Zebrafish eggs are widely used in biological experiments to study the environmental and genetic influence on embryo development. Due to the high throughput of microscopic imaging, automated analysis of zebrafish egg microscopic images is highly demanded. However, machine learning algorithms for zebrafish egg image analysis suffer from the problems of small imbalanced training dataset and subtle inter-class differences. In this study, we developed an automated zebrafish egg microscopic image analysis algorithm based on deep convolutional neural network (CNN). To tackle the problem of insufficient training data, the strategies of transfer learning and data augmentation were used. We also adopted the global averaged pooling technique to overcome the subtle phenotype differences between the fertilized and unfertilized eggs. Experimental results of a five-fold cross-validation test showed that the proposed method yielded a mean classification accuracy of 95.0% and a maximum accuracy of 98.8%. The network also demonstrated higher classification accuracy and better convergence performance than conventional CNN methods. This study extends the deep learning technique to zebrafish egg phenotype classification and paves the way for automatic bright-field microscopic image analysis.

Keywords: zebrafish egg; microscopy image processing; convolutional neural network

---

1. Introduction

Zebrafish embryos have gained popularity in biological research since they share 84% of genes associated with human disease [1] and they are nearly transparent under bright-field microscopes. Zebrafish egg is a special form of the embryo, and it is usually used to study the influence of environmental factors on embryo development. To evaluate the biological endpoints based on zebrafish eggs, microscopic screening is frequently performed [2]. By far, the analysis of zebrafish microscopic images is mostly performed by human operators. With the advances in image acquisition systems, the number of microscopic images is increasing rapidly, making manual assessments increasingly time-consuming. Therefore, automatic analysis of zebrafish microscopic image becomes an urgent demand [3].

To meet this stringent demand, a series of studies was conducted for computerized zebrafish microscopic image analysis [4,5]. Most techniques were based on traditional machine learning strategies, i.e., using texture filters to extract hand-crafted image features and then using classification algorithms (e.g., the supported vector machine and random forest) to conduct phenotype pattern recognition. The performances of these methods highly rely on the quality of hand-crafted image features, but the design and selection of hand-crafted features involve subjective human interventions, which limit the objectiveness and robustness of the method.

In the last decade, deep learning methods experienced dramatic development, leading to improvements in many pattern recognition applications, such as image processing, video analysis, and language recognition [6–11]. Compared to the conventional machine learning methods, deep learning overcomes the limitation of hand-crafted features by automatically optimizing the feature extraction and classification procedure. The core of deep learning for image analysis is the revolutionary development of the convolutional neural network (CNN) [12,13]. CNN was originally designed to recognize and classify object patterns in images. As of today, numerous CNN-based powerful image classification models are developed, including Alex Krizhevsky Network (AlexNet) [14], Visual Geometry Group (VGG) nets [15], and Residual Neural Network (ResNet) [16]. These methods were also applied to biological image analysis [17,18], leading to improvement of accuracy and robustness.

Despite the fast development of deep learning techniques, their applications in zebrafish egg microscopic image analysis are rare. A common data analysis task for zebrafish egg images analysis is to classify whether the egg is fertilized or not, in order to verify if the tested drug has impaired the fertilization process. To accomplish this task, there are several challenging problems to solve:

- Imbalanced training dataset. In biological research, it is difficult to collect a balanced number of fertilized and unfertilized egg samples as the training dataset. The imbalanced training set will result in insufficient classification ability for the category with fewer training samples, leading to unsuccessful network training.
- Small training dataset. The training of deep neural network requires no less than thousands of training samples. However, it is difficult to collect enough training data for a specific biological image analysis task. Small training sample set will lead to overfitting of the training data, hampering the generalization ability of the network.
- Subtle inter-class differences. In bright-field microscopic images, fertilized and unfertilized zebrafish eggs usually demonstrate subtle inter-class differences. This challenging problem becomes a technical bottleneck for automated zebrafish egg image analysis.

To overcome these problems, this paper proposed a deep learning algorithm for automated zebrafish egg fertilization status classification from microscopic images. Dedicated data augmentation and transfer learning strategy were used to tackle the imbalanced and small training set problem. The global average pooling scheme was used to address the subtle inter-class differences. Experimental results showed that the proposed method yielded dramatic accuracy improvement compared to traditional CNN network, and the classification accuracy for zebrafish eggs could reach up to 98.8%.

2. Materials and Methods

2.1. Data Collection

In this study, the microscopic images of zebrafish eggs were acquired using a bright-field microscopy imaging device called ImageXpress [19]. The system automatically placed three or four embryos in a U-shaped bottom transparent well plate. The image of each plate was collected using a ×2 dry objective between 3 and 3.7 h post fertilization. Transition Metal Oxide Nanoparticles were applied to the zebrafish embryos, and some of the eggs became unfertilized due to the toxicity effect of the nanoparticles. Figure 1 shows a typical sample image of the zebrafish eggs. The eggs are to be classified into two classes, fertilized and unfertilized. The fertilized eggs contain the nucleus surrounded by dark yolk membranes, whereas the unfertilized eggs have clear yolk membranes.

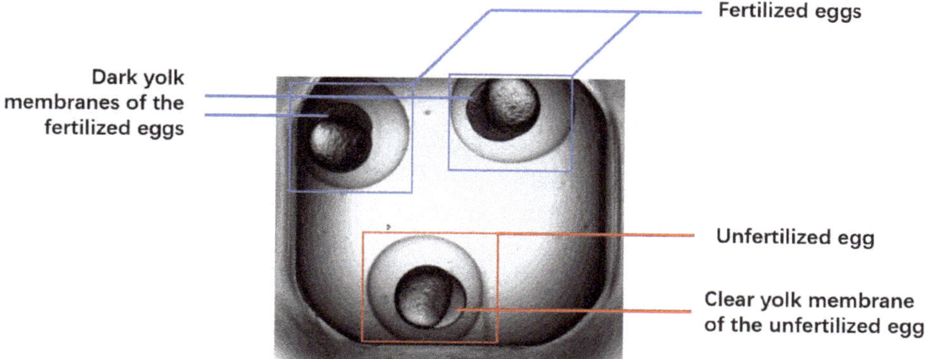

Figure 1. A typical example of the zebrafish eggs bright-field microscopic image, in which the fertilized and unfertilized eggs are marked. The fertilized eggs contain the nucleus surrounded by dark yolk membranes, whereas the unfertilized eggs have clear yolk membranes.

2.2. Method Workflow

As illustrated in Figure 2, our automatic zebrafish egg recognition and counting method consisted of three major steps. The input image is a well plate image containing three or four eggs. For the first step, each individual egg was detected and separated as a small patch. The patch of each egg was then fed into a deep convolutional neural network to calculate the classification feature vector. Finally, a global average pooling layer was used to classify the fertilization status based on the feature vector. Details of the proposed method are explained in the following subsections.

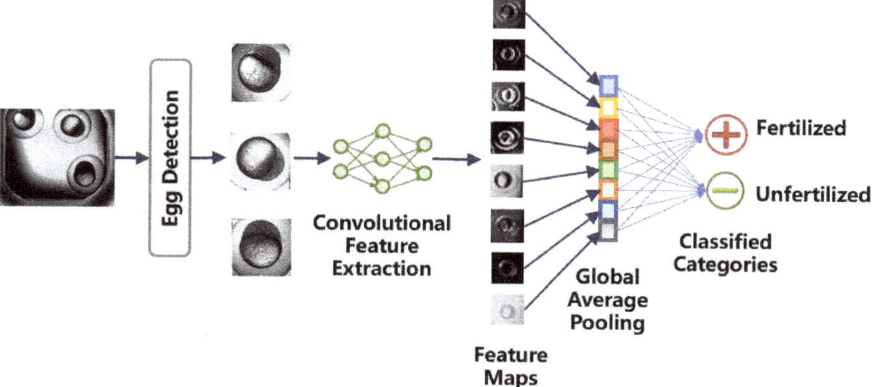

Figure 2. The workflow of the proposed method.

2.3. Egg Detection

As required by the classification task, the microscopic images were pre-segmented and cropped into square patches of single eggs. This was achieved via a template matching step, which detected the center of each egg. Figure 3 demonstrates the principle of template matching. As shown in Figure 3a, the template was constructed by manually cropping K typical egg patches of N × N pixels from the training images. The K patches were reoriented into the same direction, and an averaged template was created by calculating the average image of them. Then, the average template was rotated with 30 degrees interval to generate 12 template patches of different orientations (Figure 3b). To perform template matching, each of the 12 templates was moved with N/10 pixels interval in both x and y directions throughout the test image. For each moved position, the mutual information between the template and its covered image area was calculated as the similarity metric. The top 20% positions with the largest mutual information were maintained as the candidate egg centers. At last, candidate centers close to each other (within N/5 pixels distance) were clustered, and the mean coordinates of clustered candidates were used as the egg center. Based on the detected egg centers, a bounding box of size N × N was used to crop the egg out of the image. In this study, we found K = 10 sufficed for our needs, and a cropping size of N = 150 pixels ensured to enclose all eggs.

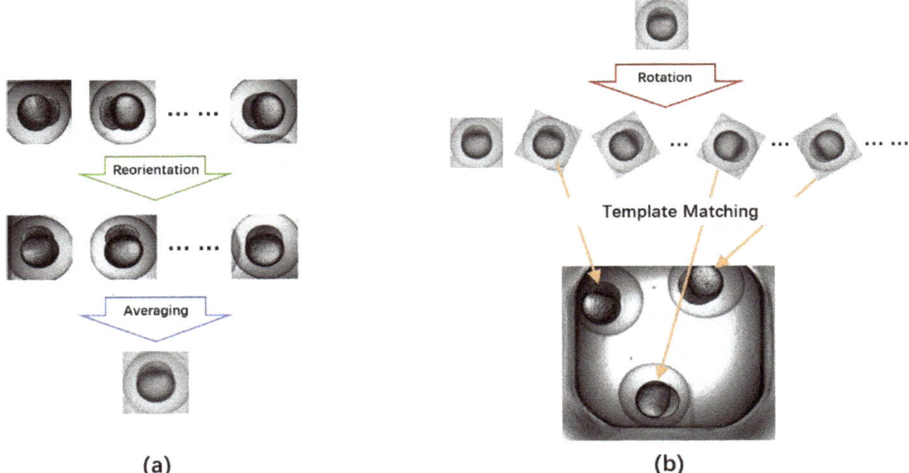

Figure 3. The workflow of egg detection. (**a**) The egg template is created by averaging K patches of egg samples; (**b**) The egg template is rotated into different directions, and each rotated patch is moved through the target image to find its matched egg.

2.4. Convolutional Feature Extraction

After the egg detection step, each cropped egg patch was fed into a convolutional neural network to extract the image features for egg classification. To train such a network, we needed to overcome the limitation of the small and imbalanced training dataset. Our study involved only a few hundreds of samples of zebrafish eggs, which were not enough for training a deep neural network. Compared to the popular ImageNet [14] dataset of over ten million sample images, the size of our datasets is at least four orders of magnitude less. When the number of weights to be trained in a neural network is far more than the number of training samples, the problem of overfitting is likely to occur, and the network will have poor generalization ability.

Another problem with our training set is that the sample numbers of different categories were seriously imbalanced. The ratio between fertilized and unfertilized eggs was almost 6:1 in our dataset. Imbalanced training data could potentially diminish the specificity of the network, making the network incompetent to recognize the relatively smaller category, i.e., the unfertilized eggs.

To overcome the limitation of the small and imbalanced training set, we used the image augmentation strategy to increase the training set size and to balance the training sample numbers of different categories. Image augmentation is the process to increase the training set by creating altered versions of the existing sample images, and it is proved to be an effective solution to prevent overfitting [14]. Typical ways of data augmentation include rotation, translation, zooming, flipping, scaling, color perturbation, and adding random noise.

In our study, image augmentation strategies were carefully chosen according to the characteristics of our datasets (as shown in Figure 4). Since the eggs were captured at the same time point, they had similar sizes. All the images were captured under the same environmental light condition so that the grayscale level of different eggs was similar. The most possible variation of the eggs is the different orientation caused by random placement. Therefore, we used image rotation and flipping to simulate possible deviation of egg orientations. In order to improve the balance of the dataset, we augmented the unfertilized eggs more than the fertilized eggs. Each fertilized patch was rotated three times with 60 degrees interval, while each unfertilized patch was rotated 18 times with 10 degrees interval. All the rotated patches were also flipped vertically to simulate the effect of different illumination orientations of the environmental light. As a result, the ratio between fertilized and unfertilized eggs was close to 1:1 after the augmentation.

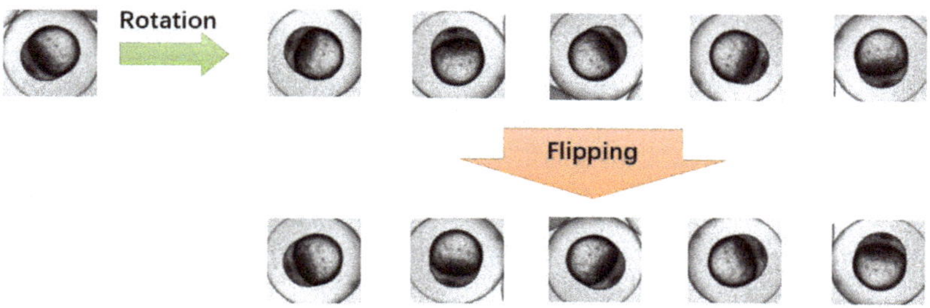

Figure 4. Examples of the one zebrafish egg (the leftmost patch) and its augmented patches, including rotational augmentation and flipping augmentation.

Based on the augmented training dataset, a convolutional neural network was trained. The network used the architecture of VGG-16 [15], the winner of the 2014 Large Scale Visual Recognition Challenge (ILSVRC). As shown in Figure 5, this architecture consisted of 5 blocks of 13 convolutional layers. For each convolutional layer, a convolution kernel of size 3×3 was convolved with the layer input to produce a tensor of outputs. The output tensor of the convolutional layer was then transferred into a finite value by an activation function of Rectified Linear Unit (ReLu), i.e., $F(x) = x$ for $x > 0$ and 0 otherwise. At the end of each block of the convolutional layers, there was a max-pooling layer to perform down-sampling by dividing the output feature map from each block into 2×2 pooling regions and computing the maximum of each region. The down-sampled feature map from each max-pooling layer was then fed into the next convolutional layers as an input.

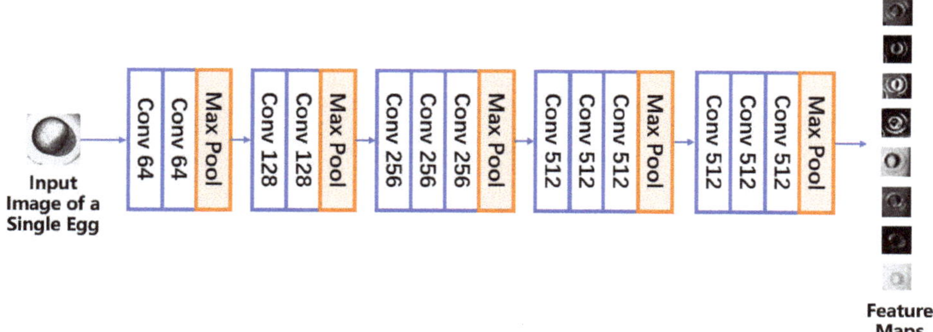

Figure 5. The architecture of the convolutional feature extraction network, where 'Conv N' stands for a 3×3 convolutional layer with N channels, 'Max Pool' stands for a 2×2 max pooling layer.

To train this deep convolutional network, the transfer learning strategy was used. The network weights pre-trained on the ImageNet dataset was adopted as the initialization. By using VGG-16 as the initial model, we were able to take advantage of deep features learned from millions of natural images [20]; therefore, the risk of overfitting was further reduced, and the convergence of the training was accelerated. During the training, the weights of the first three blocks of layers were frozen to retain the extracted simple features by VGG-16. Other two blocks of convolutional layers were fine-tuned with a small learning rate to make sure that the magnitude of the updates from each fine-tuning iteration stayed small.

For the bottleneck feature training phase, RMSprop optimizer was used for faster general localization. For the fine-tuning phase, Stochastic Gradient Descent (SGD) optimizer with momentum was chosen for better generalization ability. Choosing proper optimizers is crucial since it directly affects the convergence of the algorithm. Both optimizers we chose here originate from the optimizer of Gradient Descent. However, the basic Gradient Descent method calculates the gradient of the whole data set for performing only one update. Therefore, it is extremely slow and memory expensive for experiments with large datasets. Stochastic Gradient Descent (SGD) method was designed to rectify the above problems of the regular Gradient Descent method by performing a parameter update for each training example. To further improve convergence accuracy and reduce fluctuation, a momentum term was added to the SGD method. It restricts the oscillation in one direction during searching to improve the speed of the convergence. Based on the SGD with momentum method, RMSprop optimizer restricts the oscillations in the vertical direction. In this way, a larger learning rate could be adapted to have a larger searching pace in the horizontal direction to increase convergence speed. For bottleneck feature training phase, RMSprop optimizer was used for faster general localization at the beginning. While Stochastic Gradient Descent (SGD) optimizer with momentum was chosen for more precisely global minima localization.

Learning rate is one of the most important aspects of Gradient Descent because it determines the pace size for searching the global minima of the optimizing algorithm. Here, a small learning rate of 0.0001 was used to perform fine adjustments to weights without changing the overall weight structure. We used a small learning rate so that the features learned previously were not wrecked. For the training process of each data fold, we ran 50 epochs and saved the best result of model weights at the epoch when the validation loss was the least. The technique of reduced learning rate was used, i.e., the learning rate was multiplied with 0.2 when the training loss stopped reducing for 3 epochs.

2.5. Global Average Pooling Classifier

After features were extracted by the convolutional network module, a classification module based on global average pooling method was used instead of the traditional fully connected layer classifier. Conventionally, in a convolutional neural network, convolutional layers are usually followed by several fully connected layers to vectorize the feature extracted by convolutional layers and to accomplish the classification task via a softmax logistic regression layer. However, fully connected layers involve many weights to be trained, which increase the cost of computing and reduce the convergence speed of the network. On the other hand, the increment of weights will also increase model complexity, which may easily lead to overfitting. Effective techniques have been proposed to avoid overfitting, such as dropout [21,22]. Using global average pooling (GAP) instead of fully connected layers to classify different categories directly from feature maps is a revolutionary innovative improvement made to traditional convolutional neural networks [23]. Instead of adding fully connected layers on top of the feature maps from convolutional layers, GAP generates one feature map for each corresponding category to be classified, vectorize the features by global average pooling, and feed the vectors directly into the final softmax classifier, as shown in Figure 2. Compared to traditional fully connected layers, GAP had enforced the correspondences between feature maps and categories. Besides, the GAP didn't introduce extra weights to be optimized for the network, which had reduced the prone of overfitting.

3. Results

In this study, a total of 211 zebrafish egg microscopic images containing 638 eggs were acquired using the ImageXpress system. A human biologist with over 10 years' experience was invited to assign fertilization labels to all the eggs, resulting in 546 fertilized eggs and 92 unfertilized. The labels of human expert were used as the gold standard for method validation. The network was constructed using the Keras platform on a server with NVIDIA K4000 Graphics Processing Unit (GPU). The training process took ~60 min for each training subsample set and took less than 5 s on each test image.

3.1. Zebrafish Egg Classification Accuracy

To validate the proposed method, a five-fold cross-validation scheme was used. For K-fold cross-validation, the original dataset was randomly partitioned into K equal-sized subsample sets. The training and validation processes were repeated K times. Each time, one subsample set was retained, in turn, as the validation data, while other K−1 subsample sets were used as the training data. We chose 5-fold cross-validation according to the overall size of the dataset so that there were no less than 500 eggs in each training set. The training and validation processes were repeated five times, and the accuracy of each validation subsample set was calculated.

Table 1 reports the accuracy of each cross-validation fold. The accuracy was defined as Accuracy = $(N_{TP} + N_{TN})/N_{all}$, where N_{TP}, N_{TN}, N_{all} stand for the number of true positive, true negative, and all eggs, respectively. In this study, we considered unfertilized eggs as positive samples and fertilized eggs as negative samples, respectively. As reflected in Table 1, the third fold had the highest accuracy (98.8%), and the second fold had the lowest accuracy (93.2%). Even the lowest accuracy was higher than 93%, and the mean accuracy of all folds was 95.0%, meaning that the proposed method has a quite high classification accuracy for zebrafish egg fertilization status. The standard deviation of all the folds was also small (2.2%), meaning this method performs stably over different test datasets.

Table 1. Classification Accuracy of the Proposed Method.

Fold 1	Fold 2	Fold 3	Fold 4	Fold 5	Mean ± Std.
93.3%	93.2%	98.8%	93.7%	95.9%	95.0 ± 2.2%

3.2. Comparison between Regular Fully Connected Layers and Global Average Pooling

Our method used global averaged pooling layers instead of the regular fully connected layers to improve the classification accuracy. To verify the advantage of global averaged pooling, we calculated the classification results with regular fully connected layer to compare with the results based on global average pooling. In this experiment, the dropout technique with an experimental value of 50% dropout probability was adopted with two fully connected layers to compare with the global average pooling classifier. Fully connected layers are usually accompanied by the dropout method to promote convergence. Dropout is a regularization technique to prevent overfitting for neural network models [14,22]. Neurons are randomly selected with a given probability to be dropped out and ignored during training so that the network could learn multiple independent internal representations and improve the generalization ability.

The comparison was based on the same five-fold cross-validation dataset, as mentioned above. The highest classification accuracy of the regular fully connected layers was 97.3%, which was less than the highest accuracy of 98.8% of the global average pooling method. Moreover, we also found that the global averaged pooling method had better convergence performance for model training. Figure 6 plots the training accuracy curve and validation accuracy curve of both global averaged pooling and conventional fully connected layers. The global averaged pooling method shows more steady convergence process with less fluctuation. The global averaged pooling method also has a narrower gap between the training and validation curves, implying better generalization ability than the fully connected layers method.

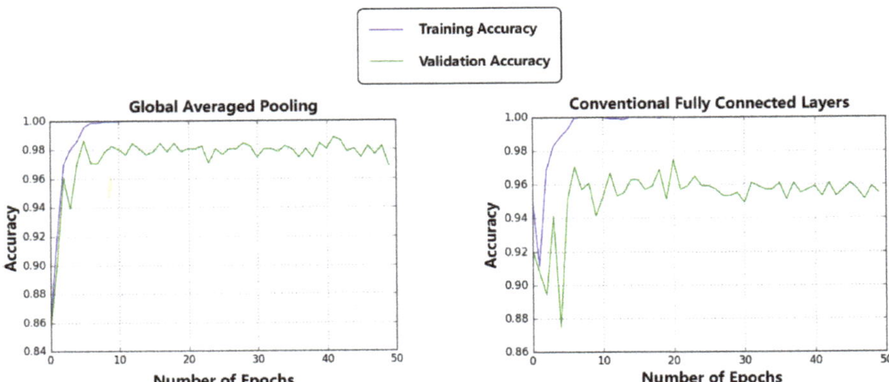

Figure 6. The training accuracy curve and validation accuracy curve of both global averaged pooling method and conventional fully connected layers method.

3.3. Comparison between Augmented Dataset and Original Dataset

Our method used data augmentation strategy to overcome the limitation of the small and imbalanced training dataset. In this Section, we compared the performances of zebrafish egg classification with and without data augmentation. Besides the training result acquired with augmented datasets of 7864 image patches, another model was trained with the original dataset of 638 patches without augmentation. The improvement of accuracy in the case of the augmented datasets against the original datasets was significant. Validation accuracy was improved from 83.8% to 98.8% after data augmentation.

To further analyze the effect of balancing the imbalanced datasets by augmentation, we had computed the metrics of sensitivity, specificity, precision, and accuracy between the methods with and without data augmentation (as shown in Table 2). The metrics were defined as Sensitivity = $N_{TP}/(N_{TP} + N_{FN})$, Specificity = $N_{TN}/(N_{TN} + N_{FP})$, Precision = $N_{TP}/(N_{TP} + N_{FP})$, where N stands for the number of samples, TP, FP, TN, FN represent true positive, false positive, true negative, false negative, respectively. From Table 2, it can be observed that data augmentation led to evident improvements in both sensitivity and accuracy, while the specificity and precision of both methods were at the same level. We also compared the convergence performance of model training between the methods with and without data augmentation. As shown in Figure 7, data augmentation led to faster convergence speed and a smaller gap between the training accuracy curve and validation accuracy curve, implying that data augmentation yielded better specificity and generalization ability.

Table 2. Comparison of the classification performance between the methods with and without data augmentation.

Method	Sensitivity	Specificity	Precision	Accuracy
with Data Augmentation	97.3%	99.2%	99.2%	98.8%
without Data Augmentation	68.0%	99.6&	99.4%	83.8%

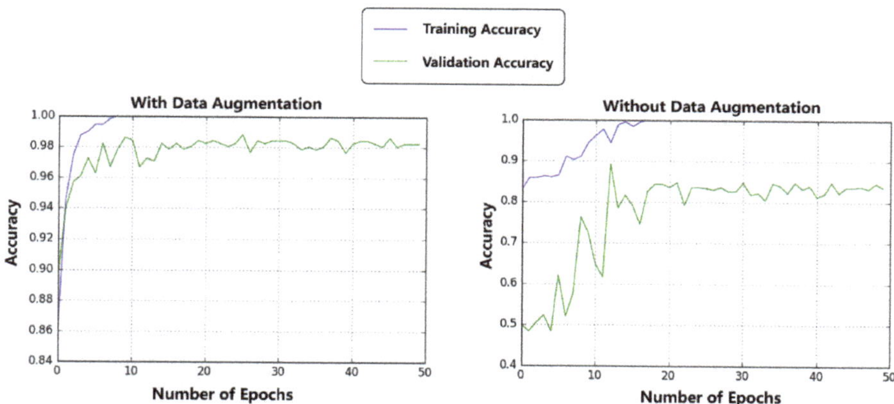

Figure 7. The training accuracy curve and validation accuracy curve of model training with and without data augmentation.

3.4. Comparison with Other Zebrafish Embryo Microscopic Image Analysis Studies

As we surveyed the existing studies, there was rarely any research on zebrafish egg fertilization status classification from microscopic images. The most similar study to ours is from Liu et al. [4] who used support vector machine (SVM) to classify zebrafish embryo hatching status based on hand-crafted image features. It is hard to rigorously compare our method with Liu's method since the application purpose is different. As a rough comparison, their method achieved average recognition accuracy of 97.4 ± 61.0%, while our method had an average accuracy of 95.0 ± 2.2%. Although the two methods have similar accuracy, the standard deviation of our method (2.2%) is much less than theirs (61.0%), meaning that our method is considerably more stable. Moreover, our method doesn't need any hand-crafted feature; thus, the cost of algorithm design and the involvement of subjective interference of our method is much less. It is evident that our deep learning approach has better stability and objectiveness than the traditional machine learning methods based on hand-crafted features.

4. Discussion

In this study, exploratory research was conducted on CNN-based zebrafish egg phenotype classification from microscopic images. Due to the particularity of zebrafish egg research, we were facing the problems of the small imbalanced dataset and subtle inter-class difference. To tackle these problems, the strategies of transfer learning, data augmentation, and global averaged pooling were used.

It is known that training a deep network from scratch with random initialization is a formidable task. It requires millions of well-annotated training images, which are difficult to obtain in our study. Transfer learning is a technique to obtain deep features that an existing model has learned from tens of thousands of natural image datasets, either as an initialization or a fixed feature extractor for the task of interest. In some studies, transfer learning has been used to analyze medical images and achieved dramatic performance improvement for classification tasks of small datasets [24,25]. In this study, we used VGG-16 model previously trained on millions of natural scene images [15]. Compared to medical images like Computed Tomography (CT) and Magnetic Resonance Imaging (MRI), bright-field microscopic images share more common image features with natural scene images; therefore, we directly used the original VGG-16 model without modifications to its network architecture. As shown in our experimental results (Table 1), a mean accuracy of 95.0% was obtained based on five-fold cross-validation, proving the effectiveness of the transfer learning.

To further address the small imbalanced dataset problem, data augmentation strategy was used in this study. The effect of data augmentation was evident. As shown in Table 2, the sensitivity and accuracy were improved dramatically after data augmentation. The model trained without data augmentation yielded quite low sensitivity (68.0%), implying that this model tended to make negative judgments (fertilized). This is because the training data without augmentation contained much less unfertilized eggs than the fertilized eggs, making the model inadequate to recognized fertilized eggs. Therefore, dedicated data augmentation is very crucial for training a network for recognizing both types of eggs.

To cope with the subtle inter-class differences, global averaged pooling was used instead of the conventional fully connected layers. Global average pooling classifier enforced the correspondences between feature maps and categories without introducing extra weights to be optimized, and thus reduced the fluctuation during the training process and promoted fast and steady convergence. As reflected from the experimental results (Figure 3), global averaged pooling not only yield improved averaged classification accuracy but also lead to faster and more stable convergence of the training and validation curves. Such an advantage is crucial for biological microscopic image classification since the genetic or biological changes usually result in quite subtle phenotype differences.

As a limitation of this study, the proposed method still used a conventional template matching scheme to locate each egg in the well-plate image. There are several state-of-the-art neural networks for fast object detection, such as Faster-RCNN, YOLO, etc. [26]. However, as we tested these models, they performed well on locating the eggs but failed to accurately distinguish between the fertilized and unfertilized eggs. Therefore, we chose to use conventional CNN structure equipped with global averaged pooling to overcome the subtle inter-class difference. In the future study, we will focus on combining the object detection networks with our network architecture so that the whole workflow (including detection and classification) can be performed with only one network.

5. Conclusions

This study applied the deep learning technique to classify fertilized and unfertilized zebrafish eggs from bright-field microscopic images. Transfer learning and data augmentation schemes were used to overcome the problem of the small imbalanced training dataset. Global averaged pooling was adopted to improve the classification accuracy over subtle inter-class differences. Our future research direction will focus on applying this method in daily zebrafish egg acquisition workflow so that the proposed algorithm can promote the research outcome of high throughput biological experiments.

Author Contributions: Conceptualization, F.C. and S.L.; methodology, S.S.; software, S.S.; validation, S.S.; formal analysis, S.S.; investigation, S.L. and L.L.; resources, S.L. and L.L.; data curation, L.L.; writing—original draft preparation, S.S.; writing—review and editing, F.C.; visualization, S.S.; supervision, F.C.; project administration, F.C. and S.L.; funding acquisition, F.C. and S.L.

Funding: This research is supported by the National Natural Science Fund of China, No. 21607115; the general program of the National Natural Science Fund of China, No. 21777116; the Xinghai Scholar Cultivating Funding of Dalian University of Technology, No. DUT14RC(3)066.

Acknowledgments: The authors would like to thank Hongkai Wang for his advice on deep neural network training.

References

1. Blaser, R.E.; Vira, D.G. Experiments on learning in zebrafish (Danio rerio): A promising model of neurocognitive function. *Neurosci. Biobehav. Rev.* **2014**, *42*, 224–231. [CrossRef] [PubMed]
2. Li, L.; LaBarbera, D.V. 3D High-Content Screening of Organoids for Drug Discovery. In *Chemistry, Molecular Sciences and Chemical Engineering*; Elsevier Inc.: Amsterdam, The Netherlands, 2017; pp. 388–415. [CrossRef]
3. Unser, M.; Sage, D.; Delgado-Gonzalo, R. Advanced image processing for biology, and the Open Bio Image Alliance (OBIA). In Proceedings of the 21st European Signal Processing Conference, Marrakech, Morocco, 9–13 September 2013; pp. 1–5.

4.   Liu, R.; Lin, S.; Rallo, R.; Zhao, Y.; Damoiseaux, R.; Xia, T.; Lin, S.; Nel, A.; Cohen, Y. Automated phenotype recognition for zebrafish embryo based in vivo high throughput toxicity screening of engineered nano-materials. *PLoS ONE* **2012**, *7*, e35014. [CrossRef] [PubMed]

5.   Jeanray, N.; Marée, R.; Pruvot, B.; Stern, O.; Geurts, P.; Wehenkel, L.; Muller, M. Phenotype classification of zebrafish embryos by supervised learning. *PLoS ONE* **2015**, *10*, e0116989. [CrossRef] [PubMed]

6.   Lawrence, S.; Giles, C.L.; Tsoi, A.C.; Back, A.D. Face recognition: A convolutional neural-network approach. *IEEE Trans. Neural Netw.* **1997**, *8*, 98–113. [CrossRef] [PubMed]

7.   Sun, Y.; Wang, X.; Tang, X. Deep convolutional network cascade for facial point detection. In Proceedings of the IEEE Conference on Computer Vision and Pattern Recognition, Portland, OR, USA, 23–28 June 2013; pp. 3476–3483.

8.   Hu, B.; Lu, Z.; Li, H.; Chen, Q. Convolutional neural network architectures for matching natural language sentences. *Adv. Neural Info. Process. Syst.* **2014**, *3*, 2042–2050.

9.   Kalchbrenner, N.; Grefenstette, E.; Blunsom, P. A convolutional neural network for modelling sentences. In Proceedings of the 52nd Annual Meeting of the Association for Computational Linguistics, Baltimore, ML, USA, 23–25 June 2014; pp. 655–665.

10.   Ji, S.; Xu, W.; Yang, M.; Yu, K. 3D convolutional neural networks for human action recognition. *IEEE Trans. Pattern Anal. Mach. Intell.* **2013**, *35*, 221–231. [CrossRef] [PubMed]

11.   Tompson, J.; Jain, A.; LeCun, Y.; Bregler, C. Joint training of a convolutional network and a graphical model for human pose estimation. In Proceedings of the Neural Information Processing Systems, Montreal, QC, Canada, 8–13 December 2014; pp. 1799–1807.

12.   Nielsen, M.A. Neural Networks and Deep Learning. 2015. Available online: http://neuralnetworksanddeeplearning.com/about.html (accessed on 15 August 2019).

13.   Lecun, Y.; Bottou, L.; Bengio, Y.; Haffner, P. Gradient-based learning applied to document recognition. *Proc. IEEE* **1998**, *86*, 2278–2324. [CrossRef]

14.   Krizhevsky, A.; Sutskever, I.; Hinton, G.E. ImageNet Classification with Deep Convolutional Neural Networks. *Adv. Neural Info. Process. Syst.* **2012**, *25*, 2012. [CrossRef]

15.   Simonyan, K.; Zisserman, A. Very Deep Convolutional Networks for Large-Scale Image Recognition. In Proceedings of the International Conference on Learning Representations (ICLR), San Diego, CA, USA, 7–9 May 2015; pp. 1–14.

16.   He, K.; Zhang, X.; Ren, S.; Sun, J. Deep Residual Learning for Image Recognition. In Proceedings of the Computer Vision and Pattern Recognition, Las Vegas, NV, USA, 26 June–July 2016; pp. 770–778.

17.   Angermueller, C.; Pärnamaa, T.; Parts, L.; Stegle, O. Deep learning for computational biology. *Mol. Syst. Biol.* **2016**, *12*, 878. [CrossRef] [PubMed]

18.   Zhang, W.; Li, R.; Zeng, T.; Sun, Q.; Kumar, S.; Ye, J.; Ji, S. Deep model based transfer and multi-task learning for biological image analysis. *IEEE Trans. Big Data* **2016**, 1475–1484. [CrossRef]

19.   Lin, S.; Zhao, Y.; Xia, T.; Meng, H.; Ji, Z.; Liu, R.; George, S.; Xiong, S.; Wang, X.; Zhang, H.; et al. High content screening in zebrafish speeds up hazard ranking of transition metal oxide nanoparticles. *ACS Nano* **2011**, *5*, 7284–7295. [CrossRef] [PubMed]

20.   Chollet, F. Xception: Deep Learning with Separable Convolutions. In Proceedings of the IEEE Conference on Computer Vision and Pattern Recognition (CVPR), Honolulu, HI, USA, 21–26 July 2017; pp. 1251–1258.

21.   Wager, S.; Wang, S.; Liang, P. Dropout training as adaptive regularization. In Proceedings of the 26th International Conference on Neural Information Processing Systems, Lake Tahoe, NV, USA, 5–10 December 2013; pp. 351–359.

22.   Srivastava, N.; Hinton, G.; Krizhevsky, A.; Sutskever, I.; Salakhutdinov, R. Dropout: A Simple Way to Prevent Neural Networks from Overfitting. *J. Mach. Learn. Res.* **2014**, *15*, 1929–1958.

23.   Min, L.; Qiang, C.; Yan, S. Network in Network. In Proceedings of the International Conference on Learning Representations, Banff, AL, Canada, 14–16 April 2014; pp. 1–8.

24.   Ravishankar, H.; Sudhakar, P.; Venkataramani, R.; Thiruvenkadam, S.; Annangi, P.; Babu, N.; Vaidya, V. Understanding the Mechanisms of Deep Transfer Learning for Medical Images. In *Deep Learning and Data Labeling for Medical Applications, Proceedings of the International Workshop on Deep Learning in Medical Image Analysis, Athens, Greece, 21 October 2016*; Carneiro, G., Mateus, D., Eds.; Springer International Publishing AG: Cham, Switzerland, 2016.

25. Zhou, Z.; Shin, J.; Lei, Z.; Gurudu, S.; Gotway, M.; Liang, J. Fine-tuning Convolutional Neural Networks for Biomedical Image Analysis: Actively and Incrementally. In Proceedings of the IEEE Conference on Computer Vision and Pattern Recognition, Honolulu, HI, USA, 22–25 July 2017; pp. 4761–4772.
26. Liu, L.; Ouyang, W.; Wang, X.; Fieguth, P.W.; Chen, J.; Liu, X.; Pietikainen, M. Deep Learning for Generic Object Detection: A Survey. *arXiv* **2018**, arXiv:1809.02165.

Article

Zebrafish Larvae Phenotype Classification from Bright-field Microscopic Images Using a Two-Tier Deep-Learning Pipeline

Shang Shang [1], Sijie Lin [2,*] and Fengyu Cong [1,*]

[1] School of Biomedical Engineering, Faculty of Electronic Information and Electrical Engineering, Dalian University of Technology, Dalian 116024, China; dlutbmehomework@163.com
[2] College of Environmental Science and Engineering, Key Laboratory of Yangtze River Environment, Shanghai Institute of Pollution Control and Ecological Security, Tongji University, Shanghai 200092, China
* Correspondence: lin.sijie@tongji.edu.cn (S.L.); cong@dlut.edu.cn (F.C.); Tel.: +86-21-6598-1217 (S.L.); +86-411-8470-6309 (F.C.)

Received: 30 December 2019; Accepted: 10 February 2020; Published: 13 February 2020

Featured Application: Phenotype classification of zebrafish larvae from brightfield microscopic images.

Abstract: Classification of different zebrafish larvae phenotypes is useful for studying the environmental influence on embryo development. However, the scarcity of well-annotated training images and fuzzy inter-phenotype differences hamper the application of machine-learning methods in phenotype classification. This study develops a deep-learning approach to address these challenging problems. A convolutional network model with compressed separable convolution kernels is adopted to address the overfitting issue caused by insufficient training data. A two-tier classification pipeline is designed to improve the classification accuracy based on fuzzy phenotype features. Our method achieved an averaged accuracy of 91% for all the phenotypes and maximum accuracy of 100% for some phenotypes (e.g., dead and chorion). We also compared our method with the state-of-the-art methods based on the same dataset. Our method obtained dramatic accuracy improvement up to 22% against the existing method. This study offers an effective deep-learning solution for classifying difficult zebrafish larvae phenotypes based on very limited training data.

Keywords: zebrafish larva; microscopy image analysis; deep neural network

---

1. Introduction

The zebrafish and its larvae are becoming prominent vertebrate models in biological and medical research [1–6]. The transparency of zebrafish larvae facilitates convenient observation of the experimental process. Compared to the observations made by naked eyes in early days, the development of modern microscopy and imaging technology has facilitated the implementation of more complicated experiments by obtaining and storing massive zebrafish phenotype imaging data digitally [7].

Early methods of phenotype image analysis mostly rely on manual work which is labour intensive and error prone. Due to the increasing amount of image data, efficient image-processing methods extracting meaningful information from a massive amount of phenotype images are increasingly important [8]. So far, automatic approaches have been proposed for classifying the stages of zebrafish embryo development [3,9,10] and recognition of adult fish behavior [11–13], following classical image-processing techniques like wavelet analysis and fractal geometry [14–16]. These studies opened a new era of computerized zebrafish microscopic image analysis, liberating biologists from tedious manual work and increasing the objectivity of the task.

Comparing to the development of image-analysis methods for zebrafish adults or eggs, the studies on zebrafish larvae are relatively rare. The analysis of the larvae image remains a challenging task due to the complexity and diversity of larvae phenotypes [17–19]. The distinctions among different phenotypes are difficult to identify even for human experts. Among the limited number of existing zebrafish larvae image-analysis studies, Nathalie Jeanray et al. proposed an automatic approach for larvae defects classification from microscopic images using a supervised machine-learning approach [20]. They used dense random sub-windows to extract hand-crafted image features for phenotype defects classification. For nine out of 11 classification categories, their method reached 90% to 100% agreement with consensus voting of biologists. This method gives new insights into automatic recognition of zebrafish larvae phenotypes by replacing human labour with the computerized method.

Even though the machine-learning method using hand-crafted features achieved comparable performance to human experts, there is still significant scope to improve the accuracy and automation of the method. The accuracy of these methods mostly depends on the quality of the extracted features. The process for feature extraction involves subjective factors and human interventions. In contrast to the classical machine-learning methods using hand-crafted features, the newly developed deep-learning (DL) methods automatically extract image features and achieved a dramatic improvement of the classification accuracy. In the last decade, DL methods, especially convolutional neural networks (CNNs), have revolutionised the field of image analysis [21–23]. Despite the success of deep learning in image processing, there are some unique difficulties of zebrafish larvae microscopic images hampering the application of DL approaches:

(i)   Scarce training data. Biological experiments usually produce small datasets far from enough for training large neural networks. It is difficult to generate a large training set of expert-annotated zebrafish larvae images. Manual annotation of the training images is labour-intensive. It is unlikely to create large annotated biological image datasets with comparable scale to the natural image datasets [24].

(ii) Fuzzy inter-class difference. The appearance difference between different zebrafish larvae phenotypes are sometimes too trivial to be identified by the naked eye, and even human experts do not agree with each other on the annotation labels.

(iii) Multi-label nature. Unlike the natural image classification tasks which assign one unique label (e.g., cat or dog) to each subject, some zebrafish larvae phenotypes may simultaneously belong to multiple classes. For example, a larva with a necrosed yolk sac may also have an up curved short tail. Existing DL models are mostly designed for single-label classification task thus cannot be directly applied to zebrafish larvae phenotype classification.

Considering these difficulties, it is necessary to develop a specific deep-learning method for zebrafish phenotype classification from brightfield microscopic images. In this work, to cope with the scarce training data problem, we adopted the idea of separable convolution from the Xception (extreme inception) network [25] to overcome the overfitting issue caused by limited training data. To tackle the fuzzy inter-class difference and multi-label problems, a two-tier classification strategy is used. The phenotypes that are easy to recognize are firstly recognized in the first-tier classification process. The remaining difficult phenotypes are then recognized with dedicated trained classifiers in the second tier. For each of the difficult phenotypes, we train one network for each single phenotype, so that the networks are more focused on the specific phenotypes. Since each test image is processed with multiple networks, multiple phenotype labels can be assigned to one image, thus the multi-label problem is tackled.

To test our algorithm and compare it fairly with the state-out-the-art solutions, we used a publicly available zebrafish larvae microscopic image dataset [20] which has corresponding human expert annotation labels. Based on the researching achievements of the predecessor, we worked further on this dataset and developed a DL-based zebrafish larvae classification model that is more robust and accurate. In Section 2, the dataset annotated elaborately by the Nathalie Jeanray et al. is described.

Section 3 introduces the detailed structure and method. Experimental results are reported in Section 4, with further discussions presented in Section 5.

2. Materials and Methods

2.1. Data Collection

To train and develop an effective model for phenotype classification, the quality of the carefully collected dataset is crucial. Ground-truth datasets acquired and annotated by domain experts with the depiction of various classes are required [26]. Research based on publicly available ground-truth datasets has the advantage of enabling better evaluation and comparison of algorithms to enable continuous progress. Therefore, we chose the dataset from the Jeanray et al. group which were produced and collected with rigorous processes [20].

To obtain a collection of different defect phenotypes, Jeanray et al. used increasing concentrations of varied chemicals for zebrafish embryo treatment, including propranolol, amiodarone, acetaminophen, valproic acid caffeine, theophylline 4,5-dichloroanilin, as well as heavy metals such as thallium, methylmercury, lead acetate and zinc sulfate. Phenotype images of control or treated embryos were captured in a lateral view with a high-resolution Olympus SZX10 microscope coupled with a camera and transmitted light illumination. The acquired images were in the size of 2575×1932 pixels.

A sophisticated workflow of manual annotation was performed to build a high-quality ground-truth dataset. Three biologists labelled images with different categories of phenotypes by observations only in the first round. For each phenotype assigned to each image, the ground-truth was calculated by majority voting. Ten categories of phenotypes are identified, i.e., "normal", "dead", "chorion", "down curved tail", "hemostasis", "necrosed yolk sac", "edema", "short tail", "up curved tail" and "up curved fish". Sample images for different phenotypes are shown in Figure 1. By contrast with a regular classification dataset, there may be more than one phenotype assigned to a single image since more than one deformity may be found on one larva. Next, all annotations were reviewed by three biologists at the same time in order to reach agreement on phenotypes assigned to each image. After repeating the annotation and voting session, consensus was reached on 870 images. In total, 529 of them acquired in five independent acquisition days were set as a training dataset and other 341 images acquired in three additional acquisition days were integrated into the test dataset. The rigorous process of building the ground-truth dataset has laid the foundation of high-performance phenotype classification algorithms. Table 1 reports the number of training and testing images in this dataset. The training images are to be used for training the machine learning model for phenotype classification. The testing images are used for validating the trained model. The testing images are not included in the training images.

Table 1. The number of training and testing zebrafish larvae microscopic images of each phenotype.

Dataset-	Normal	Dead	Necrosed Yolk Sac	Edema	Hemostasis	Short Tail	Up Curved Tail	Up Curved Fish	Chorion	Down Curved Tail
Training	160	114	167	160	57	49	32	64	18	11
Testing	82	53	11	54	83	149	17	13	5	16

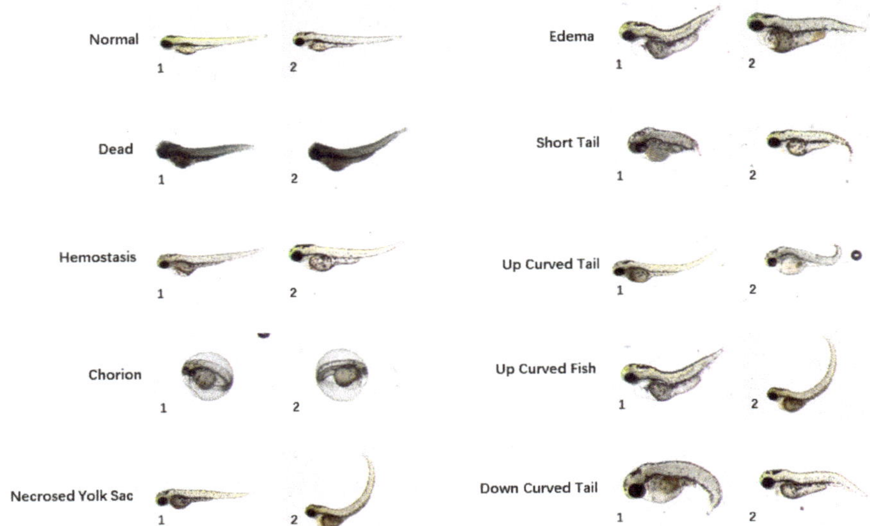

Figure 1. Example images of ten zebrafish larvae phenotypes from the publicly available dataset provided by Jeanray et al. A detailed description of these phenotypes are referred to the original paper of this dataset [20]. Note that some larvae have multiple phenotype labels, such as the larva of "edema 1" and "up curved fish 1", as well as the larva of "necrosed yolk sac 2" and "up curved fish 2".

2.2. Data Pre-Processing

One of the greatest impediments to developing deep-learning networks for biological images is that the sizes of training datasets from biological experiments are usually too small. The zebrafish larvae image dataset used in this study is the biggest dataset we could find, but its data size (hundreds of images) is still up to four orders of magnitudes less than the number of images needed to train a deep neural network. Training a deep network using such a small dataset will inevitably cause overfitting, leading to imperfect generalization ability of the network. Another obstacle is that the numbers of each category of the dataset are seriously imbalanced (as Table 1 reports). Imbalanced training samples will result in poor classification performance for those phenotypes with insufficient training samples.

To address the problem of the small and imbalanced training dataset, a data argumentation strategy is applied. Data augmentation is an effective way to extend an existing dataset by creating altered versions of the training data according to possible varied experimental situations. The trained network will be more robust against possible image variations due to the added possible diversification. Image augmentation is proved to be one of the effective solutions for preventing over-fitting. In this way, a wider variety of images is available for training the classification network to make it more robust to possible image variations.

Data-augmentation techniques were chosen based on the peculiarity of the zebrafish larvae microscopic image dataset. The most possible variations of the microscopic images were caused by random larvae placement which resulted in slight differences of the larvae location and orientation. Therefore, each training image was shifted and rotated in different directions to generate pseudo augmented images. We applied small shifting in four directions of the objects, i.e., up, down, left and right. Similarly, we also added small rotations to the original and shifted images in both clockwise and counterclockwise directions to simulate possible deviation caused by manual operation during experiments.

To cope with the imbalanced training data problem, the images of different phenotypes were augmented a different number of times to produce a more balanced dataset. The phenotypes with relatively abundant images (i.e., normal, dead, necrosed yolk sac, and edema) were shifted in each direction four times with 10 pixels step and rotated in each direction three times with 10 degrees interval, therefore they were augmented $44 \times (3 \times 2) = 1536$ times. The phenotypes with middle sample sizes (i.e., hemostasis, short tail, upcurved tail and upcurved fish) were shifted in each direction four times with 10 pixels step and rotated in each direction five times with six degrees interval, therefore they were augmented $44 \times (5 \times 2) = 2560$ times. The phenotypes with scarce training sample (i.e., chorion and down curved tail) were shifted in each direction five times with eight pixels step and rotated in each direction 10 times with three degrees interval, they were augmented $54 \times (10 \times 2) = 12{,}500$ times. Table 2 summarizes the number of training images of each phenotype before and after data augmentation.

Table 2. The number of training images of each phenotype before and after data augmentation.

	Normal	Dead	Necrosed Yolk Sac	Edema	Hemostasis	Short Tail	Up Curved Tail	Up Curved Fish	Chorion	Down Curved Tail
Before	160	114	167	160	57	49	32	64	18	11
After	2.46×10^5	1.75×10^5	2.56×10^5	2.46×10^5	1.46×10^5	1.25×10^5	8.19×10^4	1.64×10^5	2.25×10^5	1.38×10^5

2.3. Network Structure

Augmented and normalized images were then fed into our classification module to identify the malformation type of each larva. Unlike regular classification datasets of which one image belongs to one category only, images of zebrafish larvae can be classified to have more than one label. Therefore, we used a two-tier classification pipeline to address the multi-label problem. As Figure 2 shows, the first-tier classifier performs a tri-category classification, classifying the input image as edema, chorion or other phenotypes. We separate edema and chorion from other classes because they are quite easy to recognize, and it is not necessary to train individual classifiers for them. If the image is classified as others, it will go into the second-tier classification process in which eight binary classifiers are applied in parallel. Each of the classifiers is only responsible to judge whether the larva belongs to one of the eight classes. The objective of the second-tier classification is to find out whether the larva has one or more of the labels out of the eight phenotypes.

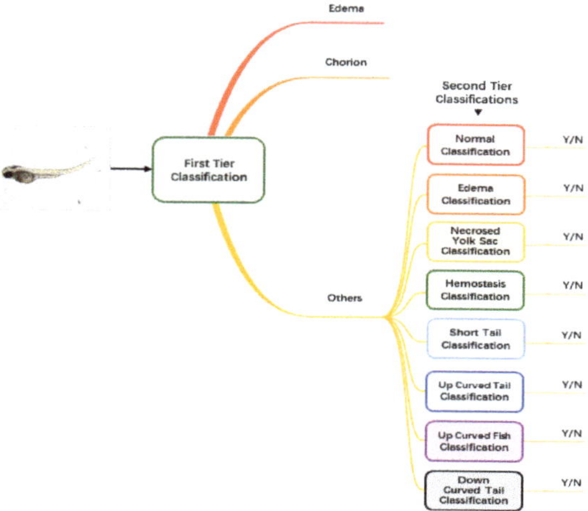

Figure 2. The pipeline of the two-tier classification strategy.

2.4. Larva Classification Network

For the entire pipeline, we use the same network structure for both the first- and second-tier classification. We adopt the Xception network as the classifier model. The network structure is illustrated in Figure 3. For conciseness of the article, we refer the readers to [25] for a detailed explanation of the network architecture. For the first-tier classification, the network outputs the probabilities of three phenotypes, i.e., edema, chorion and others. The phenotype label with the maximum probability is assigned to the target larva. For the second-tier classification, the same network is used for each binary classifier, outputting the probability of belonging to each of the eight phenotypes. If the probability of a certain phenotype is over 0.5, the larva is classified to have the label of this phenotype.

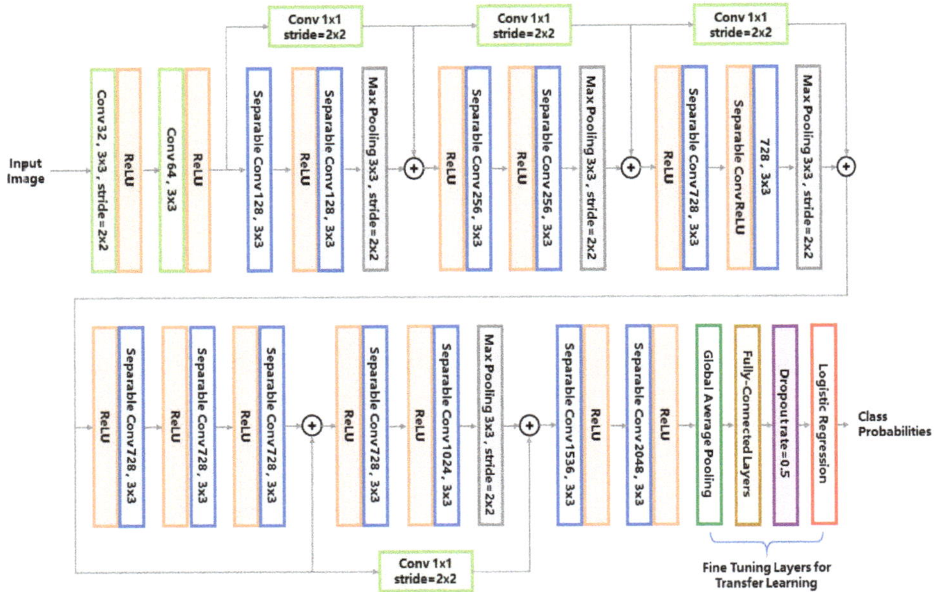

Figure 3. The classifier network architecture used in this study. The abbreviations 'conv' and 'ReLU' stands for 'convolution kernel' and 'rectified linear unit', respectively.

So far, various types of convolutional networks have been developed for image classification. The reason we chose the Xception network instead of other classical networks (e.g., VGG16 [27]) is that the Xception network uses the so-called 'separable convolution' mechanism (shown as the blue blocks in Figure 3) to reduce the number of network hyperparameters. Figure 4 compares the separable convolution architecture with conventional convolution architecture. The input of the convolutional architecture is a k-channel feature map (k = 4 in Figure 4 as an example) produced by the previous network layer. The separable convolution architecture (Figure 4a) conducts separated channel-wise and space-wise convolutions while the conventional convolution architecture (Figure 4b) conducts channel-wise and space-wise convolutions simultaneously using a k × 3 × 3 kernel containing 9k hyperparameters. The channel-wise convolution is firstly performed using a k × 1 × 1 kernel, producing a single-channel intermediate feature map which is further convoluted with a 1 × 3 × 3 kernel. Such separable convolution uses only 9 + k hyperparameters which is much less than the 9k parameters of the conventional convolution. The merit of separable convolution for zebrafish larvae image classification is that it avoids overfitting to the limited training data because the network has much fewer hyperparameters than the conventional CNN.

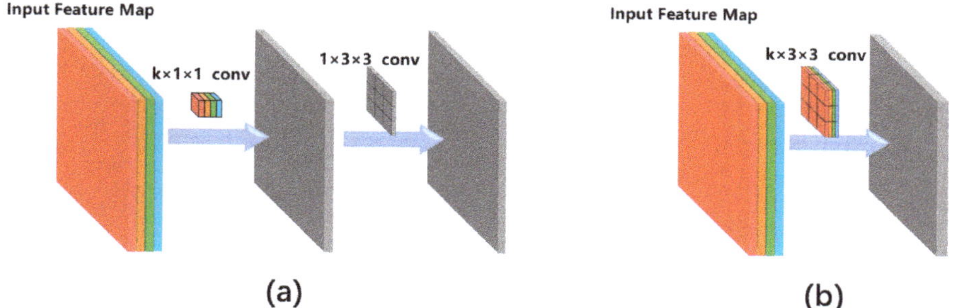

Figure 4. Comparison between (**a**) the separable convolution architecture and (**b**) the conventional convolution architecture. Different colours of the input feature map indicate different feature channels. k is the number of feature channels and this figure uses k = 4 as an example.

To further relieve the problem of scarce training data, a technique called transfer learning is used to train a robust network for complicated zebrafish larvae classification. By transfer learning, we can obtain deep features that an existing model has learned from tens of thousands of natural images, therefore the burden of training such a deep network from scratch is alleviated. We used an existing Xception network pretrained on the imagenet datasets [25] and froze the hyperparameters of the feature extraction layers, leaving only the classification layers trainable (i.e., the global average pooling layer, the fully-connected layers and the logistic regression layer as marked in Figure 3). These classification layers were trained on our augmented dataset using the stochastic gradient descent (SGD) optimizer with a slow learning rate of 0.0001 and a momentum of 0.9. We used such a slow learning rate because the network was already pretrained, the learning rate should be small enough to fine tune the network to suit the limited training data of zebrafish larvae.

3. Results

Thanks to the availability of public zebrafish larvae microscopic images, we can compare our approach with the published state-of-the-art machine learning methods based on the same training and test data. As explained in the Introduction Section, the strength of our method lies in the usage of a two-tier classification pipeline and the Xception network. To prove the advantages of the two-tier pipeline, we compare our method with a single-tier classification pipeline. To evaluate the effectiveness of the Xception network, we run a comparison with an existing method using other classical CNN models, including AlexNet, the Visual Geometry Group networks of 16 and 19 layers (VGG16 and VGG19) and GooLeNet [24,27,28]. All these comparisons were conducted using the classification accuracy as the performance metric, i.e., Accuracy = $(N_{TP} + N_{TN})/N_{all}$, where N_{TP}, N_{TN} and N_{all} denote the number of true positive, true negative, and all test samples respectively for each phenotype. Our networks were programmed with the Keras platform running on a server using NVIDIA TITAN X graphics processing unit (GPU). The training process took −200 min on the entire augmented training set and took −2 s on each test image.

3.1. Comparison with the State-of-the-Art Larva Phenotype Classification Method

In the previous study by Jeanray et al. [20], they published the zebrafish larvae image dataset along with a supervised phenotype classification algorithm using a randomized trees classifier, which is a conventional machine learning method dedicatedly adapted to the zebrafish larvae images. They also used a two-tier classification strategy quite similar to ours, and therefore the major difference between our and their method is whether the deep learning approach was used. We refer Jeanray's method as the state-of-the-art baseline method and conducted the comparison based on the same training images (as listed in Table 2) and testing images (as listed in Table 1). Figure 5 plots the classification accuracies of the two methods. The accuracy values of the baseline method were collected from Jeanray's original paper [20]. Our method achieved evident accuracy improvement over the baseline method for most phenotypes. For the phenotypes which are easy to recognize (dead and chorion), our method yielded an accuracy of 1.0 which is quite promising. Jeanray's method also obtained accuracy over 0.95 for these phenotypes but did not reach 1.0. The biggest improvement is with hemostasis, for which our method obtained an accuracy of 0.73 vs. 0.51 of the baseline method. Normal and short tail are the two phenotypes for which our method yielded slightly lower accuracies than the baseline method, but the differences (0.010 and 0.005 for normal and short tail, respectively) were too small to be considered significant.

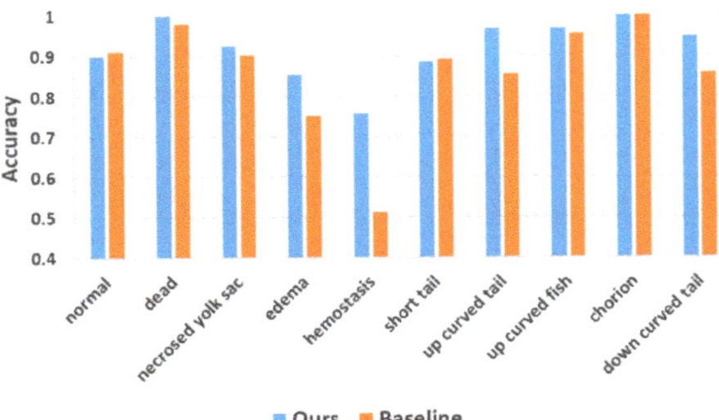

Figure 5. Comparison of classification accuracy between our method and the state-of-the-art baseline method.

3.2. Comparison with the Single-Tier Classification Pipeline

In this study, we used the two-tier pipeline to make the classifiers focus on the fuzzy phenotypes which are difficult to recognize. To evaluate the effectiveness of the two-tier pipeline, we also conducted phenotype classification based on a single-tier pipeline, in which 10 separated binary classifiers were trained for each of the 10 phenotypes. By doing so, each classifier has to recognize one phenotype from 10 phenotypes, which might be more difficult than recognizing one from eight as in the two-tier pipeline. Figure 6 compares the accuracies of the two-tier vs. single-tier pipelines. It can be observed that although the two-tier pipeline does not outperform the single-tier pipeline for all the phenotypes, and it achieved much better accuracy for those difficult phenotypes including edema, hemostasis and down curved tail whose accuracies were improved by 0.06, 0.03 and 0.09, respectively.

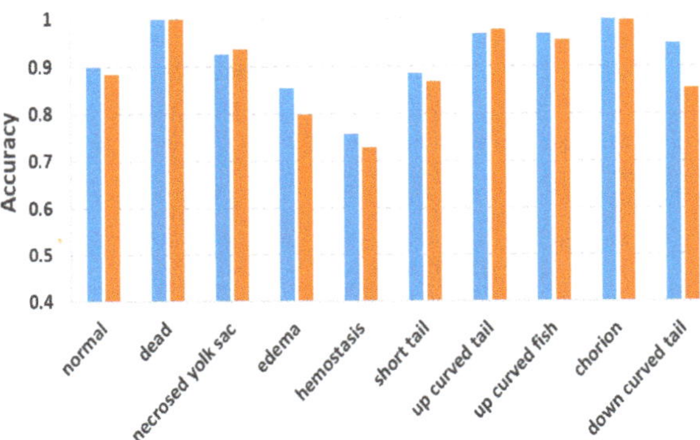

Figure 6. Comparison of classification accuracy between the two-tier and the single-tier pipelines.

3.3. Comparison with the Classical Convolutional Neural Network (CNN) Model

To validate the advantage of using the Xception network instead of a conventional CNN, we compare our method with an existing study using the classical CNN model [19]. VGG16 is a widely used CNN model for image classification [27]. Tyagi et al. fine tuned a pre-trained VGG16 model for zebrafish larvae microscopic image classification [19]. Their experiments were also based on the same ground-truth dataset from the Jeanray group. In their experiment, the network was trained to classy different combinations of five, eight and all phenotypes, respectively. They aimed to validate the classicization performance for different numbers of target phenotypes. The combination of five phenotypes included hemostasis, necrosed yolk sac, edema, short tail and normal. The combination of eight phenotypes included the normal, dead, down-curved tail, chorion, up-curved tail, short tail, edema and necrosed yolk sac.

To conduct a fair comparison, we followed the same experiment setup of Tyagi et al. [19] That is, our Xception network was also trained to classify the same combinations of five, eight and all phenotypes, respectively. The averaged classification accuracy of each combination is compared with Tyagi's accuracy in Table 3. For all the combinations, the Xception network resulted in better accuracy than the VGG16 network. When more phenotypes are included in the combination, the improvement of accuracy becomes larger. This result means that the Xception network is more suitable for multi-phenotype classification and better suits the multi-label nature of the zebrafish larvae.

Table 3. Comparison of classification accuracy between our Xception-based method with Tyagi's VGG16-based method [19] and our finetuning based on AlexNet [24], VGG19 [27], and GoogLeNet [28].

	# of Network Parameters	# of Weight Layers	Five-Phenotype Combination	Eight-Phenotype Combination	All-Phenotype Combination
Xception	22M	16	0.86	0.92	0.92
AlexNet	60M	8	0.80	0.75	0.70
VGG16	138M	16	0.85	0.87	0.84
VGG19	144M	19	0.88	0.87	0.85
GoogLeNet	5M	22	0.85	0.86	0.88

Besides the VGG16-based method of Tyagi et al., we further compared our method with transfer learning using three other popular CNN models including AlexNet [24], VGG19 [27], and GoogLeNet [28] which were all pretrained with the ImageNet data. For a fair comparison, these networks were also validated on the combinations of five, eight and all phenotypes, respectively. The validation results are reported in Table 3 as well. For the five-phenotype combination, VGG19 achieved the best accuracy (0.88) while our Xception-based method yielded suboptimal accuracy (0.86). For both the eight- and all-phenotype combinations, our method was the most accurate.

4. Discussion

In this study, we take advantage of deep learning to classify zebrafish larvae phenotypes from brightfield microscopic images. We aim to solve the classification problems especially for those phenotypes with fuzzy class features. Based on the same training and testing data, Jeanray et al. had already developed a larvae phenotype classification method using the classical machine learning method [20]. Their method provides a state-of-the-art baseline for our DL-based approach. The experimental results of Figure 5 reveals that our deep-learning method outperforms the baseline method in terms of classification accuracy. Besides the accuracy improvement, the advantages of the deep-learning method also include the avoidance of designing hand-crafted image features. Jeanray et al. used image edge and sub-window intensity distribution as the image features, but these features may not be optimal for classifying some difficult phenotypes like edema and hemostasis. Our deep-learning approach automatically learns the crucial image features for phenotype classification. The deep-learning approach not only relieves the algorithm developers from the burden of image feature design but also automatically discovers the useful features which the human designers cannot conceive of. As a result, we observe dramatic accuracy improvement for the phenotypes with fuzzy class features, such as edema and hemostasis.

To address the scarce training data problem, we performed augmentation to the training data. We augmented different phenotypes with different scales to compensate for the imbalanced training data. As Figure 5 reveals, this strategy decreased the inter-class accuracy differences compared to the baseline method using original training data. However, some phenotypes with too few original training samples (e.g., hemostasis) still have relatively lower accuracy than other phenotypes. This is because our augmentation process only implemented image shifting and rotation, its capacity to improve data variability is limited. For a future study, we will introduce more variations into the data augmentation step, such as non-linear image deformation and image contrast variations [29].

To cope with the problems of fuzzy inter-class differences and multi-label issues, we used the two-tier classification pipeline. Experiment results proved the effectiveness of the two-tier strategy. Accuracy improvements against the single-tier pipeline were observed for the difficult classes like edema, hemostasis and down-curved tail. The improvements for these phenotypes are 0.06, 0.03 and 0.09, respectively. It is worth mentioning that Jeanray et al. also compared the two-tier and single-tier pipelines using their randomized tree classifier [20]. However, their two-tier pipeline only improved the accuracy of edema and down-curved tail by 0.01 and 0.03, respectively, which are much less than our improvements. For hemostasis, their two-tier pipeline even resulted in a 0.03 decrease in accuracy. It is obvious that the advantage of the two-tier strategy is more obvious for our method, thanks to the superior classification ability of the DL models.

Currently, there are various types of DL models for image classification. We chose Xception as the classifier because of its small hyperparameter scale [25]. Training a smaller network on scarce annotated data can potentially avoid the overfitting problem because larger networks are more flexible on data fitting and tend to overfit the limited training data. The conventional VGG16 network [27] has over 138 million parameters which is much more than the 22 M parameters of the Xception network. As we compared our Xception-based approach with the VGG16-based approach [19], an obvious improvement of classification accuracy was observed. More importantly, the improvement becomes bigger when more phenotypes are involved as the target classes. That means the VGG16 network has imperfect generalization ability for more classes. The Xception network successfully alleviated the overfitting issue without losing the accuracy of classification. Our study has explored an effective way to cope with scarce annotation data for biological image classification.

As we further compare our method with more CNN networks, it seems that deeper network tends to yield more accurate results. An intuitive explanation is that deeper networks are better at learning deep features from the large ImageNet dataset and thus performs better at transfer learning. However, the Xception network that we use is not the deepest, but it yielded the most accurate results for eight- and all-phenotype combination tests. This might be attributed to the residual connections (i.e., the shortcut connections of 1 × 1 convolution in Figure 3) which does not exist in AlexNet, VGG16, VGG19 or GoogleNet. As demonstrated in the original paper of the Xception network [25], such a residual connection is essential for increasing the classification accuracy since it improved the convergence performance on a large pretraining dataset and helped the network to learn more distinctive image features for transfer learning.

As a limitation of this study, we only tested our method on a publicly available dataset. However, the appearance of zebrafish larvae microscopic images may vary between institutes, acquisition devices and sample preparation procedures. It is important to test our method with more diverse datasets. Moreover, our method also needs to be implemented in an on-site experimental environment to produce real-time classification results. Fortunately, the Xception model has small file size which can be implemented easily in embedded operating systems.

5. Conclusions

We developed a deep-learning pipeline for zebrafish larvae phenotype classification from brightfield microscopic images. Facing the difficulties of scarce training data, fuzzy inter-class difference and the multi-label problem, we used a two-tier classification pipeline and the Xception network architecture to address all these challenges. Experimental results proved the effectiveness of our solution. We look forward to applying our method in daily biological experiments and making further improvements to process more diverse datasets. The method in this paper is also transferable to other biological phenotype recognition problems for researchers who are facing similar challenges of biological image classification.

Author Contributions: Conceptualization, F.C. and S.L.; methodology, S.S.; software, S.S.; validation, S.S.; formal analysis, S.S.; investigation, S.L.; resources, S.L.; data curation, S.S.; writing—original draft preparation, S.S.; writing—review and editing, F.C.; visualization, S.S.; supervision, F.C.; project administration, F.C. and S.L.; funding acquisition, F.C. and S.L. All authors have read and agreed to the published version of the manuscript.

Funding: This research is supported by the National Key R&D Program of China (2018YFC1803100), NSFC grant No. 21607115 and No. 21777116; the Xinghai Scholar Cultivating Funding of Dalian University of Technology, No. DUT14RC(3)066.

Acknowledgments: The authors appreciate Hongkai Wang for his advices on deep neural network training.

Conflicts of Interest: The authors declare no conflict of interest.

References

1. Blaser, R.E.; Vira, D.G. Experiments on learning in zebrafish (Danio rerio): A promising model of neurocognitive function. *Neurosci. Biobehav. Rev.* **2014**, *42*, 224–231. [CrossRef] [PubMed]
2. Lin, S.; Wang, X.; Ji, Z.; Chang, C.H.; Dong, Y.; Meng, H.; Liao, Y.-P.; Wang, M.; Song, T.-B.; Kohan, S.; et al. Aspect Ratio Plays a Role in the Hazard Potential of CeO_2 Nanoparticles in Mouse Lung and Zebrafish Gastrointestinal Tract. *ACS Nano* **2014**, *8*, 4450–4464. [CrossRef] [PubMed]
3. Liu, R.; Lin, S.; Rallo, R.; Zhao, Y.; Damoiseaux, R.; Xia, T.; Lin, S.; Nel, A.; Cohen, Y. Automated phenotype recognition for zebrafish embryo based in vivo high throughput toxicity screening of engineered nano-materials. *PLoS ONE* **2012**, *7*, e35014. [CrossRef] [PubMed]
4. Mungall, C.J.; Washington, N.L.; Nguyen-Xuan, J.; Condit, C.; Smedley, D.; Köhler, S.; Groza, T.; Shefchek, K.; Hochheiser, H.; Robinson, P.N.; et al. Use of Model Organism and Disease Databases to Support Matchmaking for Human Disease Gene Discovery. *Hum. Mutat.* **2015**, *36*, 979–984. [CrossRef]
5. Nishimura, Y.; Inoue, A.; Sasagawa, S.; Koiwa, J.; Kawaguchi, K.; Kawase, R.; Maruyama, T.; Kim, S.; Tanaka, T. Using zebrafish in systems toxicology for developmental toxicity testing. *Congenit. Anom.* **2016**, *56*, 18–27. [CrossRef]
6. Li, C.; Lim, K.M.K.; Chng, K.R.; Nagarajan, N. Predicting microbial interactions through computational approaches. *Methods* **2016**, *102*, 12–19. [CrossRef]
7. Li, L.; LaBarbera, D.V. 3D High-Content Screening of Organoids for Drug Discovery. *Comprehensive Medicinal Chemistry III* **2017**, 388–415.
8. Unser, M.; Sage, D.; Delgado-Gonzalo, R. Advanced image processing for biology, and the Open Bio Image Alliance (OBIA). In Proceedings of the 21st European Signal Processing Conference (EUSIPCO 2013), Marrakech, Morocco, 9–13 September 2013; pp. 1–5.
9. Lin, S.; Zhao, Y.; Xia, T.; Meng, H.; Ji, Z.; Liu, R.; George, S.; Xiong, S.; Wang, X.; Zhang, H.; et al. High content screening in zebrafish speeds up hazard ranking of transition metal oxide nanoparticles. *ACS Nano* **2011**, *5*, 7284–7295. [CrossRef]
10. Lin, S.; Zhao, Y.; Ji, Z.; Ear, J.; Chang, C.H.; Zhang, H.; Low-Kam, C.; Yamada, K.; Meng, H.; Wang, X.; et al. Zebrafish high-throughput screening to study the impact of dissolvable metal oxide nanoparticles on the hatching enzyme, ZHE1. *Small* **2013**, *9*, 1776–1785. [CrossRef]
11. Goldsmith, P. Zebrafish as a pharmacological tool: The how, why and when. *Curr. Opin. Pharmacol.* **2004**, *4*, 504–512. [CrossRef]
12. Pylatiuk, C.; Sanchez, D.; Mikut, R.; Alshut, R.; Reischl, M.; Hirth, S.; Rottbauer, W.; Just, S. Automatic zebrafish heartbeat detection and analysis for zebrafish embryos. *Zebrafish* **2014**, *11*, 379–383. [CrossRef]

13. Ishaq, O.; Sadanandan, S.K.; Wählby, C. Deep Fish: Deep Learning-Based Classification of Zebrafish Deformation for High-Throughput Screening. *J. Biomol. Screen.* **2016**, *22*, 102–107. [CrossRef] [PubMed]

14. Guariglia, E. Entropy and Fractal Antennas. *Entropy* **2016**, *18*, 84. [CrossRef]

15. Frongillo, M.; Gennarelli, G.; Riccio, G. TD-UAPO diffracted field evaluation for penetrable wedges with acute apex angle. *J. Opt. Soc. Am. A-Opt. Image Sci. Vis.* **2015**, *32*, 1271–1275. [CrossRef] [PubMed]

16. Guariglia, E. Harmonic Sierpinski Gasket and Applications. *Entropy* **2018**, *20*, 714. [CrossRef]

17. Lussier, Y.A.; Liu, Y. Computational Approaches to Phenotyping: High-Throughput Phenomics. *Proc. Am. Thorac. Soc.* **2007**, *4*, 18–25. [CrossRef] [PubMed]

18. Mikut, R.; Dickmeis, T.; Driever, W.; Geurts, P.; Hamprecht, F.A.; Kausler, B.X.; Ledesmacarbayo, M.J.; Maree, R.; Mikula, K.; Pantazis, P.; et al. Automated processing of zebrafish imaging data: A survey. *Zebrafish* **2013**, *10*, 401–421. [CrossRef]

19. Tyagi, G.; Patel, N.; Sethi, I. A Fine-Tuned Convolution Neural Network Based Approach For Phenotype Classification Of Zebrafish Embryo. *Proc. Comput. Sci.* **2018**, *126*, 1138–1144. [CrossRef]

20. Jeanray, N.; Marée, R.; Pruvot, B.; Stern, O.; Geurts, P.; Wehenkel, L.; Muller, M. Phenotype classification of zebrafish embryos by supervised learning. *PLoS ONE* **2015**, *10*, e0116989. [CrossRef]

21. Leng, B.; Yu, K.; Qin, J. Data augmentation for unbalanced face recognition training sets. *Neurocomputing* **2017**, *235*, 10–14. [CrossRef]

22. Sun, Y.; Wang, X.; Tang, X. Deep convolutional network cascade for facial point detection. In Proceedings of the IEEE Conference on Computer Vision and Pattern Recognition, Portland, OR, USA, 25–27 June 2013; IEEE: New York, NY, USA, 2013; pp. 3476–3483.

23. Tompson, J.; Jain, A.; LeCun, Y.; Bregler, C. Joint training of a convolutional network and a graphical model for human pose estimation. In Proceedings of the Advances in Neural Information Processing Systems 27, Montréal, QC, Canada, 8–13 December 2014; Neural Information Processing Systems Foundation, 2014; pp. 1799–1807.

24. Krizhevsky, A.; Sutskever, I.; Hinton, G.E. ImageNet Classification with Deep Convolutional Neural Networks. *Adv. Neural Inf. Proc. Syst.* **2012**, *25*, 1097–1105. [CrossRef]

25. Chollet, F. Xception: Deep Learning with Separable Convolutions. In Proceedings of the 2017 IEEE Conference on Computer Vision and Pattern Recognition (CVPR), Honolulu, HI, USA, 21–26 July 2017; IEEE: New York, NY, USA, 2017; pp. 1800–1807.

26. Marée, R. The need for careful data collection for pattern recognition in digital pathology. *J. Pathol. Inf.* **2017**, *8*, 19. [CrossRef] [PubMed]

27. Zhang, X.; Zou, J.; He, K.; Sun, J. Accelerating Very Deep Convolutional Networks for Classification and Detection. *IEEE Trans. Pattern Anal. Mach. Intell.* **2015**, *8*, 1943–1955. [CrossRef] [PubMed]

28. Szegedy, C.; Liu, W.; Jia, Y.; Sermanet, P.; Reed, S.; Anguelov, D.; Erhan, D.; Vanhoucke, V.; Rabinovich, A. Going deeper with convolutions. In Proceedings of the IEEE Conference on Computer Vision and Pattern Recognition, Boston, MA, USA, 7–12 June 2015; IEEE: New York, NY, USA, 2015; pp. 1–9.

29. Zhao, A.; Balakrishnan, G.; Durand, F.; Guttag, J.V.; Dalca, A.V. Data Augmentation Using Learned Transformations for One-Shot Medical Image Segmentation. In Proceedings of the Computer Vision and Pattern Recognition, Long Beach, CA, USA, 15–20 June 2019; IEEE: New York, NY, USA, 2019; pp. 8543–8553.

Article

Unsupervised Generation and Synthesis of Facial Images via an Auto-Encoder-Based Deep Generative Adversarial Network

Jeong gi Kwak [1] and Hanseok Ko [1,*

School of Electrical Engineering, Korea University, Seoul 136-701, Korea; jgkwak@ispl.korea.ac.kr
* Correspondence: hsko@korea.ac.kr

Received: 26 December 2019; Accepted: 11 March 2020; Published: 14 March 2020

Abstract: The processing of facial images is an important task, because it is required for a large number of real-world applications. As deep-learning models evolve, they require a huge number of images for training. In reality, however, the number of images available is limited. Generative adversarial networks (GANs) have thus been utilized for database augmentation, but they suffer from unstable training, low visual quality, and a lack of diversity. In this paper, we propose an auto-encoder-based GAN with an enhanced network structure and training scheme for Database (DB) augmentation and image synthesis. Our generator and decoder are divided into two separate modules that each take input vectors for low-level and high-level features; these input vectors affect all layers within the generator and decoder. The effectiveness of the proposed method is demonstrated by comparing it with baseline methods. In addition, we introduce a new scheme that can combine two existing images without the need for extra networks based on the auto-encoder structure of the discriminator in our model. We add a novel double-constraint loss to make the encoded latent vectors equal to the input vectors.

Keywords: generative models; GAN (Generative adversarial networks); facial image; generation; database augmentation; synthesis

---

1. Introduction

In the last few years, deep neural networks (DNNs) have been successfully applied to a range of computer vision tasks, including classification [1–3], detection [4–6], segmentation [7,8], and information fusion [9,10]. However, because data augmentation is essential for the effective training of DNNs, and because there are numerous image-to-image translation and information fusion problems that need to be overcome, deep generative models have received significant attention. In this field, research on facial datasets has been particularly active, because they have a large number of real-world applications, such as facial classification and the opening of closed eyes in photos. Despite this increase in research interest, implementing generative models remains challenging because the process required to generate realistic images from low-level to high-level information is complex.

Since Goodfellow et al. [11] first proposed the generative adversarial network (GAN), which is based on adversarial learning between two networks, a generator and a discriminator, many GAN models have demonstrated excellent performance in terms of their photo-realistic output. The key principle underlying the use of a GAN is to ensure that the probability distribution of the generated data is close to that of the real data via the adversarial training of the generator and discriminator. In the early stages of training, the generator may generate poor-quality images; thus, the discriminator can easily distinguish between real and fake samples. As the generator learns more during training, its output becomes more photo-realistic and the discriminator finds it more difficult to distinguish between real and fake samples. When the training reaches convergence, the generator can generate realistic but fake images. However, many GAN models suffer from instability during the training process, leading to problems such as mode collapse and the lack of diversity.

BEGAN [12] is an auto-encoder-based GAN model with an auto-encoder architecture as the discriminator. Unlike many existing GAN models [11,13,14] that attempt to directly match the real data distribution, this model seeks to match the loss distribution of the auto-encoder. The BEGAN developers introduced an equilibrium hyperparameter to maintain the balance between the generator and the discriminator. It makes it possible for a user to control the visual quality and diversity of an image by changing the parameters. However, it suffers from the trade-off between diversity and quality, is subject to mode collapse, and occasionally fails to generate high-quality images during the training phase.

StyleGAN [15] can generate photo-realistic output images using a style-based generator that considers the scale-specific characteristics of the generated image. Each layer in the StyleGAN generator consists of several convolutional layers and adaptive instance normalization (AdaIN) [16] layers. The AdaIN layers utilize latent vectors as input and then utilize their information with an affine transform. In addition, StyleGAN can perform style mixing, in which an image generated using two different latent vectors has both characteristics. However, StyleGAN generates an image from noise; thus, it cannot mix two existing images; i.e., it does not take existing images as input. Synthesizing two existing images using the model requires the training of an additional network that can encode real images into the latent space of StyleGAN.

Motivated by StyleGAN, we propose a generator that takes two latent vectors as input based on the scale-specific role of each layer in the generator. The front layers are involved in the creation of high-level features such as the overall shape of the face, while the back layers are involved in lower-level features such as hair color and the microstructure. Our discriminator is trained to reconstruct only real images, and its decoder has the same structure as the generator. This divided structure of the generator and decoder that utilize different latent vectors to assign scale-specific roles in image generation improves the visual quality of the image.

We also adopt a training technique that differs from that used in the conventional BEGAN model. The instability of GANs usually occurs when generating high-resolution images; thus, we adopt the progressive growth concept for the generator and discriminator introduced in [17]. The size of a generated image at the beginning of the training process is small, but it becomes twice the size after several epochs. This training scheme reduces instability and consequently improves the visual quality of the output images.

In addition to generating images using random vectors, we also propose a method to synthesize two existing images by exploiting the auto-encoder structure of our discriminator. The encoder of discriminator learns to encode both real and fake samples during the training process; thus, it does not need to train an additional model. However, in order for the decoder or generator to combine real images, the encoded latent space of the real images should be similar to that of the fake images. To guarantee this, we propose the novel double-constraint loss function, which constrains the latent vectors of encoded real images. Therefore, the images are combined when the decoder decodes an image using the latent vectors obtained from the different images in an unsupervised manner.

This paper is structured as follows. Section 2 presents the theoretical background and provides a detailed description of the proposed model. We then demonstrate the superiority of our model by qualitatively and quantitatively comparing it to conventional models [12,18] in Section 3. Concluding remarks are presented in Section 4.

2. Proposed Method

This section describes our proposed model in detail by first introducing the BEGAN baseline model with a brief explanation of the auto-encoder-based GAN and then outlining the structure of our proposed model and its training strategy. Subsequently, we introduce a method for combining facial images using our model.

2.1. BEGAN Baseline Model

Conventional GANs have a generator and a discriminator; the generator creates fake images, whereas the discriminator receives both real and fake images as input and attempts to distinguish them. The goal of a GAN is to match the probability distribution of the fake samples generated by the generator to that of the real samples. Therefore, the output of the discriminator is essentially a probability score, and this is fed into the loss function. However, BEGAN has a discriminator with an auto-encoder structure, meaning that the output of the discriminator is an image of the same size as the input.

Auto-encoder-based GAN models can be optimized by reducing the Wasserstein distance between the reconstruction loss distributions of the real and fake images rather than their sample distributions directly [12,19]. The discriminator attempts to reconstruct only real images, but the generator attempts to produce an image that can be accurately reconstructed by the discriminator. Therefore, the reconstruction performance of the discriminator is crucial for the generator to be able to produce high-quality output. If the decoder within the discriminator produces poor-quality images when reconstructing the input, the generator could easily fool the discriminator with those poor-quality images.

Berthelot et al. [12] introduced the hyperparameter $\gamma \in [0,1]$ to maintain the balance between generator and discriminator loss, defined as

$$\gamma = \frac{\mathbb{E}\left[\mathcal{L}(G(z))\right]}{\mathbb{E}\left[\mathcal{L}(x)\right]}, \tag{1}$$

where $\mathcal{L}(\cdot)$ denotes the L_1 or L_2 reconstruction error from the auto-encoder; i.e., the discriminator. $\mathbb{E}(\cdot)$ denotes expectation operator. $G(z)$ denotes a fake image from the generator and x denotes a real image. This ratio (γ) enables users to control the balance between the visual quality and diversity of the output images. If γ is low, the model focuses more on reducing the reconstruction loss of the real images; i.e., the auto-encoding ability of the discriminator increases. This leads to higher visual quality and lower diversity. However, BEGAN has limitations in terms of visual quality and diversity due to the inherent structure of the generator, the lack of reconstruction ability in the discriminator, and unstable training.

2.2. The Proposed Model

2.2.1. Network Architecture

We propose the novel auto-encoder-based GAN architecture illustrated in Figure 1. Our generator takes two latent vectors and consists of several blocks, with each block handling a specific resolution. The latent vectors are fed into each block and transformed by the affine transformation layer. We use an AdaIN [16] layer that stylizes feature maps with information from the affine transformation layer. We divide the generator into front and back modules, with the front module generating feature maps of a relatively low resolution (32×32) and the back module generating the final output image. z_1 is fed into the front module, and z_2 is fed into the back module, meaning z_1 is associated with the overall structure of the image (e.g., the shape or appearance of the face), whereas z_2 is associated with the details of the image (e.g., the microcharacteristics of the face or hair color). The bottom of Figure 1 presents the details of each block. Initially, the input features are upscaled, and there are three sets of Conv-ELU-AdaIN layers. As mentioned above, the AdaIN layer normalizes the features and matches them to new statistics (i.e., the mean and variance from the affine transformation layer). AdaIN is formulated as

$$\text{AdaIN}(x, y) = \sigma(y)\left(\frac{x - \mu(x)}{\sigma(x)}\right) + \mu(y), \tag{2}$$

where x denotes the feature map, and the new mean and variance ($\sigma(y)$ and $\mu(y)$, respectively) are calculated by affine transformation with the input latent vectors. Because of the scale-specific role of each layer, the visual quality of the output images is improved.

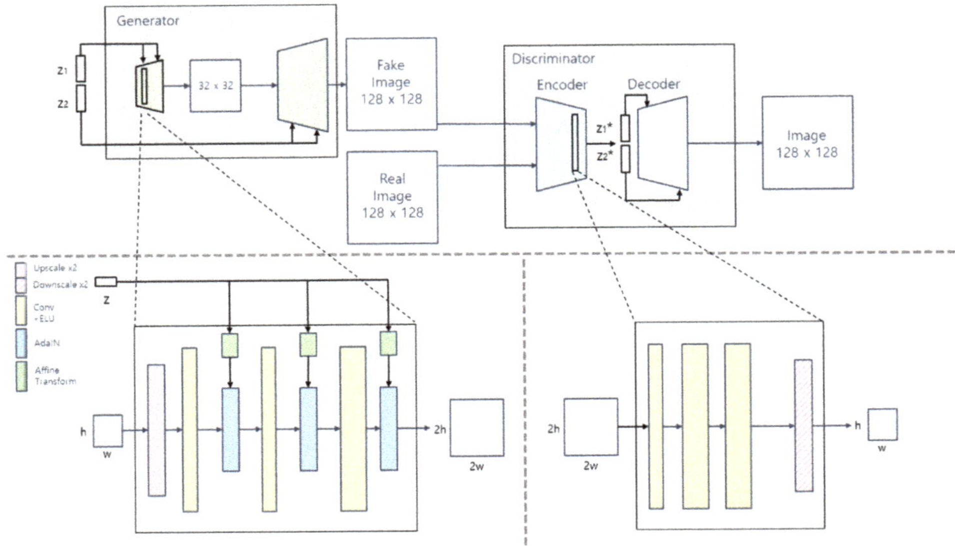

Figure 1. Overview of the proposed model, consisting of two networks; generator and discriminator. The generator takes two input vectors and generates a fake image. The discriminator takes a real or fake image and it is learned to reconstruct only real sample (Top). There are AdaIN layers which stylize feature maps with transformed input vector after convolutional layer in each block of the generator (Left bottom). The encoder down-samples input image to two latent vectors with convolutional layers and down sampling layers (Right bottom).

The discriminator of our model has an auto-encoder structure that consists of an encoder and a decoder. The encoder takes a real or fake image as input and encodes it as two latent vectors z_1^* and z_2^* of the same size as z_1 and z_2 respectively. The decoder then decodes the image with z_1^* and z_2^*. Because the decoder has the same structure as the generator, z_1^* and z_2^* affect different scale-specific characteristics.

2.2.2. Objective Function

A fake image generated from the input vector z_1, z_2 can be expressed as $G(z_1, z_2)$. The goal of the discriminator is to distinguish real image x from fake image $G(z_1, z_2)$. Therefore, the discriminator attempts to reconstruct x only, not $G(z_1, z_2)$. On the other hand, the generator attempts to produce an image that can be reconstructed well by the discriminator. As a result of the adversarial training of the generator and the discriminator, the output images from the generator become more realistic to deceive the discriminator. In other words, the generator is trained to reduce the Wasserstein distance between the loss distributions of real and fake samples in the auto-encoder. The adversarial loss of the generator and discriminator can be expressed as

$$L_D = L(x; \theta_D) - k_t L(G(z_D; \theta_G); \theta_D),$$
$$k_{t+1} = k_t - \lambda_k(\gamma L(x) - L(G(z_G))) \; for \; each \; step \; t, \tag{3}$$

and

$$L_G = L(G(z_D; \theta_G); \theta_D), \tag{4}$$

where $L(\cdot)$ denotes the L_1 loss from the auto-encoder, and k_t is the parameter that controls the proportion of generator and discriminator loss introduced in BEGAN. This is required because the discriminator cannot achieve a suitable reconstruction quality at the beginning of training. At this time, k_t has a value close to zero and gradually increases as training progresses.

As mentioned in Section 2.2.1, z_1 and z_2 are both involved in the generation of different scale-specific areas. To apply this principle to the decoder, we add a novel constraint on the encoded latent vectors, referred to as double-constraint loss. It includes the difference between the input vector and the encoded vector as defined by

$$
\begin{aligned}
L_{dc} &= \|z_1 - z_1^*\|_1 + \|z_2 - z_2^*\|_1, \\
[z_1^* \; z_2^*] &= Enc(G(z_1, z_2)),
\end{aligned}
\tag{5}
$$

where $Enc(\cdot)$ denotes the output of the encoder. The double-constraint loss is designed to stabilize training because the inputs of the generator and decoder would be similar. It can also be extended to the synthesis of existing images because real samples are mapped to a space similar to the latent space of the input. Hence, the generator loss can be modified as

$$
L_G = L(G(z_D; \theta_G); \theta_D) + \alpha \cdot L_{dc},
\tag{6}
$$

where hyperparameter α represents a weighting factor for the double-constraint loss.

2.2.3. Training Scheme

Unstable training is a major concern when using GANs, and it can occasionally result in mode collapse or low-quality output. In auto-encoder-based GAN models in particular, the reconstruction performance of the discriminator is a decisive factor in establishing the visual quality of an output image. However, excellent reconstruction performance cannot be guaranteed because the importance of the reconstruction error for a real image decreases as k_t increases, as can be seen in Equation (3). Training a discriminator on relatively large images (e.g., 128×128) is slow and difficult, and k_t becomes larger because the discriminator does not effectively function as an auto-encoder. Motivated by PGGAN [17], our model attempts to overcome this problem by starting the training process with low-resolution images. In other words, the size of the training images increases as training progresses (Figure 2). When the size of an image is larger, new layers are added to both the generator and discriminator to adjust the size of the input and output correctly. While PGGAN starts training with 4×4 images, our model begins with 32×32 images because our model performs sufficiently without threatening stability when generating images with sizes of 32×32 or lower, thereby reducing the training time. After a few epochs of training, the size of the training images is doubled, and new layers are added to the generator, encoder, and decoder while maintaining the weights in the conventional layers. By progressively training the generator and the discriminator in this manner, our model achieves better reconstruction performance than when k_t remains constant. Because the discriminator is trained to some extent in the previous stage, the training process becomes more stable and the visual quality is higher than when training with 128×128 images directly. In addition, the layers of the generator and the decoder can accurately reflect the spatial properties of their input.

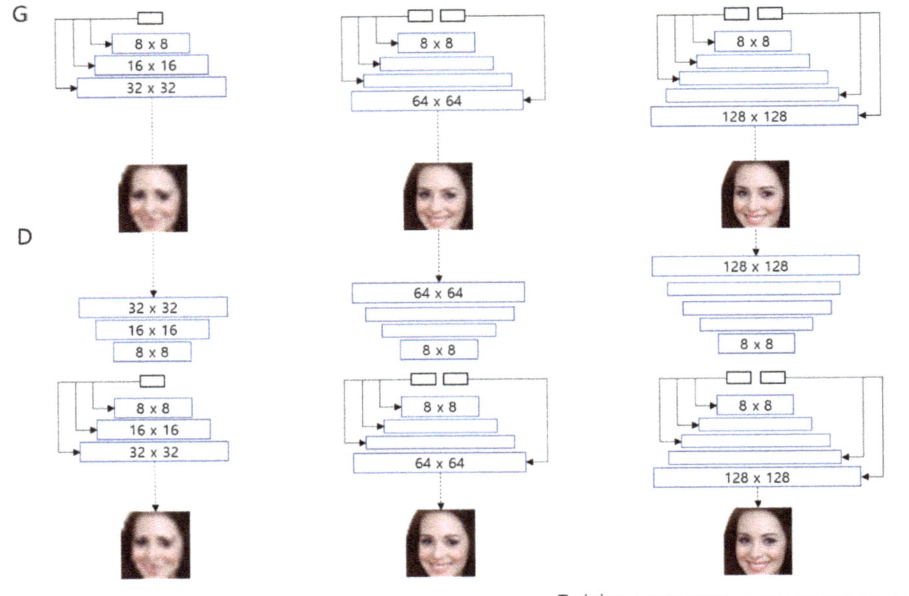

Figure 2. Progressive training of the proposed generator and discriminator. Our model starts with a 32 × 32 image in the first stage, and the size of the training images is doubled in the next stage.

2.2.4. Facial Synthesis Method

In addition to generating images from random noise, as with other unsupervised GANs, our model can also be used to synthesize two images. StyleGAN introduced style mixing, which exploits two or more input vectors, but it was used on only random noise input, not existing images. To mix two existing images, an additional encoder needs to be trained to map the images onto the latent space of the input. However, our model does not require an additional network because our discriminator already has an encoder. By taking advantage of the auto-encoder structure of our discriminator, we present a method for mixing existing images. The encoder encodes an input image as two latent vectors, and they are exploited in different layers of the decoder. Reconstruction occurs when the decoder uses the two latent vectors from a single image. However, if the decoder exploits a combination of the two latent vectors from two different images, the output of the decoder is a mixed image.

Let X and Y denote the two images to be mixed; the output of the encoder when the input is X and Y can be expressed as

$$[z^*_{X_1}, z^*_{X_2}] = Enc(X), \quad [z^*_{Y_1}, z^*_{Y_2}] = Enc(Y). \tag{7}$$

If the decoder decodes the image using $z^*_{X_1}$ and $z^*_{X_2}$, it reconstructs X, and if it uses $z^*_{Y_1}$ and $z^*_{Y_2}$, it reconstructs Y; i.e.,

$$X^* = Dec\left(z^*_{X_1}, z^*_{X_2}\right) \rightarrow Reconstruction\ of\ X, \tag{8}$$

$$Y^* = Dec\left(z^*_{Y_1}, z^*_{Y_2}\right) \rightarrow Reconstruction\ of\ Y, \tag{9}$$

where X^* and Y^* denote the reconstructed images of X and Y, respectively, and $Dec(.)$ denotes our decoder. To synthesize X and Y, the decoder needs to take the latent vectors from the two images as input. A mixed image of X and Y is acquired by exploiting a combination of the latent vectors from the two images (e.g., $z^*_{X_1}$ and $z^*_{Y_2}$), as illustrated in Figure 3. The two blue boxes in Figure 3 represent the two parts of the decoder; i.e., one is involved in generating a 32 × 32 feature map from a given latent vector, and the other is involved in generating a 128×128 image from a given 32×32 feature map. Therefore, the synthesis process can be expressed as

$$I_{X,Y} = Dec\left(z_{X_1}^*,\ z_{Y_2}^*\right) \rightarrow Synthesis\ of\ X\ and\ Y, \tag{10}$$

where $I_{X,Y}$ is an image that has the structural or coarse-scale characteristics of X and the details or fine-scale characteristics of Y.

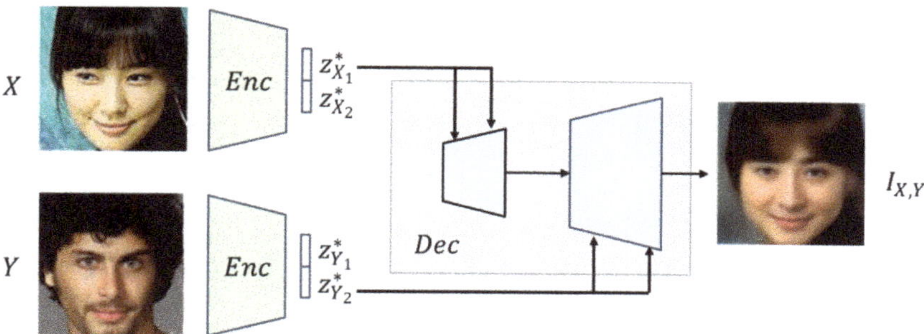

Figure 3. The facial image synthesis process for our model. The decoder takes one encoded vector from each image.

3. Experimental Results

In this section, we first explain our experimental setup and then present qualitative and quantitative comparisons of the performance of our model with those of other auto-encoder-based models.

3.1. Experimental Setup

We used the CelebFaces Attributes (CelebA) dataset (Figure 4) [20], which consists of 202,599 facial images of celebrities cropped to 178×218. We cropped each image further to 170×170, and then resized them to 128×128. For progressive training, we downsampled the images to 32×32 and used them in the first stage. The height and width of each training image were doubled every five training epochs. In our experiments, the coefficients of the objective functions in Equations (3) and (6) were set to $\gamma = 0.5$ and $\alpha = 0.1$. We used the ADAM [21] solver with $\beta_1 = 0.5$ and $\beta_2 = 0.999$, and the learning rate was set initially to 0.0005. L_1 loss was adopted as the loss function for the auto-encoder. All other parameters were the same as in BEGAN. We used Tensorflow with cuDNN as the deep-learning framework and an NVIDIA GTX 1080Ti graphics card.

Figure 4. CelebFaces Attributes (CelebA) dataset.

3.2. Qualitative Results

We conducted qualitative analysis by comparing the output of our model with those of two other auto-encoder-based GAN models, BEGAN [12] and BEGAN-CS [18]. BEGAN-CS adds a latent constraint to BEGAN. The results are shown in Figure 5. The columns (a) to (c) represent 5, 10, and 15 epochs, respectively, while each row represents the qualitative results from the compared methods. The output images are produced by the generator and the input vectors are sampled randomly from a Gaussian distribution. It should be noted that the output of our model in (a) (5 epochs) has a lower resolution than the other models because of the progressive learning strategy it employs. The visual quality of the images improves as training progresses in all three models. However, the results from BEGAN contain some artifacts, such as checkerboard patterns, while BEGAN-CS produces blurred and unstructured facial images (Figure 5d). Once the size of the training images is increased, our model produces similar visual quality in Figure 5b and clearer images than the other models after 15 epochs (Figure 5c,d).

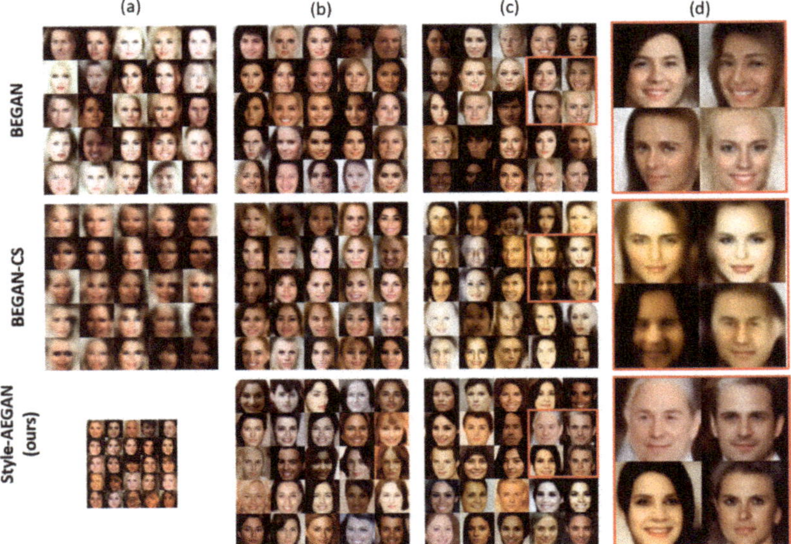

Figure 5. Qualitative results for facial image generation. The rows from top to bottom present the results of BEGAN, BEGAN-CS, and the proposed model, respectively. Each column represents five epochs; i.e., (**a**) epoch 5, (**b**) epoch 10, and (**c**) epoch 15. (**d**) An enlargement of the red box in (c).

3.3. Quantitative Results

It is difficult to verify the diversity of output images using several images. Therefore, we conducted quantitative experiments using the Fréchet inception distance (FID) [22]. The FID score can be used to measure the quality and diversity of images. The FID score is calculated using Equation (11):

$$\text{FID} = \|\mu_x - \mu_y\|_2^2 + Tr\left(\Sigma_x + \Sigma_y - 2\left(\Sigma_x \Sigma_y\right)^{\frac{1}{2}}\right), \tag{11}$$

where x and y denote the image sets x and y. In our experiments, x and y consist of real images and fake images, respectively. Because the FID score considers the mean (μ) and variance (Σ) of the images, it can represent the visual quality and diversity of the images. If the two sets of images have a similar probability distribution, the FID score is low (Equation (11)). Therefore, lower FID scores are better when comparing GAN models. We measured the FID score based on 5000 real samples and 5000 fake samples in epoch 15 for each model. The results are summarized in Table 1.

Table 1. Visual quality in terms of the Fréchet inception distance (FID) score, where a lower score is better.

	BEGAN	BEGAN-CS	Style-AEGAN (ours)
FID	47.93 ± 1.18	50.31 ± 1.01	41.88 ± 1.08

Our model produces the best results. As a result, our model can be seen as superior in terms of image quality and diversity.

3.4. Facial Synthesis Results

We test the synthesis of facial images using our model as described in Section 3.4. In Figure 6, the right-most image in each row represents the synthesis output of the two left-side images. The front module of our decoder takes a latent vector encoded from the left image, and the back module takes a latent vector from the right image. The output image has the characteristics of both images but different scale-specific features. In other words, the output has the coarse-scale characteristics of the first image (e.g., the overall structures or locations of facial attributes) and the fine-scale features of the second image (e.g., the eyes or the skin color). Note that facial synthesis is achieved without requiring additional information, such as binary attribute labels for each image.

Figure 6. Qualitative results of facial synthesis with the proposed model. The right-most image is the synthesis of the two left-side images.

4. Conclusions

In this paper, we proposed an enhanced GAN model for unsupervised facial image generation and synthesis. To overcome the limitations of GAN models (particularly auto-encoder-based models), we first introduced an enhanced generator and discriminator structure. Our generator and decoder utilize two input vectors, and every block reflects the information from the input vectors with adaptive instance normalization layers. Each layer plays a role in producing scale-specific components of the facial image. We also applied a progressive learning method to the proposed auto-encoder-based model, in which the training process was divided into several stages depending on the size of the training image. Consequently, our model can both generate and synthesize facial images via an auto-encoder structure. Our model can generate arbitrary images because it also takes noise as input and synthesizes two existing images using an encoder and decoder within the discriminator. Therefore, it does not require additional training to encode existing images or a pre-trained network. We demonstrated that the visual quality and diversity of the output images were higher than those of the baseline models using both qualitative and quantitative analysis. Additionally, we presented a method for synthesizing two existing images by exploiting the auto-encoder structure of the discriminator. Our model did not need to train a subnetwork that could encode the images for mixing. All of the networks in our model were trained in an end-to-end manner without the labeling of the images. In future research, we will further investigate this novel method from a variety of perspectives to enhance the visual quality of the output images and to ensure stable training for large-scale image generation. Furthermore, we will extend our model for use in not only unsupervised generation tasks but also conditional image generation or synthesis tasks, such as image-to-image translation.

Author Contributions: Conceptualization: J.g.K.; methodology: J.g.K.; software: J.g.K.; investigation: J.g.K.; writing—original draft preparation: J.g.K.; writing—review and editing: H.K.; supervision: H.K. All authors have read and agreed to the published version of the manuscript.

Acknowledgments: This research was supported by a National Research Foundation (NRF) grant funded by the MSIP of Korea (number 2019R1A2C2009480).

Conflicts of Interest: The authors declare no conflicts of interest.

References

1. Krizhevsky, A.; Sutskever, I.; Hinton, G.E. ImageNET classification with deep convolutional neural networks. In *Advances in Neural Information Processing Systems*; Curran Associates, Inc.: Lake Tahoe, CA, USA, 2012; pp. 1097–1105.
2. Szegedy, C.; Liu, W.; Jia, Y.; Sermanet, P.; Reed, S.; Anguelov, D.; Rabinovich, A. Going deeper with convolutions. In Proceedings of the IEEE Conference on Computer Vision and Pattern Recognition, Boston, MA, USA, 7–12 June 2015; pp. 1–9.
3. He, K.; Zhang, X.; Ren, S.; Sun, J. Deep residual learning for image recognition. In Proceedings of the IEEE Conference on Computer Vision and Pattern Recognition, Las Vegas, NV, USA, 26 June–1 July 2016; pp. 770–778.
4. Ren, S.; He, K.; Girshick, R.; Sun, J. Faster R-CNN: Towards real-time object detection with region proposal networks. In *Advances in Neural Information Processing Systems*; Curran Associates, Inc.: Montreal, QC, Canada, 2015; pp. 91–99.
5. Girshick, R. Fast r-cnn. In Proceedings of the IEEE International Conference on Computer Vision, Boston, MA, USA, 7–12 June 2015; pp. 1440–1448.
6. Redmon, J.; Divvala, S.; Girshick, R.; Farhadi, A. You only look once: Unified, real-time object detection. In Proceedings of the IEEE Conference on Computer Vision and Pattern Recognition, Las Vegas, NV, USA, 26 June–1 July 2016; pp. 779–788.
7. Long, J.; Shelhamer, E.; Darrell, T. Fully convolutional networks for semantic segmentation. In Proceedings of the IEEE Conference on Computer Vision and Pattern Recognition, Boston, MA, USA, 7–12 June 2015; pp. 3431–3440.
8. Ronneberger, O.; Fischer, P.; Brox, T. U-net: Convolutional networks for biomedical image segmentation. In Proceedings of the International Conference on Medical Image Computing and Computer-Assisted Intervention, Munich, Germany, 5–9 October 2015; pp. 234–241.
9. Chen, Z.; Li, W. Multisensor feature fusion for bearing fault diagnosis using sparse auto-encoder and deep belief network. *IEEE Trans. Instrum. Meas.* **2017**, *66*, 1693–1702. [CrossRef]
10. Liu, Y.; Chen, X.; Peng, H.; Wang, Z. Multi-focus image fusion with a deep convolutional neural network. *Inf. Fusion* **2017**, *36*, 191–207. [CrossRef]
11. Goodfellow, I.; Pouget-Abadie, J.; Mirza, M.; Xu, B.; Warde-Farley, D.; Ozair, S.; Courville, A.; Bengio, Y. Generative adversarial nets. In *Advances in Neural Information Processing Systems*; Curran Associates, Inc.: Montreal, QC, Canada, 2014; pp. 2672–2680.

12. Berthelot, D.; Schumm, T.; Metz, L. Began: Boundary equilibrium generative adversarial networks. *arXiv* **2017**, arXiv:1703.10717.
13. Mao, X.; Li, Q.; Xie, H.; Lau, R.Y.; Wang, Z.; Paul Smolley, S. Least squares generative adversarial networks. In Proceedings of the IEEE International Conference on Computer Vision, Honolulu, HI, USA, 21–26 July 2017; pp. 2794–2802.
14. Arjovsky, M.; Chintala, S.; Bottou, L. Wasserstein GAN. *arXiv* **2017**, arXiv:1701.07875.
15. Karras, T.; Laine, S.; Aila, T. A style-based generator architecture for generative adversarial networks. In Proceedings of the IEEE Conference on Computer Vision and Pattern Recognition, Long Beach, CA, USA, 16–20 June 2019; pp. 4401–4410.
16. Huang, X.; Belongie, S. Arbitrary style transfer in real-time with adaptive instance normalization. In Proceedings of the IEEE International Conference on Computer Vision, Honolulu, HI, USA, 21–26 July 2017; pp. 1501–1510.
17. Karras, T.; Aila, T.; Laine, S.; Lehtinen, J. Progressive growing of GANs for improved quality, stability, and variation. *arXiv* **2017**, arXiv:1710.10196.
18. Chang, C.C.; Hubert Lin, C.; Lee, C.R.; Juan, D.C.; Wei, W.; Chen, H.T. Escaping from collapsing modes in a constrained space. In Proceedings of the European Conference on Computer Vision, Salt Lake City, UT, USA, 18–23 June 2018; pp. 204–219.
19. Gulrajani, I.; Ahmed, F.; Arjovsky, M.; Dumoulin, V.; Courville, A.C. Improved training of Wasserstein GANs. In *Advances in Neural Information Processing Systems*; Curran Associates, Inc.: Long Beach, CA, USA, 2017; pp. 5767–5777.
20. Liu, Z.; Luo, P.; Wang, X.; Tang, X. Deep learning face attributes in the wild. In Proceedings of the IEEE international conference on computer vision, Boston, MA, USA, 7–12 June 2015; pp. 3730–3738.
21. Kingma, D.P.; Ba, J. Adam: A method for stochastic optimization. *arXiv* **2014**, arXiv:1412.6980.
22. Heusel, M.; Ramsauer, H.; Unterthiner, T.; Nessler, B.; Hochreiter, S. Gans trained by a two time-scale update rule converge to a local Nash equilibrium. In *Advances in Neural Information Processing Systems*; Curran Associates, Inc.: Long Beach, CA, USA, 2017; pp. 6626–6637.

Article

Detecting Green Mold Pathogens on Lemons Using Hyperspectral Images

Yuriy Vashpanov [1], Gwanghee Heo [1,2], Yongsuk Kim [1,3], Tetiana Venkel [4] and Jung-Young Son [1,*

[1] Public Safety Research Center, Konyang University, Nonsan 32992, Korea;
     yuriy.vashpanov@gmail.com (Y.V.); heo@konyang.ac.kr (G.H.); yongsuk@konyang.ac.kr (Y.K.)
[2] Department of Civil Engineering, Konyang University, Nonsan 32992, Korea
[3] Department of Bio-IT Engineering, Konyang University, Nonsan 32992, Korea
[4] Linguistics for Sciences Department, Chernivtsi University, 58012 Chernivtsi, Ukraine; t.venkel@chnu.edu.ua
*   Correspondence: jyson@konyang.ac.kr; Tel.: +82-41-730-5636

Received: 19 December 2019; Accepted: 6 February 2020; Published: 11 February 2020

Abstract: Hyperspectral images in the spectral wavelength range of 500 nm to 650 nm are used to detect green mold pathogens, which are parasitic on the surface of lemons. The images reveal that the spectral range of 500 nm to 560 nm is appropriate for detecting the early stage of development of the pathogen in the lemon, because the spectral intensity is proportional to the infection degree. Within the range, it was found that the dominant spectral wavelengths of the fresh lemon and the green mold pathogen are 580 nm and 550 nm, respectively, with the 550 nm being the most sensitive in detecting the pathogen with spectral imaging. The spectral intensity ratio of the infected lemon to the fresh one in the spectral range of 500 nm to 560 nm increases with the increasing degree of the infection. Therefore, the ratio can be used to effectively estimate the degree of lemons infecting by the green mold pathogens. It also shows that the sudden decrease of the spectral intensity corresponding to the dominant spectral wavelength of the fresh lemon, together with the neighboring spectral wavelengths can be used to classify fresh and contaminated lemons. The spectral intensity ratio of discriminating the fresh lemon from the infected one is calculated as 1.15.

Keywords: healthy and infected lemons; Hyperspectral image; *Penicillium digitatum* pathogen; lemon skin; dominant spectral wavelength; spectral intensity ratio

---

1. Introduction

As the volume of global production of fruits increases, the use of a spectroscopic method for the purposes of monitoring surface condition and controlling the quality of the fruits in food and agricultural industries is continuously increasing, especially in determining the ripening status [1]. This is because the method is rapid, simple, and accurate for the above-mentioned purposes, as the visual colors of the fruit skin are directly related to the wavelengths of the light reflected from objects' surfaces, especially in the visible and near infrared spectral ranges. In addition, the spectroscopic method is noninvasive. As the most fruits are soft, any form of physical forces affecting their skin, such as collisions among them and shocks from outside causes bruises and scars of the skin. The bruises and scars can induce changes in the skin color of the fruits. Moreover, the juicy components emitted to the skin in the process of forming the bruises and scars can cause various diseases in fruits. These diseases are causing further changes in the skin color of fruits. Therefore, the noninvasive method is preferable for the inspection of fruits.

Among the diseases, fungal infections induce such serious problems as asthma, cancer, and other diseases to human health [2–6]. As the fungi spores widely spread in the environment are gradually accumulated in the human body, the long-term exposure to them can induce damages to internal organs. Therefore, they are a significant threat to human health. In fact, it is known that more than a billion people suffer from the diseases, and more than 1.5 million people die each year by the diseases [7]. Among the fungi, the pathogen *Penicillium digitatum* (green mold) is the fungus that is dominantly found in the skins of citrus fruits, especially in lemons, and is very abundant in our environment [8,9]. The diseases in fruits outbreak when the biological pathogens infiltrate the fruits' surfaces. As a consequence, the fruits' skins change their colors, and other healthy fruits near the infected ones are contaminated. Thus, the infected fruits should be separated from the healthy ones. For this purpose, a quick and accurate method of identifying the infected fruits, i.e., inspecting the large amount of fruits in real-time and quantifying the degree of the infections should be devised. In this regard, the spectral imaging, which photographs an object or scene with the use of a spectral filter, has been recognized as a very promising inspection method of the qualities of agricultural products and foods [10–16]. This is because the imaging can reveal physical, chemical, geometrical, and optical parameters of the objects being inspected, whereas the spectroscopic method can provide the optical parameter only, because it can measure only the spectral intensity. However, the imaging has not been developed so far for the quantifying the degree of the infections. The separation between the fresh and infected fruits should be based on the quantification to prevent from discarding good ones with only insignificant amount of the infection. This is why the harvested citrus fruits are going through some forms of chemical treatments to protect the fruit skins from the pathogen infection [17–20]. In addition to the chemical treatments, there are several other ways of treating the fruit surfaces: Covering the surface with plant extracts and essential oils, and exposing to hot water and ultraviolet light [21]. However, these treatments are costly and can be harmful to people, especially the chemicals used for the chemical treatment. Besides, the fruits can also be damaged on the way of transporting them. This requires another separation method. This means that the surface treatment methods are convenient but costly and risky for health.

The spectral imaging is classified into three regimes such as multispectral, hyperspectral, and ultraspectral ones, depending on the transmitting spectral bandwidth of the spectral filter. When the filter can pass a spectral bandwidth of more than 10 nm or several different bandwidths, as in using color glasses, it is called multispectral imaging. When the passing bandwidths are near 1 nm to 10 nm, and not more than 1nm, they are called hyper- and ultraspectral imaging, respectively. As the spectral bandwidth of the filter becomes smaller, the physical and chemical elements contained in the object to be tested can be identified with better accuracy because each individual element has its corresponding spectral wavelength. Among the imaging, the hyper- and multispectral imaging have been used very widely for estimating the quality of fruits, vegetables, meat, grains such as surface damage (bruises, chilling, and insect biting), surface quality (firmness, moisture content, hardness, rottenness, and fungi presence), biochemical components, and quality grades in the food and agricultural industries [22–25]. However, the multispectral imaging can be hardly used for the accurate detection of the green mold pathogen infection in the fruits, as it has a spectral range too wide to trace the presence of the pathogen, though the hyperspectral imaging was used to identify the pathogen presence in tangerine.

In this paper, a spectral intensity ratio and the behavior of spectral intensity changes in a hyperspectral imaging are used to estimate the degree of the green mold pathogen infection and to discriminate between fresh and infected lemons. The presence of green mold pathogens on the surfaces of lemons is identified with the use of the dominant wavelengths of the fresh lemon and the green mold pathogen colors, which are found in this paper.

2. Materials and Methods

The pathogen infection in the lemon is initiated by a hardly visible small damaged area in the skin of a lemon. The skin color of a healthy lemon is bright yellow when it is fully ripe [26]. However, the bright yellow skin color becomes greener as the green mold infection gets deeper. To show these color changes, the pathogen infected lemon samples are prepared and the hyperspectral images of each infected sample are taken and their intensity distributions are found in the following ways.

2.1. Pathogen Infected Lemon Samples Preparation and Changes in Skin Color

A batch of the green mold pathogen infected lemon samples is prepared in the following way. Lemon samples, which are completely covered with the green mold pathogen, are prepared. Then, each of the completely pathogen infected lemon samples is put in the center of a chamber with 4 to 5 fresh lemons. This chamber is equipped with a thermometer and a humidifier, and the humidity and temperature is controlled to be kept as more than 80% and 25 °C, respectively. The fresh lemon samples are having less than 2.2% intensity differences between them. The chamber is kept under such conditions for many days and for every 24 h, then one of samples, which is not burst or unusually rotten, is selected and photographed with a camera with a spectral filter in front of its objective. 16 Hyperspectral filters in the spectral range of 500 nm to 650nm are used. Each spectral filter has the spectral bandwidth of 10 nm centered at 500 nm to 650 nm for every 10 nm interval. The camera images of the lemon samples taken without spectral filters are shown in Figure 1. There are 20 sample images of differently infected lemons. It is considered that the lemons are infected more as the sample number increases. Figure 1 clearly demonstrates that the bright yellow color of the fresh lemon skin turns more to green as the green mold infection becomes severe. The presence of the green mold pathogen is hardly recognized in images 1 to 7, whereas its presence on the surfaces of the lemons is too obvious in samples numbered 8 to 20. The samples numbered 16 to 20 also show their morphology changes as the infection goes deeper.

Figure 1. Lemon skin color changes with the increasing degree of the green mold pathogen infection. The lemons are more infected as the number increases: (**1**) healthy lemon, (2-19) infected lemons of different degrees and (**20**) lemon covered completely with the green mold pathogen.

The color changes can be quantitatively represented in histograms in 3-dimensional R (Red), G (Green), and B (Blue) color space. This is shown in Figure 2. The numbers represent 256 gray level. For the case of the undamaged lemon, i.e., fresh lemon as shown in the number 1 image of Figure 1; (1) R is the dominant color for the pixels with the gray levels up to 200, though the gray level difference between R and G is ~30, and (2) B component becomes distinct for the pixels with R and B gray levels greater than 200 but B gray level does not exceed 200. These facts indicate that the skin color of the lemon is mainly yellow but some gray colors are also there, although they are not very recognizable. The histogram as shown in Figure 2b corresponds to the sample numbered 15 of Figure 1. It shows that (1) G is the dominant color component but the gray level difference between G and B is not very large; (2) B component appears strongly for the pixels with R and G's gray level greater than 150, though the B's gray level does not exceed 200; and (3) only few pixels are colored by R and G with very low B component. These facts indicate that the color of the lemon skin shown in the sample numbered 15 of Figure 1 is shifted to the green side and more gray is added to it compared with the fresh lemon. This means that the lemon is now covered by green molds. Figure 2c is the RGB histogram of the image numbered 20 of Figure 1. It shows that (1) the gray levels of R, G, and B color components are almost the same to each other, though B population is much larger than (2) Figure 2b and the highest gray level of B is the same as those of R and G, and (3) a very few pixels have R and G's gray levels less than 100. These facts indicate that the skin color is changed mostly to the brighter gray. This means that the histogram can implicitly inform the presence of green and blue molds as demonstrated by the sample numbered 20 but cannot explicitly. As shown in Figure 2, the histograms of RGB color components for different stages of the infections are distinctively different from each other. The histogram in 3-dimensional color space informs implicitly the evolving stage of green and blue molds in lemon. However, it is difficult to define the pixel population difference of between different colors at each stage and between stages, which can serve as a criterion for segregating the not damaged lemons from the damaged ones. However, the spectral image, which will be shown in the next section, can provide the criterion.

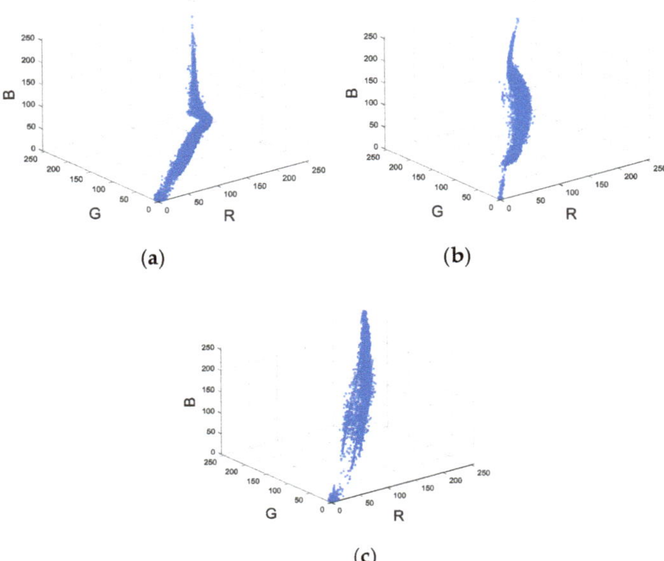

Figure 2. RGB histogram in 3-dimensional color space: (**a**) image 1 (**b**) image 15, and (**c**) image 20 of Figure 1.

2.2. Experimental Set-Up for Taking Spectral Images and Their Intensity Distributions

As earlier mentioned, each color of an object corresponds to a specific wavelength in near-UV (ultraviolet) to near-infrared wavelength regime. This means that if there is a spectral filter which passes only a narrow wavelength range within the above wavelength regime, it will be possible to take an image of the object, which has the specific color corresponding to the wavelength range. When a hyperspectral filter is represented by its center wavelength as in this paper, it indicates that the filter can transmit the light in the wavelength range of $\lambda_k \lambda_k - 5$ nm to $\lambda_k + 5$ nm. In the spectral imaging, it is important to know that the spectral intensity distribution of the illumination source and the spectral transparency of the filter because the intensity distribution of each spectral image is linearly dependent on them.

Figure 3 shows the experimental set-up to take the hyperspectral images of lemons. It consists of a camera (2), a halogen lamp (4), two polarizers (5,6), a power meter (7) with a detector (8), and 16 hyperspectral filters in the spectral range of 500 nm to 650 nm for 10 nm interval (9). Nikon D810 with the detector resolution of 7360 × 4912 [27] is used as the camera. Each spectral filter has a diameter of 25 mm and a spectral bandwidth of 10 nm, and is produced by Asahi spectra [28]. The transmittance of the filters in the spectral range of 500 nm to 650 nm is in the range of 79% to 80.5%. The images with the spectral filters with less than 500 nm and greater than 650 nm have much lower brightness than those in 500 nm to 650 nm. This is why they are not analyzed in this paper. Each spectral filter (9) is placed in front of the camera objective with its center matched on the camera optical axis. The objective is opened only through the filter. For each object, the spectral filters are changed 16 times. We used the data of the halogen lamp spectral power distribution [29] for calibrations of the optical system.

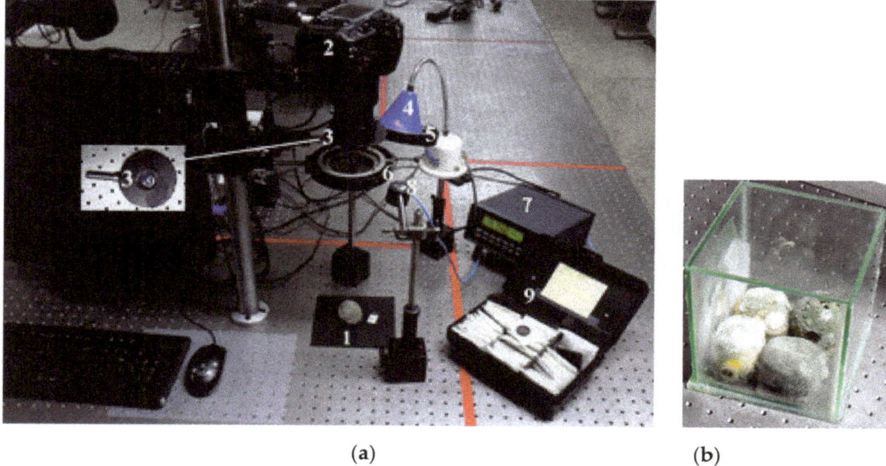

(a)	(b)

Figure 3. (a) Experimental set-up: (**1**) Object (Lemon), (**2**) Nikon Camera D810, (**3**) Spectral Filter, (**4**) a halogen lamp, (**5**) two polarizers, (**6**) analyzer, (**7**) a power meter, (**8**) a detector and (**9**) hyperspectral filters in the spectral range of 500 nm to 650 nm for 10 nm interval; (**b**) fully rotted lemons as source of green mold infection in special chamber.

Light intensity can be calibrated with the spectral power of the lamp for the lamp's spectral wavelengths. The accuracy of calibration can be in the range of 2% to 7% depending on the wavelength. The angle of the lamp beam to the normal direction of the optical table is 30° (Figure 3). The illumination direction of the lamp is adjusted in such a way that the beam center is directed to the mid position of the object (1). Each of the two polarizers is located in front of the spectral filter. The filter is for taking only the surface reflected lights from the object [30]. Therefore the polarizer in front of the camera works as an analyzer. The object is at ~350 mm distance from both polarizers. It is located so that its image appears at the mid area of the camera's viewfinder. The power meter is New Port 1830C [31], which has full-scale accuracy of 99.6%. In this experiment, the illumination and photographing conditions are kept the same for each object. As described in Section 2.1, fully infected lemons are a source of green mold infection.

As the lemon shape is changing as the infected area grows, the objects' positions are carefully adjusted to be located at the mid-position of the viewfinder of the camera to keep the illumination condition the same for all cases. In the experiment, 16 objects with different degrees of infection are investigated. The objects can be 17 when the fresh lemon is included. For each object, 16 different spectral images are taken to observe the spectral property changes of each object's skin due to the pathogen infection. The spectral property is represented by the intensity average of each spectral image of the same object.

Figures 4–7 present the 16 spectral images of the objects shown in samples numbered 1, 12, 14, and 20 of Figure 1, respectively. In Figure 4a, the number in each spectral image represents the central wavelength λ_k of the hyperspectral filter used. In images in Figures 4b and 5, Figures 6 and 7, each lemon follows the spectral image orders as in Figure 4a. This means that the 1st row images are in the blue-green color region (500–530 nm), the 2nd row—in the green color region (540–570 nm), the 3rd row—in the yellow to orange color region (580–610 nm), and the 4th row—in the orange to red color region (620–650 nm) from left to right. Figure b of Figures 4–7 represents the grayscale version of the spectral images in the same order as the spectral images. When Figure a of Figures 4–7 is converted into a 16-bit image in grayscale using the program OriginLab 2017, they are shown in Figures 8–11. The intensity value of each pixel of the image corresponds to the total light intensity passed through a spectral filter after reflecting from the lemon surface that is imaged to the pixel. The color scale shows the intensity level of each pixel. The gray scale images are intended to normalize the brightness of the spectral image because eye response to color brightness is different for different colors [32] These images inform that the fresh lemon, i.e., the leftmost image at 3rd row has the highest brightness at 580 nm and other yellow-orange images of the spectral range 590 nm to 610 nm look brighter than other spectral range images. Figure 5 distinguishes the yellow-orange colors. However, for the sample 12 as shown in Figure 6, the highest brightness still appears at 580 nm. However, the images at the 2nd and the 3rd in the 2nd row, which correspond to 550 nm and 560 nm, respectively, become brighter than other images in the 3rd row. Now the darkest images are in the 4th row. For the sample 14 as shown in Figure 7, the 2nd row images are brighter than other rows images and the 1st row images become brighter than those in Figure 6. The darkest images are still in the 4th row. For the case of Figure 7, the brightest images are the 2nd and 3rd ones in the 2nd row. The 1st row images look brighter than the images in the 3rd and 4th rows. These spectral images indicate that the skin color of the lemon turns more green as the pathogen infected areas increase, and the green becomes the dominant color of the lemon, when the skin is completely covered by the pathogens. The blue-green colors are also more enhanced than yellow to red colors. Therefore, the spectral image clearly informs the presence of green and blue molds. Figures 5 and 6 clearly inform that the infected lemon can be fully characterized by the green colors, even the infected area is too small to be neglected. To see more clearly the brightness distribution of the skin surface and brightness changes in different spectral ranges due to the infection, the intensity distributions of the grayscale images in Figures 5–8 are plotted in a 3-dimensional space for the quantitative comparison between them.

Figures 8–11 represent the grayscale plotting of gray scale images in Figures 4–7, respectively, by considering lamp spectral intensity distribution, camera's spectral response, and the transparency of the spectral filters. The distribution represents only the lemon surface. As the lemons are having an oval shape, the intensity distribution over the lemon surface will not be uniform because of different heights.

The x- and y-axes in Figures 8–11 correspond to images in columns and rows of Figures 4–7, respectively. The numbers 1–4, which specify the scales of y-axis, represent the 1st, 2nd, 3rd, and 4th rows, respectively, whereas those of x-axis—the 1st, 2nd, 3rd, and 4th columns of Figures 4–7 respectively. Figures 8–11 allow quantitative comparison of different spectral images in differently affected lemons. The intensity distributions reflect mostly the surface shapes of the lemon. Figure 8 shows that (1) the intensity levels at 580 nm and 590 nm are higher than those at other spectral wavelengths, but the intensity distributions are more uniform over the image at 580 nm than at 590 nm, (2) the areas having uniform intensity distribution are shrunk to the central region of the image for 590 nm and the uniformity is almost disappeared at 600 nm and 610 nm, and (3) the spikes appearing at other spectral wavelengths are probably caused by the lemon's uneven surface. The light reflected from each surface area will be different from others. Figure 9 informs that (1) the spectral intensities of the spectral range above 570 nm are still dominant for other spectral ranges, but the areas are reduced compared with Figure 8, and (2) the intensity distributions of below 570 nm are gained more intensities over the large surface areas. This means that the green molds' encroaching the surface area is still going on.

Figure 10 reveals that (1) the intensity distribution of each spectral image in the spectral range of 540 to 570 nm is almost uniform and their intensities are close to the maximum gray level, (2) for the spectral ranges of 580 to 610 nm, a large part of each image lost its spectral intensity if compared with Figure 9, but the central part of the image still keeps the maximum intensity level for 580 nm; (3) for 500 nm, the intensity distribution is almost uniform for the entire image area, except the spikes; and (4) the image lost much of its intensity for the spectral range greater than 610 nm. This means that the green molds cover the entire surface of the lemon and blue molds are also growing in large parts of the lemon skin. Further, the color of the fresh lemon is much reduced. Figure 11 shows that (1) the intensity distributions of the spectral images in the spectral range smaller than 580 nm are very spiky for all their image areas, and their intensities are significantly higher than those of the images in spectral range greater than 570 nm, and (2) the intensity distributions at 520 nm and 550 nm are higher than those of other spectral wavelengths. This means that there will be a significant amount of the blue mold along with the green mold pathogens and the lemon's skin is completely covered by both molds. The intensity spikes suggests that the molds have a hair shape.

Figures 8–11 inform that even in sample 14, a part of the lemon still retains its original color. This means that part of the lemon surface still remains untouched by the green and blue mold pathogens, at least for sample 14. Figures 8–11 enable comparison of absolute spectral intensities between different spectral images of differently infected lemons. However, the comparison is still difficult because they show intensity distributions of the images. To make the comparison easy, the intensity distribution of each spectral image at each infected state should be averaged and then normalized with the averaged spectral intensity distribution of the corresponding spectral image of the fresh/undamaged lemon. This will be shown in the next section.

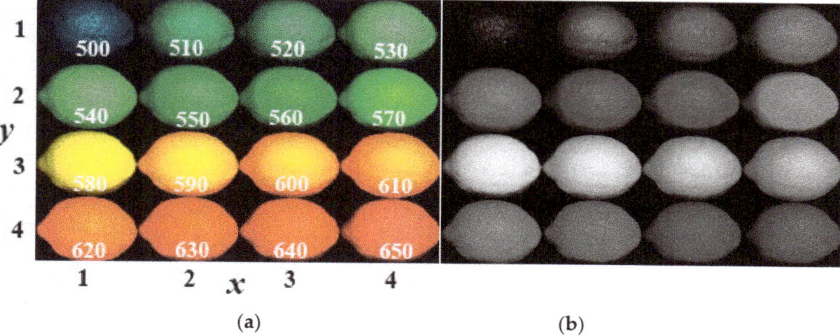

(a)   (b)

Figure 4. Sixteen spectral images of the lemon sample numbered 1 in Figure 1: (**a**) Spectral images from 500 nm at top left to 650 nm at bottom right in the order of left to right and top to bottom; (**b**) grayscale images of the spectral images in panel (a).

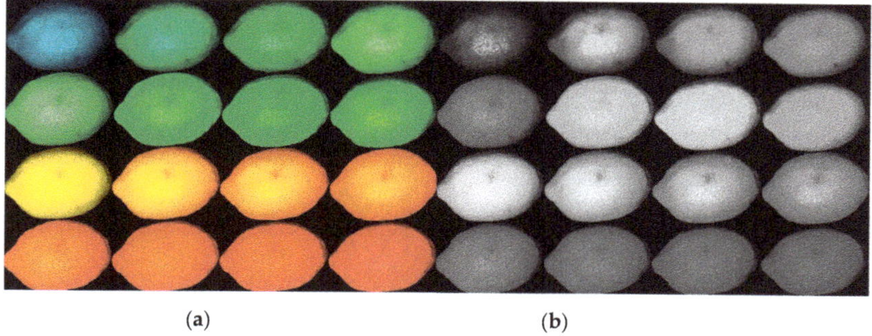

(a)   (b)

Figure 5. Sixteen spectral images of the sample lemon numbered 12 in Figure 1: (**a**) Spectral images from 500 nm at top left to 650 nm at bottom right in the order of left to right and top to bottom; (**b**) grayscale images of the spectral images in panel (a).

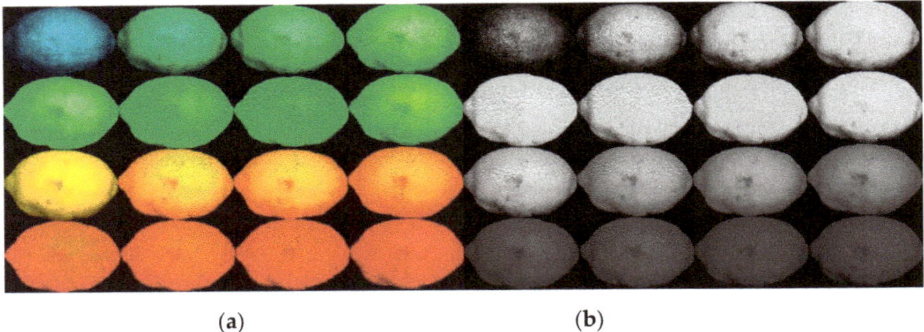

(a)   (b)

Figure 6. Sixteen spectral images of the sample lemon numbered 14 in Figure 1: (**a**) Spectral images from 500 nm at top left to 650 nm at bottom right in the order of left to right and top to bottom, (**b**) grayscale images of the spectral images in panel (**a**).

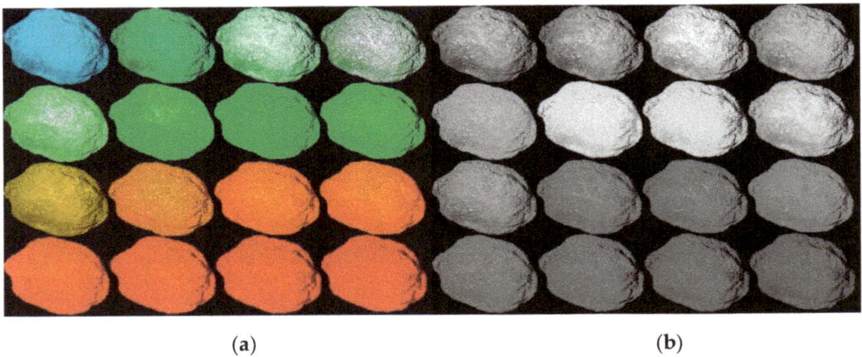

(a)   (b)

Figure 7. Sixteen spectral images of the sample lemon numbered 20 in Figure 1: (**a**) Spectral images from 500 nm at top left to 650 nm at bottom right in the order of left to right and top to bottom; (**b**) grayscale images of the spectral images in (**a**).

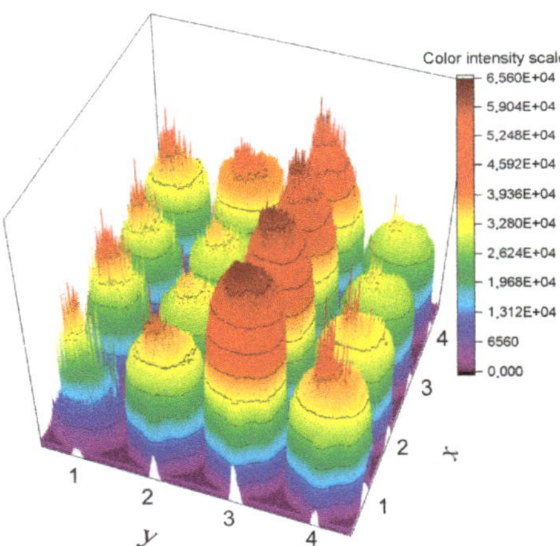

Figure 8. Intensity distribution of Figure 4b.

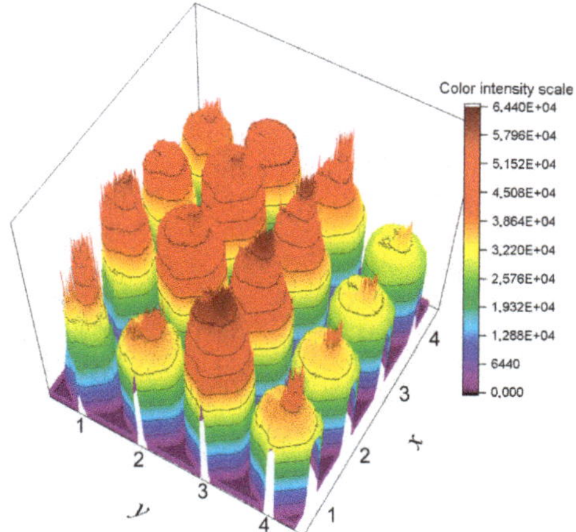

Figure 9. Intensity distribution of Figure 5b.

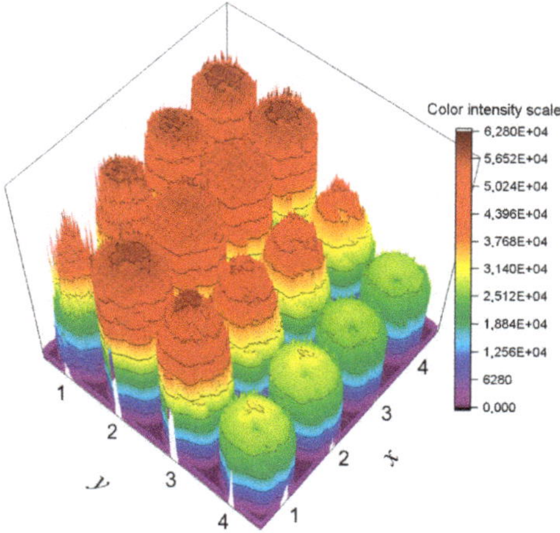

Figure 10. Intensity distribution of Figure 6b.

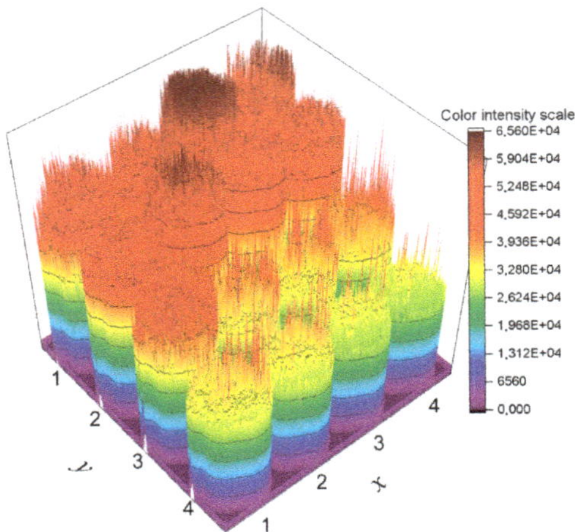

Figure 11. Intensity distribution of Figure 7b.

3. Results and Discussion

For the normalization, the intensity distribution of each spectral image from all infected states is averaged by adding up the intensities of all pixels making up the lemon image and then dividing it by the added number of pixels. These averaged intensities are divided by that of corresponding wavelength spectral image of the fresh lemon to normalize the intensities. This normalization eliminates the calibration problem incurred by the spectral intensity differences between different spectral ranges of the illumination lamp. The averaged spectral intensity $I^j_{\lambda_k}$ of jth infected image at the spectral wavelength λ_k, where $j = 1, 2, 3, 4, \cdots\cdots, 17$, including the fresh lemon, $k = 1, 2, 3, 4, \cdots\cdots, 16$, can be expressed as

$$I^j_{\lambda_k} = \frac{\sum\limits_{i=1}^{n^j_{\lambda_k}} P_i}{n^j_{\lambda_k}} \tag{1}$$

where P_i and $n^j_{\lambda_k}$ represent ith pixel intensity and the number of pixels making up each gray scale lemon image of jth infected image at the spectral wavelength λ_k, respectively. The normalization can be expressed as

$$\alpha^j_{\lambda_k} = \frac{I^j_{\lambda_k}}{I^0_{\lambda_k}} \tag{2}$$

where, $\alpha^j_{\lambda_k}$ and $I^j_{\lambda_k}$ are the normalized coefficient and averaged intensity of jth infected image at the spectral wavelength λ_k, respectively, and $I^0_{\lambda_k}$ the averaged intensity of not damaged lemon image at λ_k.

When the molds taken from the completely infected lemon, as in image 20 of Figure 1, are grafted on the skin surface of healthy lemons, the grafted area is hardly identified for the spectral filter of 540 nm, 550 nm, and 560 nm, as shown in Figure 12. Figure 12 presents photographic images of the grafted area taken with 16 spectral filters used for Figures 5–8. The numbers in this figure represent the central wavelengths of the spectral filters. The grafted area appears to be of a gray color with no filter as shown in the leftmost image of Figure 12. This means that the molds are probably having RGB color components with almost the same intensity. When the spectral filters are used, the molds are hardly identified in the spectral range of 540 nm to 560 nm. This means that the spectral wavelength of the green molds is in the range of 540 nm to 560 nm. For the case of red color, the molds are still identified up to 650 nm. Therefore, a spectral filter in the range of 535 nm to 565 nm will be enough for the accurate detection of the green mold pathogen because the spectral range is the most sensitive to the pathogens.

Figure 12. Photos of a lemon surface grafted with a green mold for the spectral range of 500 to 650 nm. The 1st image is without filter.

The averaged spectral intensities of the 20 spectral images, as shown in Figure 1, are plotted in Figure 13 at each of the spectral wavelengths. The number of each curve corresponds to that of each sample in Figure 1. The intensity has the same relative scale as in Figures 8–11. Figure 13 shows that the spectral intensities at 570 nm have the minimum difference between samples.

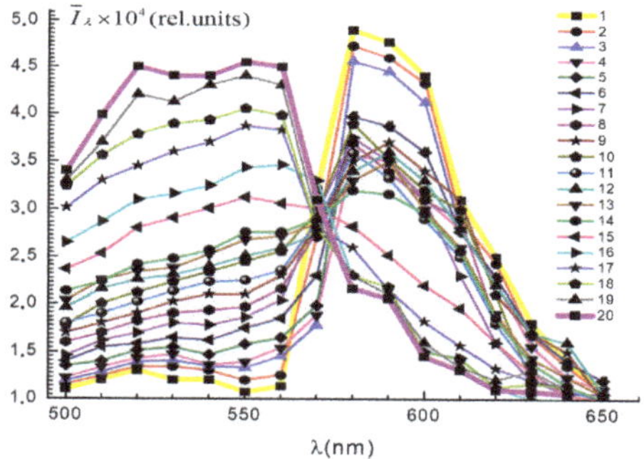

Figure 13. Spectral intensity distributions of 20 lemon samples shown in Figure 1, calculated with Equation (1).

This is expected because the dominant wavelengths of the green molds and the fresh lemon are in the range of 535 nm to 565 nm and 580 nm, respectively. As sample 1 has the highest spectral intensity at 580 nm and very low spectral intensities at below 570 nm, the spectral intensity at 570 nm has an intermediate value between these two values. Sample 2 behaves the same way as sample 1. However, starting from sample 3, the intensity at 570 nm becomes lower than that of samples 1 and 2 and then increases as the sample number increase. For a few samples, the values can slightly exceed those of samples 1and 2. Along with this feature, Figure 13 also shows other features: (1) the fresh lemon has the lowest spectral intensity at 550 nm, and (2) the intensity increases with sample numbers increase in the spectral ranges 510 nm to 560 nm though the increment between adjacent numbers is not the same. Also note that the spectral intensities at 550 nm and 560 nm are higher than those of the spectral range of 500 nm to 540 nm for most of the samples but for the samples 1 and 2, they are lower than those of 510 nm to 540 nm. (3) The intensity decrease in the spectral ranges greater than 570 nm is not orderly because the intensities of other number samples are mixed together. (4) The spectral intensity is the smallest (largest) for the fresh (the completely covered with molds) in the spectral range of 500 nm to 560 nm and the largest (smallest) in the 570 nm to 650 nm, except at 570 nm. (5) The spectral intensity difference between samples numbered 15 and 20 is higher than that between 1 to 14. (6) The spectral intensities of samples 4 to 14 entangled themselves, i.e., the intensities no longer decrease with increasing sample numbers. (7) The spectral intensity of the fresh lemon has the highest intensity at 580 nm and the intensities decrease with increasing sample numbers for samples 1 to 3, but they no longer decrease in corresponding to the increasing sample numbers, and the intensity at 590 nm is not too different from that at 580nm and shows the same behavior as the 580 nm. (8) Samples 2 and 3 keep their spectral features as the same way as that of the sample 1, i.e., the fresh lemon, though the intensity ratio of sample 1:2:3 = 1.0:0.972:0.935, 1.0:0.971:0.943 and 1.0:0.985:0.938 for 580 nm, 590 nm and 600 nm, respectively. (9) Samples 4–20 have spectral features completely different from those of sample 1 as manifested by the spectral intensity difference 0.5 between samples 1 to 3 and 4 to 20 at the spectral range 580 nm to 590 nm. (10) The intensity difference at 600 nm is slightly smaller than that at 580 nm and 590 nm but the samples 1 to 3 are still distinctively separated from other samples. These features indicate that (1) the fresh lemon has the lowest spectral intensities at 550 nm, but the intensities become higher as the green mold infection increases confirming again that the dominant spectral wavelength of the green mold is close to 550 nm and the lemons becomes more green due to the pathogen accumulations. The range is within 540 nm to 560 nm as demonstrated by Figure 12; (2) the spectral image can identify the infected states of the lemons, which are unrecognized with the naked eye; (3) the higher spectral intensities of the spectral range 550 nm insures that the spectral range is the most sensitive spectral range in quantifying the green mold pathogen infection in lemon and possibly for other citrus fruits too; and (4) the dominant spectral range of the fresh lemon is 580 nm to 590 nm and the samples 1–3 retain their spectral features but with other samples, their spectral intensity distributions in the spectral range exceeding 570 nm become completely different from those of samples 1 to 3, though the spectral intensities in the spectral range not exceeding 570 nm are slightly increasing as the sample number increases. The spectral intensity difference between samples 1 and 2 is less than 3% for the spectral range 580 nm to 600 nm, but this difference is slightly higher than the spectral intensity difference 2.2% between fresh lemons. This is why the spectral intensity of sample 2 at 570 nm is very similar to that of sample 1. However, for samples 1 and 3, the differences are close to 7% for the spectral range 580 nm to 600 nm. This difference is bigger than the spectral intensity difference 2.2% between fresh lemons. This informs that the pathogens are already on the skin surface of sample 3. This can be the reason why the spectral intensity of sample 3 at 570 nm becomes much lower than that of samples 1 and 2. This means that sample 2 can also be considered a fresh lemon.

Figure 13 shows that the normalized spectral intensity distribution defined by Equation (2) for the spectral range of 500 nm to 560 nm is derived as shown in Figure 14. In Figure 14, 19 curves are presented because the curve for the fresh lemon is used as the reference one. The numbers 1 to 19 correspond to the samples numbered 2 to 20, respectively. Figure 14 shows that (1) the difference between samples is more distinctive than that in Figure 13; (2) α values are increasing with increasing sample numbers for the specified spectral range and the normalized spectral intensity has the highest value at 550 nm for the samples numbered 9–20; (3) the α values of sample 2 and 3 at 550 nm are 1.15 and 1.3, respectively; and (4) among samples 2–14, sample 2 to 5 and 6 to 14 shows the highest α value differences for the spectral range 540 nm to 560nm. The difference is more than 0.15. The coefficient α characterizes the degree of spectral characteristic changes in the lemon skin, i.e., the degree of pathogen infections on the lemon samples. The increasing α values with increasing lemon sample numbers indicate that the pathogens are growing more as the sample number increases. In Figure 14, the α value of the sample 20 at 550 nm is close to 4.2. This value represents the completely pathogen infected lemon. For the samples 17 to 20, the α values are in the range of 3 to 4.2. Therefore, the degree of pathogen infections on lemons can be quantified by calculating α values only at 550 nm. According to Figure 13, the α value that can be considered as a fresh lemon will be up to 1.15 to <1.3.

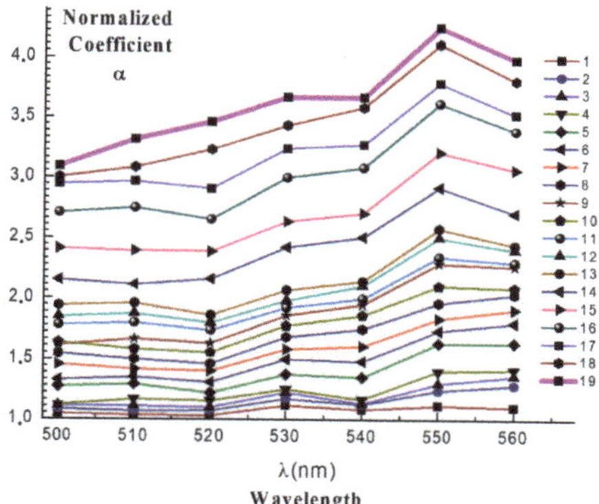

Figure 14. Normalized spectral intensity distribution of Figure 13 in the spectral range of 500 to 560 nm.

The skins of lemons viewed with a microscope with the magnification of 200 are shown in Figure 15. Figure 15 compares the surface states of samples 1 and 6. The fresh lemon surface specified by number 1 have no stains, i.e., it is clean but the sample 6 specified by number 2 is mottled by the pathogens. The thinner color areas represent the pathogen infected areas. When the infection goes deeper, all surfaces are densely covered by pathogens, as shown by sample 20 in Figure 1. The α value of the sample 6 is 1.51. However, it is noticed that the surface is greatly damaged, though it is hard to find the pathogen presence in sample 6 in Figure 1 with naked eye. Thus it can be said that it is not safe to use the lemon with α value of 1.51. It is considered that the α is useful to quantify the degree of pathogen infection on lemon skin.

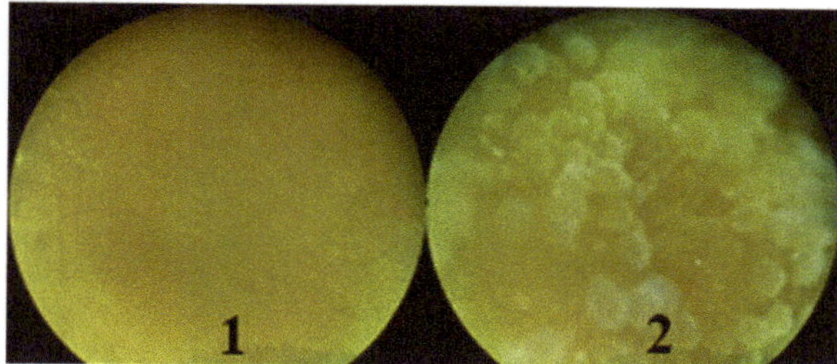

Figure 15. Microscope image of the surfaces with 200 times magnification: **1**: Fresh lemon (Sample 1), and **2**: Sample 6 with the $\alpha = 1.51$.

4. Conclusions

Spectral imaging is not only a noninvasive, but also a very efficient method, of identifying and quantifying the lemons infected by green mold pathogens. The spectral range of 540 nm to 560 nm and 570 nm to 600 nm are good for identifying the states of lemon infected by the pathogens. The spectral range of 550 nm and 580 nm, which is found to be the dominant wavelengths of the green mold pathogen and the lemon's skin color, is especially sensitive for quantifying the infection degrees, and discriminating between the fresh and infected lemons, respectively: the quantification can be done by calculating the normalized coefficient α at 550 nm and discriminating by spectral intensity variations at 570 nm and 580 nm. The α value at which the lemon can be considered fresh will be less than 1.3.

The proposed method for determining the parameter α and calculating the spectral intensity variations can be useful in practice for detecting pathogens in other citrus fruits and foods too. Defining the α value range, which can help to discriminate the fresh and the infected lemons more accurately, and comparing this range with those for other citrus fruits will be the subject of our further research.

The determined value range will enable the automatization of the process of inspecting the large amounts of citrus fruits in real-time.

Author Contributions: Y.V., Setting experimental setup and data collection; G.H., Total Planning and sample preparation; Y.K., Software design; T.V., Formatting, assisting manuscript preparation and English corrections; J.-Y.S., Full manuscript preparation and corresponding. All authors have read and agreed to the published version of the manuscript.

Funding: This work was supported by Priority Research Centers Program through the National Research Foundation of Korea (NRF) funded by the Ministry of Education (Grant Number: NRF-2018R1A6A1A03025542).

Conflicts of Interest: The authors declare that they have no conflict of interest.

References

1. Lakshmi, S.; Pandey, A.K.; Ravi, N.; Chauhan, O.P.; Gopalan, N.; Sharma, R.K. Non-destructive Quality Monitoring of Fresh Fruits and Vegetables. *Def. Life Sci. J.* **2017**, *2*, 103–110. [CrossRef]
2. Green Mold on Wood, Bread, Wall. This is How to Get Rid of This. Available online: https://healthida.com/green-mold (accessed on 14 August 2019).
3. Bongomin, F.; Gago, S.; Oladele, R.; Denning, D. Global and Multi-National Prevalence of Fungal Diseases-Estimate Precision. *J. Fungi.* **2017**, *3*, 57. [CrossRef]
4. Chrdle, A.; Mallátová, N.; Vašáková, M.; Haber, J.; Denning, D.W. Burden of serious fungal infections in the Czech Republic. *Mycoses* **2015**, *58*, 6–14. [CrossRef] [PubMed]
5. Huh, K.; Ha, Y.E.; Denning, D.W.; Peck, K.R. Serious fungal infections in Korea. *Eur. J. Clin. Microbiol. Infect. Dis.* **2017**, *36*, 957–963. [CrossRef]
6. Ghosh, P.N.; Fisher, M.C.; Bates, K.A. Diagnosing emerging fungal threats: A one health perspective. *Front. Genet.* **2018**, *9*, 376. [CrossRef]
7. Garcia, M.D.; Chua, S.; Low, Y.; Lee, Y.; Francis, K.A.; Wang, J.; Nouwens, A.; Lonhienne, T.; Williams, C.M.; Fraser, J.A.; et al. Commercial AHAS-inhibiting herbicides are promising drug leads for the treatment of human fungal pathogenic infections. *PNAS* **2018**, *115*, E9649–E9658. [CrossRef]
8. Lluís, P. Penicillium digitatum, Penicillium italicum(Green Mold, Blue Mold). In *Postharvest Decay. Control Strategies*; Academic Press: Cambridge, MA, USA, 2014; chapter 2; pp. 45–102. [CrossRef]
9. Yahia, E.M. *Postharvest Biology and Technology of Tropical and Subtropical Fruits*; Woodhead Publishing: Cambridge, UK, 2011; pp. 437–516.
10. Su, W.; Sun, D. Multispectral Imaging for Plant Food Quality Analysis and Visualization. *Compr. Rev. Food Sci. Food Saf.* **2018**, *17*, 220–239. [CrossRef]
11. Adão, T.; Hruška, J.; Pádua, L.; Bessa, J.; Peres, E.; Morais, R.; Sousa, J.J. Hyperspectral Imaging: A Review on UAV-Based Sensors, Data Processing and Applications for Agriculture and Forestry. *Remote Sens.* **2017**, *9*, 1110. [CrossRef]
12. Report: Hyperspectral Imaging Market to Reach $11.34 Billion in Five Years. Zion Market Research. 2017. Available online: https://www.zionmarketresearch.com/marketanalysis/hyper-spectral-imaging-system-market (accessed on 14 August 2019).
13. Gaurav, S.G. Hyperspectral Imaging: Technologies and Global Markets. BCC Research. 2016. Available online: https://www.bccresearch.com/marketresearch/instrumentationand-sensors/hyperspectral-imaging-tech-markets-report-IAS106A.html (accessed on 14 August 2019).
14. Ahmed, M.R.; Yasmin, J.; Mo, C.; Lee, H.; Kim, M.S.; Hong, S.J.; Cho, B.K. Outdoor Applications of Hyperspectral Imaging Technology for Monitoring Agricultural Crops: A Review. *J. Biosyst. Eng.* **2016**, *41*, 396–407. [CrossRef]
15. VetriDeepika, K.; Johnson, A.; LafrulHudha, M.; Ramya, M. Enhancement of Precision Agriculture Using Hyperspectral Imaging. *IJTRD* **2016**, *Special issue ETEIAC-16.* 19–21.
16. Dale, L.M.; Thewis, A.; Boudry, C.; Rotar, I.; Dardenne, P.; Baeten, V.; Pierna, J.A.F. Hyperspectral Imaging Applications in Agriculture and Agro-Food Product Quality and Safety Control: A Review. *J. Appl. Spectrosc. Rev.* **2013**, *48*, 142–159. [CrossRef]
17. Mohammadi, P.; Tozlu, E.; Kotan, R.; Kotan Şenol, M. Potential of some bacteria for biological control of postharvest citrus green mold caused by Penicillium digitatum. *Plant Protect. Sci.* **2017**, *53*, 134–144. [CrossRef]

18. El-Gali, Z.I. Control of Penicillium Digitatum on Orange Fruits with Calcium Chloride Dipping. *J. Microbiol. Res. Rev.* **2014**, *2*, 54–61.

19. Wang, W.; Liu, S.; Deng, L.; Ming, J.; Yao, S.; Zeng, K. Control of citrus post-harvest green molds, blue molds, and sour rot by the cecropin a-Melittin hybrid peptide BP21. *Front. Microbiol.* **2018**, *9*, 2455. [CrossRef]

20. Wang, W.; Deng, L.; Yao, S.; Zeng, K. Control of green and blue mold and sour rot in citrus fruits by the cationic antimicrobial peptide PAF56. *Postharvest Biol. Technol.* **2018**, *136*, 132–138. [CrossRef]

21. Papoutsis, K.; Mathioudakis, M.M.; Hasperué, J.H.; Ziogas, V. Non-chemical treatments for preventing the postharvest fungal rotting of citrus caused by Penicillium digitatum (green mold) and Penicillium italicum (blue mold). *Trends Food Sci. Technol.* **2019**, *86*, 479–491. [CrossRef]

22. Zhang, X.; Liu, F.; He, Y.; Li, X. Application of hyperspectral imaging and chemo metric calibrations for variety discrimination of maize seeds. *Sensors* **2012**, *12*, 17234–17246. [CrossRef]

23. Pu, Y.Y.; Feng, Y.Z.; Sun, D.-W. Recent progress of hyperspectral imaging on quality and safety inspection of fruits and vegetables: A review. *Compr. Rev. Food Sci. Food Saf.* **2015**, *14*, 176–188. [CrossRef]

24. El Masry, G.M.; Nakauchi, S. Image analysis operations applied to hyperspectral images for non-invasive sensing of food quality. A comprehensive review. *J. Biosyst. Eng.* **2016**, *142*, 53–82. [CrossRef]

25. Feng, C.; Makino, Y.; Oshita, S.; Francisco, J.; Martín, G. Hyperspectral imaging and multispectral imaging as the novel techniques for detecting defects in raw and processed meat products: Current state-of-the-art research advances. *Food Control* **2018**, *84*, 165–176. [CrossRef]

26. BBC Good Food. Available online: https://www.bbcgoodfood.com/glossary/lemon (accessed on 14 August 2019).

27. Digital Camera Nikon D810, User's Manual. Nikon Corporation USA. Available online: https://www.nikon-usa.com/en/nikon-products/product/dslr-cameras/d810.html (accessed on 14 August 2019).

28. Asahi Spectra Inc. Available online: https://www.asahi-spectra.com (accessed on 14 August 2019).

29. MacDonald, L.; Giacometti, A.; Campagnolo, A.; Robson, S.; Weyrich, T.; Terras, M.; Gibson, A. Multispectral Imaging of Degraded Parchment. In *International Workshop on Computational Color Imaging*; Springer: Berlin/Heidelberg, Germany, 2013; pp. 143–157.

30. Son, J.Y.; Vashpanov, Y.; Jung, D.H.; Lee, D.S.; Kwack, K.D.; Kim, S.H. Polarized light for measuring a human skin feature indicating aging. *Jpn. J. Appl. Phys.* **2009**, *48*, 09LD06. [CrossRef]

31. Picowatt Digital Optical Power Meter Model 1830-C Newport. Available online: https://www.newport.com (accessed on 14 August 2019).

32. Valberg, A. *Light Vision Color*; John Wiley & Sons Ltd.: Chichester, West Sussex, UK, 2005.

Review

Review on Computer Aided Weld Defect Detection from Radiography Images

Wenhui Hou [1,2], Dashan Zhang [1,2], Ye Wei [3] and Jie Guo [3] and Xiaolong Zhang [1,2,*]

[1] School of Engineering, Anhui Agriculture University, No. 130 West Changjiang Road, Hefei 230026, China; hwh303@mail.ustc.edu.cn (W.H.); zhangds@mail.ustc.edu.cn (D.Z.)
[2] Intelligent Agricultural Machinery Laboratory of Anhui Province, Hefei 230026, China
[3] Department of Precision Machinery and Precision Instrumentation, University of Science and Technology of China, Hefei 230027, China; weiye128@mail.ustc.edu.cn (Y.W.); guojiegj@mail.ustc.edu.cn (J.G.)
* Correspondence: xlzhang@ahau.edu.cn; Tel.: +86-133-8569-0949

Received: 31 December 2019; Accepted: 2 March 2020; Published: 10 March 2020

Abstract: The weld defects inspection from radiography films is critical for assuring the serviceability and safety of weld joints. The various limitations of human interpretation made the development of innovative computer-aided techniques for automatic detection from radiography images an interest point of recent studies. The studies of automatic defect inspection are synthetically concluded from three aspects: pre-processing, defect segmentation and defect classification. The achievement and limitations of traditional defect classification method based on the feature extraction, selection and classifier are summarized. Then the applications of novel models based on learning(especially deep learning) were introduced. Finally, the achievement of automation methods were discussed and the challenges of current technology are presented for future research for both weld quality management and computer science researchers.

Keywords: radiographic image; image processing; feature extraction; classifier; deep learning; defect detection

---

1. Introduction

Welded structures have been widely used in many areas, such as construction, vehicle, aerospace, railway, petrochemical and machinery electrical. The weld defects are inevitable due to the different environmental conditions and welding technology in the welding process. It is critical to check the quality of welded joints to assure the reliability and safety of the structure, especially for those critical applications where weld failure can be catastrophic. As the most commonly used methods to detect the quality of welding, nondestructive testing techniques (NDT) include radiographic, ultrasonic, magnetic particle and liquid penetrant testing methods. In this paper, we mainly pay attention to radiography testing technology commonly used to inspect the inner defects of welds. The X-ray and Gamma-ray sources are usually used to produce the radiographic weld images by penetrating the weld structure and exposing photographic films.

Weld flaws are described by the variation of intensity in the radiographic films. These films should be checked by certified inspectors to evaluate and interpret the quality of welds called human interpretation. However, the radiogram quality, the welding over-thickness, the bad contrast, the noise and the weak sizes of defects make difficult the task [1]. There are some drawbacks for human interpretation. Firstly, the inspectors are generally trained and have relevant expertise and experience. However, it is still difficult for the skilled inspector to recognize the small flaws within a short time. Secondly, the human interpretation is usually short of objectivity, consistency and intelligence. Finally, The labor intensity of human interpretation is large because lots of films are produced each day due to the improvement of production efficiency in modern industry. In addition, human visual inspection is, at best, 80% effective, and this effectiveness can only be achieved if a rigidly structured set of inspection checks is implemented [2]. Thus many researchers began to build the intelligent systems based on computer which help human on evaluating the quality of welds before the 1990s. Such computer-aided systems typically take the digital images as the object to extract the welds and detect the flaws in the images by various algorithms. Thus, for the conventional films, the digitization should be necessary. Fortunately, digital radiography systems (digitizers) are currently available for digitizing radiographic films without losing the useful information of the original radiograph. Unlike the conventional films which can only be evaluated manually, the digitized radiographic images not only enable the storage, management and analysis of radiographic inspection data easier, but also make the more intelligent inspection of welds possible.

The Advanced Quality Technology Group of Lockheed Martin Manned Space Systems had been supporting three projects which contribute to the building of computer-assisted X-ray film interpretation system, development of weld flaw detector based on image processing and using of Geometric Arithmetic Parallel Processor (GAPP) chips [3–5]. Automatic detection method for weld flaws have rapidly advanced in recent decades. This benefits from the development of technologies such as image processing, computer vision, pattern recognition and deep learning which improved the analysis capability for images.

In the initial study, many researchers took the intensity plot of the line image as the object, and processed the 2D image line-by-line. These methods are based on observation that the weld defect would destroy the bell shape possessed by the line image of good weld. Thus the detection is to find the abnormity of intensity plot. The features used for detection and classification are often defined in the intensity plot. The defects can be discriminated accurately. However, these methods are often time-consuming due to their processing style for images. In addition, it is difficult to recognize diverse types of weld defects. Then most detection systems based on 2D images relied on the image processing, feature extraction and classification. Various image processing technologies were successfully applied to improve the quality of images and remove background to highlight defect region. The geometric features, texture features or combination of both features were applied to characterize the shape, size and texture of defects for further classification. The MFCCs together with polynomial features were also used for defect identification due to their robust to noise and time shifts in signals. The feature selection is usually used between feature extraction and classifier for reducing the number of features to save the computational costs. Furthermore, developments in computer hardware technology and representation learning has provided the perfect conditions for automation of weld defects inspection. Especially, with recent advances of deep learning theory, in optical image recognition domain, considerable effort has been made to design multistage architectures which learn the hierarchical features from images automatically.

This paper aims to review the common practices for weld defects detection and classification based on the digitized radiographic images. The radiation involved in these studies is X-ray (or sometimes Gamma-ray). The two radiographic sources are used in different occasions which are not distinguished in this paper. The paper focuses on the summary of analysis methods for digitized radiographic images. It gives a detailed and comprehensive summary of literatures from image pre-processing, defect segmentation and defect classification. It elaborates from four aspects: (1) the quality improvement of weld images; (2) traditional techniques for defect detection and classification; (3) the application of novel models based on learning; and (4) the achievements and challenges of current methods.

2. Data Collection

In order to review the relevant literatures in weld detection, we searched accessible databases. The publications collected were papers published between 1982 and 2019 from peer-reviewed journals and conference proceedings. This review was carried out only on automatic technologies for weld defects based on digital radiographic images.

The peer-reviewed journals from a variety of fields including Expert Systems with Applications, NDT&E International, Fuzzy Sets and Systems, Information Sciences, Journal of Manufacturing Systems and so on. Weld defect detection automation has always been a topic of interest for NDT&E International. After conducting a comprehensive literature analysis through the literatures, it can be summarized that the automatic detection systems for weld defect mainly involve several technologies: image pre-processing, defect segmentation, feature extraction and selection, classification. Figure 1 shows a classical procedure welding defect detection system involving these aspects [6]. In our paper, the literatures are summarized through a detailed analysis of each section. The block diagram indicating the structure of this paper is shown in Figure 2. All the abbreviation in Figure 2 have full names in the text.

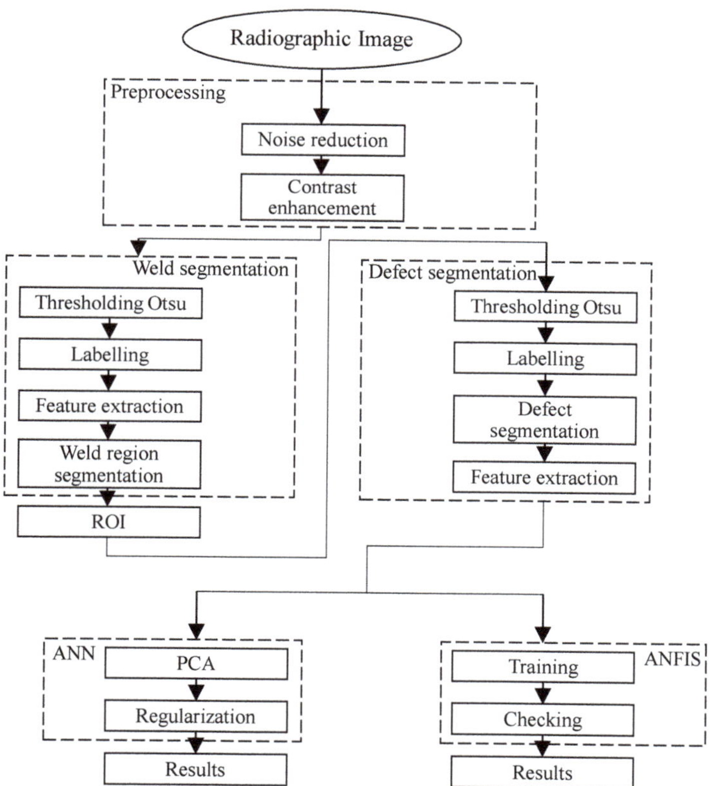

Figure 1. Procedure of automatic welding defect detection system [6].

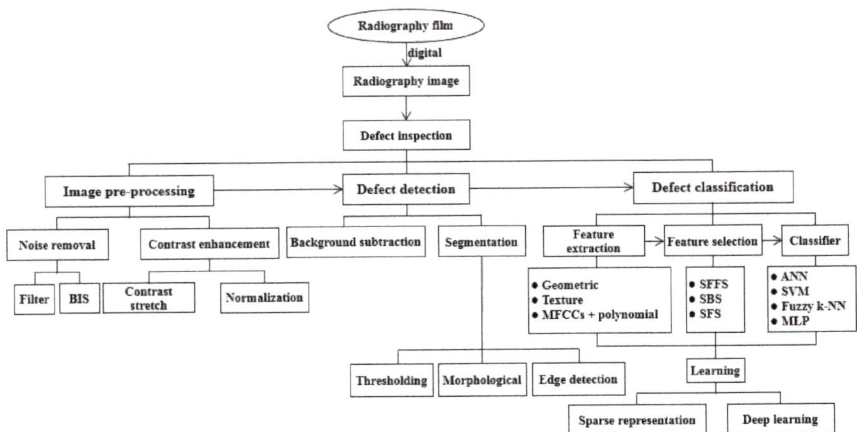

Figure 2. Flowchart of the review for studies on automatic welding defect detection.

3. Image Preprocessing

The digital radiographic images often show low contrast, existence of noise, inconsistent distribution of grays. However, the quality of images influenced the detection of weld defects to a large extent especially for those small defects which can be easily drown in the noise. Many processing methods were applied to eliminate or relieve the existence of such problems. It should be mentioned that the processing must be carefully so that the important information is not lost. For instance, the original shape and brightness distribution of the defect which is important for the discrimination of the defect types may be lost when the image enhancing methods, such as normalization and histogram equalization are applied. The pre-processing methods mainly involve noise reduction and contrast enhancement. There are several key tasks in this phase: noise filtering, object highlighting and vision improvement.

3.1. Noise Removal

The noise pixels are usually distributed irregularly in the image. Their gray level values pixels are different from those of their surrounding pixels. Filtering methods are commonly used to remove the noise pixels based on the fact that the noise is characterized as high frequency values. Zscherpel designed a one-dimensional FFT-filter for detecting the crack flaw. This filter including a column-wise FFT high-pass Bessel operation can distinguish between undercuts and cracks [7]. The noise can be easily eliminated and the cracks can be recognized clearly when a row-oriented low-pass is used to the output of this filter. Strang used a wavelet filter to transform the image in order to restrain the noise with a simple threshold operation [8]. The median filter and adaptive Wiener filter were applied successfully for removing the noise from images in many literatures [9–13]. Median filter is a nonlinear, low-pass filter. It usually deploys a template and replaces the pixel value with the median of the neighboring pixels. Wang and Liao presented that the Median filter is adequate to be applied to radiographs of continuous welds [9]. Zapata applied an adaptive Wiener filter and Gaussian low-pass filter for eliminating noise. The Wiener filters carried out the smoothing with different degrees. This adaptive filter retained the edges and other high-frequency information. Gaussian low-pass filter smooth the image in the frequency domain by alternating a specified range of high-frequency components [6]. The performance of the filter depends on the size of the filter. The defect may not be filtered out when the size is too small, while the background is estimated inaccurately when the size is large. Moreover, finding a filter which can be used on all radiographic images is difficult.

El-Tokhy used the blind image separation (BIS) instead of the filtering method. The method firstly selected the appropriate method for source extraction, then identified multiple-input multiple-output finite impulse response, lastly eliminated ambiguities of convolutive blind source separation by a correlation method. Based on the procedure, the noise is separated from the gamma radiography image [14].

3.2. Contrast Enhancement

Radiographic images are usually low-contrast and short of detail due to the limitation of intensities range accommodated by the capture device. The objective of contrast enhancement is to adjust the contrast value for highlighting the important parts without the unimportant information lost [9,15]. El-Tokhy used the contrast stretch and normalization algorithm to improve the image. The method first normalize the image with the low and high threshold values, then find the closest value to the minimum and maximum values, and finally do the contrast stretching according to determined range of contrast values [14]. Shafeek applied the histogram stretch and the histogram equalization to obtain the optimum image before the segmentation process. The histogram stretch algorithm aims to increase the contrast of images. The objective of histogram equalization is to obtain an image with equally distributed brightness levels over the whole brightness scale [2]. Ye improved both the contrast of welding seam area and background area with the method of sin function intensification. After that, the grayscales of background area and welding seam area was concentrated on high and low grayscale, and the curve was double-peak [16].

4. Defect Segmentation

The weld image contains not only the weld but also the background information. The background is defined to be the regions in an image that are not important to the analyst. In welding defect detection, the regions with defects in the weld bead is the target of recognition. Thus these regions are the focus of analyst, while the background regions without defects should be removed. The background subtraction method tries to recognize the defects by subtracting the background from the original image.

The defects can be inspected by subtracting the background from the original image due to their superimposing on the other image structures. This inspection method is popular in the early research of weld seam detection. Hyatt designed a multiscale method for removing the background structure in the digital radiographs while reserving the defect details [17]. The gray level in defect part usually changed with high spatial frequencies, meanwhile that in normal part varied gradually, namely with low spatial frequencies [18]. Many methods are built on this basis. Liao pointed that the line profile of a good weld has a bell shape, while the existence of defects will destroy the bell. Based on this, they first scaled line image so that each profile has approximately the same size, then chosen the suitable threshold value by observing the histograms of the scaled images to remove the background, finally detected whether there are anomalies in the profiles and generated a two-dimensional (2D) flaw-map [19]. Furthermore, Wang and Liao simulated a 2D background model of normal welding bead which can be subtracted from the original image. This method can be applied to defect all types of flaws [9]. These methods taken the line profile as the object of detection and processed each weld image line by line. Aoki constructed the background subtraction method for abstracting the defect image. In the process of subtracting the background, a special points connection method is proposed to persist the background distribution of defective parts [20]. Carrasco and Mery used a bottom-hat filter to separate the majority of defects from the background. This filter is made up of two stages: first a background image without flaws was produced by a morphological closure operator, secondly, the defective regions were identified by subtracting the background image from the original image [21].

Kazantsev took the weld detection problem as a problem of hypothesis testing. They applied the statistical techniques for weld defect detection in radiography images. The results showed that the defects are segmented from the original radiography images based on nonparametric statistics successfully [22].

The other image processing methods involving thresholding operation [23], watershed transform [24], morphological operations [25], and edge detection [26] are used for segmenting the defects by searching the significant variation of the pixels which may be the defects. Carrasco and Mery used the binary thresholding method to segment the defective regions after the noise reduction and background subtraction. In order to eliminate over-segmentation, filters taken from morphological mathematics were next used. The watershed transform was finally used to separate internal defective regions. The watershed transform coming from the field of mathematical morphology is a well established tool for the segmentation of images. Generally, the watershed transform is computed on the gradient of the original image, the boundaries are located at high gradient points. It can produce a complete division of the image in separated regions. The result image of each step was shown in Figure 3. The result generated an area of 0.9358 underneath the receiver operation characteristic (ROC) curve [21]. Anand applied the morphological image processing to detect suspected defect regions. The approach first detected the important edges by applying the Canny operator (a computational approach to edge detection) [26] with an appropriate threshold value. In order to obtain a closed contour, morphological image processing approach was used which dilated few similar boundaries and eroded some irrelevant boundaries [27]. Nafaa applied Artificial Neural Networks (ANN) in the edge detection of X-ray images containing defects of welding replacing the application of filtering techniques. The results show that the directly contours of welding defects in radiograms is successfully delivered. It is also proved that the proposed neural segmentation technique is robust to the noise and variable luminance. However, in this type of application, the powerful and rapid computers are needed due to the slow execution speed of this technique [1]. Shafeek implemented the segmentation process using a suitable threshold to convert the image in the specified window to binary image for separating the defects from the surrounding areas. Then they applied the eight-neighborhoods boundary chain code algorithm to identify the contours of defects. As a result, the coordinates of the boundary edges of all defects can be extracted and stored. The boundary edge codes marked can be used to calculate area, perimeter, width and height of the defects. This useful information could be used to assessment the defect and to classify the welding defects in the future [2].

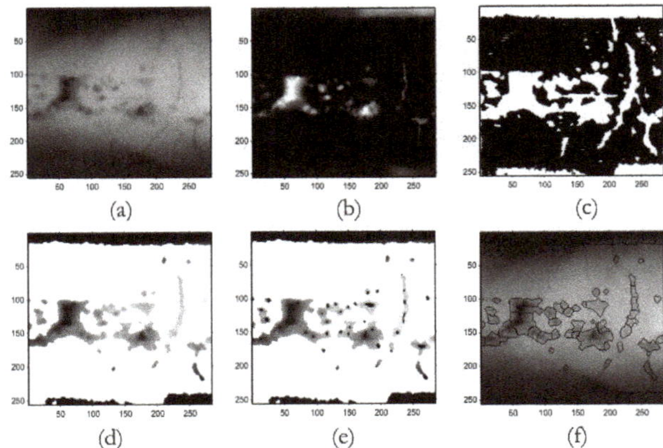

Figure 3. Summary of the proposed segmentation process: (**a**) image-altering the application of the median filter. (**b**) Application of the Bottom-Hat filter. (**c**) Application of binary thresholding. (**d**) Application sectioning process. (**e**) Modification of minima. (**f**) Application of the Watershed transformation [21].

Valavanis used graph-based segmentation algorithms [28] to extract segments that correspond to the defects or false positives. The method observed the global image characteristics other than local variation. Thus it can identify the distinct regions, even in the case that there is large variability in their interior [29].

5. Defect Classification

The classification of defects can be treated as two tasks: binary (defect, non-defect) and multi-class (gas cavity, lack of penetration, porosity, slag inclusion, crack, lack of fusion, wormhole and undercut etc.) problem. The problem involved the feature extraction, feature selection and the choice of classifier.

5.1. Feature Extraction

The raw data of the defect image cannot be analyzed directly for further classification. Some measures or descriptors should be used to extract the characteristics of defects. This process is often called feature extraction. The classification method based on features of the defects is one of the most widely used techniques. The features with a lower dimension space than the raw data are usually human-defined according to the expertise or experience. The choice of proper features called feature engineering is a key factor for recognition of the defects in the intelligent system. This feature engineering is similar to the human interpretation, which recognizes a kind of weld defect according to visual information.

At the beginning of the study, many scholars defined the features on a one-dimensional grayscale curve which are simple. Then more and more scholars began to focus on the shape, size, location, and the intensity of corresponding pixels of the defects in 2D images. Thus the geometric features and texture features were widely applied. The two kind of features defined by different scholars were also not identical. Furthermore, to synthesize the information obtained from the image, several scholars combined different kinds of features. After that, someone tried to apply features whose performance is good in other fields to weld detection.

Liao et al. processed each image line by line. For each line, they extracted 25 features, such as the degree of symmetricity (DOS), the median DOS and the goodness of fitting (GOF) etc. as the inputs of classifiers [30]. Furthermore, they extract three new features for each object in the line image for classification. These features are the width, the mean square error (MSE) between the object and its Gaussian, and the peak intensity (gray level) [31,32]. Perner calculated various parameters of profile plot as features for classification [33]. All the features mentioned above are based on the gray lever curve of line profile.

The geometric features and texture features have been the most commonly used for weld defect classification in recent decades [6,34]. The geometric features usually describe the shape, size, location, and intensity information of welding defects, while texture features can provide very useful visual cues commonly used in pattern recognition of image. Wang extracted 12 numeric features from the segmented binary defect image, such as distance from center, radius mean, standard deviation, and circularity etc. [35]. Four geometric features including position, aspect ratio, ratio, roundness were extracted to build the inputs of nonlinear pattern classifier [36]. Three new geometric features were defined to be added the features for classification [37]. Shen defined four new features: roughness of defect edge, roughness of defect region, skewness and kurtosis, which are closely related to the defect types but cannot be detected by human eyes [38]. The expert vision system proposed by Shafeek was based on the features estimating the shape, orientation and location of the defect [39]. Zhang selected eight parameters regarding to weld center, symmetry, filling degree index and relative gray-scale as defect features, such as edge flatness and ratio of perimeter to area etc. [40]. Mery extracted 28 texture features based on co-occurrence matrix for three distances, and 64 texture features based on Gabor functions for classification [41]. Valavanis proposed the multimodal feature definition by combining the geometric features and texture features for capturing all visual attributes [29]. Kumar extracted a set of 8, 64 and 44 texture features vectors based on gray level co-occurrence matrix for classifying various defects [15]. Further, they combined the geometric features and texture features for classification, and compared the performance of classifiers with different feature sets. The results show that the classifier with the combined features perform better than with only geometric features or only texture features [12].

Mel-Frequency Cepstral Coefficients (MFCCs) as low-dimensional, fixed dimension feature vectors are very effective in speech processing [42]. They are already successfully tested as s damage sensitive features for mechanical systems [43] and applied in damage detection for structural health monitoring (SHM) [44]. Kasban used them with the polynomial features for weld defect identification. These features extracted from the 1D lexicographically ordered signals or their power density spectra are suitable for defect detection in the existence of noise. [11,45].

These features are defined on the gray-lever curve of line image or 2D images. In general, the appropriate feature is decided by the inspector or researcher who judges according to his experience and knowledge. Thus, different inspectors are inclined to diverse types of features. It is difficult to say what kind of feature is best.

5.2. Feature Selection

Sometimes, the number of features extracted is large, thus the dimension of feature vector is high, which lead to excessive computational burden. Feature selection as one of the mainstream approaches for processing high-dimensional data is essential to reduce the difficulty of classification task and computational costs. It tries to find an optimum subset of original features which can provide useful information. The work is too complicated for people, even impossible. There are a variety of methods for evaluating the quality of the extracted features. Jain compared the performance of feature selection [46]. Many feature selection methods found the optimum subset by evaluation function which assessing the quality of feature subsets. Liao used a simple feature selection approach based on the correlation coefficients between independent variable and the dependent variable to meet 7 ± 2 rule of fuzzy expert system. The method reduced the dimension of features from 12 to 7 or 9 without too much loss of accuracy [35]. However, to some extent, using of feature subset reduced the accuracy. Garcı'a-Allende presented sequential forward floating selection (SFFS) as the substitution of principal component analysis (PCA) for weld detection. The algorithm performed the dimensionality reduction better than PCA, and improved the computational performance [47]. In the work of Mery, 148 texture features were extracted for each segmented region which is too many to be input of classifier. In order to reduce the computational time, they used a sequential forward selection (SFS) method which requires an objective function obtained from the Fisher discriminant to evaluate the performance of classifier with m feature. The method began with one feature (m = 1), then added one feature in each calculation, finally searched the features that maximize the objective function until the optimal n features were obtained [41]. Valavanis used a sequential backward selection (SBS) method with classifier and compared the performance of classification using different feature sets (43 features and its subset with seven features) [29]. The results showed that the SBS saves about 80% feature computation. The accuracies of the artificial neural network (ANN) classifier using 7 features were almost as high as those using 43 features. However, the situation is not the same when using the support vector machine (SVM). The feature selection procedure favored the ANN but not SVM in classification perform.

Many selection methods applied the classification performance (classification accuracy or error) of classifier as the evaluation function. In this situation, the effectiveness of feature selection method is related to the choice of classifier.

5.3. Classifier

The choice of classifier is another key factor influencing the performance of defect classification. It is a more complex problem to choose an appropriate classifier. The classification task was finished by fuzzy K-NN algorithm and fuzzy c-means algorithm according to extracted features. It was concluded that the classification performance of fuzzy K-NN algorithm is superior to that of fuzzy c-means [30,31]. Mery and Berti used three statistical classifiers which conducting the classification by using the concept of similarity: polynomial classifier, Mahalanobis classifier and nearest neighbor classifier [48]. They compared the performance of these classifiers with 7 selected features using the true positives, false positives, false negatives, true negatives, sensibility, and 1-specificity [41]. Liao used fuzzy expert systems for classification and compared it with the fuzzy k-nearest neighbor algorithm and multi-layer perceptron neural networks. It is concluded that the proposed fuzzy system is more transparent and easier understood by human through the evaluation by bootstrap method [35]. Zhang found that multi-SVM is almost independent of the impact of the training samples reduction. Thus it had higher accuracy than fuzzy neural network under the condition of small samples [40]. Shen used direct multiclass SVM (DMSVM) to classify the defects through the features defined by themselves. DMSVM yields a direct method for training multiclass predictors instead of constructing the classifier according to the samples to be classified [38]. Silva implemented a study of nonlinear classifier using ANN and proved that the quality of the extracted features is more important than the quantity of the features [36]. In the last ten years, ANN has been widely used in the classification of welding defects [6,12,49]. Vilar applied ANN to classify the defect candidates. They used three different regularization methods (i.e., regularization with modified performance function, automatically setting of the regularization parameters and early stopping or bootstrap) to improve the network generalization. The network with best performance is obtained through several tests [49].

Perner compared the performances of neural networks and decision trees. In this work, they pointed out the error rate cannot be the only criterion for comparing the different classifiers. More criteria, such as generalization, representation ability, classification quality and classification cost etc. were proposed [33]. They noted that the RBF neural net and the back-propagation network perform better than decision tree on the error rate for test data set while the unpruned decision tree shows the best error rate for the design data set. The representation and generalization ability of the neural networks is more balanced while the unpruned decision tree overfits the data. However, the neural nets need the feature selection before learning. Otherwise, the neural nets will be more complex and time-consuming. Moreover, the explanation capability of decision trees whose rules can be understood and controlled by human is strong.

The selection of classifiers is a complex problem. Their accuracies are closely related not only to the extracted features, but also to the characteristics of the original data. For example, the complexity of some classifiers depends on the number of input features. Some classifiers need an additional feature selection operation when the number of the input features is large.

Table 1 shows the primary coverage of most literatures for reader guidelines. The table shows the traditional computer vision technologies including the above-mentioned aspects. The results of various technologies show the only best accuracy or false alarm rate obtained by the classifiers with better performance. The results of different literature are generated through the different data using the different evaluation criteria. Thus it is not appropriate for the comparison. The values are only used to understand the performance of classifiers. The word "complex" present the performance of the classifier is evaluated by complex style (more criteria) which is partly introduced in preceding text.

Through the summary of the above-mentioned literatures, it can be concluded the research on the classification of welding defects based on radiographic images has been focusing on feature extraction, feature selection and classifier design. These three aspects are closely connected which all influence the classification performance. However, the contribution of most of the works is to optimize one or both of them. The human-defining of features cannot be updated on-line. Feature selection involves prior selection of hyperparameters such as latent dimension. Thus the system including feature definition/extraction, feature selection and classifier training cannot be jointly optimized which may hinder the final performance of the whole system [50].

Table 1. Main techniques used in the papers cited for readers guideline.

Ref	Base	Pre-Processing	Feature Number; Type	Feature Selection	Classifier	Results	Evaluation
[30]	Line profile	-	25 profile measurements	-	Fuzzy k-NN; Fuzzy C-means	6.01% 18.67%	Missing rate False alarm
[31]	Line profile	-	3 profile measurements	-	Fuzzy k-NN; Fuzzy C-means	-	False alarm
[33]	Line profile	-	36 profile measurements	-	NN; Decision-tree	complex	Generalization; Representation; Quality; Cost; etc.
[32]	Line profile	-	3 profile measurements	-	Fuzzy reasoning	100%	Accuracy
[9]	2D image	Noise removal; Contrast improve; Defect segment	12 numeric	-	Fuzzy k-NN; MLP	92.39%	Bootstrap accuracy
[41]	2D image	Potential defect segment	148 texture	SFS	Polynomial; Mahalanobis; Nearest neighbor	90.91%	Area under the ROC
[35]	2D image	-	12 geometric	Filter methods	Fuzzy expert; Fuzzy k-NN; MLP	0.9205	Bootstrap accuracy
[36]	2D image	Noise removal; Contrast improve;	4 geometric	-	Nonlinear pattern classifiers using NN	complex	Classification performance; Relevance criterion; Principal components
[40]	2D image	Noise removal; Enhancement; Segmentation	8 geometric	-	SVM; Fuzzy NN	83.3%	Accuracy rate
[37]	2D image	-	7 geometric	-	Nonlinear classifier	92%	Bootstrap accuracy
[29]	2D image	Defect segment	43 geometric+ texture	SBS	SVM; ANN; k-NN	98.51%	3-fold cross validation accuracy
[15]	2D image	Noise removal; Contrast improve	8,64,44 texture	-	ANN	86.1%	Classification accuracy
[12]	2D image	Noise removal; Contrast improve; Image segment	16 texture 8geometric 72 geometric + texture	-	ANN	87.34%	Classification accuracy
[45]	1D signal	-	13MFCCs+ 26polynomial features	-	ANN	100%	Recognition rates
[6]	2D image	Noise removal; Contrast improve; Defect segment	12 geometrical	-	ANN ANFIS	100%	Classification accuracy
[11]	Power Density Spectra	Image enhancement; Image segmentation	MFCCs+ polynomial features	-	ANN	100%	Probability of detection; False alarm rate
[14]	2D image	Noise removal; Contrast improve; Image segment	Energy of the wavelet coefficients	-	SVM	99.5%	Classification rate
[34]	2D image	Location of the weld bead region	8 geometrical	-	MLP	88.6% 87.5%	Accuracy F-score

5.4. New Methods

Considering that most of the defect features are designed manually and lack of intelligence, many researchers began to focus on the automatic extraction method based on learning. As the classical learning-based approaches, deep learning [51] and sparse representation [52] supplied new idea in recognizing the object automatically from the optical images. Olshausen pointed out the classical image is sparse and can be compressed [53]. Deep learning can learn hierarchical feature replacing the handcrafted features. Their excellent performance in feature learning from image urged people to use them in automatic defect detection [54–56]. For automatic detection of welding defects, the application of two methods begin to rise, especially the application of convolutional neural network (CNN) which is a typical deep learning model [57–59].

In the focus of this paper, Chen developed a system for recognizing diverse types of defects from images inspired by the human visual inspection mechanism. The method which is an unsupervised algorithm learned a dictionary from lots of normal images replacing the experienced workers. The dictionary can be used to sparsely reconstruct the background and the weld region of test images without the defective regions. Thus the defects would stand out in difference image between the test image and the reconstructed image [60]. Yang proposed a model-based on CNN to classify the X-ray weld images by improving the convolution kernel and the activation function. The method doesn't need noise reduction, the feature extracting and enhancing [61]. Based on the principle of visual perception, Li constructed a deep learning network with 10 layers to directly determine the type of suspected defect. The network can judge whether the defect is linear or circular without extracted features [62]. Ye designed online segmentation and recognition method based on the compressed sensing for weld defect. The method firstly established offline database of defects images, then segmented the defect using clustering method, lastly determined the type of defects using the optimal dictionary whose atoms are characteristic values of defects. Hou developed a model based on a deep convolutional neural network (DCNN) to extract the deep features directly from the X-ray images without any preprocessing operation. The performance of deep feature was compared with those of traditional designed feature, such as texture features and histogram of oriented gradients features. The results showed that the separability of deep features is better [63]. Figure 4 shows the difference of traditional computer vision technology based on the above-mentioned technologies and the deep learning approach.

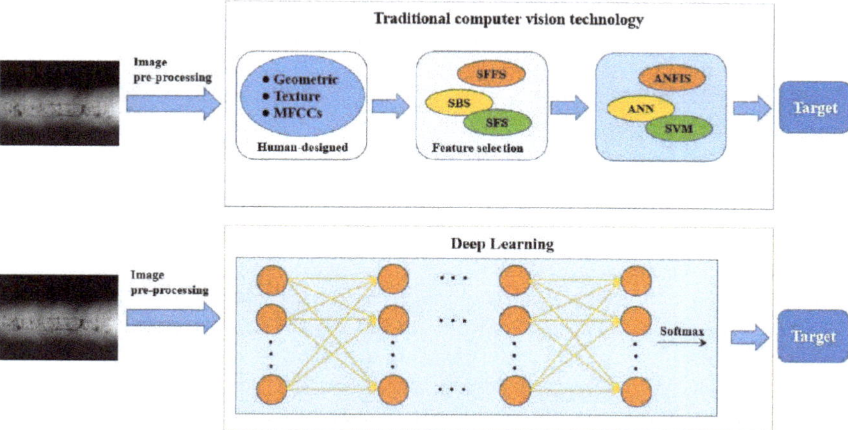

Figure 4. Traditional computer vision technology and deep learning approach comparison.

The deep models can directly learn the high-level structural features from the original images avoiding the segment of the weld region. Thus unlike previous algorithms, the performance of these models does not depend on accurate weld region extraction. Moreover, the deep models embracing the feature extraction, selection and classification can be jointly optimized.

6. Discussions

6.1. Achievements

Inspection of weld defects from the radiographic images has always been an important and challenging task. Limitations of human evaluation persuade the researchers to develop new technologies for automatic inspections. The developments in image analysis and computer vision techniques have made the perfect conditions for proposing methods for automation of weld defect inspection. Looking into the accomplishments introduced in previous sections, developments in detection methods provide enough data for segmenting weld, extracting the features and classification in order to detect and classify nearly all defects.

The defect segmentation methods based on traditional image processing can separate the defects from the original the images. The process can be regarded as the recognition of defect. The location of defects can be obtained from the segmented results. This is similar to the second level of Rytter classification [64], which is often used in damage detection and assessment in structural health monitoring (SHM) [65]. Detection and localization of defects have been automated to some extent. At the same time, the important parameters of defects can also be calculated from the segmented defect region which are often used as the features in the future to classify the welding defects. Thus the quality of defect segmentation would affect the performance of classification system. Most of the defect classification systems based on feature extraction, selection and classifier introduced in Section 5 achieved high accuracy. However, the features introduced are defined or designed by human, which hinder the automation of system. The innovative technologies such as deep neural networks have made it possible to learn the features automatically from the image. Deep neural networks represented remarkable performance in recent years according to their recent application in weld defect detection. If the X-ray images are from thick steel pipes, the edge of image will be blurry and the gray scale distribution of the weld region will be uneven. Deep neural networks can deal with the situation well-avoiding noise reduction, extracting and enhancing features. As the end-to-end detection system, the networks can directly determine whether the suspected defect image is a linear defect, circular defect or noise.

6.2. Challenges

Contrary to the development mentioned in the previous section, there are still challenges in the automation of weld defect inspection. The main contribution made in the last decade is the development of efficient and well-organized methods for feature extraction for constructing a robust classifier. The features providing enough discrimination should be selected. It is an important step in classifier design. Previous studies have mainly focused on the geometric and texture features while there are infinite unknown patterns and shapes for each type of weld defect.

Although the deep neural networks introduced are dominant in learning the hierarchical feature from the weld images, the previous research show the drawbacks in developing a generalized defect inspection model based on these networks. There are mainly two aspects: the model training and the preparation of the datasets. Due to the huge model complexity behind deep learning methods, the training time of models is usually long which made it difficult to realize the real time defect detection. However, we believe that the implementation an on-line, real-time welding monitor system to prevent the possible defects from happening will be one of the key concerns in future. Moreover, the performance of deep models heavily depend on the scale and quality of datasets. For weld defect inspection, it is difficult to obtain a good dataset (a lots of weld images with defects being labeled by human). We believe that unsupervised representation learning provides good ideas for learning a layer of feature representations from unlabeled images. However, the literature involving the application of unsupervised representation learning in welding defect detection is still very few. Its effectiveness has yet to be supported and proven by more work.

There is an intrinsic problem in the domain of weld inspection as in many other domains such as fraud detection, oil spill detection. It is just the class imbalance problem, namely the number of examples of one class is much higher than the others. The classifier tends to obtain high accuracy over the majority class, but poor accuracy over the minority class. Figure 5 presents that class imbalance leads to poorer performance when classifying minority class. The weld defect detection is just the problem of classification for minority class. Boaretto carried out the identification of the defects successfully. However, they failed on the attempt of classifying the defects due to the unbalanced data generated by the few samples of each defect type [34]. Liao studied the imbalanced data problem in the classification of different types of weld flaws. He used eight evaluation criteria to study the effectiveness of 22 data preprocessing methods for dealing with imbalanced data. The results indicated that some preprocessing methods do not improve any criterion, and they vary from one classifier to another [66]. These preprocessing methods solved the problem from only the perspective of data. The literature involving improving algorithms to deal with the imbalance learning for weld defect inspection is still few.

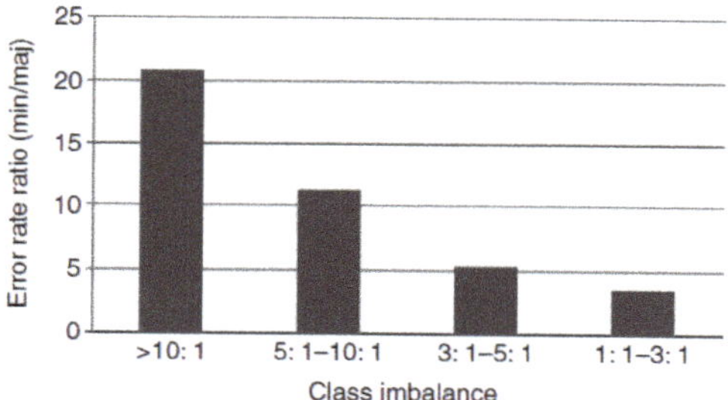

Figure 5. Impact of class imbalance on classification performance of minority class [67].

7. Conclusions

To assure the reliability of the weld structure, it is critical to inspect and assess the quality of the weld joints from the radiographic films according to NDT test. The evaluation of films performed by certified operators is time-consuming, subjective, and error prone. Therefore, many researchers have tried to automate the evaluation using computer assistance. This article introduced available practices in weld defects inspection from digital radiographic images. The technologies were introduced by stage in weld defect inspection. This review supplied a comprehensive literature review on automatic models based on traditional computer vision techniques falling into three categories: image pre-processing, feature extraction and classification. The paper firstly found that some image pre-processing tools like morphological operation and thresholding operation which improved the quality and segmented the defect region. Following this, it concluded three aspects: feature extraction, selection and classifier and pointed out the limitations of these traditional methods. Lastly, the advantage of sparse representation and deep models were analyzed and their applications in weld defects inspection were introduced. It was concluded with the analysis of achievements of automating weld defect detection and challenges summarized as open questions for future research in the field.

Author Contributions: W.H., X.Z. and J.G. conceived and designed the field survey; W.H. and Y.W. processed and analyzed the data; D.Z. and X.Z. contributed to the survey and reviewed the submitted manuscript. All authors have read and agreed to the published version of the manuscript.

Funding: This project was funded by National Natural Science Foundation of China (No. 51805006 and No. 51675005).

Acknowledgments: The authors would like to thank the anonymous reviewers for their valuable comments and suggestions.

Conflicts of Interest: The authors declare no conflict of interest.

References

1. Nafaa, N.; Redouane, D.; Amar, B. Weld defect extraction and classification in radiographic testing based artificial neural networks. In Proceedings of the 15th World Conference on Non Destructive Testing, Rome, Italy, 15–21 October 2000.
2. Shafeek, H.I.; Gadelmawla, E.S.; Abdel-Shafy, A.A. Assessment of welding defects for gas pipeline radiographs using computer vision. *NDT E Int.* **2004**, *37*, 291–299.
3. Yeh, P.S.; Le Moigne, J.; Fong, C.B. Computer-aided X-ray film interpretation. In *MML TR 88–76*; Martin Marietta Laboratories: Baltimorem, MD, USA, 1988.
4. Basart, J.P.; Xu, J. Automatic detection of flaws in welds. In *Final Report to Martin Marietta MSS Contract# A71445*; Center for NDE, Iowa State University: Ames, IA, USA, 1991.
5. Cloud, E.; Fraser, K.; Krywick, S. Automated examination of X-ray welds. In *Final Report to Martin Marietta MSS*; Martin Marietta Electronics, Information & Missiles Group: Orlando, FL, USA, 1992.
6. Zapata, J.; Vilar, R.; Ruiz, R. Performance evaluation of an automatic inspection system of weld defects in radiographic images based on neuro-classifiers. *Expert Syst. Appl.* **2011**, *38*, 8812–8824.
7. Zscherpel, U.; Nockemann, C.; Mattis, A.; Heinrich, W. Neue Entwicklungen bei der Filmdigitalisierung. In Proceedings of the DGZfP-Jahrestagung in Aachen, Tagungsband, Aachen, Germany, 22–24 March 1995.
8. Strang, G. Wavelets and dilation equations: A brief introduction. *SIAM Rev.* **1989**, *31*, 614–627.
9. Wang, G.; Liao, T.W. Automatic identification of different types of welding defects in radiographic images. *NDT E Int.* **2002**, *35*, 519–528.
10. Aoki K; Suga, Y. Application of artificial neural network to discrimination of defect type in automatic radiographic testing of welds. *ISIJ Int.* **1999**, *39*, 1081–1087.
11. Zahran, O.; Kasban, H.; El-Kordy, M.; Abd El-Samie, F.E. Automatic weld defect identification from radiographic images. *NDT E Int.* **2013**, *57*, 26–35.
12. Kumar, J.; Anand, R.S.; Srivastava, S.P. Flaws classification using ANN for radiographic weld images. In Proceedings of the IEEE International Conference on Signal Processing and Integrated Networks (SPIN), Noida, India, 20–21 February 2014; pp. 145–150.
13. Li, L.; Xiao, L.; Liao, H. Welding quality monitoring of high frequency straight seam pipe based on image feature. *J. Mater. Process. Technol.* **2017**, *246*, 285–290.
14. El-Tokhy, M.S.; Mahmoud, I.I. Classification of welding flaws in gamma radiography images based on multi-scale wavelet packet feature extraction using support vector machine. *J. Nondestruct. Eval.* **2015**, *34*. [CrossRef]
15. Kumar, J.; Anand, R.S.; Srivastava, S.P. Multi-class welding flaws classification using texture feature for radiographic images. In Proceedings of the IEEE International Conference on Advances in Electrical Engineering (ICAEE), Vellore, India, 9–11 January 2014; pp. 1–4.
16. Ye, H.; Juefei, L.; Huijun, L. Detection and recognition of defects in X-ray images of welding seams under compressed sensing. *J. Phys. Conf. Ser.* **2019**, *1314*, 012064.
17. Hyatt, R.; Kechter, G.E.; Nagashima, S. A method for defect segmentation in digital radiographs of pipeline girth welds. *Mater. Eval.* **1996**, *54*, 379793.
18. Daum, W.; Rose, P.; Heidt, H.; Builtjes, J.H. Automatic recognition of weld defects in x-ray inspection. *Br. J. Nondestruct. Test.* **1987**, *29*, 79–81.
19. Liao, T.W.; Li, Y. An automated radiographic NDT system for weld inspection: Part II—Flaw detection. *NDT E Int.* **1998**, *31*, 183–192.
20. Aoki, K.; Suga, Y. Intelligent image processing for abstraction and discrimination of defect image in radiographic film. In Proceedings of the Seventh International Offshore and Polar Engineering Conference, Honolulu, HI, USA, 25–30 May 1997.
21. Carrasco, M.A.; Mery, D. Segmentation of welding defects using a robust algorithm. *Mater. Eval.* **2004**, *62*, 1142–1147.
22. Kazantsev, I.; Lemahieu, I.; Salov, G.I.; Denys, R. Statistical detection of defects in radiographic images in nondestructive testing. *Signal Process.* **2002**, *82*, 791–801.
23. Murakami, K. Image processing for non-destructive testing. *Weld. Int.* **1990**, *4*, 144–149.
24. Grau, V.; Mewes, A.U.J.; Alcaniz, M.; Kikinis, R.; Warfield, S.K. Improved watershed transform for medical image segmentation using prior information. *IEEE Trans. Med. Imaging* **2004**, *23*, 447–458. [PubMed]

25. Sofia, M.; Redouane, D. Shapes recognition system applied to the non destructive testing. In Proceedings of the 8th European Conference on Non-Destructive Testing, Barcelona, Spain, 17–21 June 2002.
26. Canny, J. A computational approach to edge detection. *IEEE Trans. Pattern Anal. Mach. Intell.* **1986**, *8*, 679–698.
27. Anand, R.S.; Kumar, P. Flaw detection in radiographic weld images using morphological approach. *Ndt E Int.* **2006**, *39*, 29–33.
28. Felzenszwalb, P.F.; Huttenlocher, D.P. Efficient graph-based image segmentation. *Int. J. Comput. Vis.* **2004**, *59*, 167–181.
29. Valavanis, I.; Kosmopoulos, D. Multiclass defect detection and classification in weld radiographic images using geometric and texture features. *Expert Syst. Appl.* **2010**, *37*, 7606–7614.
30. Liao, T.W.; Li, D.M.; Li, Y.M. Detection of welding flaws from radiographic images with fuzzy clustering methods. *Fuzzy Sets Syst.* **1999**, *108*, 145–158.
31. Liao, T.W.; Li, D.; Li, Y. Extraction of welds from radiographic images using fuzzy classifiers. *Inf. Sci.* **2000**, *126*, 21–40.
32. Liao, T.W. Fuzzy reasoning based automatic inspection of radiographic welds: weld recognition. *J. Intell. Manuf.* **2004**, *15*, 69–85.
33. Perner, P.; Zscherpel, U.; Jacobsen, C. A comparison between neural networks and decision trees based on data from industrial radiographic testing. *Pattern Recognit. Lett.* **2001**, *22*, 47–54.
34. Boaretto, N.; Centeno, T.M. Automated detection of welding defects in pipelines from radiographic images DWDI. *NDT E Int.* **2017**, *86*, 7–13.
35. Liao, T.W. Classification of welding flaw types with fuzzy expert systems. *Expert Syst. Appl.* **2003**, *25*, 101–111.
36. da Silva, R.R.; Calôba, L.P.; Siqueira, M.H.S.; Rebello, J.M.A. Pattern recognition of weld defects detected by radiographic test. *NDT E Int.* **2004**, *37*, 461–470.
37. Da Silva, R.R.; Siqueira, M.H.S.; de Souza, M.P.V.; Rebello, J.M.A. Estimated accuracy of classification of defects detected in welded joints by radiographic tests. *NDT E Int.* **2005**, *38*, 335–343.
38. Shen, Q.; Gao, J.; Li, C. Automatic classification of weld defects in radiographic images. In *Insight—Non-Destructive Testing and Condition Monitoring*; The British Institute of Non-Destructive Testing: Northampton, UK, 2010; Volume 52, pp. 134–139.
39. Shafeek, H.I.; Gadelmawla, E.S.; Abdel-Shafy, A.A.; Elewa, I.M. Automatic inspection of gas pipeline welding defects using an expert vision system. *NDT E Int.* **2004**, *37*, 301–307.
40. Zhang, X.G.; Xu, J.J.; Ge, G.Y. Defects recognition on X-ray images for weld inspection using SVM. In Proceedings of the IEEE International Conference on Machine Learning and Cybernetics, Shanghai, China, 26–29 August 2004; Volume 6, pp. 3721–3725.
41. Mery, D.; Berti, M.A. Automatic detection of welding defects using texture features. In *Insight—Non-Destructive Testing and Condition Monitoring*; The British Institute of Non-Destructive Testing: Northampton, UK, 2003; Volume 45, pp. 676–681.
42. Civera, M.; Ferraris, M.; Ceravolo, R.; Surace, C. The Teager-Kaiser Energy Cepstral Coefficients as an Effective Structural Health Monitoring Tool. *Appl. Sci.* **2019**, *9*, 5064.
43. Balsamo, L.; Betti, R.; Beigi, H. A structural health monitoring strategy using cepstral features. *J. Sound Vib.* **2014**, *333*, 4526–4542.
44. Ferraris, M.; Civera, M.; Ceravolo, R.; Surace, C.; Betti, R. Using enhanced cepstral analysis for structural health monitoring. In Proceedings of the 13th International Conference on Damage Assessment of Structures, Porto, Portugal, 9–10 July 2019; Springer: Berlin, Germany, 2020; pp. 150–165.
45. Kasban, H.; Zahran, O.; Arafa, H. Welding defect detection from radiography images with a cepstral approach. *NDT E Int.* **2011**, *44*, 226–231.
46. Jain, A.; Zongker, D. Feature selection: Evaluation, application, and small sample performance. *IEEE Trans. Pattern Anal. Mach. Intell.* **1997**, *19*, 153–158.
47. Garcia-Allende, P.B.; Mirapeix, J.; Conde, O.M.; Cobo, A.; Lopez-Higuera, J.M. Spectral processing technique based on feature selection and artificial neural networks for arc-welding quality monitoring. *NDT E Int.* **2009**, *42*, 56–63.
48. Mery, D.; da Silva, R.R.; Calôba, L.P.; Rebello, J.M.A. Pattern recognition in the automatic inspection of aluminium castings. In *Insight—Non-Destructive Testing and Condition Monitoring*; The British Institute of Non-Destructive Testing: Northampton, UK, 2003; Volume 45, pp. 475–483.

49. Vilar, R.; Zapata, J.; Ruiz, R. An automatic system of classification of weld defects in radiographic images. *NDT E Int.* **2009**, *42*, 467–476.

50. Zhao, R.;Yan, R.; Chen, Z.; Mao, K. Deep learning and its applications to machine health monitoring. *Mech. Syst. Signal Process.* **2019**, *115*, 213–237.

51. LeCun, Y.; Bengio, Y.; Hinton, G. Deep learning. *Nature* **2015**, *521*, 436–444.

52. Tosic, I.; Frossard, P. Dictionary learning: What is the right representation for my signal? *IEEE Signal Process. Mag.* **2011**, *28*, 27–38.

53. Olshausen, B.A.; Field, D.J. Emergence of simple-cell receptive field properties by learning a sparse code for natural images. *Nature* **1996**, *381*, 607.

54. Xiang, Y.; Zhang, C.; Guo, Q. A dictionary-based method for tire defect detection. In Proceeidngs of the IEEE International Conference on Information and Automation, Hailar, China, 28–30 July 2014; pp. 519–523.

55. Nhat-Duc, H.; Nguyen, Q.L.; Tran, V.D. Automatic recognition of asphalt pavement cracks using metaheuristic optimized edge detection algorithms and convolution neural network. *Autom. Constr.* **2018**, *94*, 203–213.

56. Cha, Y.J.; Choi, W.; Büyüköztürk, O. Deep learning-based crack damage detection using convolutional neural networks. *Comput.-Aided Civ. Infrastruct. Eng.* **2017**, *32*, 361–378.

57. Sassi, P.; Tripicchio, P.; Avizzano, C.A. A smart monitoring system for automatic welding defect detection. *IEEE Trans. Ind. Electron.* **2019**, doi:10.1109/TIE.2019.2896165. [CrossRef]

58. Zhang, Y.; You, D.; Gao, X.; Zhang, N.; Gao, P.P. Welding defects detection based on deep learning with multiple optical sensors during disk laser welding of thick plates. *J. Manuf. Syst.* **2019**, *51*, 87–94.

59. Zhang, Z.; Wen, G.; Chen, S. Weld image deep learning-based on-line defects detection using convolutional neural networks for Al alloy in robotic arc welding. *J. Manuf. Process.* **2019**, *45*, 208–216.

60. Chen, B.; Fang, Z.; Xia, Y. Accurate defect detection via sparsity reconstruction for weld radiographs. *NDT E Int.* **2018**, *94*, 62–69.

61. Yang, N.; Niu, H.; Chen, L.; Mi, G. X-ray weld image classification using improved convolutional neural network. In Proceedings of the 2018 International Symposium on Mechanics, Structures and Materials Science, Tianjin, China, 9–10 June 2018; Volume 1995, p. 020035.

62. Yaping, L.; Weixin, G. Research on X-ray welding image defect detection based on convolution neural network. *J. Phys. Conf. Ser.* **2019**, *1237*, 032005.

63. Hou, W.; Wei, Y.; Jin, Y.; Zhu, C. Deep Features Based A DCNN Model Classifying Imbalanced Weld Flaw Types. *Measurement* **2019**, *131*, 482–489.

64. Rytter, A. Vibrational Based Inspection of Civil Engineering Structures. Ph.D. Thesis, Aalborg University, Aalborg, Denmark, 1993.

65. Civera, M.; Zanotti Fragonara, L.; Surace, C. An experimental study of the feasibility of phase-based video magnification for damage detection and localisation in operational deflection shapes. *Strain* **2020**, *56*, e12336.

66. Liao, T.W. Classification of weld flaws with imbalanced class data. *Expert Syst. Appl.* **2008**, *35*, 1041–1052.

67. He, H.; Ma, Y. *Imbalanced Learning: Foundations, Algorithms, and Applications*; John Wiley & Sons: Hoboken, NJ, USA, 2013.

 *applied sciences*

Article

Feature Extraction with Discrete Non-Separable Shearlet Transform and Its Application to Surface Inspection of Continuous Casting Slabs

Xiaoming Liu [1], Ke Xu [2,*], Peng Zhou [1] and Huajie Liu [2]

1   National Engineering Research Center for Advanced Rolling Technology, University of Science and Technology Beijing, Beijing 100083, China; jklxm333@e-mail.com (X.L.); zhoupeng@nercar.ustb.edu.cn (P.Z.)
2   Collaborative Innovation Center of Steel Technology, University of Science and Technology Beijing, Beijing 100083, China; lhjk2s@163.com
*   Correspondence: xuke@ustb.edu.cn; Tel.: +86-10-62332159

Received: 17 September 2019; Accepted: 25 October 2019; Published: 1 November 2019

Featured Application: First, the surface defect inspection characteristics of continuous casting slab is that the slab moves slowly on the production line, so the feature extraction method does not need too fast a calculation speed. Second, the inspection difficulties of continuous casting slabs are the defect with complex backgrounds, so some common feature extraction methods cannot meet these needs. DNST (discrete non-separable shearlet transform) is a new multiresolution analysis method with moderate computing speed. It can extract images information from multiple scales and directions. Therefore, this paper proposed a DNST-GLCM-KSR (discrete non-separable shearlet transform-gray-level co-occurrence matrix-kernel spectral regression) feature extraction method. The method is suitable for surface defects inspection with complex background and moderate running speed of production line.

Abstract: A new feature extraction technique called DNST-GLCM-KSR (discrete non-separable shearlet transform-gray-level co-occurrence matrix-kernel spectral regression) is presented according to the direction and texture information of surface defects of continuous casting slabs with complex backgrounds. The discrete non-separable shearlet transform (DNST) is a new multi-scale geometric analysis method that provides excellent localization properties and directional selectivity. The gray-level co-occurrence matrix (GLCM) is a texture feature extraction technology. We combine DNST features with GLCM features to characterize defects of the continuous casting slabs. Since the combination feature is high-dimensional and redundant, kernel spectral regression (KSR) algorithm was used to remove redundancy. The low-dimension features obtained and labels data were inputted to a support vector machine (SVM) for classification. The samples collected from the continuous casting slab industrial production line—including cracks, scales, lighting variation, and slag marks—and the proposed scheme were tested. The test results show that the scheme can improve the classification accuracy to 96.37%, which provides a new approach for surface defect recognition of continuous casting slabs.

Keywords: continuous casting slabs; surface defect classification; discrete non-separable shearlet transform; gray-level co-occurrence matrix; kernel spectral regression

---

1. Introduction

At present, machine vision–based surface inspection technology has been widely used in the detection and identification of surface defects of various industrial products due to its non-contact and real-time detection properties [1]. The machine vision–based detection method is to collect the image of the industrial product under the irradiation of the high-intensity light source and use the image processing and pattern recognition algorithm to analyze the surface image [2]. For different industrial products, one need consider defect image features of the products themselves and then adopt appropriate recognition methods.

In the production process of continuous casting slabs, defects often occur due to various factors like raw material, preprocessing technologies, etc. The defects will have a negative impact on the next rolling process, and severe defects will even lead to the scrapping of entire slabs [3,4]. The defect feature extraction method plays an important role in defect inspection, which is one of the hotspots in the research on surface defect recognition algorithms. The most important characteristic of surface defects of continuous casting slab are complex backgrounds, which make recognition difficult.

At present, research is more active on strip steel products with a simple background image [5–7]. However, the defect recognition of continuous casting slabs with complex backgrounds has received comparatively little attention. Wei SY et al. [3] extracted the shape feature values of the image to classify and recognition defects. Yun [8] proposed a surface defect recognition algorithm based on Gabor wavelet that can detect the fine cracks and angular cracks on the surface of the slabs by minimizing the cost function of the energy separation criterion for the defect area and the defect-free area. Pan E [9] proposed an engineering-driven rule-based detection (ERD) method according to the mechanism of deep longitudinal crack and transverse crack on slabs. Xu K et al. [10] used non-sampled wavelet to decompose the surface image by calculating the scale co-occurrence matrix and grayscale co-occurrence matrix, and used AdaBoosting classifier to identify cracks from water marks, slag marks, scales, and vibration marks. Subsequently, the author proposed combining the discrete shearlet transform (DST) and kernel local preservation projection (KLPP) algorithm to extract surface defect features [2]. Y. Ai utilized the combination of Contourlet transform and kernel local preservation projection (KLPP) algorithm to extract the defect features [1], then Xu K [11] improved the above method by introducing a texture feature. Si Yang [12] improved the local binary pattern and proposed a multi-block local binary pattern (MB-LBP) feature extraction method.

Of the above mentioned methods, the wavelet-based feature extraction (for example, references [1,2,8,10,11]) is the more effective and more studied technology. Although these methods have achieved some results, the recognition accuracy of surface defects of continuous casting slabs needs to be further improved with the increasingly strict quality requirements of users.

Discrete nonseparable shearlet transform (DNST) [13,14] is a new kind of wavelet-based method. It is a compactly supported shearlet transform with excellent localization properties in the spatial domain and excellent directional selectivity. DNST has been successfully introduced in the fields of compressed sensing magnetic resonance imaging [15,16]. According to defect images of continuous casting slabs with the scale and directionality traits, DNST was introduced into the surface defect feature extraction of continuous casting slabs. The gray-level co-occurrence matrix (GLCM) is an effective texture feature extraction method that can reflect the comprehensive information of the image gray in direction, adjacent pixel interval, and gray level variation [17]. Some defects images of continuous casting slabs also have texture traits, thus we consider introducing GLCM into the feature extraction of continuous casting slabs. Since the features extracted by DNST and GLCM are redundant, we use kernel spectral regression (KSR) [18,19] technology to remove redundant features. KSR is a kind of manifold learning dimensionality reduction technology, and it casts the problem of learning an embedding function into a regression framework that facilitates efficient computations. The proposed feature extraction approach is named discrete nonseparable shearlet transform gray-level co-occurrence matrix kernel spectral regression (DNST-GLCM-KSR), which combines multi-scale and multi-directional features of DNST with texture features of GLCM and uses KSR to remove redundant features. The DNST-GLCM-KSR approach can improve the defect recognition accuracy of continuous casting slabs and achieved better performance than traditional methods.

The novelty of our work lies in introducing DNST into the surface defect recognition of continuous casting slabs with the complex backgrounds, fusing GLCM texture features, and using a suitable dimensionality reduction algorithm KSR, which makes defect recognition easier and more effective. The rest of this paper is organized as follows. In Section 2, the surface defects information of continuous casting slabs is depicted. Section 3 introduces the basic principles of the DNST-GLCM-KSR feature extraction approach. The surface defect recognition algorithm is presented in Section 4. Section 5 describes the experimental results and discussions, followed by conclusions in Section 6.

2. The Characteristics of Defect Images

The surface temperature of continuous casting slabs is very high during the production process. The temperature can reach about 800~900 °C, which results in the surface being oxidized and forming a large number of various shapes scales [1]. The scales seriously interfere with the detection and recognition of defects. At the same time, due to insufficient illumination in the production site and the rough surface of continuous casting slabs, the slab surface appears uneven. Also, in the continuous casting slab production process, the quality of surface images collected is decreased due to the splash of cooling water, rolling mill vibrations, and other factors. Figure 1 shows several common surface images of continuous casting slabs, including cracks, scales, lighting variations, and slag marks. The cracks are true defect, while the scales, lighting variations, and slag marks are the interference factors that may lead to misclassification. The interference factors are labeled as false defects and also as a type of recognition object. The main task of continuous casting slabs is to recognize crack defects in the complicated background and interference factors.

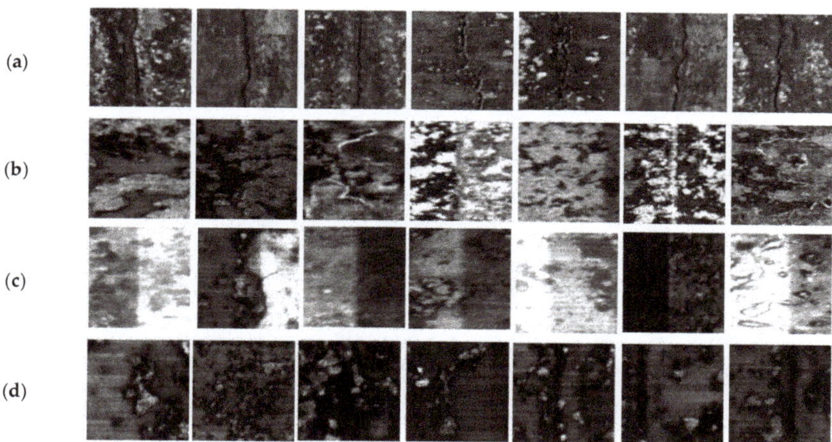

Figure 1. Surface images of continuous casting slabs. (**a**) Cracks; (**b**) Scales; (**c**) Lighting variations; (**d**) Slag marks.

Figure 1a is longitudinal crack defect sample, abbreviated as cracks. The cracks are mainly along the longitudinal distribution of the slabs, the shape is curved, and the length ranges from a few centimeters to dozens of centimeters. The cracks usually have a certain depth, which shows a different grayscale value than the surrounding pixels in the intense light irradiation. The occurrence of cracks is mostly related to the high-temperature steel and various mechanical behaviors in the solidification process. A crack is a very serious defect.

Figure 1b is a scales sample. The shape of scales is uncertain and varies greatly. Sometimes, some of the scales are warped, but most scales are attached to the surface of the slabs and show texture features. Due to the scales covering the surface of the slabs, some fine crack defects are hard to identify.

Figure 1c is the lighting variations. The bright and dark areas are caused by the irradiation of multiple light sources or changes in illumination intensity. The boundaries of the bright and dark areas shcw straight line shapes, and the gray values on both sides are noticeably different. Therefore, these boundaries are sometimes misclassified as cracks.

Figure 1d is the slag marks. They are mainly formed by the residual slag. Slag marks are also distributed along the longitudinal direction of the slabs with a certain width, and their gray values are lower than that of the surrounding pixels. These images are easily misclassified as crack defects.

3. Basic Principles of the Proposed Method

The feature extraction method is the core of the defect recognition algorithm. The quality of feature extraction directly affects the results of defect recognition. The proposed feature extraction scheme DNST-GLCM-KSR in this paper utilizes three technologies, including DNST, GLCM, and KSR, which are described in detail as follows.

3.1. Discrete Nonseparable Shearlet Transform

The wavelet analysis method has been applied in many fields due to its advantages of multi-scale decomposition and fast computation speed. However, the drawback of the wavelet is that the direction decomposition is insufficient. It can only decompose the horizontal, vertical, and diagonal directions. In order to compensate for the deficiency of directional decomposition, the multi-scale geometric analysis (MGA) method was proposed. Typical multi-scale geometric analysis methods include Ridgelet transform [20], Curvelet transform [21], Contourlet transform [22], Shearlet transform [23], and Bandelet transform [24]. The most widely used ones are Contourlet transform and Shearlet transform, while the other methods are limited by their slow computational speed. The advantage of Contourlet is fast computational speed, but its direction representation is limited. The computational speed of Shearlet transform is slower than that of Contourlet, but the direction representation is more flexible. The continuous casting slabs run very slowly on the track, with a speed less than 1 m/s, so the computational speed of the defect recognition algorithm is not required to be fast. The discrete nonseparable shearlet transform (DNSTS) [13,14] is a new kind of shearlet transform.

W.Q. Lim [13] proposed discrete nonseparable shearlet transform (DNST) based on the discrete frame. DNST is constructed from a 2D nonseparable fan filter (improved directional selectivity) and a separable compactly supported shearlet generator (excellent localization properties). It is a direction representation system that extends the wavelet frame. DNST exhibits the same advantages wavelet, namely a unified treatment of the continuum and digital situation.

Two-dimensional discrete shearlet transform is usually obtained using a cone adapted discrete shearlet system defined by scaling functions ϕ_m and shearlet functions $\psi_{j,k,m}$, $\widetilde{\psi}_{j,k,m}$ (by swapping the order of two variables of $\psi_{j,k,m}$).

$$\left\{ \phi_m : m \in \mathbb{Z}^2 \right\} \cup \left\{ \psi_{j,k,m}, \widetilde{\psi}_{j,k,m} : j \in \mathbb{Z}, j \geq 0, |k| \leq 2^{\frac{j}{2}}; m \in \mathbb{Z}^2 \right\} \tag{1}$$

where j is scaling parameter, m is translation parameter, and k is shear (direction) parameter.

Non-separable generator ψ^{non} is defined as follows:

$$\hat{\psi}^{non}(\xi) = P(\xi_1/2, \xi_2)\hat{\psi}(\xi) \tag{2}$$

where the trigonometric polynomial P is a 2D fan filter, which can improve directional selectivity in the frequency domain at each scale. ψ is the 2D separable shearlet generator. which can provide excellent localization properties. $\hat{\psi}$ is the Fourier transform of ψ. The nonseparable shearlets $\psi^{non}_{j,k,m}(x)$ generated by ψ^{non} by setting

$$\psi^{non}_{j,k,m}(x) = 2^{\frac{3}{4}j}\psi^{non}\left(S_k A_{2^j} x - M_{c_j} m\right) \tag{3}$$

where A_{2^j} is parabolic scaling matrix, S_k is shear matrix, and M_{c_j} is a sampling matrix given by $M_{c_j} = \mathrm{diag}(c_1^j, c_2^j)$. We only discuss the case of the shearlet coefficients associated with A_{2^j} and S_k; the same procedure can be applied to compute the shearlet coefficients associated with $\widetilde{A}_{2^j}$ and $\widetilde{S}_k$. After faithfully digitizing $\psi^{non}_{j,k,m}$, the digital formulation of the discrete nonseparable shearlet transform (DNST) is given by

$$DNST_{j,k,m}(f_J) = \left(f_J * \overline{\psi}^{d}_{j,k}\right)\left(2^J A^{-1}_{2^j} M_{c_j} m\right) \text{ for } j = 0, \ldots, J-1 \tag{4}$$

where f_J is the scaling coefficients. $\psi^{d}_{j,k}$ is digital shearlet filters, $\psi^{d}_{j,k} = S^{d}_{k2^{-j/2}}(p_j * W_j)$, $S^{d}_{k2^{-j/2}}$ is the discrete shear operator, p_j is the Fourier coefficients of P, and $W_j = g_{J-j} \otimes h_{J-j/2}$, g_{J-j} and $h_{J-j/2}$ are 1D filters. Please refer to reference [13] for details.

The frequency tiling induced by such discrete shearlet system is shown in Figure 2a, where $\hat{\phi}$, $\hat{\psi}^{non}$ and $\hat{\psi}^{non}$ are associated with the square in the center, the horizontal cone (white), and the vertical cone (yellow), respectively. Each scale corresponds to a ring of tiles and shear is associated with a pair of tiles in a certain direction within the ring. With a proper choice of the parameters associated with the translation, the DNST is obtained as a series of filtering operations. Each shearlet function has two symmetric tiles. The magnitude of the shearlet filter is shown in Figure 2b,c. Figure 2b is the frequency tiles of a DNST filter corresponding to the first scale. Figure 2c is the frequency tiles of a DNST filter corresponding to the second scale.

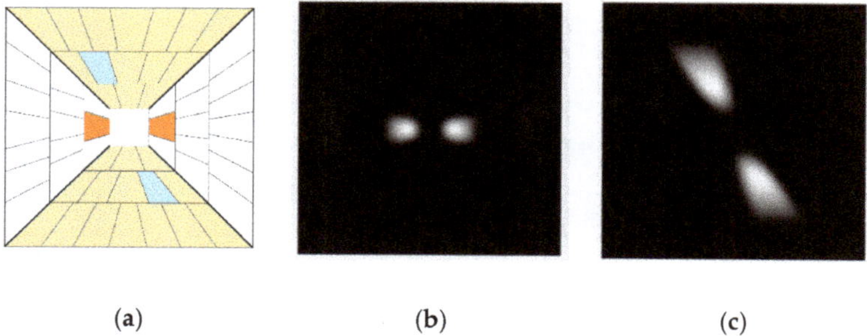

(a)   (b)   (c)

Figure 2. (**a**) The tiling of the ideal frequency plane by a cone adapted shearlet system. (**b**) The frequency tiles of a discrete nonseparable shearlet transform (DNST) filter corresponding to the first scale. (**c**) The frequency tiles of a DNST filter corresponding to the second scale.

3.2. Gray-Level Co-Occurrence Matrix

Gray-level co-occurrence matrix (GLCM) is an effective texture feature extraction approach. GLCM considers not only the distribution of intensities but also the relative positions of pixels in an image [25]. Let Q be an operator that defines the position of two pixels relative to each other, and consider an image, I_f, with possible intensity levels. Let G be a matrix whose element is the number of times that pixel pairs with intensities z_v and z_h occur in I_f in the position specified by Q. A matrix formed in this manner is referred to as a gray-level co-occurrence matrix. Generally, GLCM is not directly regarded as a texture feature, but it is represented by some descriptors such as energy, contrast, entropy, homogeneity, and correlation.

3.3. Kernel Spectral Regression

Kernel spectral regression (KSR) is a dimensionality reduction method based on manifold learning and subspace [18]. The KSR assumes that the original data is embedded in the low-dimensional manifold of the high-dimensional observation space, and each sample is kept adjacent to it by the manifold learning algorithm, so as to mine the low-dimensional manifold structure contained in the high-dimensional data. KSR only needs to solve a set of regularized least squares problems, which results in huge savings of both time and memory. KSR can make efficient use of label and local neighborhood information to discover the intrinsic discriminant structure in the data. The algorithmic procedure is stated below.

(1) Constructing the adjacency graph. Let G denote a graph with m nodes. The *i-th* node corresponds to the sample x_i. If x_i shares the same label with x_j, put an edge between nodes i and j.

(2) Choosing the weights:

$$w_{ij} = \begin{cases} \frac{1}{l_k}, & \textit{if } x_i \textit{ and } x_j \textit{ both belong to the } k - th \textit{ class} \\ 0 & \textit{otherwise} \end{cases} \tag{5}$$

where W_{ij} is the weight of the edge joining vertices i and j.

(3) Responses generation. Find $y_0, y_1, \cdots, y_{c-1}$, the largest c generalized eigenvectors of eigenproblem.

$$\mathbf{Wy} = \lambda \mathbf{Dy} \tag{6}$$

where D is a diagonal matrix whose (i, i)-*th* element equals to the sum of the *i-th* column of W, c is the number of classes.

(4) Regularized kernel least squares. Find $c - 1$ vectors $\alpha_1, \dots, \alpha_{c-1} \in R^m$. $\alpha_k (k = 1, \cdots c - 1)$ is the solution the linear equations system.

$$(\mathbf{K} + \alpha \mathbf{I}) \alpha_k = \mathbf{y}_k \tag{7}$$

where $K_{ij} = K(x_i, x_j)$. It can be easily verified that function $f(x) = \sum_{i=1}^{m} a_i^k K(x, x_i)$ is the solution of the following regularized kernel least square problem:

$$\min_{f \in H_K} \sum_{i=1}^{m} \left(f(x_i) - y_i^k \right)^2 + \alpha \| f \|_k^2 \tag{8}$$

(5) KSR Embedding: Let $\Theta = [\alpha_1, \alpha_2, \cdots, \alpha_{c-1}]$, Θ is a $m \times (c - 1)$ transformation matrix. The samples can be embedded into $c - 1$ dimensional subspace by

$$x \to z = \Theta^T K(:, x) \tag{9}$$

where $K(:, x) = [K(x_1, x), K(x_2, x), \cdots, K(x_m, x)]^T$.

4. Defect Recognition Algorithm

The defect recognition algorithm is the core of the surface quality inspection system. Generally, defect recognition algorithm consists of image preprocessing, image feature extraction, and image classification. In order to obtain more comprehensive information of the surface images of continuous casting slabs, we do not carry out image preprocess. Figure 3 is the schematic diagram of the defect recognition algorithm. The details are as follows.

(1) Extract DNST features. All sample images are decomposed into multiple subbands at different scales and different directions by DNST, and the coefficients matrices of subbands of horizontal and vertical cones are obtained. Then calculate the mean and variance statistics of each subband for all samples. The mean reflects the average value of the subband coefficient matrix, and the variance reflects the deviation between the subband coefficient and the average value of the subband coefficients. The DNST feature of each sample image is a one-dimensional long vector.

(2) Extract texture features. The gray level of the original image is compressed to reduce the computational complexity. The directions choose $0°$, $45°$, $90°$, and $135°$. The texture descriptors energy, contrast, entropy, homogeneity, and correlation are calculated for four directions,

respectively. Then, the mean and variance of each texture descriptor are calculated. Thus, each sample image can get a 10-dimensional GLCM texture feature vector.

(3) Feature combination. The DNST and GLCM feature vector of each sample are spliced into a new one-dimensional long vector. Then the feature vector of each sample are arranged from top to bottom to form the feature matrix. Finally, the feature matrix is normalized to [−1, 1].

(4) Dimensionality reduction. Since the extracted features are high-dimensional and redundant, which is not conducive to subsequent classification. Therefore, we first use the KSR to reduce the dimensionality of the feature matrix of the training set and use the projection matrix obtained from the training set to reduce the dimensionality of the feature matrix of the test set. The high-dimensional feature matrix reduced to $c − 1$ dimensional subspace, where c is the number of classes.

(5) Defect classification. First, the low-dimensional feature matrix is normalized to [−1, 1]. Then, the low dimensional features and labels data are input into the SVM [26] for training and classification. Finally, the surface defects of the continuous casting slabs are recognized.

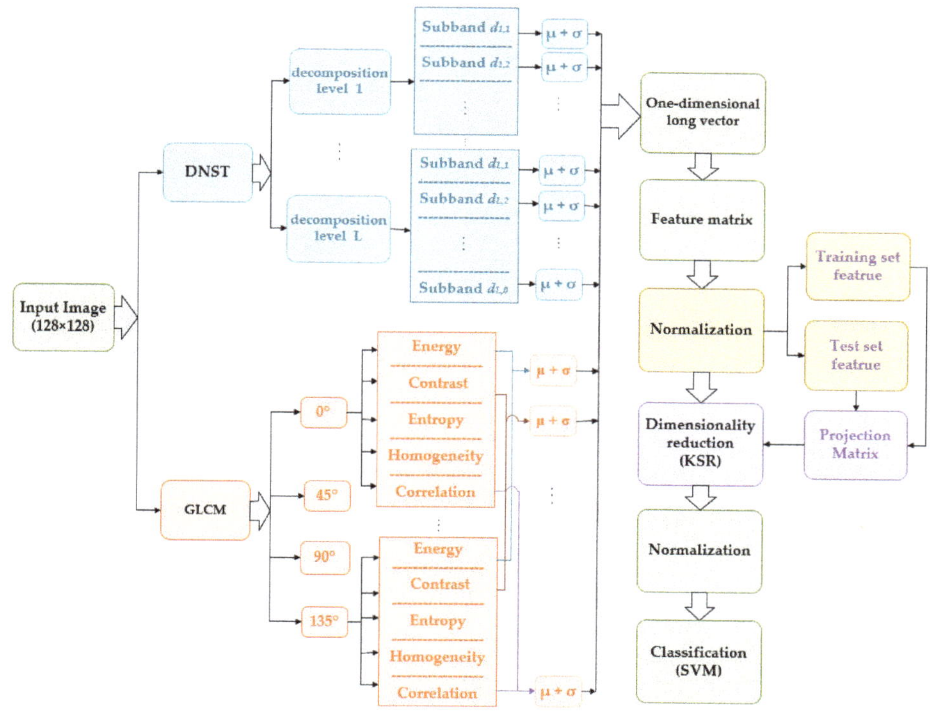

Figure 3. Schematic diagram of defect recognition algorithm.

5. Experiments and Discussions

In this section, the authors introduce the sample database in the experiment in Section 5.1. Some important parameters settings are explained in Section 5.2. In Section 5.3, the feature extraction results of DNST are presented and compared with three commonly used multi-scale methods and one texture extraction method. In Section 5.4, the experimental results of the proposed scheme are presented and compared with other feature combination schemes. The advantages of our proposed scheme in classification time and accuracy are discussed in detail in Section 5.5. Finally, we analyze the specific classification and visualization results of the proposed schemes.

5.1. Sample Database

The samples were collected by an online surface inspection system on a continuous casting slabs production line in a steel plant. The defects database consists of 496 samples, which are divided into two types—positive samples and negative samples. The positive samples have crack defects, with 222 samples. The negative samples include three types of images—scales, lighting variations, and slag marks, with 274 samples. The cracks are defects, and the other three types of samples are pseudo-defects. The pseudo-defects are the main factor of false classification, so the pseudo-defects are labeled as a type of samples. The odd numbers of samples were used for the training set, and the even numbers of samples were used for the test set. All sample images are cropped to 128 by 128 pixels for classification.

5.2. Parameter Setting

To test the feature extraction performance of the proposed scheme, the proposed scheme is compared with wavelet, Contourlet, DST, GLCM, etc. Some important parameters are listed as follows.

DNST: The parameters chose [1 2], [0 1 2], etc. Take an example to explain parameters. When the parameter is set to [1 2], the first scale has eight shearlet direction filters, and the second scale has 16 direction filters, which produced a total of 24 high-frequency subbands and one low frequency subband. The number of DNST features is $(8 + 16 + 1) \times 2 = 50$.

GLCM: The gray level of the original image was compressed to 8, 16, 24 levels, etc. The distance parameter was set to 1.

Wavelet: The wavelet type chose "Haar," "db2," etc. The decomposition level was set to 2,3,4.

Contourlet: The Laplacian filter chose "9-7," and the directional filter chose "pkva." The direction parameter chose [2 3 4], [3 4 5], etc.

DST: The scale filter chose "Symmlet" with the fourth-order vanishing moment, scale parameter was set to [2 1 1], [2 1 1 0], etc. The directional parameters were set to [2 2 2], [2 2 3 3], etc.

KSR: The kernel type chose "Gaussian," and the kernel parameter was set to 0.001.

We chose the radial basis kernel (RBF) as the kernel function of the SVM classifier, and the kernel parameter gamma γ was set to iterate through all values from 0 to 4 with step length 0.01. The other parameters took the default values. Our experiments were based on using a ThinkPad E440 PC equipped with a 2.29 GHz Intel i7 processor and 8GB of RAM. The application software is MATLAB published by MathWork company.

5.3. Comparison of DNST Feature

Numerous experiments on each method were carried out and took the best value and average value as the objective basis of the comparison. The experimental results are shown in Figure 4. From Figure 4, the DNST feature achieves the highest classification results. The best accuracy is 89.92%, and the average accuracy is 89.36%. The Contourlet and DST schemes obtained the same recognition result. The average accuracy of DNST is 1.34% higher than that of DST and Contourlet, 3.15% higher than that of wavelet, and 5.97% higher than that of GLCM. This is due to the fact that DNST has excellent spatial localization properties and directional selectivity, and it can capture defects features more accurately for continuous casting slabs. The classification accuracy obtained by GLCM is the lowest, which indicates that only extracting texture information is not sufficient to represent features of the continuous casting slabs. Figure 4 also shows that the four multi-scale multi-directional feature extraction methods are superior to GLCM. Besides, the difference between the best accuracy and the average accuracy of the wavelet is $88.31\% - 86.21\% = 2.1\%$, which indicates that the wavelet is less robust to parameter changes. The difference between the best accuracy and average accuracy of DNST is 89.92% − 89.36% = 0.56%, the value is lowest, which shows the DNST has better robustness for parameter changes.

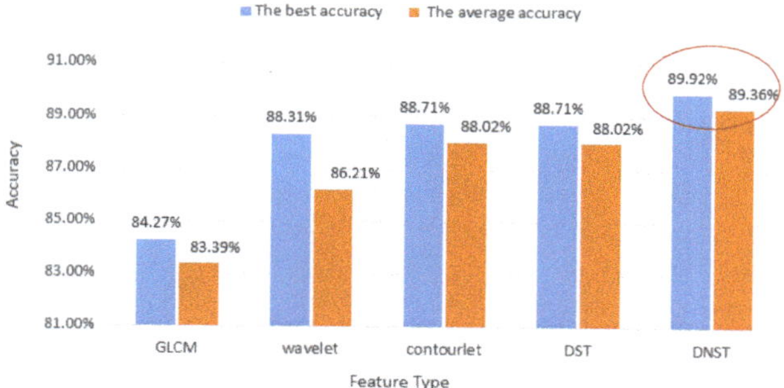

Figure 4. Classification results of different features.

5.4. Comparison of Feature Combination

In addition to the classification accuracy evaluation metrics, the other commonly used evaluation metrics include recall, false positive rate (FPR), F-measurement, precision, the area under the receiver operating characteristic (ROC) curve (AUC), etc. [27]. The lower value of the FPR metrics indicates the better feature extraction performance, while the higher value of the other metrics indicates the better feature extraction performance.

In order to verify the superiority of the proposed scheme DNST-GLCM-KSR, we compared it with the Contourlet-KLPP scheme in reference [1], the DST-KLPP scheme in reference [2], Contourlet-GLCM-KLPP in reference [11], as well as with several similar combination feature schemes. Table 1 lists classification results of some evaluation metrics. From Table 1, the DNST-GLCM-KSR scheme achieves the lowest value of FPR, the highest values of precision, F-measure, AUC, and accuracy, indicating that the proposed scheme obtained the best comprehensive performance in the seven schemes. Besides, when only extracting DNST features, using the KLPP algorithm to reduce dimension can achieve better metrics than that of using KSR except AUC metrics. When extracting DNST-GLCM features, using KSR algorithm to reduce dimension can achieve better metrics than that of using KLPP. The above shows that both dimensionality reduction technologies are effective, and which one is better to use depends on experiments. The principle of the two technologies is the same, but the calculation technic is different.

Table 1. Results of different evaluation metrics.

Feature	FPR (%)	Precision (%)	F-Measure (%)	AUC	Accuracy (%)
Contourlet-KLPP	6.57	92.04	92.79	0.961	93.55
Contourlet-GLCM-KLPP	5.11	93.69	94.09	0.967	94.35
DST-KLPP	5.11	93.69	94.09	0.972	94.35
DNST-KSR	6.57	92.17	93.25	**0.976**	94.35
DNST-KLPP	5.84	93.04	94.09	**0.973**	95.16
DNST-GLCM-KLPP	3.65	95.41	95.15	0.979	95.16
DNST-GLCM-KSR	**2.92**	**96.36**	**96.37**	**0.980**	**96.37**

Taking the accuracy metrics as an example, DNST-GLCM-KSR scheme achieved the highest accuracy of 96.37%, which is 2.82% higher than that of reference [1] and 2.02% higher than that of [2] and [11], indicating that the proposed scheme is superior to the traditional ones. When using the same dimensionality reduction technology, the accuracy of Contourlet-KLPP is 93.55%, DST-KLPP is 94.35%, and DNST-KLPP is 95.16%, which indicates that the DNST can extract more discriminant features of continuous casting slabs than that of Contourlet and DST. The above results show that DNST-GLCM-KSR is an excellent feature fusion approach for continuous casting slabs.

5.5. Comparisons of Dimensionality Reduction

Table 2 lists the results of different combined features of DNST. It can be seen that the accuracy of the training set and the test set of DNST-GLCM is 96.77% and 90.73% respectively, the accuracy of the training set and the test set of DNST is 93.55% and 89.92% respectively, and both results of DNST-GLCM are higher than those of DNST, which indicates DNST feature combined with GLCM texture features can improve the recognition accuracy. In addition, the accuracy of DNST-KSR is 94.35% − 89.92% = 4.43% higher than that of DNST, and the accuracy of DNST-GLCM-KSR is 96.37% − 90.73% = 5.64% higher than that of DNST-GLCM. The above shows that KSR can effectively remove redundancy and interference features and improve recognition accuracy. At the same time, we noticed that the classification time was shortened from tens of seconds to several seconds by KSR. This is because KSR reduces feature dimensionality to $c - 1$, where c is the number of classes. The continuous casting slabs samples include positive samples and negative samples—that is to say, the number of classes is 2. The feature number was reduced to 1 dimensionality. Finally, it should be noted that the number of subbands by DNST decomposition is different when different feature combinations achieve the highest recognition accuracy.

Table 2. Result of feature combination.

Feature Type	The Number of Features	Time Cost (ms)	Accuracy (%)	
		Classification	Training Set	Test Set
DNST	66	33.10	**93.55**	89.92
DNST-GLCM	36	25.04	**96.77**	90.73
DNST-KSR	1	3.39	96.37	94.35
DNST-GLCM-KSR	**1**	**3.13**	**97.58**	**96.37**

Figure 5 shows the recognition accuracy of the training set and test set when using KSR and not using KSR. With the increase of SVM kernel parameter γ, the accuracy of the DNST-GLCM training set gradually increases and is finally close to 100%, while the accuracy of the test set first increases and then decreases, which indicates that DNST-GLCM feature data is sensitive to SVM kernel parameter; in other words, the feature is complex and not conducive to classifier learning. For DNST-GLCM-KSR, the accuracy of test set and training set are both high, and the curve fluctuation are small. The above results show that the DNST-GLCM-KSR scheme makes the extracted features more discriminative, easier to learn and classify, and it has a strong robustness to the kernel parameter of SVM.

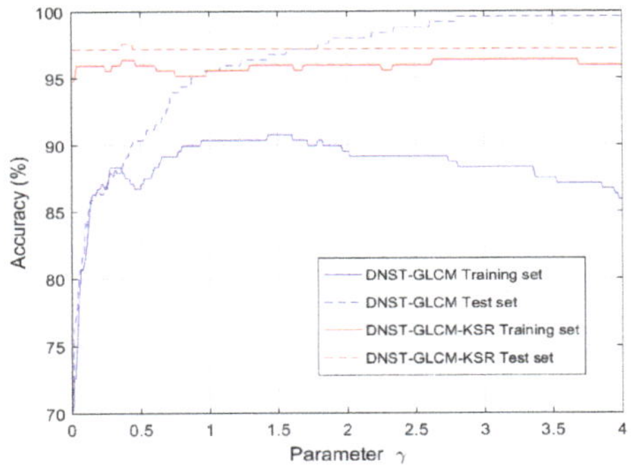

Figure 5. The recognition accuracy of training set and test set.

5.6. Confusion Matrix and Visualization

Table 3 lists the detailed classification of test set of DNST-GLCM-KSR scheme. The accuracy of positive sample (crack defect) is 95.50%; the accuracy of negative sample is 97.08%. There are four pseudo-defects misclassified as crack defects, and five crack defects are misclassified as pseudo-defects. This was due to the fact that inter-class defects have similar aspects in appearance. The false alarm rate of cracks is 4/110 = 3.64%, and the missing alarm rate is 5/111 = 4.5%. The above metrics meet the needs of engineering application of continuous casting slabs.

Table 3. The confusion matrix of discrete nonseparable shearlet transform gray-level co-occurrence matrix kernel spectral regression (DNST-GLCM-KSR) scheme.

Sample Type	Positive Sample	Negative Sample	Correct Number	Sample Total	Accuracy
Positive sample	106	5	106	111	95.50%
Negative sample	4	133	133	137	97.08%
total	110	138	239	248	96.37%

Figure 6 shows the visualization of the DNST-GLCM-KSR test set. The result is a straight line, because the number of the feature dimensionality is 1. The pink circle represents the positive sample, the blue one represents the negative sample, and the upper left corner is a local enlargement image. The result of Figure 6 is consistent with that of Table 3, namely that four pseudo-defects are misclassified as crack and five crack defects are misclassified as pseudo-defects. From the graph, we can see that the DNST-GLCM-KSR features truly reflect the similarity between defect images. The intra-class scatter is small, and the inter-class scatter is large. The above analysis shows the method proposed can effectively recognize crack defects of continuous casting slabs in complex background images and interference factors.

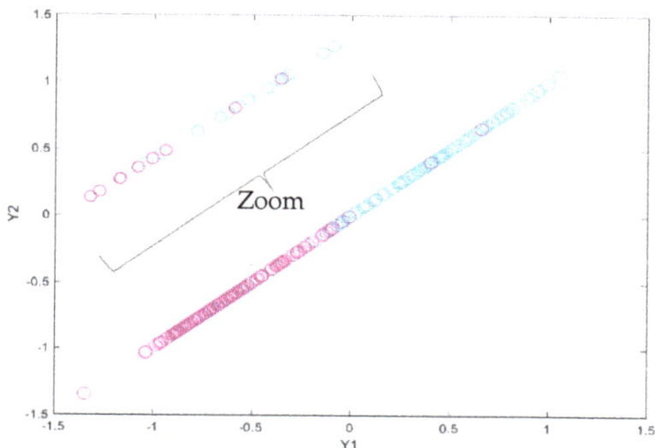

Figure 6. Visualization results.

6. Conclusions

According to the direction and texture information of surface defects of the continuous casting slabs with complex backgrounds, a new feature extraction approach DNST-GLCM-KSR is proposed, which combines multi-scale and multi-directional DNST features with GLCM texture features and uses KSR technology to reduce dimensionality. The experimental results are as follows.

(1)   The DNST feature obtained the highest average accuracy and the best accuracy. It can better characterize defects of continuous casting slabs than that of Contourlet, DST, wavelet, and GLCM.

(2)   The accuracy of the training set and test set of the DNST-GLCM feature were 96.77% and 90.73%, respectively. Both results were higher than those of DNST feature. The recognition accuracy of continuous casting slabs can be improved by combining the features of DNST and GLCM.

(3)   The recognition accuracy of the DNST-GLCM-KSR scheme is 5.64% higher than that of DNST-GLCM, and the classification time of DNST-GLCM-KSR was shorter than that of DNST-GLCM. Using KSR technology can improve recognition accuracy and shorten classification time.

(4)   The proposed scheme can extract more discriminative features of defects and make the recognition accuracy of crack defect up to 95.50% and the total accuracy up to 96.37%. The new scheme provides a new method for the surface defect recognition of continuous casting slabs.

(5)   Future work should collect more defect samples, establish a complete sample database, and improve the recognition accuracy of crack defect.

Author Contributions: K.X. contributed to the methodology of the study and investigation. X.L. contributed significantly to formal analysis and manuscript writing. P.Z. and H.L. helped perform the analysis with constructive discussions.

Funding: This research was funded by the National Natural Science Foundation of China (grant number 51674031 and 51874022) and the Fundamental Research Finds for the Central Universities under Grant FRF-TP-18-009A2.

Conflicts of Interest: The authors declare there is no conflicts of interest regarding the publication of this paper.

References

1. Ai, Y.; Xu, K. Feature extraction based on contourlet transform and its application to surface inspection of metals. *Opt. Eng.* **2012**, *51*, 113605. [CrossRef]
2. Xu, K.; Liu, S.; Ai, Y. Application of shearlet transform to classification of surface defects for metals. *Image Vis. Comput.* **2015**, *35*, 23–30. [CrossRef]
3. Wei, S.; Zheng, S.; Gong, H.; Tang, H. Study on On-line Detection of the Cast-billet Surface Defect. *J. Wuhan Univ. Technol. (Inf. Manag. Eng.)* **2009**, *31*, 433–436.
4. Cai, K.; Sun, Y.; Han, C. The "Zero Defect" Philosophy of Controlling Strand Quality for Steel Continuous Casting. *Contin. Cast.* **2011**, *s1*, 288–298.
5. Zhang, Y.; Cheng, W.S.; Zhao, J. Classification of surface defects of strips based on invariable moment functions. *Opt. Electron. Eng.* **2008**, *35*, 90.
6. Xu, K.; Yang, C.L.; Zhou, P. Technology of on-line surface inspection for hot-rolled steel strips and its industrial application. *J. Mech. Eng.* **2009**, *45*, 111. [CrossRef]
7. Song, K.; Yan, Y. A noise robust method based on completed local binary patterns for hot-rolled steel strip surface defects. *Appl. Surf. Sci.* **2013**, *285*, 858–864. [CrossRef]
8. Ban, J.; Seo, M.; Goh, T.; Jeong, H.; Kim, S.W.; Koo, G.; Shin, C.; Choi, H.; Lee, J.-H.; Kim, S.W.; et al. Automatic detection of cracks in raw steel block using Gabor filter optimized by univariate dynamic encoding algorithm for searches (uDEAS). *NDT E Int.* **2009**, *42*, 389.
9. Pan, E.; Ye, L.; Shi, J.; Chang, T.-S. On-line bleeds detection in continuous casting processes using engineering-driven rule-based algorithm. *J. Manuf. Sci. Eng.* **2009**, *131*, 061008. [CrossRef]
10. Xu, K.; Yang, C.L.; Zhou, P.; Yang, C. On-line detection technique of surface cracks for continuous casting slabs based on linear lasers. *J. Univ. Sci. Technol. Beijing* **2009**, *31*, 1620.
11. Ke, X.U.; Yang, C.L.; Peng, Z.; Yang, C. Recognition of surface deffects in continuous casting slabs based on Contourlet transform. *J. Univ. Sci. Technol. Beijing* **2013**, *35*, 1195–1200.
12. Tian, S.-Y.; Xu, K.; Guo, H.-Z. Application of local binary patterns to surface defect recognition of continuous casting slabs. *Chin. J. Eng.* **2016**. [CrossRef]
13. Lim, W.-Q. Nonseparable shearlet transform. *IEEE Trans. Image Process.* **2013**, *22*, 2056–2065. [PubMed]
14. Kutyniok, G.; Lim, W.Q.; Reisenhofer, R. ShearLab 3D: Faithful Digital Shearlet Transforms based on Compactly Supported Shearlets. *arXiv* **2014**, arXiv:1402.5670. [CrossRef]
15. Kang, R.; Cao, B.; Wang, B.; Yan, L.; Qu, G. Compressed sensing longitudinal MRI based on wavelet and shearlet. *J. Beijing Jiaotong Univ.* **2017**, *41*, 127–132.

16. Pejoski, S.; Kafedziski, V.; Gleich, D. Compressed Sensing MRI Using Discrete Nonseparable Shearlet Transform and FISTA. *IEEE Signal Process. Lett.* **2015**, *22*, 1566–1570. [CrossRef]

17. Liu, L.; Kuang, G.Y. Overview of image textural feature extraction methods. *J. Image Graph.* **2009**, *4*, 622–635.

18. Cai, D.; He, X.; Han, J. *Spectral Regression for Dimensionality Reduction*; Technical Report; UIUC: Champaign, IL, USA, 2007.

19. Cai, D. Spectral Regression: A Regression Framework for Efficient Regularized Subspace Learning. Ph.D. Thesis, UIUC, Champaign, IL, USA, 2009.

20. Donoho, D.L. Orthonormal ridgelets and linear singularities. *Siam J. Math. Anal.* **2000**, *31*, 1062–1099. [CrossRef]

21. Candes, E.; Demanet, L.; Donoho, D.; Ying, L. Fast discrete Curvelet transforms. *Multiscale Model. Simul.* **2006**, *5*, 861. [CrossRef]

22. Po, D.; Do, M. Directional multiscale modeling of images using the contourlet transform. *IEEE Trans. Image Process.* **2006**, *15*, 1620. [CrossRef]

23. Lim, W.Q. The discrete shearlet transform: A new directional transform and compactly supported shearlet frames. *IEEE Trans. Image Process.* **2010**, *19*, 1166–1180. [PubMed]

24. Pennec, E.L.; Mallat, S. Sparse geometric image representations with bandelets. *IEEE Trans. Image Process.* **2005**, *14*, 423–438. [CrossRef] [PubMed]

25. Gonzalez, R.C.; Woods, R.E. *Digital Image Processing*, 3rd ed.; Prentice-Hall, Inc.: Upper Saddle River, NJ, USA, 2007.

26. Chang, C.C.; Lin, C.J. LIBSVM: A library for support vector machines. *ACM Trans. Intell. Syst. Technol.* **2011**, *2*, 1–27. [CrossRef]

27. Zhou, Z. *Machine Learning*; Tsinghua University Press: Beijing, China, 2016.

Article

A Novel Extraction Method for Wildlife Monitoring Images with Wireless Multimedia Sensor Networks (WMSNs)

Wending Liu, Hanxing Liu, Yuan Wang, Xiaorui Zheng and Junguo Zhang *

School of Technology, Beijing Forestry University, Beijing 100083, China; Liu_wending@163.com (W.L.);
liuhanxing@bjfu.edu.cn (H.L.); wangyuan@bjfu.edu.cn (Y.W.); zheng_xiaorui@126.com (X.Z.)
* Correspondence: zhangjunguo@bjfu.edu.cn; Tel.: +86-10-6233-6398

Received: 26 April 2019; Accepted: 30 May 2019; Published: 2 June 2019

Featured Application: The novel extraction method for wildlife monitoring images can achieve the extraction of the foreground region and provide effective support for image progressive transmission in Wireless Multimedia Sensor Networks (WMSNs).

Abstract: In remote areas, wireless multimedia sensor networks (WMSNs) have limited energy, and the data processing of wildlife monitoring images always suffers from energy consumption limitations. Generally, only part of each wildlife image is valuable. Therefore, the above mentioned issue could be avoided by transmitting the target area. Inspired by this transport strategy, in this paper, we propose an image extraction method with a low computational complexity, which can be adapted to extract the target area (i.e., the animal) and its background area according to the characteristics of the image pixels. Specifically, we first reconstruct a color space model via a CIELUV (LUV) color space framework to extract the color parameters. Next, according to the importance of the Hermite polynomial, a Hermite filter is utilized to extract the texture features, which ensures the accuracy of the split extraction of wildlife images. Then, an adaptive mean-shift algorithm is introduced to cluster texture features and color space information, realizing the extraction of the foreground area in the monitoring image. To verify the performance of the algorithm, a demonstration of the extraction of field-captured wildlife images is presented. Further, we conduct a comparative experiment with N-cuts (N-cuts), the existing aggregating super-pixels (SAS) algorithm, and the histogram contrast saliency detection (HCS) algorithm. A comparison of the results shows that the proposed algorithm for monitoring image target area extraction increased the average pixel accuracy by 11.25%, 5.46%, and 10.39%, respectively; improved the relative limit measurement accuracy by 1.83%, 5.28%, and 12.05%, respectively; and increased the average mean intersection over the union by 7.09%, 14.96%, and 19.14%, respectively.

Keywords: wireless multimedia sensor networks; wildlife monitoring image; extraction; Hermite; adaptive mean-shift

---

1. Introduction

As an important part of ecological environment protection, wildlife protection is crucial for maintaining the balance and stability of the whole ecosystem [1]. Presently, the widely used wildlife monitoring methods are mainly based on GPS collar systems [2,3], infrared camera technology [4,5], and remote sensing monitoring technology [6,7]. Although GPS collar systems can obtain wildlife locations with a great accuracy and precision, this monitoring method cannot obtain wildlife images, which poses a problem for researchers. The monitoring image, which can accurately estimate species diversity, population size, and habitat distribution, is an important part of wildlife protection. Thus, the monitoring image can provide a scientific basis for wildlife resource conservation [8]. Infrared camera technology and remote sensing monitoring technology can capture wildlife images, but these methods also have limitations. Infrared camera images need to be saved on secure digital (SD) memory cards, thus resulting in long monitoring delays and high labor costs. Remote sensing systems are capable of obtaining species diversity over a large geographic area, but cannot monitor wildlife at an individual level. In order to avoid these problems, the use of wireless multimedia sensor networks (WMSNs), which are used as monitoring carriers, has received extensive attention from researchers [9].

Wireless sensor nodes usually rely on batteries for their power supply, but this setup is difficult to recharge in remote areas. Reducing the consumed energy and extending the life cycle of sensor networks has recently become a hot topic of research [10]. Meanwhile, the energy consumption of the sensor node is mainly concentrated in the data transmission process. Therefore, the progressive transmission of monitoring images in WMSNs can effectively reduce the energy consumption and improve the life cycle of nodes. At the same time, extraction of the image is the basis of progressive transmission [11,12]. The current research work cannot achieve the extraction of wildlife monitoring images with a complex background and uneven illumination. Against this background, this study proposes an image extraction algorithm with a high accuracy for wildlife monitoring images. In this paper, we propose an extraction method for wildlife monitoring images based on the combination of color space construction and Hermite transform for WMSNs. Firstly, we reconstruct a color space model for utilizing the novel color enhancement method to extract the color parameter. The color enhancement method utilizes a bilateral filter with different kernels to process the luminance components based on the Retinex framework. Then, we construct a filter through Hermite transform to acquire texture information in the wildlife monitoring images. Finally, according to the characteristics of the pixel, an adaptive mean-shift algorithm is presented as an ideal clustering model to implement the extraction of the foreground area in the image. In this work, we took the wildlife of the Saihan Ula Nature Reserve in Inner Mongolia province as targets, and applied the proposed method to extract the wildlife images captured by the WMSN monitoring system. The experimental results showed that the proposed method achieved effective extraction results for the wildlife monitoring images and provided effective support for reducing the power consumption when transmitting wildlife monitoring images.

2. Related Work

Currently, image extraction methods mainly include edge detection, threshold extraction [13,14], the clustering algorithm [15,16], saliency detection [17], and semantic segmentation [18]. Image extraction based on edge detection preserves the edge information of the input image by calculating the derivative between different pixels [19]. However, the algorithm is affected by noise, which might be misjudged as a boundary, thus reducing the edge position accuracy; this is a general disadvantage of the above methods. Threshold extraction, one regional extraction technique, divides pixels into several categories, with the advantage of a low computational complexity [20]. However, this algorithm is applicable to images in which the foreground and background are in different grayscale ranges; the cluster algorithm divides similar features of color space information into a specific group [21]. The saliency detection algorithm extracts the region of the foreground by simulating the visual characteristics of human beings. Considering the characteristics of a high computational complexity, the algorithm may not be applied to real-time applications [22]. Semantic segmentation describes the process of associating each pixel of an image with a class label [23]. The algorithm, based on semantic segmentation, has a high accuracy, but requires a large number of wildlife monitoring images to be marked, resulting in high labor costs. Meanwhile, the high computational complexity of the algorithm makes it unsuitable for WMSNs. It is difficult to extract wildlife images due to complex backgrounds and uneven illumination, which varie significantly in non-uniform illumination in different seasons. Due to the disadvantages of the above extraction methods, Shehu A et al. [24] proposed an edge detection algorithm for the pixel detection of wildlife images captured by sensor nodes through the calculation of the gray threshold and gradient amplitude. To enhance the accuracy of edge detection,

Feng et al. [25] defined an ideology of saliency detection by introducing a positional saliency map. Combined with the edge detection method to extract the images captured by the wireless sensors, this method optimizes the extraction effect and realizes the extraction of wildlife images. Tian et al. [26] segmented multi-colored wildlife images into watershed regions based on watershed transformation. Then, they used the traditional mean-shift algorithm to cluster the watershed regions, which preserves the edge information and effectively suppresses the occurrence of over-segmentation; nevertheless, the traditional mean-shift method based on color space information cannot accurately segment images captured in a complex environment, because it cannot consider the texture parameters in the foreground and background regions. The traditional mean-shift image segmentation algorithm was improved by Akbulut Y et al. [27] in 2018. This method combines texture parameters with color space information and then uses edge-preserving filtering to preserve as much edge information as possible, thus using the mean-shift algorithm to achieve superior image extraction. The algorithm is successful in image extraction within a certain range, but a fixed bandwidth must be set manually, which is not suitable for the real-time extraction of images, and the high computational complexity of the algorithm makes it unsuitable for use in WMSNs with a limited power consumption. We try to utilize the adaptive mean-shift algorithm combined with color information and texture parameters to extract the WMSN wildlife monitoring images.

3. Materials

In this study, a WMSN monitoring system designed by our laboratory was used to perform the task of capturing images of wildlife. The system was deployed in the Saihan Ula National Nature Reserve in the Inner Mongolia province, which is located in the southern mountainous area of the Greater Khingan Range. It is a forest ecological nature reserve, which has a medium-temperate, semi-humid, and warm climate. This reserve is home to 37 primary species of wildlife, including three kinds of secondary, nationally protected mammals, such as *Cervus elaphus*, *Naemorhedus goral*, and *Lynx lynx*.

The WMSN wildlife monitoring system is used to capture wildlife images using industrial-grade cameras with terminal node equipment embedded. The system, which is mostly composed of coordination nodes, terminal nodes, gateway nodes, and servers, achieves real-time, remote monitoring images. The detailed distributions are depicted in Figure 1.

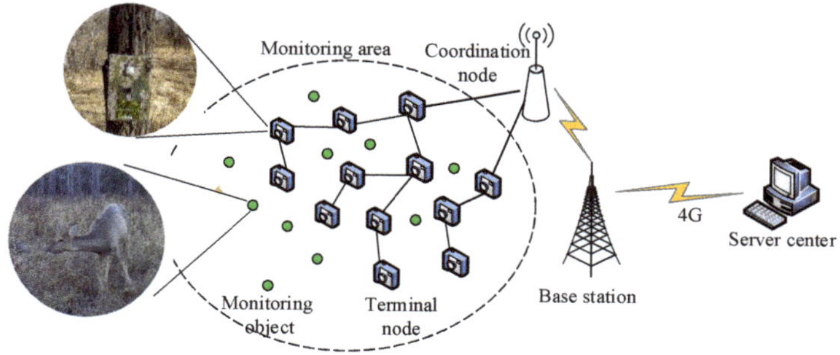

Figure 1. Wildlife monitoring system.

The WMSN node establishes a sensor network in a self-organizing manner by using the ZigBee network protocol. Detailed parameters are given in Table 1. When the wild animals enter the monitoring range, the camera is triggered by the infrared sensor of the terminal node to capture the images. The captured images are then sent to the coordination nodes in a multi-hop manner. After the coordinating nodes successfully receive the image data information, the information is transmitted to the server center through the gateway node in the form of a 4G signal. If there is no target shown in the monitoring region, the nodes stop working to reduce the energy consumption.

Table 1. Parameters of the wireless multimedia sensor network (WMSN) node.

Monitoring Node	Parameter
Camera	OV7725 QVGA 30 fps
Pixel	640 × 480
Memory card	SD 16G
Controller	STM32
Monitoring range	120°/Radius 10 m
Maximum transmission distance	1200 m

More than 20 sensor nodes were deployed in Saihan Ula National Nature Reserve, and the distance between every two sensor nodes was 150 m. In this study, more than 2000 images of 12 species of wildlife were collected in the nature reserve using the monitoring system, including *Cervus elaphus* and *Lynx lynx*, which are nationally protected animals. The monitoring images are shown below in Figure 2.

(a)   (b)   (c)

Figure 2. Wildlife monitoring images: (**a**) *Capreolus pygargus*; (**b**) *Sus scrofa*; (**c**) *Cervus elaphus*.

4. Experimental Methods

A novel image extraction method is proposed to process the wildlife monitoring images captured by the WMSN monitoring system, as depicted in Figure 3. The target region that contains the wildlife is the major object of interest, whereas the background regions only provide reference information. The steps of the algorithm are as follows:

1. Color information is extracted by constructing a color space, which is utilized to solve the problem of uneven illumination in the monitoring images;
2. Texture information is extracted based on Hermite transform to ensure the target region; in other words, the extracted texture information is not affected by illumination or the shooting angle, even under conditions including a complex background and uneven illumination;
3. To guarantee the extraction efficiency, the adaptive mean-shift algorithm is utilized to extract the foreground region when the color and texture information is received.

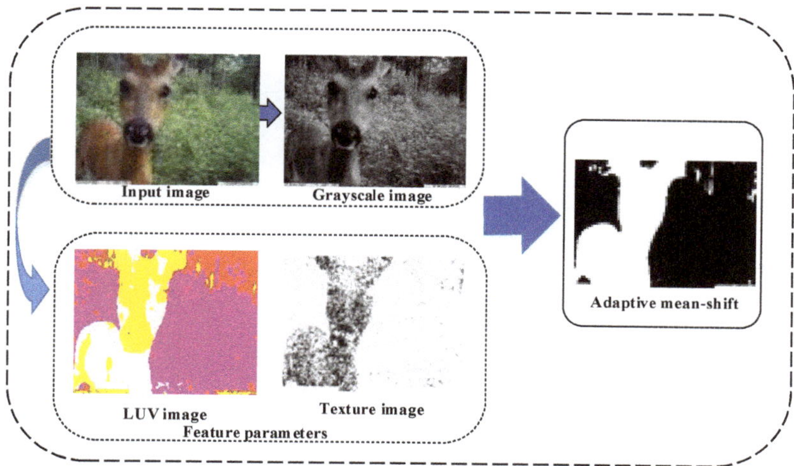

Figure 3. Flow-process diagram of the proposed method; LUV:CIELUV.

4.1. Color Space Information Extraction

A traditional red-green-blue (RGB) color space may not extract the desired color parameters, as RGB color space components always have strong correlations in wildlife monitoring images. Therefore, constructing the color space model to extract color parameters is a significant procedure in our proposed method. However, the weakened quality of the acquired wildlife monitoring images aggravates the difficulty of completing the color space model construction due to the different illumination variations in wild environments.

In our algorithm, the CIELUV (LUV) color space is applied in the output of luminance and chrominance components step, as it has two distinguished advantages over other color spaces. One is that it has non-correlation between color components, and the other is that it has been validated to extract detailed edge regions in the color image [28].

To obtain more detailed color parameters, novel color enhancement based on the Retinex method was introduced into the color space model. The classic Retinex model decomposes images into reflectance and illumination:

$$log(I(x,y)) = log(R(x,y)) + log(L(x,y)) \tag{1}$$

where $I(x,y)$ is the observed pixel in the monitoring image at the location of (x,y) [29,30], $R(x,y)$ is the reflectance, and $L(x,y)$ denotes the illumination of the image.

In the theory of multi-scale Retinex, a plurality of individual convolutions with different Gaussian kernels can be applied to the original $I(x,y)$ to approximate the component of $L(x,y)$ by using different weights, as shown in Equation (2), where σ_i is the Gaussian Kernel modulus, and the sum of the weights w_i is equal to 1.

$$R_{MSR}(x,y) = \sum_{i=1}^{n} w_i[log(I(x,y)) - log(g(\sigma_i) \times I(x,y))]$$
$$g(\sigma_i) = \frac{1}{2\pi\sigma_i^2} e^{-\frac{((x-\mu)^2+(y-\mu)^2)}{2\pi\sigma_i^2}} \tag{2}$$

The novel color enhancement was inspired by the traditional Retinex framework, and it processes the wildlife monitoring images in the LUV color space. The color shift occurs when the traditional Retinex with a Gaussian Filter simultaneously changes the luminance and chrominance. Therefore, the proposed method utilizes a bilateral filter in the luminance channel to avoid color shift, while the traditional Retinex algorithm utilizes Gaussian filtering in different color space channels. The bilateral filters with different kernels are used to process the L channel by the same weights considering the uneven distribution in the spatial domain, as shown in Equation (3), where σ_i is the kernel coefficient.

$$R_{Bilateral-MSR}(x,y) = \sum_{i=1}^{n} w_i[log(I(x,y)) - log(g_{Bilateral}(\sigma_i) \times I(x,y))] \tag{3}$$

The workflow of the proposed method for reconstructing the color space model is denoted in Figure 4, which can be divided into three main parts: (a) color space transformation, (b) the color enhancement with the filter in the middle, and (c) the extraction of the color parameters. The whole process of the method consists of the following steps:

1. Convert the wildlife monitoring images from the RGB color space to the LUV color space;
2. Apply convolutions with a bilateral filter to the L channel of the motoring image to obtain the luminance component $\hat{L}$. The σ_i values of the bilateral filter represent different scales, which are 10, 50, and 100 for three scales;
3. Preserve the original chrominance parameters with the *U* and *V* channel;
4. Reconstruct the color space model with $\hat{L}UV$.

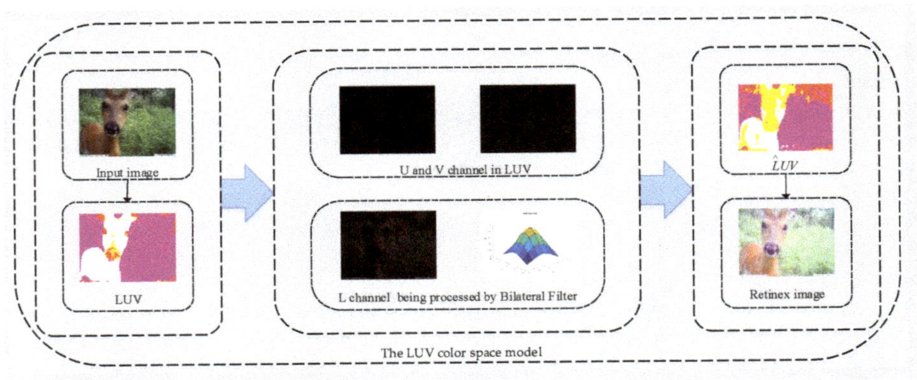

Figure 4. Flow-process diagram of the proposed method.

4.2. Texture Information Extraction

After extracting the color parameters of the wildlife image, we can determine that different areas of the same texture are susceptible to color changes in the image. Here, the texture information is obtained to ensure the integrity of the extraction of the foreground area. In this study, the texture information of the images is extracted by Hermite transform, and the texture information is used as a vital parameter component of the mean-shift algorithm. The continuous Hermite function is defined as follows [31,32]:

$$H_n(x) = (-1)^n e^{x^2/2} \frac{d^n}{dx^n} e^{-x^2/2}, n = 0, 1, 2, \ldots \tag{4}$$

Hermite transform of a signal is defined as

$$f_n(t_0) = \int f(t)H_n(t_0 - t)V_n^2(t_0 - t)dt \tag{5}$$

where $V(t)$ is a Gaussian window function defined as follows:

$$V(t) = \frac{1}{\sqrt{\sigma\sqrt{2\pi}}}e^{-\frac{(t-\mu)^2}{2\sigma^2}} \tag{6}$$

$f_n(t_0)$ is obtained by convolution with the Hermite analytic function $d_n(t)$ by the input signal $f(t)$. This is described in terms of the window and the Hermite polynomials as:

$$d_n(t) = \frac{1}{\sigma\sqrt{2\pi}}H_n(t)e^{-\frac{(t-\mu)^2}{2\sigma^2}} \tag{7}$$

Because the function is rotational symmetry and separable, the one-dimensional Hermite space can be transformed into a two-dimensional space. The formula is as follows:

$$d_{n-m,m}(x,y) = d_{n-m}(x)d_m(y) \tag{8}$$

where $n-m$ and m denote the order in x and y directions, respectively [33]. The filter of $d_{n-m,m}(x,y)$ has the characteristics of continuous attenuation, and is not as steep and discontinuous as the ideal filter, which can effectively preserve the edge information of the image. The perspective of the Hermite analytic function is shown in Figure 5.

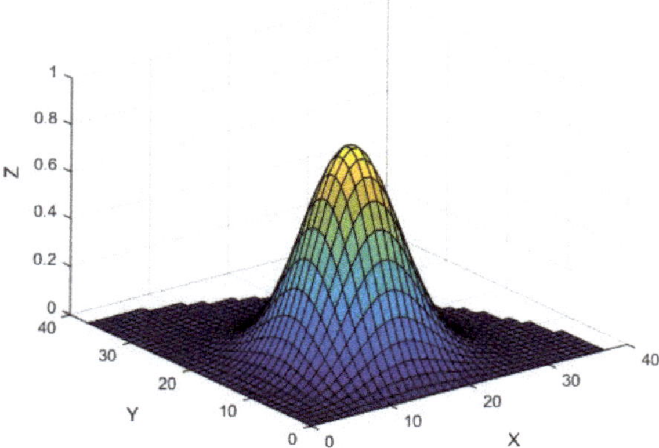

Figure 5. Perspective of Hermite analytic function H_{12}.

Finally, the input image $I(x,y)$ can be transformed into $d_{n-m,m}(x,y)$ as [34]

$$I_{n-m,m}(x_0,y_0) = \iint I(x,y)d_{n-m,m}(x-x_0,y-y_0)dxdy \tag{9}$$

Since all the Hermite analytic functions were obtained by multiplication of the Gaussian window function and Hermite polynomial, in order to reduce the computational complexity of the method, several polynomials were chosen as some of the polynomials were not capable of extracting valuable texture parameters. The steps were as follows:

1. In order to reduce the amount of data calculated, we set the value relationship between m and n as $m+n \leq 5$;
2. Since H_{mn} and H_{nm} are mutually converted matrices, one is arbitrarily selected as the convolution kernel;
3. The polynomial H_{11} was not chosen due to the fact it did not extract any texture information.

Taking the input image of Figure 4 as an example, the image was convolved with different polynomials to obtain different texture parameters, as shown in Figure 6.

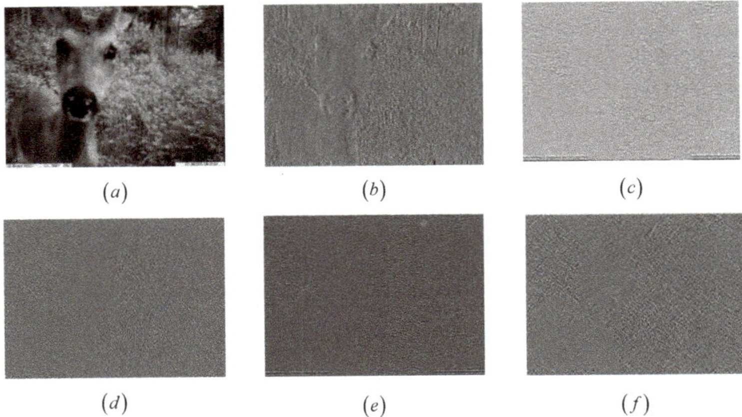

Figure 6. Polynomial convolution texture information; (**a**) H_{11}; (**b**) H_{12}; (**c**) H_{31}; (**d**) H_{23}; (**e**) H_{41}; and (**f**) H_{22}.

As shown in Figure 6, when these polynomials were convolved with the input image, the polynomials H_{12}, H_{22}, H_{23}, H_{31}, and H_{41} were determined to extract useful texture parameters in different directions, in which the texture parameters directed by H_{12}, H_{22}, and H_{31} in Figure 6 were particularly conspicuous; however, H_{11} did not extract any texture information. Instead, a parameter image, which is similar to the grayscale image of the input image, was generated. Thus, we next constructed the filter to extract the texture parameter. The architecture of the proposed method for extracting the Hermite texture parameter is demonstrated in Figure 7.

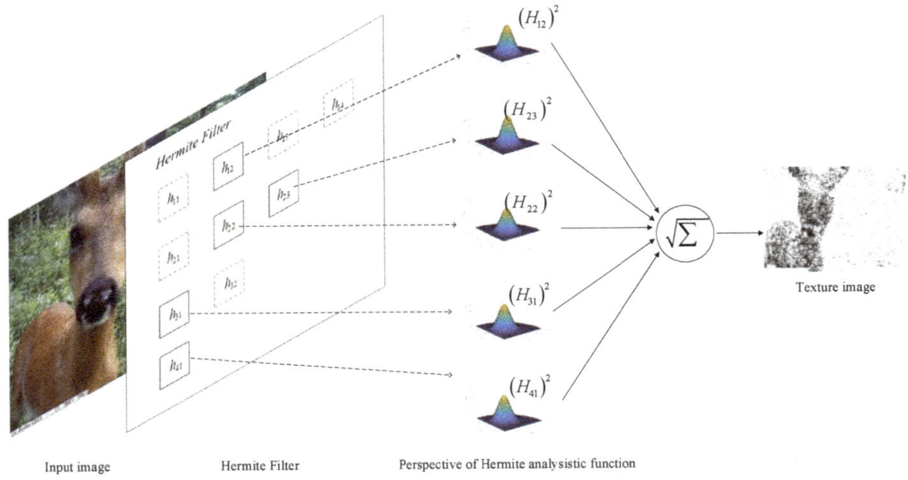

Figure 7. Flow-process diagram of texture extraction.

The extracted texture parameter was saved for further processing to obtain a texture image, as shown in Figure 8.

$$(a) \qquad\qquad\qquad (b)$$

Figure 8. Texture parameters extraction; (**a**) Input image; (**b**) texture image.

4.3. Adaptive Mean-Shift Algorithm

Wildlife monitoring images of different species have different backgrounds and different light intensities. Therefore, after the color and texture parameters were received, we utilized the adaptive mean-shift algorithm to adaptively select the bandwidth according to the pixel characteristics of the wildlife images, which guarantees the extraction quality of the foreground region.

The kernel density estimated with a fixed bandwidth for a set of data points $\{x_i, i = 1, 2, \ldots n\}$ [35,36] in the mean-shift algorithm is defined as

$$p(x) = \frac{1}{nh} \sum_{i=1}^{n} K\left(\frac{x - x_i}{h}\right) \tag{10}$$

where $K(x)$ is a symmetric kernel function with respect to the origin, and the integration of its domain is 1 [37]. The mean shift algorithm usually uses a Gaussian function as a kernel function in which h represents the fixed bandwidth of the core.

In this study, the adaptive mean-shift algorithm was used to cluster pixel data, which means that different sampling data x_i adopted different bandwidths $h = h(x_i)$. The variable bandwidth kernel function density estimate [38] is defined as

$$p(x_i) = \frac{1}{n} \sum_{i=1}^{n} \frac{1}{h(x_i)} K\left(\frac{x - x_i}{h(x_i)}\right) \tag{11}$$

$$h(x_i) = h_0 \times \sqrt{\frac{r}{f(x_i)}} \tag{12}$$

In the above function, the pixel point at the center of the grayscale image was taken as the initial center point, and the bandwidth $h(x_1)$ was calculated from this point.

$$h_0 = \frac{1}{n \times n} \sum_{x=1}^{n} \sum_{y=1}^{n} |M - I(x, y)| \tag{13}$$

where h_0 is the average offset of all pixel values and median M in the image. The probability that a pixel has a gray level of x_i is $f(x_i)$. Then, the scale factor r is defined as

$$logr = \frac{1}{m} \sum log(f(x_i)) \tag{14}$$

where m is the number of gray levels of the image. The center point iteration of the kernel function is given in Figure 9, where two-dimensional Gaussian data was randomly generated as coordinates of the data points. The mean-shift vector shifts to where the sample point changes most, and is also the direction of the density gradient.

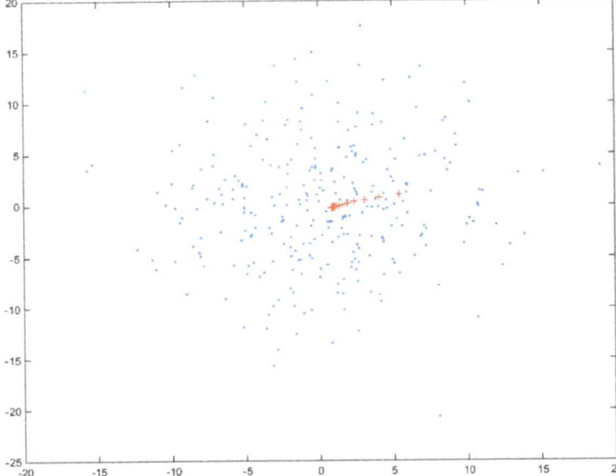

Figure 9. The center point iteration of the kernel function.

Segmentation and extraction results with different mean-shift algorithms are given in Figure 10. The first column of Figure 10 represents the original image. Figure 10b,f are the best segmentation results obtained by multiple experiments for Figure 10a,e using the traditional mean-shift algorithm, while Figure 10c,g, and Figure 10d,h show the segmentation result and extracted foreground area of the proposed algorithm, respectively. As shown in Figure 10, the results of the proposed algorithm are very close to the best segmentation results obtained by the traditional mean-shift algorithm. The proposed method can control over-segmentation to a small extent and reduce the mean-shift algorithm debugging time. The algorithm comparison parameters are shown in Table 2.

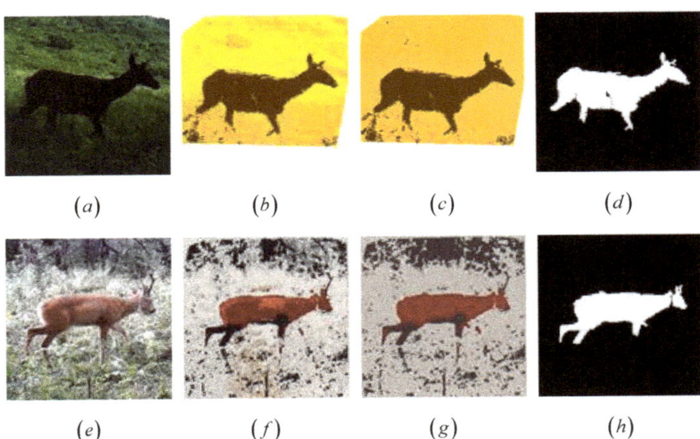

(a) $\qquad$ (b) $\qquad$ (c) $\qquad$ (d)

(e) $\qquad$ (f) $\qquad$ (g) $\qquad$ (h)

Figure 10. Visual comparison of wildlife image segmentation: (**a**,**e**) the original image; (**b**,**f**) the best segmentation results of the traditional mean-shift algorithm; (**c**,**g**) the segmentation result of the proposed algorithm; and (**d**,**h**) the extraction result of the proposed algorithm.

Table 2. Comparison of image algorithm parameters.

Parameters	Figure 10b	Figure 10c	Figure 10f	Figure 10g
h	16	18	24	34

Time complexity of the mean-shift is defined as $O(Tn^2)$, where T is the iterations number of the data sets, and n is the number of all sample data points [39]. The larger the value of the bandwidth parameter is, the less number of iterations required to traverse all the data sets, which means that the time complexity $O(Tn^2)$ is reduced. Choosing a larger bandwidth value means ignoring the detail information of the image, which can directly affect the quality of segmentation. The parameters of the algorithm proposed by this study are larger than the parameters of the best results in the comparison experiments, which means the proposed method reduces the debugging time and time complexity for image segmentation.

5. Experimental Results and Discussion

In order to verify the adaptability and effectiveness of our proposed algorithm, an extraction analysis of the captured wildlife images was conducted. The result was evaluated by several evaluation criteria, and it was compared with other conventional algorithms for image extraction.

5.1. Evaluation Criteria

The pixel accuracy, relative limit measurement accuracy, and mean intersection over the union were utilized as objective criteria to evaluate the quality of image extraction [40].

The pixel accuracy rate PA was used to calculate the ratio of the number of correctly segmented pixels to the number of pixels in the image:

$$PA = \frac{\sum_{i=1}^{n} n_{ii}}{\sum_{i=1}^{n} t_i} \times 100\% \tag{15}$$

where t_i is the number of pixels belonging to the division category i in the original picture, n_{ii} represents the number of pixels whose actual category is i, and the prediction category is also i.

The relative limit measurement accuracy $RLMA$ indicates the deviation value between the actual value of the segmented image and the true value of the foreground region:

$$RLMA = \frac{|\alpha - \beta|}{\alpha} \times 100\% \tag{16}$$

where α is the actual number of pixels in the image to be segmented, and β is the number of pixels in the foreground region obtained by segmentation. The smaller the $RLMA$, the better the segmentation effect.

Mean intersection over union $MIoU$ is used as the intersection ratio calculation of the segmentation result and the true value, which can reflect the accuracy and completeness of the segmentation result, and is the most commonly used evaluation index:

$$MIoU = \frac{1}{n+1} \sum_{i=0}^{n} \frac{n_{ii}}{t_i + \sum_{j=0}^{n} n_{ji} - n_{ii}} \times 100\% \tag{17}$$

where n_{ji} denotes the number of pixels whose actual category is j, the prediction category is i, and t_i is the number of pixels belonging to category i.

5.2. Experiment and Analysis

We compared the experimental results of our algorithm with three other extraction algorithms, including N-cuts (N-cuts) [41], the aggregating super-pixels (SAS) algorithm [42], and the histogram contrast saliency detection (HCS) algorithm [25]. Experimental samples were selected from wildlife monitoring images of different species with different backgrounds and different light intensities in the Saihan Ula Nature Reserve due to seasonal variations. The comparison results are shown in Figure 11. There are wildlife images in Figure 11(1–3) with diverse background complexity. Figure 11(1) has a simple background and shows an extreme difference between the foreground and background color.

Figure 11(2) has a higher background complexity with a similar color between the grass and trees in the background, whereas the image of Figure 11(3) has only a single background, and the shadows region under light conditions is similar to those of the wildlife. There are three typical images of wildlife with different intensities of light in Figure 11(4–6). Figure 11(4,5) show wildlife captured under the normal lighting conditions and weak illumination, in which the overall color of the latter is dark and the details are not obvious; there is a bright and dark mutation area under the condition of non-uniform strong illumination in Figure 11(6). All the experiments were performed using MATLAB (2014b, The MathWorks, Natick, MA, USA, 1984) in a workstation with Intel (R) Core (TM) i5-4590 and 8GB RAM.

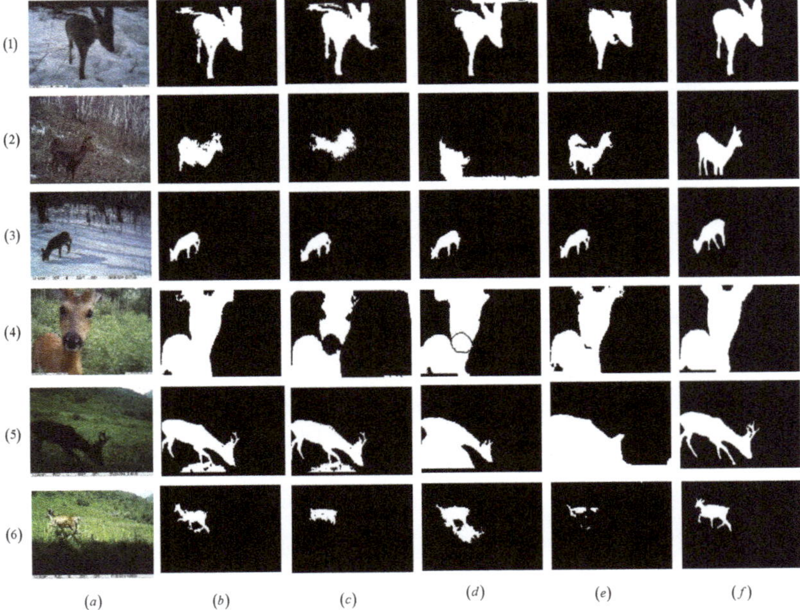

Figure 11. Vision comparison of wildlife image extraction: (**a**) the original image; (**b**) the extraction results of the proposed algorithm; (**c**) the extraction results of the N-cuts algorithm; (**d**) the extraction results of the aggregating super-pixels (SAS) algorithm; (**e**) the extraction results of the histogram contrast saliency detection (HCS) algorithm; and (**f**) ground-truth.

The above extraction results show that the method proposed in this paper has a superior performance and that its extraction of wildlife regions is more accurate than those of the other three methods. For example, the proposed algorithm, N-cuts, SAS algorithm, and HCS algorithm have a better effect on the extraction of the images of Figure 11(1,3) with simple backgrounds, compared with the extraction of the image of Figure 11(2) with a higher background complexity and a foreground in which grass and trees are very similar in color with slightly different texture features. Therefore, we proposed a method based on adaptive mean-shift and Hermite transform which could effectively segment the image and obtain satisfactory extraction results, whereas the N-cuts algorithm, SAS algorithm, and the HCS algorithm show problems of over-segmentation and even incorrect segmentation. In the surveillance images of wildlife under different illumination conditions, the SAS algorithm and the HCS algorithm caused over-segmentation in weak illumination due to the influence of the shadows, as shown in Figure 11(5). Under the conditions of non-uniform strong illumination, the N-cuts algorithm caused incorrect segmentation in the head and legs of the wildlife in bright and dark mutations, as shown in Figure 11(6).

For the six wildlife monitoring images described in Figure 11, the wildlife regions segmented by hand were manually labeled through an image splitter with reference to the true value. The pixel accuracy, relative limit measurement accuracy, and mean intersection over the union of the extraction results of the proposed algorithm, N-cuts algorithm, SAS algorithm, and HCS algorithm extraction results are shown in Figure 12.

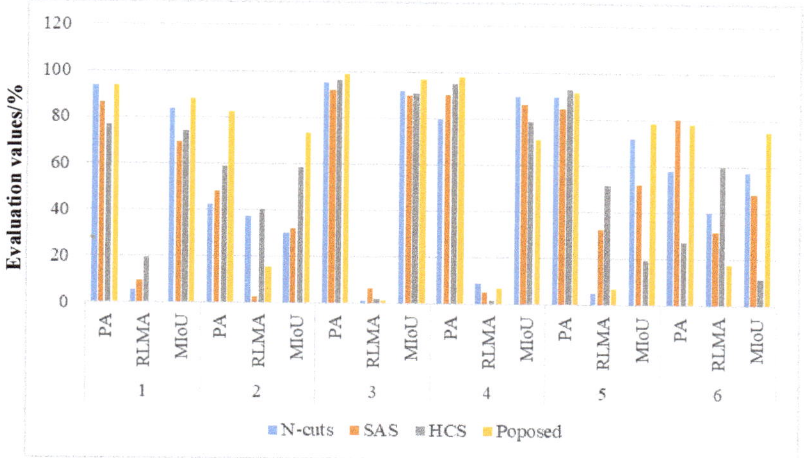

Figure 12. Comparative results of each algorithm. PA: Pixel accuracy; RLMA: Relative limit measurement accuracy; MIoU: Mean intersection over the union.

Analysis of the image algorithms corresponding to the data in Figure 12 led to four main findings: (1) Compared with the N-cuts, SAS, and HCS algorithm, the relative limit measurement accuracy *RLMA* of the proposed algorithm was the lowest, which indicates that the foreground region extracted by the proposed algorithm had the least deviation from the reference's true value; (2) an accurate extraction method produces a *PA* value that is close to 100%. Thus, our proposed method yields better extraction than the other methods for all the images except Figure 11(5). Although the HCS algorithm produced the best extraction in Figure 11(5), the foreground and background images could not be effectively extracted; (3) the mean intersection over union *MIoU* of the proposed algorithm was remarkably higher than those of the SAS and HCS algorithms, and slightly higher than that of the N-cuts algorithm, which indicates that the proposed algorithm has the best accuracy and completeness; (4) the mean intersection over union *MIoU* of the proposed algorithm was more than 70, which proves that the proposed algorithm has a high applicability in wildlife monitoring images. The above results show that our algorithm is more suitable for the extraction of wildlife images captured by WMSNs.

In order to further verify the performance of the proposed algorithm, this experiment randomly selected 120 images from the wildlife monitoring images and calculated the performance evaluation index values of the different algorithms. As shown in Figure 13, the proposed algorithm improved the pixel accuracy compared with the N-cuts, SAS, and HCS algorithm. By comparing the mean of the experimental data, the pixel accuracy of the proposed algorithm increased by 11.25%, 5.46%, and 10.39%, respectively; the relative limit measurement accuracy improved by 1.83%, 5.28%, and 12.05%, respectively; and the average mean intersection over the union increased by 7.09%, 14.96%, and 19.14%, respectively. The above results show that the proposed algorithm consistently outperforms other algorithms with respect to both pixel accuracy and average mean intersection over the union.

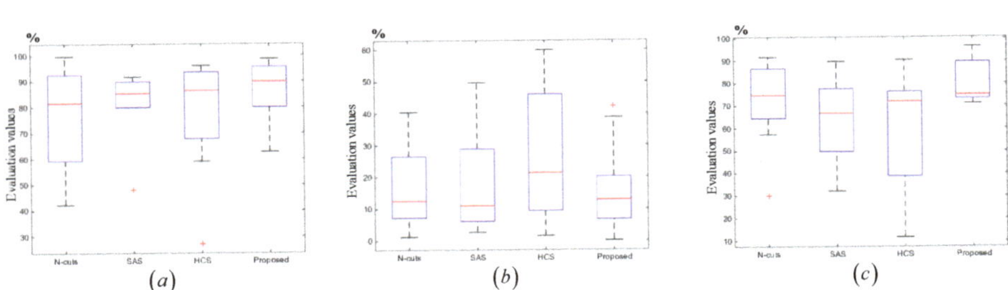

Figure 13. Comparative results of different algorithms. (**a**) Pixel accuracy (PA) experimental result; (**b**) relative limit measurement accuracy (RLMA) experimental result; (**c**) mean intersection over the union (MIoU) experimental result.

6. Conclusions

In this paper, we proposed a novel extraction method for wildlife images, which can achieve extraction of the foreground region and reduce the energy loss of sensor nodes in WMSNs. The method uses Hermite transform to extract image texture information and combine color information obtained by the color space to achieve adaptive mean-shift clustering. This study used wildlife images captured by a WMSN monitoring system, which was developed by our laboratory, in the Saihan Ula Nature Reserve, as an experimental sample. The proposed method was compared with the N-cuts algorithm, SAS algorithm, and HCS algorithm considering four criteria, including the extraction effect, pixel accuracy, relative limit measurement accuracy, and mean intersection over the union. The experimental results confirmed that the algorithm proposed in this paper was superior to the other three algorithms. The experimental data and results show that the proposed method can realize more accurate extraction of wildlife monitoring images and provide effective support for image transmission in WMSNs. However, uncertainty still remains as to the accurate extraction of the foreground by the threshold segmentation method in the case of irregular numbers of clusters. In future work, we are planning to construct a method for extraction based on gray histogram estimation and regional mergers for each input image.

Author Contributions: J.Z. proposed the algorithm; W.L. and H.L. conceived and designed the experiments; W.L., H.L., Y.W., X.Z. and J.Z. performed the experiments; Y.W. and X.Z. analyzed the data; H.L. wrote the paper.

Funding: This study was financially supported by the National Natural Science Foundation of China (Grant No.31670553), Fundamental Research Funds for the Central Universities (Grant No.2016ZCQ08).

Conflicts of Interest: The authors declare no conflict of interest.

References

1. Hori, B.; Petrell, R.J.; Fernlund, G.; Trites, A. Mechanical reliability of devices subdermally implanted into the young of long-lived and endangered wildlife. *J. Mater. Eng. Perform.* **2012**, *21*, 1924–1931. [CrossRef]
2. Handcock, R.N.; Swain, D.L.; Bishophurley, G.J. Monitoring animal behaviour and environmental interactions using wireless sensor networks, GPS collars and satellite remote sensing. *Sensors* **2009**, *9*, 3586–3603. [CrossRef] [PubMed]
3. Evans, M.N.; Guerrero-Sanchez, S.; Bakar, M.S.A. First known satellite collaring of a viverrid species: Preliminary performance and implications of GPS tracking Malay civets (*Viverra tangalunga*). *Ecol. Res.* **2016**, *31*, 475–481. [CrossRef]
4. Gomez, A.; Salazar, A.; Vargas, F. Towards Automatic Wild Animal Monitoring: Identification of Animal Species in Camera-trap Images using Very Deep Convolutional Neural Networks. *Ecol. Inform.* **2017**, *41*, 24–32. [CrossRef]
5. Rovero, F.; Zimmermann, F. Camera Trapping for Wildlife Research. *J. Wildl. Manag.* **2017**, *81*, 1125.
6. Pettorelli, N.; Laurance, W.F.; O'Brien, T.G. Satellite remote sensing for applied ecologists: Opportunities and challenges. *J. Appl. Ecol.* **2014**, *51*, 839–848. [CrossRef]

7.   Feng, G.; Mi, X.; Yan, H. CFor Bio: A network monitoring Chinese forest biodiversity. *Sci. Bull.* **2016**, *61*, 1163–1170. [CrossRef]
8.   Luis, G.; Glen, M.; Eduard, P. Unmanned Aerial Vehicles (UAVs) and Artificial Intelligence Revolutionizing Wildlife Monitoring and Conservation. *Sensors* **2016**, *16*, 97.
9.   Wang, H.; Fapojuwo, A.O.; Davies, R.J. A Wireless Sensor Network for Feedlot Animal Health Monitoring. *IEEE Sens. J.* **2016**, *16*, 6433–6446. [CrossRef]
10.   Al-Turjman, F.; Radwan, A. Data Delivery in Wireless Multimedia Sensor Networks: Challenging and Defying in the IoT Era. *IEEE Wirel. Commun.* **2017**, *24*, 126–131. [CrossRef]
11.   Zhang, J.; Xiang, Q.; Yin, Y.; Chen, C.; Luo, X. Adaptive compressed sensing for wireless image sensor networks. *Multimed. Tools Appl.* **2017**, *76*, 4227–4242. [CrossRef]
12.   Banerjee, R.; Bit, S.D. An energy efficient image compression scheme for wireless multimedia sensor network using curve fitting technique. *Wirel. Netw.* **2017**, *3*, 1–17. [CrossRef]
13.   Shen, X.; Xiang, L.; Chen, H. Fast Computation of Threshold Based on Multi-threshold Otsu Criterion. *J. Electron. Inf. Technol.* **2017**, *39*, 144–149.
14.   Han, J.; Yang, C.; Zhou, X. A new multi-threshold image segmentation approach using state transition algorithm. *Appl. Math. Model.* **2017**, *44*, 588–601. [CrossRef]
15.   Yeom, S. Infrared image segmentation based on region of interest extraction with Gaussian mixture modeling. In Proceedings of the Automatic Target Recognition XXVII, SPIE Defense + Security, Anaheim, CA, USA, 1 May 2017; p. 102020.
16.   Fang, Z.; Yu, X.; Wu, C.; Chen, D.; Jia, T. Superpixel Segmentation Using Weighted Coplanar Feature Clustering on RGBD Images. *Appl. Sci.* **2018**, *8*, 902. [CrossRef]
17.   Goferman, S.; Zelnikmanor, L.; Tal, A. Context-aware saliency detection. *IEEE Trans. Pattern Anal. Mach. Intell.* **2012**, *34*, 1915–1926. [CrossRef] [PubMed]
18.   Long, J.; Shelhamer, E.; Darrell, T. Fully Convolutional Networks for Semantic Segmentation. In Proceedings of the IEEE Conference on Computer Vision and Pattern Recognition (CVPR), Boston, MA, USA, 8–10 June 2015; pp. 3431–3440.
19.   Zhu, J.; Du, X.; Fan, H. Image multi-scale edge detection and image multi-scale segmentation. *Geogr. Geo-Inf. Sci.* **2013**, *29*, 45–48.
20.   Jiang, Y.; Hao, Z.; Yang, Z. A cooperative honey bee mating algorithm and its application in multi-threshold image segmentation. *Inf. Sci.* **2016**, *369*, 171–183. [CrossRef]
21.   Souza, É.L.; Pazzi, R.W.; Nakamura, E.F. A prediction-based clustering algorithm for tracking targets in quantized areas for wireless sensor networks. *Wirel. Netw.* **2015**, *21*, 2263–2278. [CrossRef]
22.   Zou, B.; Liu, Q.; Chen, Z. Saliency detection using boundary information. *Multimed. Syst.* **2016**, *2*, 245–253. [CrossRef]
23.   Wei, Z.; Yi, F.; Xiaosong, W. An Improved Image Semantic Segmentation Method Based on Superpixels and Conditional Random Fields. *Appl. Sci.* **2018**, *8*, 837.
24.   Shehu, A.; Hulaj, A.; Bajrami, X. An Algorithm for Edge Detection of the Image for Application in WSN. In Proceedings of the 2nd International Conference on Applied Physics, System Science and Computers, Dubrovnik, Croatia, 27–29 September 2017; pp. 207–213.
25.   Feng, W.; Zhang, J.; Hu, C. A Novel Saliency Detection Method for Wild Animal Monitoring Images with WMSN. *J. Sens.* **2018**, *2018*. [CrossRef]
26.   Tian, X.; Yu, W. Color image segmentation based on watershed transform and feature clustering. In Proceedings of the IEEE Advanced Information Management, Communicates, Electronic and Automation Control Conference, Xi'an, China, 3–5 October 2016; pp. 1830–1833.
27.   Akbulut, Y.; Guo, Y.; Şengür, A. An effective color texture image segmentation algorithm based on hermite transform. *Appl. Soft Comput.* **2018**, *67*, 494–504. [CrossRef]
28.   Shinde, S.R.; Sabale, S.; Kulkarni, S. Experiments on content based image classification using Color feature extraction. In Proceedings of the 2015 International Conference on Communication, Information & Computing Technology, Mumbai, India, 15–17 January 2015; pp. 1–6.
29.   Rahman, Z.; Jobson, D.D.; Woodell, G.A. Retinex processing for automatic image enhancement. *J. Electron. Imaging* **2004**, *13*, 100–110.
30.   Zhang, S.; Wang, T.; Dong, J. Underwater image enhancement via extended multi-scale Retinex. *Neurocomputing* **2017**, *245*, 1–9. [CrossRef]

31. Neto, J.R.D.O.; Lima, J.B. Discrete Fractional Fourier Transforms Based on Closed-Form Hermite–Gaussian-Like DFT Eigenvectors. *IEEE Trans. Signal Process.* **2017**, *65*, 6171–6184. [CrossRef]

32. Yun, L.; Shu, S.; Gang, H. A Weighted Measurement Fusion Particle Filter for Nonlinear Multisensory Systems Based on Gauss–Hermite Approximatio. *Sensors* **2017**, *17*, 2222–2235.

33. Leibon, G.; Rockmore, D.N.; Park, W. A Fast Hermite Transform. *Theor. Comput. Sci.* **2008**, *409*, 211–228. [CrossRef]

34. Luo, Z.Z. Hermite Interpolation-Based Wavelet Transform Modulus Maxima Reconstruction Algorithm's Application to EMG De-noising. *J. Electron. Inf. Technol.* **2009**, *31*, 857–860.

35. Craciun, S.; Kirchgessner, R.; George, A.D.; Lam, H.; Principe, J.C. A real-time, power-efficient architecture for mean-shift image segmentation. *J. Real-Time Image Process.* **2018**, *14*, 379–394. [CrossRef]

36. Lang, F.; Yang, J.; Yan, S.; Qin, F. Superpixel Segmentation of Polarimetric Synthetic Aperture Radar (SAR) Images Based on Generalized Mean Shift. *Remote Sens.* **2018**, *10*, 1592. [CrossRef]

37. Liu, C.; Zhou, A.; Zhang, Q. Adaptive image segmentation by using mean-shift and evolutionary optimization. *IET Image Process.* **2014**, *8*, 327–333. [CrossRef]

38. Wang, G.; Wang, G.; Yang, H. Mean shift segmentation algorithm based on fused color-texture model. *Sci. Surv. Mapp.* **2015**, *40*, 108–112.

39. Daniel, F.; Pavel, K. Fast mean shift by compact density representation. In Proceedings of the IEEE Conference on Computer Vision and Pattern Recognition, Miami, FL, USA, 20–25 June 2009; pp. 1818–1825.

40. Mesas-Carrascosa, F.J.; Rumbao, I.C.; Berrocal, J.A.B.; Porras, A.G.-F. Positional Quality Assessment of Orthophotos Obtained from Sensors Onboard Multi-Rotor UAV Platforms. *Sensors* **2014**, *14*, 22394–22407. [CrossRef] [PubMed]

41. Shi, J.; Malik, J. Normalized cuts and image segmentation. *IEEE Trans. Pattern Anal. Mach. Intell.* **2000**, *22*, 888–905.

42. Ban, Z.; Liu, J.; Cao, L. Superpixel Segmentation Using Gaussian Mixture Model. *IEEE Trans. Image Process.* **2018**, *27*, 4105–4117. [CrossRef] [PubMed]

Article

IMU-Aided High-Frequency Lidar Odometry for Autonomous Driving

Hanzhang Xue [1], Hao Fu [1] and Bin Dai [1,2,*]

[1] College of Intelligence Science and Technology, National University of Defense Technology, Changsha 410073, China; xuehanzhang13@nudt.edu.cn (H.X.); fuhao@nudt.edu.cn (H.F.)

[2] Unmanned Systems Research Center, National Innovation Institute of Defense Technology, Beijing 100071, China

* Correspondence: bindai.cs@gmail.com

Received: 14 March 2019; Accepted: 6 April 2019; Published: 11 April 2019

Abstract: For autonomous driving, it is important to obtain precise and high-frequency localization information. This paper proposes a novel method in which the Inertial Measurement Unit (IMU), wheel encoder, and lidar odometry are utilized together to estimate the ego-motion of an unmanned ground vehicle. The IMU is fused with the wheel encoder to obtain the motion prior, and it is involved in three levels of the lidar odometry: Firstly, we use the IMU information to rectify the intra-frame distortion of the lidar scan, which is caused by the vehicle's own movement; secondly, the IMU provides a better initial guess for the lidar odometry; and thirdly, the IMU is fused with the lidar odometry in an Extended Kalman filter framework. In addition, an efficient method for hand–eye calibration between the IMU and the lidar is proposed. To evaluate the performance of our method, extensive experiments are performed and our system can output stable, accurate, and high-frequency localization results in diverse environment without any prior information.

Keywords: ego-motion estimation; hand-eye calibration; IMU; lidar odometry; sensor fusion

1. Introduction

Precise and high-frequency localization is one of the key problems for autonomous vehicles. In recent years, the fusion of Global Navigation Satellite System (GNSS) and Inertial Measurement Unit (IMU) has been the most popular localization method. However, the GNSS/IMU system will fail in some environments where GNSS signals suffer from satellite blockage or multipath propagation [1], such when an autonomous vehicle is driving in a tunnel, in the mountains, or in an environment with electromagnetic interference. Thus, an alternative localization method is necessary when the GNSS signal is unavailable or is of poor quality.

Currently, most autonomous vehicles are equipped with light detection and ranging (lidar) device, which is a promising sensor that could accurately calculate the range measurements of the surroundings. Some recent navigation approaches have begun to use lidar as a complementary sensor for the GNSS/IMU localization system. A typical lidar odometry algorithm could roughly be divided into three steps: a pre-processing step, which tries to compensate the intra-frame distortion caused by the vehicle's own movement; an intermediate step, which is the main body of the lidar odometry algorithm, and a last step, which outputs the localization result.

As the lidar mostly spins at 10 Hz, its output frequency is thus less than 10 Hz, which might not be fast enough to meet the needs of other modules, like planning or control. To remedy this, some approaches try to combine the lidar with other sensors, such as IMU. In these approaches, the IMU and the lidar odometry are usually combined in an Extended Kalman Filter (EKF) framework, or by using a factor graph representation [2–7] (as shown in the left of Figure 1). In this paper, we claim that IMU information can not only be fused with the lidar odometry at the output level, but it can also aid lidar odometry in the pre-processing level and the intermediate level (as shown in the right of Figure 1).

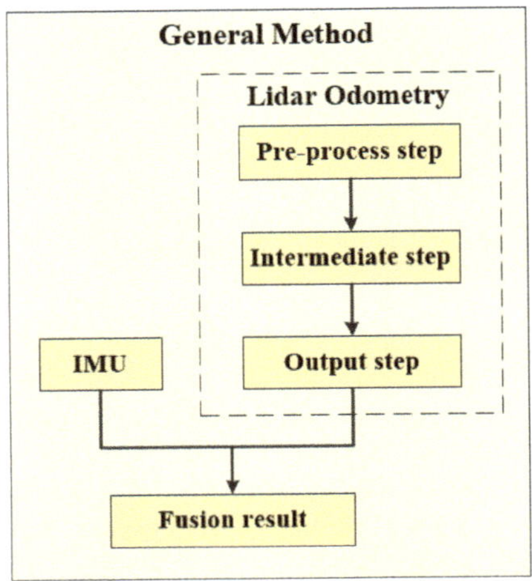

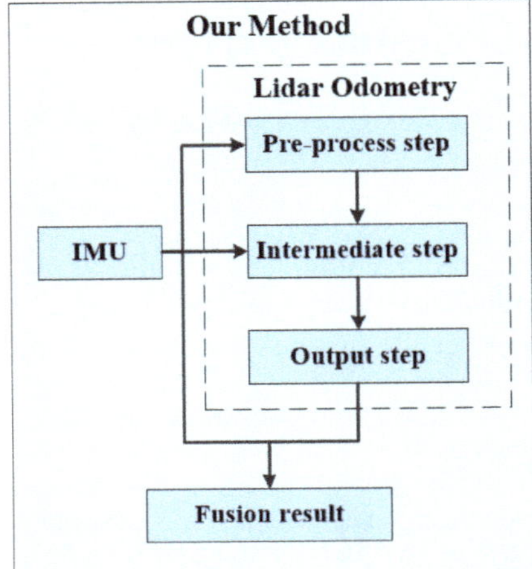

Figure 1. A typical lidar odometry algorithm can be roughly divided into three steps: a pre-processing step, an intermediate step, and an output step. Most existing lidar/Inertial Measurement Unit (IMU) fusion approaches try to fuse lidar odometry and the IMU in the output level [2–7]. In this paper, we try to utilize the IMU information in all the three steps.

In the pre-processing step, as the vehicle itself is moving during the lidar spin, the generated point clouds will be distorted. This intra-frame distortion is usually ignored. In this paper, we show that the high-frequency IMU information could naturally help correct this distortion, as long as the IMU and lidar are well-calibrated. We also claim that **the problem of intra-frame correction is in fact equivalent to inter-frame registration**, and propose an interesting two-step correction procedure which does not rely on the IMU. In the intermediate step, the IMU could also provide a good initial guess for the lidar odometry algorithm. Past works usually use the registration result from the previous two scans as the initial guess. We will do experiments to compare these two approaches. In the output step, we try to combine the lidar odometry with the IMU in the EKF framework.

In summary, we make the following contributions in this paper:

- We propose an efficient method for hand–eye calibration between the IMU and the lidar, which is a pre-requisite for making good use of the IMU.
- In the pre-processing step of the lidar odometry algorithm, we emphasize that the intra-frame correction is equivalent to inter-frame registration, and propose an efficient method for utilizing the IMU information to rectify the distorted point clouds.
- In the intermediate step, we use IMU to obtain the initial guess for the lidar odometry algorithm, and compare this approach with previous methods which usually use the registration results from the previous two scans as the initial guess.
- By combining the lidar odometry with the IMU information, we obtain a complete localization system which can output stable, accurate, and high-frequency localization results at a frequency of around 40 Hz.

The rest of this paper is organized as follows. In Section 2, we introduce some related works. Section 3.1 provides an overview of the proposed approach, and introduces the coordinate systems and notations in our approach. The details of the proposed approach are presented from Sections 3.2–3.6. The experimental results and comparison with state-of-the-art approaches are presented in Section 4. Finally, the conclusions are summarized in Section 5.

2. Related Works

Lidar odometry has been a hot topic in recent years. Lidar odometry is, in essence, equivalent to scan matching, which tries to estimate the relative transform between two frames of point clouds. The scan matching methods can be classified into three categories: point-based, mathematical property-based, and feature-based. The Iterative Closest Point (ICP) algorithm [8] and its variants [9] are the most popular point-based methods which estimate the relative transform through finding the nearest points between the two scans. For the mathematical property-based method, such as the Normal Distribution Transform (NDT) algorithm and its variants (D2D-NDT [10] and P2D-NDT [11]), the observed points are modeled as a set of Gaussian probability distribution. For the feature-based method, there have been various features applied, such as the line segment feature [12], corner feature [13], and grid feature [14]. Zhang et al. proposed a method called Lidar Odometry And Mapping (LOAM) [15,16], which extracts feature points located on sharp edges or planar surfaces, and matches these feature points using either the point-to-line distance or the point-to-plane distance.

All these lidar odometry algorithms can only output localization results at a frequency of less than 10 Hz, which might not be fast enough. Some recent approaches have started to fuse the lidar odometry with the IMU using either the Kalman Filter framework [2–5,17,18] or the factor graph representation [6,7].

In [2], the authors proposed a localization system which combines a 2D Simultaneous Localization and Mapping (SLAM) subsystem with a 3D navigation subsystem based on IMU. This system can estimate the accurate state of an unmanned ground vehicle (UGV), but the usage of 2D SLAM is not suitable for the off-road environment. In [3], a hybrid lidar odometry algorithm was proposed which combines the feature-based scan matching method with the ICP-based scan matching method, and the lidar odometry result was then fused with the IMU in an EKF framework. In [6], a bunch of IMU measurements were considered as a relative constraint using the pre-integration theory, and the results of lidar odometry were fused with IMU measurements in a factor graph framework. However, in all these previous works, the IMU information was only fused with the lidar odometry in the result level; whilst in this paper, we try to utilize the IMU information to aid the whole process of the lidar odometry algorithm.

3. The Proposed Approach

3.1. Coordinate Systems and Notations

The aim of this work is to accurately estimate the pose of the vehicle. Three coordinate systems are used in our approach, which are defined as follows:

- World Coordinate System $\{W\}$. In $\{W\}$, we define the coordinates of a point by latitude, longitude, and altitude, the x-axis points to the east, the y-axis points to the north, and the z-axis points to the sky, following the right-hand rule. This coordinate system is defined in a global scale and will never change.
- Body Coordinate System $\{B\}$. This coordinate is fixed with the IMU, which is installed in the center of two rear wheels of the vehicle. The x-axis points to the right, the y-axis points to the forward, and the z-axis points to the upward, following the right-hand rule.
- Lidar coordinate system $\{L\}$. The lidar is mounted on top of the vehicle, and the origin of $\{L\}$ is located at the center of the lidar. The three axes of this coordinate system are the same as in $\{B\}$.

Figure 2 shows the coordinate system $\{W\}$ and $\{B\}$ in the 2D plane and the relationship between them. The body coordinate system $\{B\}$ changes with the movement of the vehicle. T_k^{WB} represents the transformation between the coordinate system $\{W\}$ and $\{B\}$ at time t_k, which can also denote the pose of the vehicle at time t_k. On the contrary, T_k^{BW} transforms the coordinate system $\{B\}$ to $\{W\}$ and $T_k^{BW} = \left(T_k^{WB}\right)^{-1}$. Moreover, $\triangle T_{k-1,k}^{B}$, which could be obtained from the IMU, represents the relative transformation of the vehicle between time t_{k-1} and t_k. The relationship between $\triangle T^{B}_{k-1,k}$ and T^{WB}_k can be expressed as:

$$\triangle T^{B}_{k-1,k} = \left(T^{WB}_{k-1}\right)^{-1} T^{WB}_k .$$ (1)

In addition, the relationship between the coordinate system $\{B\}$ and $\{L\}$ is shown in Figure 3.

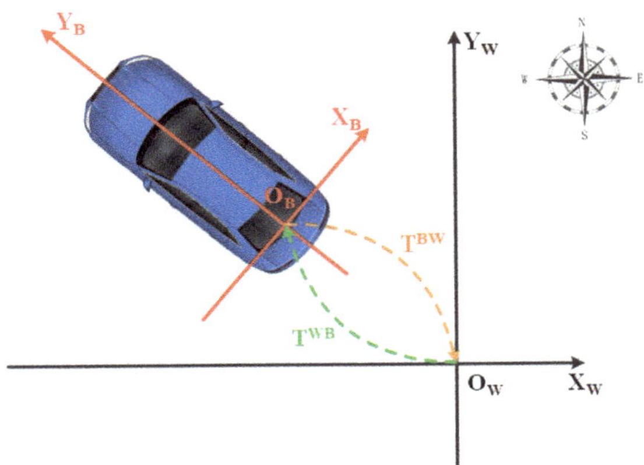

Figure 2. The relative relationship between the world coordinate system $\{W\}$ and the body coordinate system $\{B\}$ in the 2D plane, where the black axis indicates $\{W\}$ and the red axis indicates $\{B\}$.

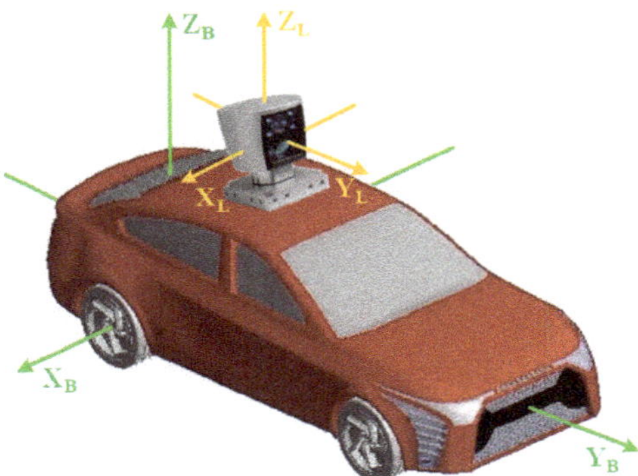

Figure 3. The relationship between the body coordinate system $\{L\}$ and the lidar coordinate system $\{B\}$, where the yellow axis indicates $\{L\}$ and the green axis indicates $\{B\}$.

We use T^{WL}_k to denote the pose of the lidar at time t_k. $\triangle T^{L}_{k-1,k}$ represents the relative transformation of the lidar between time t_{k-1} and t_k. The transformation T^{BL}, which transforms the coordinate system $\{B\}$ to $\{L\}$, can be obtained from the hand–eye calibration.

The transformation T can be expressed by:

$$T = \begin{bmatrix} R & p \\ 0^T & 1 \end{bmatrix},$$ (2)

where $R \in \mathbb{R}^{3\times3}$ represents the rotational matrix and $p \in \mathbb{R}^3$ is the translational vector.

In *Lie Group* [19], T belongs to the *Special Euclidean group* SE(3) and the rotational matrix R belongs to the *Special Orthogonal group* SO(3). Correspondingly, the $\mathfrak{se}(3)$ and $\mathfrak{so}(3)$ are two kinds of *Lie Algebra*. $\theta \in \mathfrak{so}(3)$ indicates the rotation angle which can be expressed as $\theta = \theta n$, where θ represents the magnitude of the rotation angle and n denotes the rotation axis. The pose can thus be represented by $\xi = \begin{bmatrix} \theta & p \end{bmatrix}^T \in \mathfrak{se}(3)$.

The exponential map and logarithmic map can be used to derive the relationship between the *Lie Group* and the *Lie Algebra*. For example, T can be represented by the exponential map of the ξ, namely, $T = \exp(\xi^\wedge)$. Conversely, ξ can be represented by the logarithmic map of the T, namely, $\xi = \log(T)^\vee$.

The exponential map between θ and R can be formulated by using the Rodriguez formula:

$$R = \cos\theta \cdot I + (1 - \cos\theta) \cdot n \cdot n^T + \sin\theta \cdot n^\wedge . \tag{3}$$

The operator $\wedge$ maps a vector to a skew-symmetric matrix as Equation (4), and the operator $\vee$ maps a skew symmetric matrix back to a vector.

$$\theta^\wedge = \begin{bmatrix} \theta^x \\ \theta^y \\ \theta^z \end{bmatrix}^\wedge = \begin{bmatrix} 0 & -\theta^z & \theta^y \\ \theta^z & 0 & -\theta^x \\ -\theta^y & \theta^x & 0 \end{bmatrix} . \tag{4}$$

The notation of the data used in our system are specified as follows. Each type of data used in our system is composed of the timestamp and the data itself. The timestamp is defined as the time when the data is generated. t^B denotes the timestamp of the IMU measurement and t^L represents the timestamp of the lidar point clouds. f_k^L indicates all points generated from the lidar during $[t_{k-1}^L, t_k^L]$. In addition, the i-th point generated by the b-th laser beam is denoted as $X_{(i,b,k)}^L$ in the lidar coordinate system $\{L\}$. The number of points received per laser beam is denoted as S_b and $i \in [1, S_b]$.

3.2. Motion Prior from IMU and Wheel Encoder

It is well-known that the accelerometer in the IMU is affected by the gravity, and it needs to be integrated twice to obtain the translational vector [20]. In contrast, the gyro in the IMU directly measures the angular velocity, and it only needs to be integrated once to obtain the rotation angle. It is reasonable to believe that the angular output from the IMU is much more accurate than the translational measurement.

For the ground vehicle, besides IMU, it is also equipped with the wheel encoder. The wheel encoder directly measures the rotation of the wheel, and thus provides a much more accurate translational measurement.

In this paper, we try to combine the angular output from the IMU and the translational output from the wheel encoder to obtain the motion prior.

The motion model of the vehicle can be simplified as Figure 4. We mount the IMU at the center of two rear wheels, and mount two wheel encoders on the left and right rear wheels, respectively. The black car denotes the position of the vehicle at the previous moment t_{k-1}^B, and the blue car indicates its position at the current time t_k^B. $\triangle sl$ and $\triangle sr$ represent the traveled distances of the left rear wheel and the right rear wheel during the short period of time.

Firstly, we compute the rotation matrix R_k^{WB} by triaxial angular velocities of the IMU. We introduce the following kinematic model [21]:

$$\frac{dR_\tau^{BW}}{dt} = R_\tau^{BW} \begin{bmatrix} \omega_\tau^x \\ \omega_\tau^y \\ \omega_\tau^z \end{bmatrix}^\wedge . \tag{5}$$

In Equation (5), R_τ^{BW} indicates the rotation matrix from the coordinate system $\{B\}$ to $\{W\}$. ω_τ^x, ω_τ^y, and ω_τ^z represent the instantaneous angular velocities of the IMU.

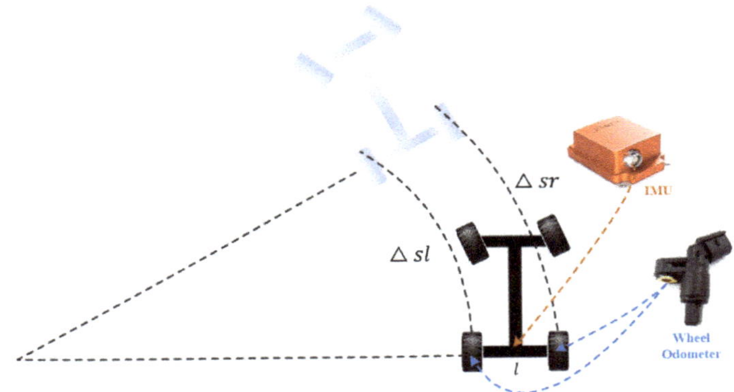

Figure 4. Schematic diagram of the simplified vehicle motion model. The black car represents the previous position of the vehicle, and the blue car indicates its current position. During a time period, the left rear wheel travels a distance of $\triangle sl$, and the right rear wheel travels $\triangle sr$.

By assuming the angular velocity of the IMU constant during $[t_{k-1}^B, t_k^B]$, we integrate Equation (5) from t_{k-1}^B to t_k^B and solve the differential equation. The rotation matrix R_k^{BW} can be obtained by:

$$R_k^{BW} = R_{k-1}^{BW} \operatorname{Exp}\left((\omega_k \, \triangle t)^\wedge \right). \tag{6}$$

To compute the translational component p_k^{WB} of the pose T_k^{WB}, we approximate the traveled distance $\triangle s$ of the vehicle to be the average of the two rear wheels. $\triangle s$ indicates the relative displacement of the vehicle during $[t_{k-1}^B, t_k^B]$ in the body coordinate system $\{B\}$ and the instantaneous speed of the vehicle can be denoted as $v = \triangle s / (t_k^B - t_{k-1}^B)$. We assume the velocity of the vehicle to be constant during $[t_{k-1}^B, t_k^B]$, where the wheel encoder always records the distance traveled along the heading direction of the vehicle. Thus, the instantaneous velocity vector can be denoted as $v_k^B = [v \ 0 \ 0]^T$. The relationship between the velocity vector v_k^W in the coordinate system $\{W\}$ and the velocity vector v_k^B in the coordinate system $\{B\}$ is:

$$v_k^W = R_k^{WB} v_k^B. \tag{7}$$

Using p_{k-1}^{WB}, R_k^{WB}, and v_k^W, the translational component p_k^{WB} can be computed as:

$$\begin{aligned}
p_k^{WB} &= p_{k-1}^{WB} + v_k^W \, (t_k - t_{k-1}) \\
&= p_{k-1}^{WB} + R_k^{WB} \begin{bmatrix} \triangle s & 0 & 0 \end{bmatrix}^T.
\end{aligned} \tag{8}$$

Since both the IMU and the wheel encoder output data at a frequency of 100 Hz, we use two separate buffers to cache these two kinds of data. These two sources of information are then fused by Equations (6) and (8). This fusion process only involves a few operators, and usually takes less than 1 millisecond. Therefore, the fusion output, which we term as the motion prior data, is also generated at 100 Hz.

3.3. Pre-Processing Step 1: Lidar-IMU Hand–Eye Calibration

As described before, the lidar point clouds are represented in the lidar coordinate system $\{L\}$. Taking the lidar point clouds of two frames f_{k-1}^L and f_k^L as inputs, we can calculate $\triangle T_{k-1,k}^L$ by the lidar odometry algorithm. $\triangle T_{k-1,k}^L$ indicates the relative transformation in the coordinate system $\{L\}$. However, the $\triangle T_{k-1,k}^B$ obtained from the IMU is expressed in the body coordinate system $\{B\}$.

Therefore, in order to utilize the IMU information, we need to establish the relationship between $\{L\}$ and $\{B\}$ first.

The calculation of the relative transformation T^{BL} between $\{L\}$ and $\{B\}$ is a typical hand–eye calibration problem. Based on T^{BL}, we can transform a point from the lidar coordinate system $\{L\}$ to the body coordinate system $\{B\}$:

$$X^{\mathrm{B}}_{(i,b,k)} = T^{\mathrm{BL}} X^{\mathrm{L}}_{(i,b,k)} \cdot \tag{9}$$

According to [22], we can formulate the hand–eye calibration problem between the coordinate system $\{L\}$ and $\{B\}$ as:

$$\triangle T^{\mathrm{B}}_{k,k-1}\, T^{\mathrm{BL}} = T^{\mathrm{BL}}\, \triangle T^{\mathrm{L}}_{k,k-1} \cdot \tag{10}$$

$\triangle T^{\mathrm{B}}_{k-1,k}$ and $\triangle T^{\mathrm{L}}_{k-1,k}$ are inputs of Equation (10). Let the timestamps of frames f^{L}_{k} and f^{L}_{k-1} be τ^{L}_{k} and τ^{L}_{k-1}, respectively. We firstly find the nearest timestamps τ^{B}_{k} and τ^{B}_{k-1} from the IMU, then T^{WB}_{k} and T^{WB}_{k-1} can be obtained. The relative transformation $\triangle T^{\mathrm{B}}_{k-1,k}$ can be computed by Equation (1). We then use $\triangle T^{\mathrm{B}}_{k-1,k}$ as the initial guess and $\triangle T^{\mathrm{L}}_{k-1,k}$ can be obtained from the lidar odometry.

Equation (10) can be solved by calculating the rotational and translational component separately. The rotational component is:

$$\triangle R^{\mathrm{B}}_{k,k-1}\, R^{\mathrm{BL}} = R^{\mathrm{BL}}\, \triangle R^{\mathrm{L}}_{k,k-1}\, , \tag{11}$$

and the translational component is:

$$\triangle R^{\mathrm{B}}_{k,k-1}\, p^{\mathrm{BL}} + \triangle p^{\mathrm{B}}_{k,k-1} = R^{\mathrm{BL}}\, \triangle p^{\mathrm{L}}_{k,k-1} + p^{\mathrm{BL}} \cdot \tag{12}$$

There are many methods to solve Equations (11) and (12). According to [23], the method proposed by Tsai et al. [24] exhibits the best performance. We therefore chose to use this method.

In our approach, multiple pairs of transformation $\triangle T^{\mathrm{L}}_{i,i-1}$ and $\triangle T^{\mathrm{B}}_{i,i-1}$ are needed. To reduce the error of hand–eye calibration [25,26], the following requirements need to be met:

- To avoid the intra-frame distortion caused by the vehicle's own movement, the frames should be chosen when the vehicle is stationary.
- The relative rotation angle of the two frames needs to be greater than 90 degrees, and the relative translation distance should be small.
- The rotational axes for these multiple transformation pairs should not be parallel.

After completing the hand–eye calibration, each point of the frame f^{L}_{k} can be transformed by Equation (9) and be represented in the coordinate system $\{B\}$.

3.4. Pre-Processing Step 2: Intra-Frame Correction

As mentioned in Section 3.1, it usually takes 100 ms for the lidar to spin a full circle. During this time, the vehicle also keeps moving. Some of these data are received at the previous moment $t^{\mathrm{L}}_{k} - t_j\,(t_j \in (0, 100)\,(\mathrm{ms}))$ and their coordinates are represented in the coordinate system corresponding to the moment $t^{\mathrm{L}}_{k} - t_j$. These points can be denoted as $X^{\mathrm{B}}_{(i,b,k-j)}$. If we want to represent these points in the coordinate system corresponding to time t^{L}_{k}, we need to transform the point $X^{\mathrm{B}}_{(i,b,k-j)}$ to $X^{\mathrm{B}}_{(i,b,k)}$. This transformation is usually called "correction" or "rectification" of the lidar scan.

There are two common ways to rectify the distorted point clouds. One utilizes the information of IMU [16,27]. We make some improvements in this method which is described in Section 3.4.1. The other one uses the previous registration results [15], which is presented in Section 3.4.2. Furthermore, we also propose an interesting two-step correction procedure which does not rely on the IMU in Section 3.4.3.

3.4.1. Rectification with the Aid of IMU

In this method, we try to use IMU to rectify the raw points. Let the pose at time t^{L}_{k} and t^{L}_{k-1} be T^{WB}_{q} and T^{WB}_{p}. Since the IMU data is generated at a frequency of 100 Hz and the lidar point clouds are generated at a frequency of 10 Hz, there are always nine frames of IMU data between the pose T_p^{WB} and T_q^{WB}. Therefore, ten relative transformations $\triangle T_{m-1,m}^B$ ($m \in (p,q]$) can be obtained by Equation (1) and each $\triangle T_{m-1,m}^B$ represents the relative transformation during $[t_{m-1}^B, t_m^B]$, where $t_m^B - t_{m-1}^B$ equals to 10 ms.

Since the rotational velocity of the lidar is constant during $[t_{k-1}^L, t_k^L]$, the rotation angle of the lidar per 10 ms is also fixed. We can separate all points of a frame f_k^L into ten sets, and the ID of points in each set can be denoted as i $\in [\frac{m-1-p}{q-p} \times S_b, \frac{m-p}{q-p} \times S_b]$. Let the total number of points in each set be S_e, which is equal to $S_b/10$. We further assume that the motion of the vehicle is constant during $[t_{m-1}^B, t_m^B]$. Each set of points now correspond to a unique relative transformation $\triangle T_{m,m-1}^B$ which can also be denoted as $exp\left(\triangle \xi_{m-1,m}^\wedge\right)$.

We linearly interpolated $exp\left(\triangle \xi_{m-1,m}^\wedge\right)$ to transform point $X_{(i,b,k-j)}^B$ to $X_{(i,b,m-1)}^B$. Then, $X_{(i,b,m-1)}^B$ will be transformed to $X_{(i,b,k)}^B$ by $\triangle T_{q,m-1}^B$. The whole rectification process can be denoted as:

$$\hat{X}_{(i,b,k)}^B = \triangle T_{q,m-1}^B \cdot exp\left(\triangle \xi_{m-1,m} \cdot \frac{i\%S_e}{S_e}\right)^\wedge \cdot X_{(i,b,k-j)}^B , \tag{13}$$

where $A\%B$ is used to compute the remainder when A is divided by B, and $\hat{X}_{(i,b,k)}^B$ represents the undistorted point after rectification.

3.4.2. Rectification by Previous Registration Results

This approach assumes that the motion of the vehicle is constant during $\left[t_{k-2}^L, t_k^L\right]$. Thus, it can assume that the relative transformation $\triangle T_{k-1,k-2}^B$ between frame f_{k-1}^L and f_{k-2}^L is the same as the relative transformation $\triangle T_{k,k-1}^B$ between frame f_k^L and f_{k-1}^L. Therefore, we can linearly interpolate the relative transformation $exp\left(\triangle \xi_{k-1,k-2}^\wedge\right)$ to rectify the raw distorted points in frame f_k^L as:

$$\hat{X}_{(i,b,k)}^B = exp\left(\triangle \xi_{k-1,k-2} \cdot \frac{i}{S_b}\right)^\wedge \cdot X_{(i,b,k-j)}^B . \tag{14}$$

3.4.3. Rectification by Registering Two Distorted Point Clouds

In this approach, we note that intra-frame correction is in fact quite related to inter-frame registration. As the task of intra-frame correction is to estimate the lidar motion between 0 ms and 100 ms, inter-frame registration tries to estimate the motion between 100 ms and 200 ms. Both frame *a* (corresponding to the lidar date generated between 0 ms and 100 ms) and frame *b* (generated between 100 ms and 200 ms) are distorted point clouds. It is reasonable to assume that the degree of distortion of these two frames are quite similar. We could thus directly register these two frames without any intra-frame correction first, and then use the registration result to correct each frame.

Let the relative transformation obtained from registering two distorted point clouds be $\triangle \tilde{T}_{k,k-1}^B$. We assumed that $\triangle \tilde{T}_{k,k-1}^B$ can approximate the lidar motion between time $\left[t_k^L, t_{k-1}^L\right]$ and linearly interpolate $exp\left(\triangle \tilde{\xi}_{k,k-1}^\wedge\right)$ to rectify the raw distorted points of f_k^L, as in Equation (15).

$$\hat{X}_{(i,b,k)}^B = exp\left(\triangle \tilde{\xi}_{k,k-1} \cdot \frac{i}{S_b}\right)^\wedge \cdot X_{(i,b,k-j)}^B . \tag{15}$$

The undistorted frame after rectification is denoted as $\hat{f}_k^L$. In Section 4, we do experiments to compare these three approaches.

3.5. Intermediate Step: Inter-Frame Registration

To register two frames of point clouds, we could use any off-the-shelf scan registration approaches, such as ICP, NDT, and LOAM. To successfully apply these approaches, it always requires an accurate initial guess for the estimated transformation. Just as mentioned previously, the problem of inter-frame registration is, in fact, quite related to intra-frame correction. Therefore, the approaches mentioned in Section 3.4 can also be applied here. We can either use the registration result of the previous two frames, or use the IMU data directly. We undertake experiments to compare these two approaches, presented in Section 4.

3.6. Output Step: Fusion in an EKF Framework

3.6.1. Basic Concepts

The Extended Kalman filter (EKF) [28] is one of the most commonly used data fusion algorithms. It is a recursive state estimator for the nonlinear system, and the state is assumed to satisfy a Gaussian distribution. A typical EKF algorithm consists of two steps: a prediction step, and an update step.

In the prediction step, the previous state x_{k-1} is used to compute a priori estimate of the current state by:

$$\begin{aligned}
\bar{x}_k &= f\left(\hat{x}_{k-1}\right), \\
\bar{P}_k &= F\,\hat{P}_{k-1}\,F^T + R,
\end{aligned} \tag{16}$$

where R is the covariance of the process noise. $\bar{x}_k$ and $\bar{P}_k$ are the mean and covariance matrices of the prior distribution of the state x_k, respectively. $\hat{x}_{k-1}$ and $\hat{P}_{k-1}$ are the mean and covariance matrices of the posterior Gaussian distribution of the state x_{k-1}. By performing a first-order Taylor expansion to the function f at the mean of the state x_{k-1}, the first-order partial derivative can be denoted as the Jacobian matrix F:

$$F = \left.\frac{\partial f}{\partial x_{k-1}}\right|_{\hat{x}_{k-1}}. \tag{17}$$

In the update step, the priori estimate of the current state is combined with the current observation z_k to refine the state estimate and obtain the posteriori state estimate x_k, which can be formulated as:

$$\begin{aligned}
K_k &= \bar{P}_k\,H^T\left(H\,\bar{P}_k\,H^T + Q\right)^{-1}, \\
\hat{x}_k &= \bar{x}_k + K_k\left(z_k - h\left(\bar{x}_k\right)\right), \\
\hat{P}_k &= \left(I - K_k H\right)\bar{P}_k,
\end{aligned} \tag{18}$$

where Q represents the covariance of the measurement noise. K_k is the Kalman gain and the posteriori state estimation $x_k \sim \mathcal{N}\left(\hat{x}_k, \hat{P}_k\right)$. Similarly, a first-order Taylor expansion to the function h can be performed at the mean of the prior distribution of the state x_k. The first-order partial derivative could be defined as the Jacobian matrix H:

$$H = \left.\frac{\partial h}{\partial x_k}\right|_{\bar{x}_k}. \tag{19}$$

In this section, We use an EKF framework to fuse the IMU and the results obtained from the lidar odometry, as shown in Figure 5. Let the motion prior model presented in Section 3.2 be the system model and the results from the lidar odometry (which is denoted as the red star in Figure 5) be the measurements. The system model and measurement model are described as follows.

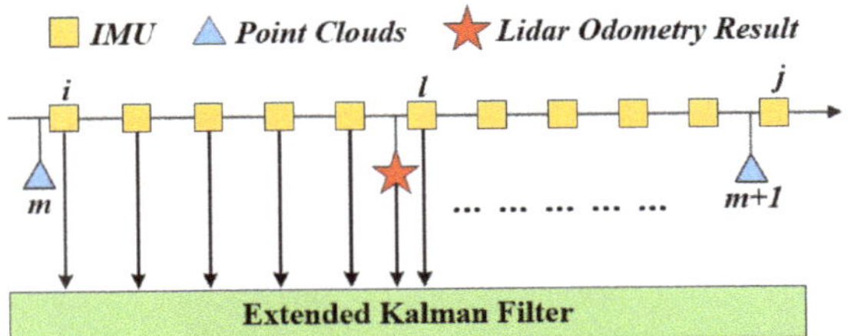

Figure 5. The diagram of the proposed fusion framework. The yellow rectangle represents the IMU data, and the blue triangle indicates the lidar point clouds. The red star denotes the localization result output from the lidar odometry.

3.6.2. System Model

Since the main purpose of our system is to obtain the pose of the vehicle, we could use the *Lie Algebra ξ* to define the state vector x_k at time t_k as:

$$x_k = \begin{bmatrix} \theta_k & p_k \end{bmatrix}^T \in \mathfrak{se}(3).$$

$$(20)$$

By utilizing the logarithmic map and the approximate Baker-Campbell-Hausdorff (BCH) formula [21], the R_k^{WB} obtained from the Equation (6) can be converted to:

$$\theta_k = J_l^{-1}(\theta_{k-1}) \cdot (\omega_k \triangle t) + \theta_{k-1},$$

$$(21)$$

where $J_l^{-1}(\theta_k)$ denotes the right BCH Jacobian of SO(3) and it equals to:

$$J_l^{-1}(\theta_k) = \frac{\theta_k}{2}\cot\frac{\theta_k}{2}I + \left(1 - \frac{\theta_k}{2}\cot\frac{\theta_k}{2}\right)n_k n_k^T - \frac{\theta_k}{2}n_k^\wedge,$$

$$(22)$$

and $J_l^{-1}(\theta_k)$ can be denoted as:

$$J_l^{-1}(\theta_k) = \begin{bmatrix} J_1^{-1}(\theta_k) & J_2^{-1}(\theta_k) & J_3^{-1}(\theta_k) \\ J_4^{-1}(\theta_k) & J_5^{-1}(\theta_k) & J_6^{-1}(\theta_k) \\ J_7^{-1}(\theta_k) & J_8^{-1}(\theta_k) & J_9^{-1}(\theta_k) \end{bmatrix}.$$

$$(23)$$

Combining the Equations (8), (21) and (23), the state transition equation could be defined as:

$$x_k = x_{k-1} + \begin{bmatrix} J_l^{-1}(\theta_{k-1}) \cdot \omega_k \triangle t \\ R_k^{WB}(1:3,1) \cdot \triangle s \end{bmatrix},$$

$$(24)$$

where R_k^{WB} is computed by θ_k^x, θ_k^y and θ_k^z according to Equation (3) and $R_k^{WB}(1:3,1)$ means the first column of the R_k^{WB}.

According to Equations (17) and (24), F can be derived as:

$$F = \begin{bmatrix} F_{\theta\theta} & 0_{3\times3} \\ F_{p\theta} & I_{3\times3} \end{bmatrix} \in \mathbb{R}^{6\times6},$$

$$(25)$$

where $F_{\theta\theta} \in \mathbb{R}^{3\times3}$, and it can be computed by:

$$F_{\theta\theta} = I_{3\times3} + \begin{bmatrix} \left(\frac{\partial J_l^{-1}(\theta_{k-1})}{\partial \theta_{k-1}^x} \cdot \omega_k \triangle t\right)^T \\ \left(\frac{\partial J_l^{-1}(\theta_{k-1})}{\partial \theta_{k-1}^y} \cdot \omega_k \triangle t\right)^T \\ \left(\frac{\partial J_l^{-1}(\theta_{k-1})}{\partial \theta_{k-1}^z} \cdot \omega_k \triangle t\right)^T \end{bmatrix}^T . \tag{26}$$

$F_{p\theta}$ can be computed by:

$$F_{p\theta} = \begin{bmatrix} \frac{\partial R_k^{WB}(1:3,1)}{\partial \theta_{k-1}^x} & \frac{\partial R_k^{WB}(1:3,1)}{\partial \theta_{k-1}^y} & \frac{\partial R_k^{WB}(1:3,1)}{\partial \theta_{k-1}^z} \end{bmatrix} . \tag{27}$$

The prior probability distribution of the state x_k could then be obtained by the Equation (16).

3.6.3. Measurement Model

The measurement vector z_k is denoted as:

$$z_k = \begin{bmatrix} \theta_{z_k}^x & \theta_{z_k}^y & \theta_{z_k}^z & p_{z_k}^x & p_{z_k}^z & p_{z_k}^z \end{bmatrix}^T . \tag{28}$$

According to Figure 5, the lidar odometry algorithm starts to run at time t_m^L, but the result is obtained at time t_k, which is several tens of milliseconds after t_m^L. We assume that the pose of the vehicle at time t_m^L and t_k were approximately equal to T_i^{WB} and T_l^{WB}. Both T_i^{WB} and T_l^{WB} can be obtained from the IMU data.

The relationship between the measurement z_k and the real state x_k of the vehicle at time t_k can be indicated by $\triangle T_{i,l}^B$, which can also be denoted as $\triangle \xi_{i,l} = \begin{bmatrix} \triangle \theta_{i,l} & \triangle p_{i,l} \end{bmatrix}^T$. The observation equation could then be represented as:

$$z_k = \begin{bmatrix} \theta_k \\ \triangle p_{i,l} \end{bmatrix} + \begin{bmatrix} J_l^{-1}(\theta_k) \cdot \triangle \theta_{i,l} \\ \triangle R_{i,l}^B \cdot p_k \end{bmatrix} . \tag{29}$$

According to Equations (19) and (29), H can be derived as:

$$H = \begin{bmatrix} H_{\theta\theta} & 0_{3\times3} \\ 0_{3\times3} & H_{pp} \end{bmatrix} , \tag{30}$$

where $H_{pp} = \triangle R_{i,l}^{WB} \in \mathbb{R}^{3\times3}$, $H_{\theta\theta} \in \mathbb{R}^{3\times3}$ and $H_{\theta\theta}$ can be computed by:

$$H_{\theta\theta} = \begin{bmatrix} \left(\frac{\partial J_l^{-1}(\theta_k)}{\partial \theta_k^x} \cdot \triangle \theta_{i,l}\right)^T \\ \left(\frac{\partial J_l^{-1}(\theta_k)}{\partial \theta_k^y} \cdot \triangle \theta_{i,l}\right)^T \\ \left(\frac{\partial J_l^{-1}(\theta_k)}{\partial \theta_k^z} \cdot \triangle \theta_{i,l}\right)^T \end{bmatrix}^T . \tag{31}$$

Based on the covariance matrix $\bar{P}_k$ from Equation (16) and the Jacobian matrix H from Equation (30), the Kalman gain K_k and the posteriori state estimation could be computed by the Equation (18).

This fusion framework is summarized in Algorithm 1.

Algorithm 1: The fusion framework of our system.

---

input :The history pose set $\{T_i^{WB}\}$ and pose $T_{B_k}^{WB}$ at time t_k^B from the IMU; pose $T_{L_k}^{WB}$ at time t_k^L from the lidar odometry; previous pose T_{k-1}^{WB}

output:Current pose T_k^{WB}

1 **begin**

2   use $T_{B_k}^{WB}$ and T_{k-1}^{WB} to do the state transition by Equation (16) and obtain the prior distribution $\bar{x}_k$;

3   **if** *have* $T_{L_k}^{WB}$ **then**

4   use $\bar{x}_k$, $T_{L_k}^{WB}$ and $\{T_i^{WB}\}$ to compute the posterior distribution $\hat{x}_k$ by Equation (18)

5   output the T_k^{WB} by $\hat{x}_k$

6   **end**

7   **else**

8   output the T_k^{WB} by $\bar{x}_k$

9   **end**

10 **end**

---

4. Experimental Results

The platform used in the following experiments is shown in Figure 6. The vehicle is equipped with a Velodyne HDL-64E lidar, a Xsens MTI300 IMU, two photoelectric encoders, a NovAtel SPAN-CPT GNSS/INS system, and a Nova 5100 industrial computer. The lidar is mounted on the top of the car, and it has 64 beams which could generate over 1.3 million points per second with a high range accuracy. The horizontal Field of View (FOV) is 360°, the vertical FOV is 26.8°, and the maximal detection distance is 120 m. The NovAtel GNSS/INS system provides localization results with centimeter-level accuracy, which can be served as the ground truth.

Figure 6. Our experimental platform. It is equipped with a Velodyne HDL-64E Lidar, a Xsens MTI300 IMU, two photoelectric encoders, a NovAtel SPAN-CPT GNSS/INS system, and a Nova 5100 industrial computer.

4.1. Experiments on Hand–Eye Calibration

As described in Section 3.3, in order to perform the lidar-IMU hand–eye calibration, we need to utilize multiple pairs of relative transformations $\{\triangle T_{k-1,k}^B, \triangle T_{k-1,k}^L\}$.

We choose the frame pairs $\{f_i^L, f_{i-1}^L\}$ before and after the vehicle makes a turn. In addition, all the frames are collected when the vehicle is stationary. Figure 7 shows several transformation pairs we have chosen. In each sub-figure, the left part is the registration result before the hand–eye calibration, whilst the right half is the result after hand–eye calibration. It can be seen that f_i^L and f_{i-1}^L are much better-registered in the right half of Figure 7a–d than the left half, which indicates the importance of hand–eye calibration.

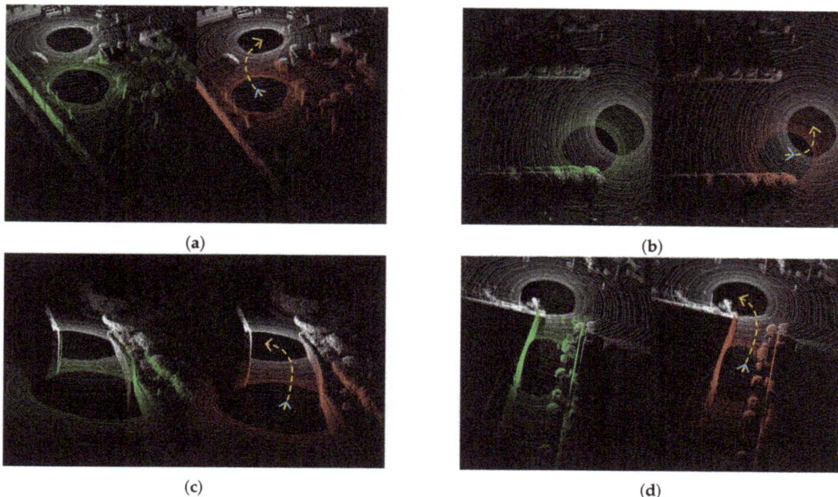

Figure 7. Four typical lidar pairs chosen to perform the hand–eye calibration. In each sub-figure, the left part is the registration result before the hand–eye calibration, whilst the right half is the result after hand–eye calibration. The yellow dotted line indicates the movement trajectory of the vehicle, while the yellow arrow and the blue arrow denote the orientation of the vehicle.

Since the ground truth of T^{BL} is impossible to be obtained, we could not verify the calibration result directly. However, we could use the IMU data to assemble different frames together, and the quality of the assembled scene can be used to evaluate the accuracy of the hand–eye calibration result. Figure 8 shows the assembled scene when the vehicle makes a turn at the intersection. Figure 8a is the assembled result before hand–eye calibration, which is very vague, while the result after the hand–eye calibration in Figure 8b is much clearer. This suggests that the result of our hand–eye calibration is accurate.

To quantitatively test the effect of hand–eye calibration, we choose the LOAM algorithm as the lidar odometry algorithm, and compare the estimated trajectories with the ground-truth trajectory generated by the GNSS/INS system. Figure 9 illustrates the localization result before and after hand–eye calibration. The translation error and rotation error are shown in Figure 9b,c, where the horizontal axis represents the traveled distance and the vertical axis shows the errors. It is obvious that the translation error and the rotation error are significantly reduced after the hand–eye calibration.

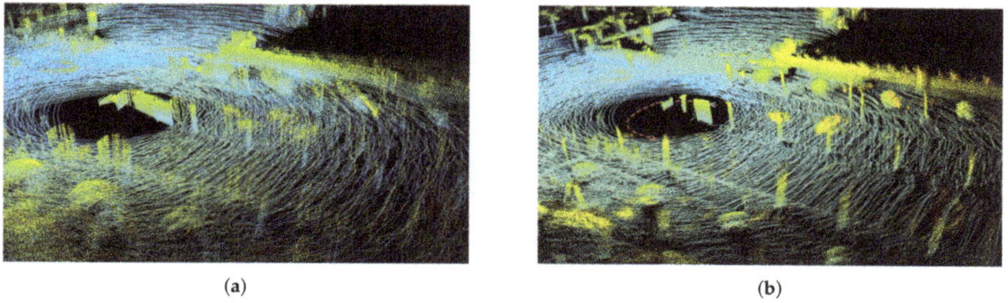

Figure 8. The 3D reconstruction result when the vehicle turns around at the intersection. The left (a) is the result before hand–eye calibration, and the right (b) is the result after hand–eye calibration.

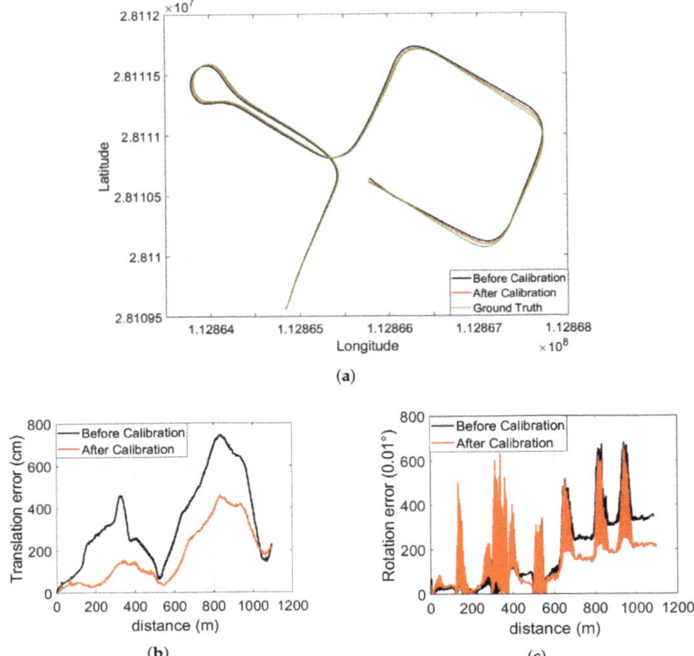

Figure 9. The localization results before and after hand–eye calibration, where (**a**) compares the trajectories before/after hand–eye calibration with the ground truth; (**b**) compares the translation error before/after calibration, and rotation error is shown in (**c**). In this experiment, the UGV drives at a speed around 25 km/h, and the overall distance is about 1.1 km.

4.2. Experiments on Intra-Frame Correction

Based on the hand–eye calibration results, we could also use the 3D reconstruction results to qualitatively test the three rectification methods introduced in Section 3.4. In order to test the result of the rectification, the vehicle should run fast to amplify the intra-frame distortion. Therefore, we carry out this experiment when the vehicle drives at around 60 km/h, and the results are shown in Figure 10.

Figure 10a–d represent different 3D reconstruction results when the vehicle is making a turn. Figure 10a is the result without any rectification, and the reconstructed scene is quite blurry. Figure 10b–d show the results when the distorted point clouds are rectified by the previous registration result, by registering two distorted point clouds and by the IMU data. The results obtained by these three approaches are all much better than the one without rectification. Moreover, the average processing time for these three methods applied on 2000 frames of point clouds is displayed in Table 1. It is seen that the approach of registering two distorted point clouds is the most computational expensive approach.

To quantitatively test the effect of intra-frame correction, LOAM is also used as the lidar odometry algorithm, and the localization accuracy is compared with the ground-truth trajectory. Figure 11 compares the localization results with/without rectification. The trajectories generated by the three rectification methods and without rectification are compared with the ground truth in Figure 11a, where "Method 1" represents rectification by previous registration, "Method 2" is the approach of registering two distorted point clouds, and "Method 3" indicates rectification by IMU data. In addition, Figure 11b,c compare the translation error and rotation error of these approaches.

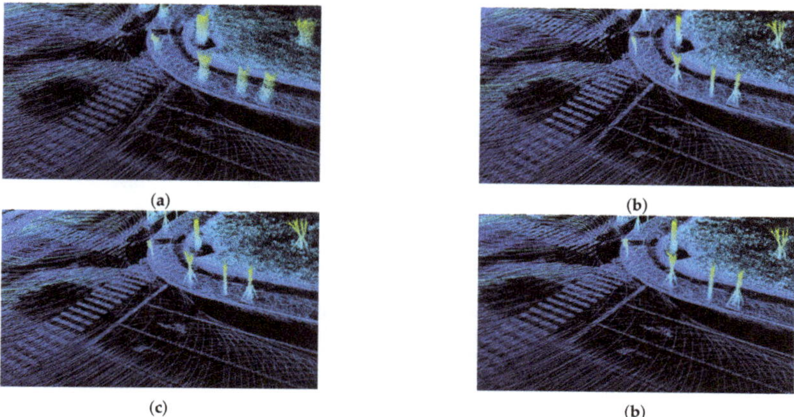

Figure 10. The 3D reconstruction results to test the three rectification methods. (**a**) represents the result without any rectification, and (**b–d**) show the results when the distorted point clouds are rectified by the previous registration result, by registering two distorted point clouds and by the IMU data, respectively.

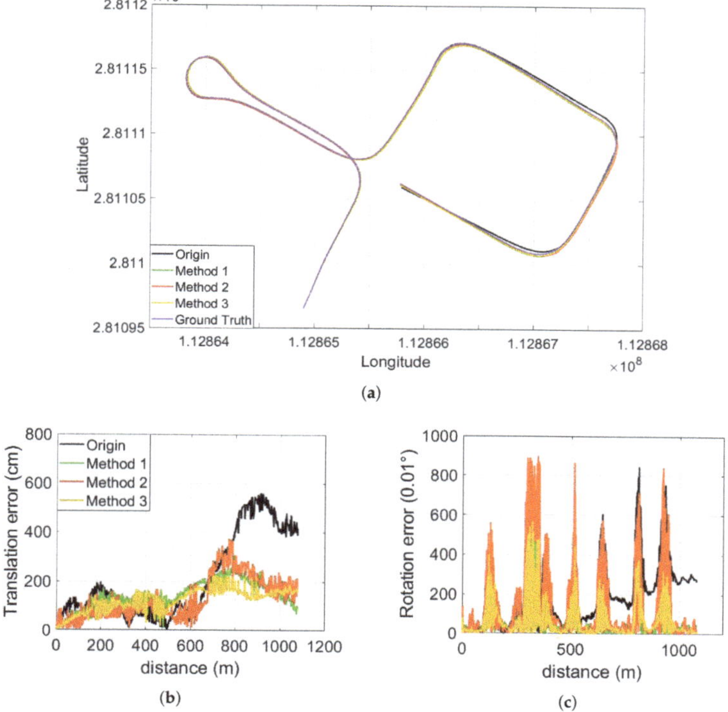

Figure 11. The localization results with/without rectification, where (**a**) compares the trajectories generated by the three rectification methods (Methods 1–3 represent rectification by the previous registration result, by registering two distorted point clouds and by the IMU data, respectively), the trajectory generated without rectification (Origin) and the ground truth; and (**b**,**c**) compare the translation error and rotation error among them, respectively. In this experiment, the UGV drives at a speed of around 60 km/h.

It can be seen that both the translation error and the rotation error are significantly reduced after rectification, and "Method 3" exhibits the best performance, followed by "Method 1" and "Method 2". A detailed error comparison is presented in Table 2.

Table 1. Computational efficiency of three rectification methods.

Rectification Methods	Average Processing Time (ms)
By previous registration	2.565
By registering two distorted point clouds	68.504
By IMU data	4.546

Table 2. Error comparison of three rectification methods.

Rectification Methods	Mean Relative Position Error
Method 1	0.79%
Method 2	0.94%
Method 3	0.57%
Without Rectification	2.28%

4.3. Experiments on Inter-Frame Registration

4.3.1. Tests with Different Methods of Assigning Initial Guess

In this experiment, two methods of assigning initial guess for the lidar odometry algorithm are compared—namely, using the IMU data or the registration result from the previous two scans. In both approaches, the raw distorted point clouds are rectified by the IMU data, and LOAM is used as the lidar odometry algorithm.

The tests are conducted when the vehicle drives at different speeds: a relatively slow speed of around 25 km/h, and a faster speed of around 60 km/h. A detailed error comparison is shown in Table 3. It can be seen that using the IMU data as the initial guess always produces a better result than using the registration result from the previous two scans, especially when the vehicle drives at a faster speed.

Table 3. Error comparison of different methods at the intermediate step.

Speed (km/h)	Methods of Assigning Initial Guess	Mean Relative Position Error
25	IMU	0.31%
	Previous regist	0.46%
60	IMU	0.57%
	Previous regist	1.32%

4.3.2. Tests with Different Lidar Odometry Algorithms

Here, we compare three popular lidar odometry algorithms: ICP, NDT, and LOAM. During the experiments, the point clouds are rectified by the IMU data. The vehicle drives at a speed around 25 km/h, and the overall distance is about 1.1 km. A detailed error comparison is shown in Table 4. It is obvious that LOAM performs much better than NDT and ICP.

Table 4. Error comparison of three lidar odometry algorithms.

Lidar Odometry Methods	Mean Relative Position Error
ICP	6.28%
NDT	3.85%
LOAM	0.31%

4.3.3. Tests in Both Urban and Off-Road Environments

Our algorithm is tested in both urban and off-road environment. We choose to use LOAM as the lidar odometry algorithm. The point clouds are rectified by the IMU data. In the urban environment, the vehicle drives at a speed of around 25 km/h, and the overall distance is about 1.1 km. In the off-road environment, the vehicle drives at a speed of around 15 km/h and the overall distance is about 1.0 km. The trajectories generated by our method are compared with the ground truth in Figure 12. A detailed error comparison is shown in Table 5. It is seen that our approach is applicable for both the urban and off-road environments, whilst the error in the off-road scenario is slightly higher than the urban environment.

Table 5. Error comparison of diverse environments.

Environment	Distance (km)	Mean Relative Position Error
Urban	1.097	0.31%
Off-road	0.983	0.93%

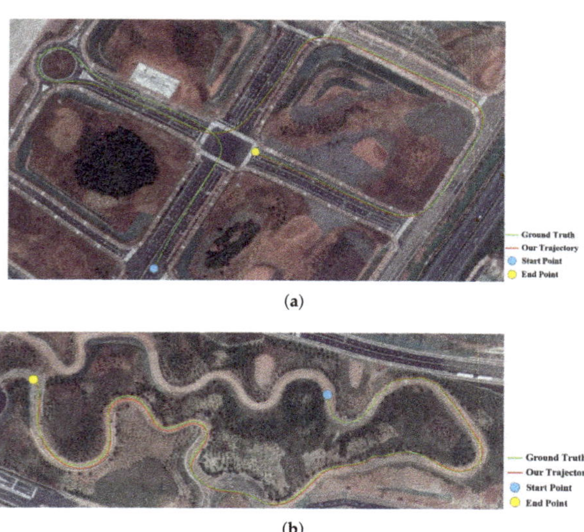

(a)

(b)

Figure 12. The results in the urban environment and off-road environment. (**a**) shows the trajectory generated in the urban environment when the UGV drives at a speed of around 25 km/h, and (**b**) shows the trajectory generated in the off-road environment when the UGV drives at a speed of around 15 km/h.

4.4. Experiments in GNSS-Denied Environment

Finally, we test our whole localization system in the GNSS-Denied environment. Figure 13a shows the trajectory generated by our localization system when the UGV drives at a speed of 20 km/h. By projecting the trajectory to the satellite image, we find that the position drift is quite small. Figure 13b compares the trajectories generated by our localization system and by the disturbed GNSS/INS system. Figure 13c shows the output frequency of our localization system. The average output frequency is around 39.7 Hz. In summary, no matter whether the GNSS signal is available or not, our system can output precise and high-frequency localization results.

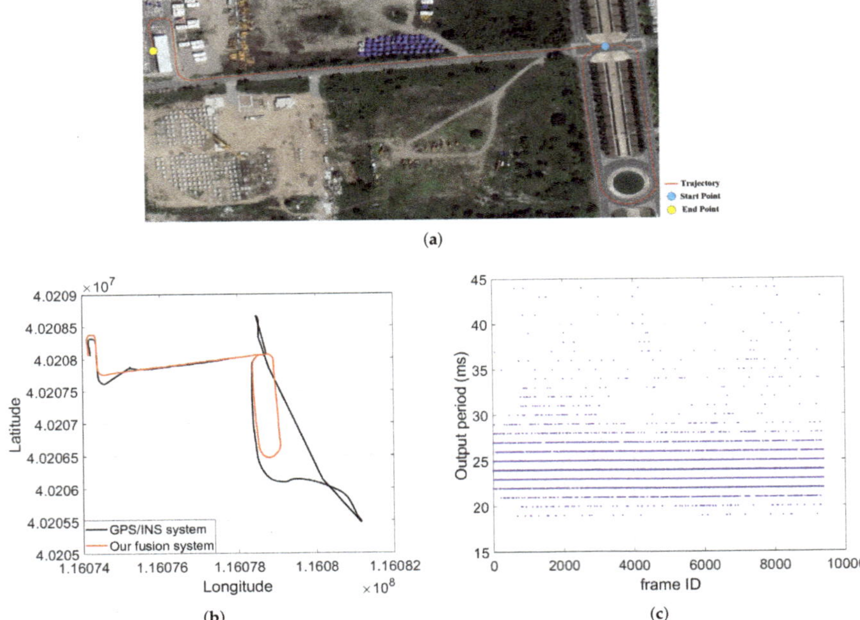

Figure 13. The results in a GNSS-deniedenvironment. (**a**) shows the trajectory generated by our system when the UGV drives at a speed around 20 km/h and the overall distance is about 1.3 km. (**b**) compares the trajectories generated by our system and by the disturbed GNSS/INS system. (**c**) shows the output period of our localization system during the whole process, and the average output frequency is about 39.7 Hz.

5. Conclusions

In this paper, we present a novel approach for combining the IMU data with lidar odometry. The proposed localization system could generate stable, accurate, and high-frequency localization results. Since our system does not recognize the loop closure, we can further extend our work to enable the ability of loop-closure detection.

Author Contributions: Conceptualization, H.F.; Data curation, H.X.; Formal analysis, H.X.; Funding acquisition, B.D.; Investigation, H.X.; Methodology, H.F. and H.X.; Project administration, B.D.; Resources, H.F. and B.D.; Software, H.F. and H.X.; Supervision, B.D.; Validation, H.F.; Original draft, H.X.; Review and editing, H.F.

Funding: This work was supported by the National Natural Science Foundation of China under No. 61790565.

Conflicts of Interest: The authors declare no conflict of interest.

References

1. Kos, T.; Markezic, I.; Pokrajcic, J. Effects of multipath reception on GPS positioning performance. In Proceedings of the Elmar-2010, Zadar, Croatia, 15–17 September 2010; pp. 399–402.

2. Kohlbrecher, S.; von Stryk, O.; Meyer, J.; Klingauf, U. A flexible and scalable SLAM system with full 3D motion estimation. In Proceedings of the 2011 IEEE International Symposium on Safety, Security, and Rescue Robotics, Kyoto, Japan, 1–5 November 2011; pp. 155–160.

3. Gao, Y.; Liu, S.; Atia, M.M.; Noureldin, A. INS/GPS/LiDAR integrated navigation system for urban and indoor environments using hybrid scan matching algorithm. *Sensors* **2015**, *15*, 23286–23302. [CrossRef] [PubMed]

4. Tang, J.; Chen, Y.; Niu, X.; Wang, L.; Chen, L.; Liu, J.; Shi, C.; Hyyppä, J. LiDAR scan matching aided inertial navigation system in GNSS-denied environments. *Sensors* **2015**, *15*, 16710–16728. [CrossRef] [PubMed]

5. Liu, S.; Atia, M.M.; Gao, Y.; Givigi, S.; Noureldin, A. An inertial-aided LiDAR scan matching algorithm for multisensor land-based navigation. In Proceedings of the 27th International Technical Meeting of the Satellite Division of the Institute of Navigation, Tampa, FL, USA, 8–12 September 2014; pp. 2089–2096.

6. Ye, H.; Liu, M. LiDAR and Inertial Fusion for Pose Estimation by Non-linear Optimization. *arXiv* **2017**, arXiv:1710.07104.

7. Pierzchala, M.; Giguere, P.; Astrup, R. Mapping forests using an unmanned ground vehicle with 3D LiDAR and graph-SLAM. *Comput. Electron. Agric.* **2018**, *145*, 217–225. [CrossRef]

8. Besl, P.J.; McKay, N.D. A Method for Registration of 3-D Shapes. *IEEE Trans. Pattern Anal. Mach. Intell.* **1992**, *14*, 239–256. [CrossRef]

9. Pomerleau, F.; Colas, F.; Siegwart, R.; Magnenat, S. Comparing ICP variants on real-world data sets. *Auton. Robots* **2013**, *34*, 133–148. [CrossRef]

10. Stoyanov, T.; Magnusson, M.; Andreasson, H.; Lilienthal, A.J. Fast and accurate scan registration through minimization of the distance between compact 3D NDT representations. *Int. J. Robot. Res.* **2012**, *31*, 1377–1393. [CrossRef]

11. Magnusson, M.; Lilienthal, A.; Duckett, T. Scan registration for autonomous mining vehicles using 3D-NDT. *J. Field Robot.* **2007**, *24*, 803–827. [CrossRef]

12. Nguyen, V.; Gachter, S.; Martinelli, A.; Tomatis, N.; Siegwart, R. A comparison of line extraction algorithms using 2D range data for indoor mobile robotics. *Auton. Robots* **2007**, *23*, 97–111. [CrossRef]

13. Aghamohammadi, A.; Taghirad, H. Feature-Based Laser Scan Matching for Accurate and High Speed Mobile Robot Localization. In Proceedings of the EMCR, Freiburg, Germany, 19–21 September 2007.

14. Furukawa, T.; Dantanarayana, L.; Ziglar, J.; Ranasinghe, R.; Dissanayake, G. Fast global scan matching for high-speed vehicle navigation. In Proceedings of the 2015 IEEE International Conference on Multisensor Fusion and Integration for Intelligent Systems (MFI), San Diego, CA, USA, 14–16 September 2015; pp. 37–42.

15. Zhang, J.; Singh, S. LOAM: Lidar Odometry and Mapping in Real-time. In Proceedings of the Robotics: Science and Systems, Berkeley, CA, USA, 12–16 July 2014; pp. 109–111.

16. Zhang, J.; Singh, S. Low-drift and real-time lidar odometry and mapping. *Auton. Robots* **2017**, *41*, 401–416. [CrossRef]

17. Hemann, G.; Singh, S.; Kaess, M. Long-range GPS-denied aerial inertial navigation with LIDAR localization. In Proceedings of the 2016 IEEE/RSJ International Conference on Intelligent Robots and Systems (IROS), Daejeon, Korea, 9–14 October 2016; pp. 1659–1666.

18. Wan, G.; Yang, X.; Cai, R.; Li, H.; Zhou, Y.; Wang, H.; Song, S. Robust and Precise Vehicle Localization Based on Multi-Sensor Fusion in Diverse City Scenes. In Proceedings of the 2018 IEEE International Conference on Robotics and Automation (ICRA), Brisbane, QLD, Australia, 21–25 May 2018; pp. 4670–4677.

19. Barfoot, T.D. *State Estimation for Robotics*; Cambridge University Press: Cambridge, UK, 2017; pp. 231–234.

20. Kepper, J.H.; Claus, B.C.; Kinsey, J.C. A Navigation Solution Using a MEMS IMU, Model-Based Dead-Reckoning, and One-Way-Travel-Time Acoustic Range Measurements for Autonomous Underwater Vehicles. *IEEE J. Ocean. Eng.* **2018**, 1–19. [CrossRef]

21. Forster, C.; Carlone, L.; Dellaert, F.; Scaramuzza, D. On-Manifold Preintegration for Real-Time Visual–Inertial Odometry. *IEEE Trans. Robot.* **2017**, *33*, 1–21. [CrossRef]

22. Wu, L.; Ren, H. Finding the Kinematic Base Frame of a Robot by Hand-Eye Calibration Using 3D Position Data. *IEEE Trans. Autom. Sci. Eng.* **2017**, *14*, 314–324. [CrossRef]

23. Shah, M.; Eastman, R.D.; Hong, T. An overview of robot-sensor calibration methods for evaluation of perception systems. In Proceedings of the Workshop on Performance Metrics for Intelligent Systems, College Park, MD, USA, 20–22 March 2012; pp. 15–20.

24. Tsai, R.Y.; Lenz, R.K. A new technique for fully autonomous and efficient 3D robotics hand/eye calibration. *IEEE Trans. Robot. Autom.* **1989**, *5*, 345–358. [CrossRef]

25. Dornaika, F.; Horaud, R. Simultaneous robot-world and hand-eye calibration. *IEEE Trans. Robot. Autom.* **1998**, *14*, 617–622. [CrossRef]

26. Huang, K.; Stachniss, C. On Geometric Models and Their Accuracy for Extrinsic Sensor Calibration. In Proceedings of the 2018 IEEE International Conference on Robotics and Automation, Brisbane, QLD, Australia, 21–25 May 2018; pp. 1–9.

27. Scherer, S.; Rehder, J.; Achar, S.; Cover, H.; Chambers, A.; Nuske, S.; Singh, S. River mapping from a flying robot: State estimation, river detection, and obstacle mapping. *Auton. Robots* **2012**, *31*, 189–214. [CrossRef]
28. Cadena, C.; Carlone, L.; Carrillo, H.; Latif, Y.; Scaramuzza, D.; Neira, J.; Reid, I.; Leonard, J.J. Past, Present, and Future of Simultaneous Localization and Mapping: Toward the Robust-Perception Age. *IEEE Trans. Robot.* **2016**, *32*, 1309–1332. [CrossRef]

Article

Determination of the Optimal State of Dough Fermentation in Bread Production by Using Optical Sensors and Deep Learning

Lino Antoni Giefer [1,2,*], **Michael Lütjen** [2], **Ann-Kathrin Rohde** [2] and **Michael Freitag** [1,2]

[1] Faculty of Production Engineering, University of Bremen, Badgasteiner Str. 1, 28359 Bremen, Germany; fre@biba.uni-bremen.de
[2] BIBA—Bremer Institut für Produktion und Logistik GmbH, University of Bremen, Hochschulring 20, 28359 Bremen, Germany; ltj@biba.uni-bremen.de (M.L.); rod@biba.uni-bremen.de (A.-K.R.)
* Correspondence: gif@biba.uni-bremen.de; Tel.: +49-(0)421-218-50147

Received: 24 July 2019; Accepted: 8 October 2019; Published: 11 October 2019

Abstract: Dough fermentation plays an essential role in the bread production process, and its success is critical to producing high-quality products. In Germany, the number of stores per bakery chain has increased within the last years as well as the trend to finish the bakery products local at the stores. There is an unsatisfied demand for skilled workers, which leads to an increasing number of untrained and inexperienced employees at the stores. This paper proposes a method for the automatic monitoring of the fermentation process based on optical techniques. By using a combination of machine learning and superellipsoid model fitting, we have developed an instance segmentation and parameter estimation method for dough objects that are positioned inside a fermentation chamber. In our method we measure the given topography at discrete points in time using a movable laser sensor system that is located at the back of the fermentation chamber. By applying the superellipsoid model fitting method, we estimated the volume of each object and achieved results with a deviation of approximately 10% on average. Thereby, the volume gradient is monitored continuously and represents the progress of the fermentation state. Exploratory tests show the reliability and the potential of our method, which is particularly suitable for local stores but also for high volume production in bakery plants.

Keywords: fermentation monitoring; quality inspection; process automation; deep learning; superellipsoid model fitting; optical sensor

---

1. Introduction

The bakery trade has experienced significant changes in recent years. While the amount of sales locations in Germany remained at a constant level, the diversity of bakeries decreased significantly [1]. That effect reflects a sharply increasing branch network structure of the whole industry, which is directly related to an increase of fermentation and baking at the sales location. Thereby, the bakeries compete directly with the fast food economy of McDonald's and Burger King [2]. A bread manufacturing process of branch network structures roughly follows the steps that are illustrated in Figure 1, whereby the primary products are produced at a central production bakery and then transported to the branches.

Figure 1. Bread manufacturing process.

In detail, the preparation of the dough represents the start of the bread manufacturing process chain, where the raw materials are mixed up and finished dough is divided into rations. In the next step, the shape of the desired product is given to each ration, followed by its cooling down to approximately −4 °C, which interrupts the fermentation. In that way, the dough is prepared for its cold storage and transportation inside of special refrigerated vans to the particular stores. This process is called proofing retardation [3]. In the branches, the fermentation process is initiated at defined conditions (e.g., discrete-time control of warmth and humidity) in special fermentation chambers. Usually, this process step ends after a defined time and the baking step follows, before the final product is produced. The proposed method is applied within the process step 'fermentation' and helps to achieve a higher process quality by using automation techniques.

Chemically, fermentation means the process of the production of carbon dioxide and ethyl alcohol by the transformation of assimilable carbohydrates and amino acids induced by the metabolism of yeast [4]. This process leads to a volume increase of the dough due to the development of gas cells. Figure 2 illustrates the expected volume development of fermenting dough schematically [5]. Herein, the area of optimal mellowness may be identified.

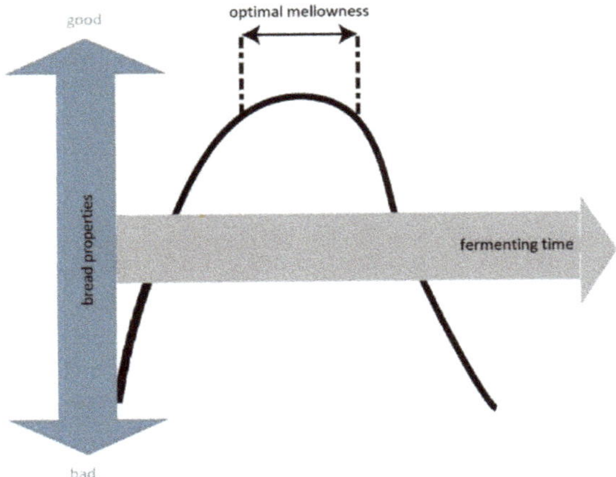

Figure 2. Volume development during fermentation (own illustration based on [5]).

Typically, when the volume gradient approaches zero, the optimal fermenting state is reached, which has to be detected ideally by the staff. Traditionally, the staff evaluates the fermenting state of the dough and estimates the perfect time for ending the fermentation phase. That requires long experience, intuition, and time from the staff. However, today there is an unsatisfied demand for skilled workers, which has lead to an increasing number of untrained and inexperienced employees at the stores. The result is a fixed standard time for the dough, which can be easily programmed and helps to get operationalizable processes. Since the main component of the dough is flour and as its fermentation ability is based on the special cultivation and growing conditions of the grain, this component massively affects the fermentation process and leads consequently to sub-optimal product qualities. By assuming a normal distribution, nearly 15% of the sold products have not only sub-optimal product qualities in taste but are also too small.

Currently, a pre-derivation of the flour properties using effective analysis is not feasible. Additionally, changes in the uniform circulation of temperature or humidity caused by the volume increase of the doughs induce an unpredictable change of the process parameters. The detection and counteracting of these impingements are requirements for high quality. In summary, it is generally not possible to reach the optimal fermenting state only by controlling the time and complying with the machine parameters. Computer-vision based-systems have become increasingly reliant on the production and logistics sector and represent one of the core concepts of the industry 4.0. The usage of image processing and machine learning techniques in the food industry is a relatively new field and offers a vast potential to control processes that are traditionally based on human observation [6]. The following paper covers the detection and control of the temporal volume growth of pieces of dough out of three-dimensional point clouds. The motivation is to create a method that is not only capable of predicting the optimal fermenting state based on the volume gradient but can also serve as the basis for other systems that control a change in the volume of objects. An additional process parameter that is used in some fermentation chambers is the insertion of aerosol mist, which consists of water drops with a diameter of few micrometers. Using that technique, humidity of nearly 100% can be achieved without, unlike with the use of steam, the problem of condensation, which can lead to hygienic risk due to the growth of mold [7]. Our proposed system should be able to perform a proper measurement despite the aerosol, which is not possible for the human eye.

A patent application for the system for the automatic capturing of the fermentation chamber topology and the determination of the volume of dough pieces has been lodged [8].

2. State of the Art

2.1. Fermentation Monitoring

Few studies deal with the topic of monitoring the fermentation state of dough pieces. Elmehdi et al. propose a non-destructive method for real-time information gaining of changes in the structure of dough during the fermentation using low-intensity ultrasonic waves [9]. Utilizing that, a correlation between the fermentation time and the ultrasonic velocity and attenuation can be observed. Increasing fermentation time leads to an increasing attenuation due to the density change of the dough, and hence leads to a decreasing speed of the ultrasonic waves. Skaf et al. describe a sensor that generates low-frequency acoustic waves through an oscillating piezoelectric element to validate the kinetics of bread dough during fermentation [10]. An emitter and a receiver are brought into contact with a piece of dough, and the attenuation of the emitted acoustic signal that changes due to the formation and growth of gas bubbles during the fermentation is measured. By means of that method, the influence of different process parameters such as temperature and humidity can be observed. Both of these proposed methods have the disadvantage of being restricted to only one dough sample at a time and hence being inappropriate for the control of a whole batch of fermenting bread without human supervision. Bajd et al. use magnetic resonance microscopy for the continuous control of dough fermentation and bread baking [11]. The proposed method delivers good results for the monitoring of dough pore distribution and dough volume regarding one dough piece but can theoretically be scaled up to more dough pieces at a time laying in one plane. A significant disadvantage is the complex measurement setup and the missing possibility for being used to upgrade existing fermentation chambers. Ivorra et al. propose an optical method of continuous fermentation state monitoring using a 3D vision system composed of a line laser and a camera and installed inside a fermentation chamber [4]. By means of this method, the height and transversal area of only one dough sample can be measured, and thus the fermentation state controlled. Pour-Damanab et al. use a digital imaging method to monitor the dynamic density of dough during fermentation [12]. The fermenting dough is taken out of the fermentation chamber to take a picture with a camera positioned orthogonal to the object. With that method, conclusions about the density by means of the calculated volume of the dough can be drawn. The technique is invasive and not practicable for multiple samples.

A restriction of all existing methods is the limitation to be only applicable in one plane. Standard fermentation chambers consist of multiple layers of metal sheets containing many dough pieces. The monitoring of one sample and even of one layer representatively is not sufficient because parameters like temperature and humidity can vary at different areas within the fermentation environment, leading to a different fermentation mellowness.

2.2. Robust Object Recognition

The approach used of monitoring multiple layers requires more robust object recognition techniques because the metal surfaces of the fermentation chambers. Several publications deal with object detection and recognition of three-dimensional point clouds. Scholz-Reiter and Thamer present a simulation platform for multi-view sensor fusion of synthetic time-of-flight (ToF) images to serve as the base for following object recognition tasks [13]. The sensor outputs are fused to one three-dimensional point cloud to obtain a suitable field of view. Qi et al. propose a novel structure of a neural network called PointNet, which allows a direct object recognition and segmentation of three-dimensional point clouds [14]. The generation of training data is very time-consuming because the ground-truth data has to be generated manually, which means that every point belonging to a particular object has to be marked as such. In comparison to the mentioned methods, we use an instance segmentation network Mask Region-based Convolutional Neural Network (Mask R-CNN), proposed by He et al. and originally developed for RGB-images [15]. To our knowledge, we are among the first who propose an approach of using this network structure for the instance segmentation of depth images for a tangible application.

Different kinds of shape representations have been developed for object recovery, such as the extruded generalized cylinder [16], the recognition-by-components based on so-called geons [17], or the representation by superquadrics [18,19]. Due to the flexibility and simplicity, superquadrics are currently used most frequently in computer graphics and computer vision. In Thamer and Scholz-Reiter, a method for the segmentation and object recognition of point clouds generated by laser scanners is proposed [20]. The segmentation is based on the local curvature using the surface normal. By fitting superquadrics to the segmented data, the type of the object is detected based on the shape. Vezzani et al. present a method for a superquadric fitting of objects to serve as a base for robot grasping applications and achieve promising results. We chose the approach of superquadric fitting to estimate the shape parameters and hence the volume of detected objects.

3. Materials and Methods

3.1. Dough Volume Monitoring

To create a topology of the inside of the fermentation chamber, we developed a three-dimensional measuring system consisting of a linear motor (Figure 3: Linear motor), a holder with adjustable angle around the x-axis (Figure 3: Holder), a two-dimensional light detection and ranging (LIDAR) scanner (RPLidar A1M8) (Figure 3: LIDAR), and a time of flight distance sensor (Adafruit VL53L0X) (Figure 3: Time of flight distance sensor) to extend the LIDAR data to the third dimension. The system needed to be as rigid as possible to obtain highly accurate measurements. We integrated the measurement system to the back of the chamber interior. Figure 3 illustrates the system schematically on the left and shows the real module on the right side.

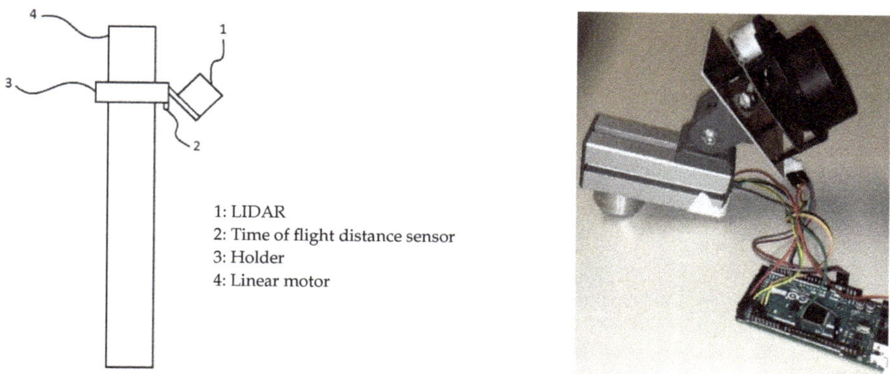

Figure 3. Schematical illustration of the measurement system (**left**) and real measurement module (**right**).

The two sensor outputs were fused by means of an Arduino microcontroller to create a single data stream containing the X, Y, and Z-coordinate of a measurement point. Therefore, the polar coordinates that are generated by the LIDAR and represented by the particular angle θ and the corresponding distance r are transformed into Cartesian coordinates by means of the transformation:

$$x = r * cos(\theta) \tag{1}$$

$$y = r * sin(\theta) \tag{2}$$

The distance sensor directing vertically to the bottom of the fermentation chamber returns measurements of the current height of the measurement sensor system. By moving the system along the Z-axis, a complete topology of one row of the observed situation is captured and serves as the basis for further image processing that detects and investigates the volume of each dough piece. The tilting of the LIDAR scanner by the angle α enables the possibility of monitoring multiple rows of dough pieces laying behind each other. In Figure 4, the general setup of the measurement system is illustrated. This represents the real setup of bread dough pieces within a typical fermentation setup on the left and the schematic overview of the measurement process on the right side.

Figure 4. Test setup (**left**) and schematic (**right**).

In the example of one object (Figure 4 right), the measurement can be described as follows. Due to the known height of the sensor system z and the tilt angle α, the height of an object z_{dough} at a certain position can be calculated as equation 3 by using simple trigonometric relationships:

$$z_{dough} = z - r \times \cos(90° - \alpha) \tag{3}$$

The motion of the linear motor in the Z-direction is performed automatically by means of a continuous loop moving between the maximal top and bottom positions with a step speed of $2\frac{mm}{s}$ to generate a sufficiently dense point cloud. In order to analyze the gradient of the volume development of the dough pieces, the topology of the fermentation chamber is measured continuously in five-minute steps. The underlying setup of the measurements in this paper generates point cloud files of approximately 6500 three-dimensional coordinate points in that way with an accuracy of 0.5-1% of the range, respectively 0.5 mm when measuring within 1.5 m, which is mostly the case in our setup. In comparison to the use of a camera, we are able to obtain depth information of the whole fermentation chamber interior, which allowed us to perform three-dimensional object recognition directly.

3.2. Data Processing

The information extraction of the measured topography of the setting is performed by executing the procedure illustrated in Figure 5. The difficulty is in obtaining a highly accurate segmentation with the number of outliers reduced to a minimum. Otherwise, false models are fitted to the segmented points leading to an inaccurate volume estimation.

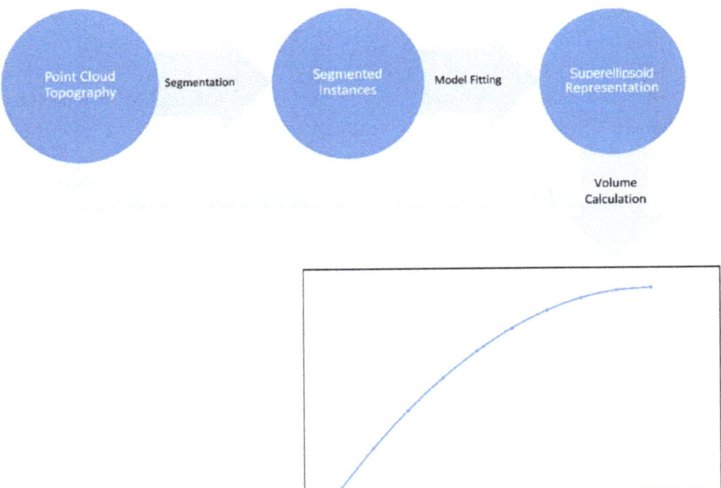

Figure 5. Processing Sequence.

Initially, the point cloud is segmented by means of a machine learning approach using a Mask Region-based convolutional neural network [15]. This neural network simultaneously performs the tasks of classification, bounding box regression, and mask estimation. The architecture of the Mask R-CNN is illustrated in Figure 6 in an abstracted form. For detailed information the reader is referred to [15].

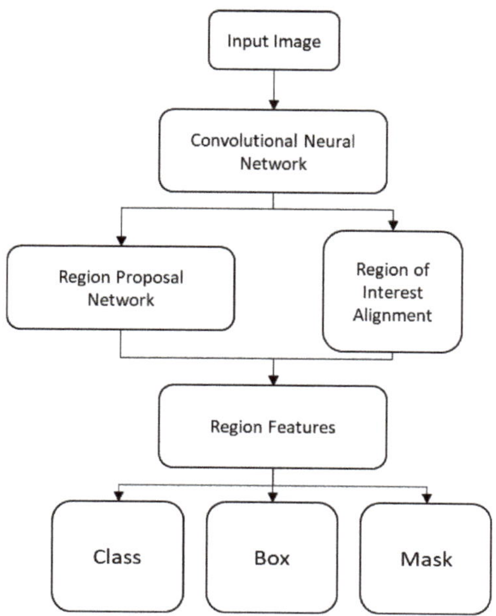

Figure 6. Architecture of Mask R-CNN (own illustration based on [15]).

Classification means the ability to recognize and understand different objects. In this paper, we use this function to predict if any bread dough is present in the current topography or not. The bounding box regression determines the rectangle that frames an object, in our case one piece of dough. By means of the estimated mask, the bounding of the particular object is precisely detected. To fit the neural network to our task, a training dataset containing example topographies that were generated during a real fermentation process was built-up and projected into depth maps by taking the x- and z-coordinates as the particular pixel positions and by normalizing the y-value to an 8-bit linear scale representing the depth. These depth maps were masked with an image annotation tool by defining the regions, and hence all pixels that contain bread. In that way, a dataset of 50 images with the corresponding ground-truth was generated. Due to the small amount of training data, the so-called 'transfer learning method' was applied [21]. Therefore, a weight file that was pre-trained on the huge COCO-dataset (Common Objects in Context dataset) containing 2.5 million labeled instances in 328.000 images was used [22]. With this status, the neural network was already able to recognize common image features. Afterward, the dataset was fed with training material for the bread instance segmentation. The output of the network was a set of images containing the particular segmented bread instances. The images were transformed back to a point cloud by converting the depth value back to the y-coordinate representation. The results of this process step are the segmented points belonging to a particular instance that can be used for the further processing steps.

After the segmentation, model fitting by means of superellipsoids is applied to achieve an implicit representation of the object that allows an analytical volume calculation. The surfaces of superellipsoids fulfill the equation

$$F(X,Y,Z) = \left[\left(\frac{X}{A} \right)^{\frac{2}{\varepsilon_2}} + \left(\frac{Y}{B} \right)^{\frac{2}{\varepsilon_2}} \right]^{\frac{\varepsilon_2}{\varepsilon_1}} + \left(\frac{Z}{C} \right)^{\frac{2}{\varepsilon_1}} = 1 \tag{4}$$

where A, B and C define the axis-scaling and ε_1 and ε_2 the deformation parameters. By means of that equation a large amount of different bodies can be represented. To achieve a complete pose estimation, the translation and rotation by quaternions are added which leads to [23]:

$$
\begin{bmatrix} x \\ y \\ z \end{bmatrix} = R_{Quat} \begin{bmatrix} X - X_0 \\ Y - Y_0 \\ Z - Z_0 \end{bmatrix}
$$
$$
= \begin{bmatrix} a^2 + b^2 - c^2 - d^2 & 2bc - 2ad & 2bd + 2ac \\ 2ad + 2bc & a^2 - b^2 + c^2 - d^2 & -2ab + 2cd \\ -2ac + 2bd & 2ab + 2cd & a^2 - b^2 - c^2 + d^2 \end{bmatrix} \begin{bmatrix} X - X_0 \\ Y - Y_0 \\ Z - Z_0 \end{bmatrix}
\tag{5}
$$

The parameters a, b, c and d represent the quaternion components. In order to find the superellipsoid that fits best regarding a given point cloud, we minimize the objective function

$$
\min_{\lambda} \sum_{i=1}^{n} \left(\sqrt{\lambda_1 \lambda_2 \lambda_3} \left(F^{\lambda_4}(p_i, \lambda) - 1 \right) \right)^2
\tag{6}
$$

where the set λ defines the parameters A, B, C, ε_1, ε_2, X_0, Y_0, Z_0, a, b, c, d of the 12-dimensional optimization problem and p_i with $i = 1 \ldots, n$ a particular point of the point cloud. The factor $\sqrt{\lambda_1 \lambda_2 \lambda_3}$ accelerates the convergence of the minimization algorithm by increasing the gradients around deep minima of the fitting function parameter space [19]. To achieve better numerical value stability, the fitting function is raised to the power of ε_1 as Whaite and Ferrie propose in [24]. In contrast to most of the other researches regarding superellisoid-, or the more general superquadric-model-fitting, where the Levenberg-Marquardt algorithm is used for solving the nonlinear optimization problem ([25–27]), we chose the interior point optimizer (IPOPT) software library. The underlying method applies a primal-dual interior-point algorithm with a filter line-search method and is specially designed for solving large-scale problems [28]. For a detailed explanation of the mathematical background, reference to the literature is provided.

On the basis of the model parameters determined, the volume of the particular superellipsoid is calculated using the equation:

$$
V = 2ABC\varepsilon_1\varepsilon_2\beta\left(\frac{\varepsilon_1}{2} + 1, \varepsilon_1\right)\beta\left(\frac{\varepsilon_2}{2}, \frac{\varepsilon_2}{2}\right)
\tag{7}
$$

where $\beta(x, y)$ represents the beta-function defined by [29]:

$$
\beta(x, y) = 2 \int_0^{\frac{\pi}{2}} \sin^{2x-1}\phi \cos^{2y-1}\phi \, d\phi
\tag{8}
$$

3.3. Preparation of the Dough Pieces and Fermentation Process

We produced standard pieces of dough made of wheat flour. It is of great importance in this process to produce equal pieces of dough with the same rheological properties. Accordingly, the dough production parameters have to be followed accurately. The following percentages of the particular ingredients were used for our recipe: 61% wheat flour Type 550, 34% water, 1.8% gold malt, 1.8% yeast, and 1.4% salt. Wheat flour Type 550 has a mineral content between 0.51% and 0.63% in dry mass and is well suited for fine-pored dough. All ingredients were combined and kneaded with a kneading machine (Diosna Dierks & Söhne GmbH Type SP24F/TEU) for two minutes at first and after a short break for an additional six minutes at room temperature. After kneading, the temperature of the dough was 24 °C, and it was left to rest for ten minutes. The dough was divided into equal dough pieces and set to the intermediate proofing for ten minutes before the final molding of the single dough pieces that were performed by using a dough press (Fortuna). Inside a fermentation chamber (MIWE GR) the temperature was set to 35 °C and the humidity to 95%, which are common process parameters to optimize the growth of the yeast species *Saccharomyces cerevisiae* [30].

4. Results

We evaluated the functionality of our proposed method firstly by means of an experimental setup consisting of a measurement chamber, a CNC (Computerized Numerical Control) positioning unit and a compressor nebulizer (Hangsun CN560) to simulate the real ambient conditions existing in real fermentation chambers. To obtain static measurement objects, we produced salt dough pieces, which were hardened and hence keep their shape during the whole measurement procedure. We measured the topology of the chamber interior containing different dough shapes for two minutes continuously, first without the addition of aerosol, and segmented the points belonging to a certain dough object by means of our instance segmentation neural network. Figure 7 shows the extracted dough surface measurements of two different dough shapes laid over images taken manually from the same perspective as from the sensor system. For better visualization, we reconstructed the surface structure by means of a subdivision surface modifier to obtain a denser mesh.

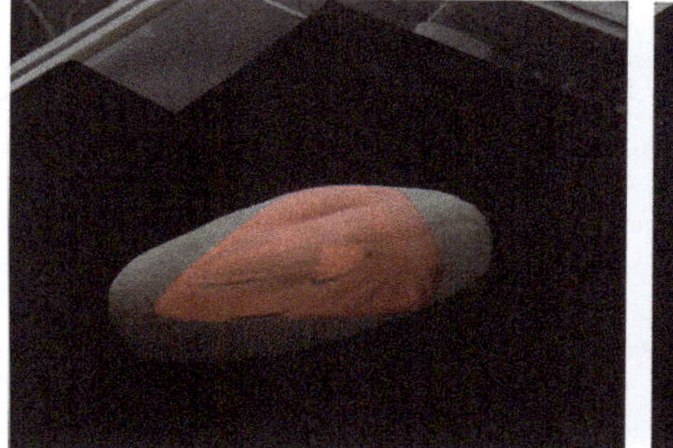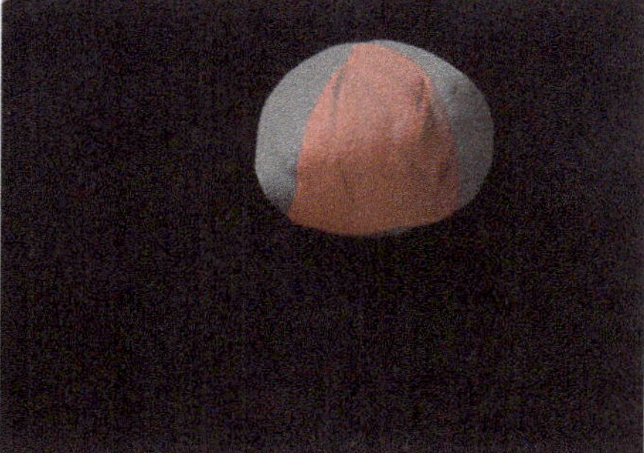

Figure 7. Surface measurement laid over scene images.

The illustration clarifies the limitations of our measurement system regarding the measurable surface part of the objects. Due to the surface curvature of the dough pieces, the reflections of the emitted light pulses outside of the red area do not reach the sensors receiving optics and hence do not generate a measurement point.

To analyze the impact of a present aerosol atmosphere, we measured dough objects with different shapes ones with and ones without induced aerosol. We again segmented the measured dough surfaces and apply a subdivision surface modifier whereby we increased the number of points with a factor of approximately 100. The resulting surface meshes allowed a superimposition to examine the deviations. Figure 8 shows the meshes of three different dough pieces.

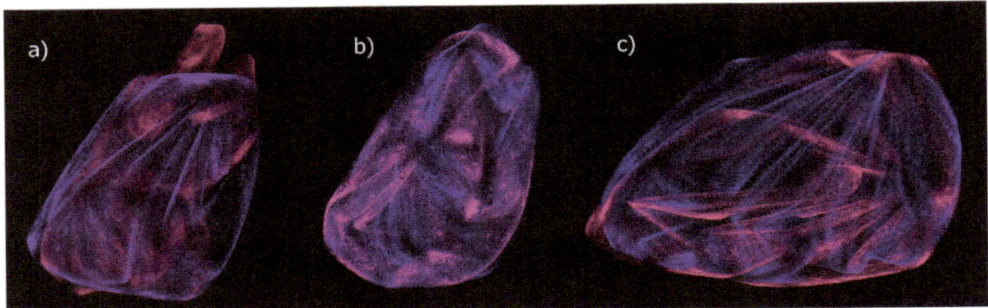

Figure 8. Superimposition of dough surface with (pink) and without induced aerosol (purple) of three different dough objects (**a–c**).

It is observable that the aerosol particles do not have an impact on the quality of the measurement. Both measurements map a comparable surface area of the dough pieces without large deviations. By applying the iterative closest point algorithm of the point clouds generated with and the one without aerosol addition, we obtained a quality measure of the similarity. All point cloud pairs achieved a Euclidean fitness score smaller than one, which proves the high congruence of measurements with and without added aerosol. Due to measurement inaccuracies of our measurement system, a certain degree of variance is not avoidable. So-called Mie scattering does not occur because the wavelength of the used LIDAR is, with ~785 nm, much smaller than the aerosol particle size of around 10 µm [31]. The effect of optical scattering is not observable in the measurements, which could stem from the weak signal strength of the beams reflected from the particles.

In comparison to our proposed method, the existing monitoring methods using optical measurement techniques are not able to work properly under the influence of aerosol. For example, the method used in [12] would not be able to catch an image by a digital camera because the light of the object would not reach the camera sensor. The structured light technique described in [4] might be able to obtain some information due to the strength of the line laser but only when being used on a short distance.

After proving the suitability of our measurement system, we evaluated the quality of our model fitting algorithm to determine the volume of dough. Therefore, we performed ten measurements of our chamber interior with ten different hardened dough pieces, each a member of either the class round or long, and determine the real volume values by means of the water displacement method to obtain a reference value. We fit superellipsoidal models to the segmented dough surface parts and compared the accuracies to the determined reference values. Thereby, the rheological properties of dough have to be taken into consideration. The dough represents a non-Newtonian fluid and dough pieces, that are placed on a surface, show a flow behavior which results in a flat base area [32]. Measurements on the ten dough pieces prepared show that the relation between height and width of the minor-axis is approximately 0.5. Because the surface area that is measured by our measurement system does not contain information about that change of shape, our algorithm would fit a model of a whole superellipsoid to the points. Therefore, we multiplied the calculated volume with the factor 0.5 to obtain the final value. Table 1 represents the reference and the calculated volume values of the ten examined dough pieces and, additionally, the deviation between the two values. In Figure 9, we compare the statistical distributions of the volume deviations of the round and the long dough test pieces by means of boxplots.

Table 1. Volume calculation.

Test dough Piece	Shape	Real Dough Volume	Calculated Dough Volume	Deviation [%]
1	Round	70	78	11.4
2	Round	70	76	8.6
3	Long	140	119	15.0
4	Long	140	122	12.9
5	Long	90	79	12.2
6	Round	70	73	4.3
7	Long	90	81	10.0
8	Long	70	62	11.4
9	Round	140	143	2.1
10	Round	70	76	8.6

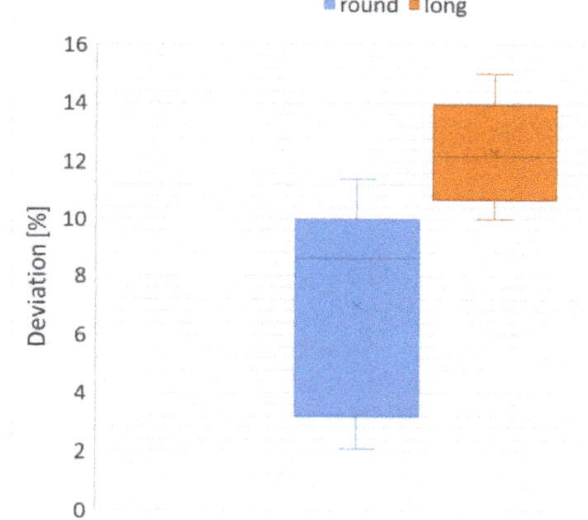

Figure 9. Boxplots for different dough shapes.

One can observe the higher median volume deviation of the elongated dough pieces (12.2% for the long compared to 8.6% for the round dough pieces), although the interquartile and the absolute ranges are smaller. The reason for that is that the surface area, which is captured by our measurement system is not unambiguous and too small to allow proper and accurate modeling. The fact that the volume of long dough test pieces is always calculated lower than the reference values supports this assumption. Nevertheless, the volumes determined represent a good approximation for our purpose.

After verifying the applicability of our volume calculation algorithm, we initiated the fermentation of three dough pieces of round shape inside a fermentation chamber and captured the topology of the interior continuously for 90 min. We determined a time step of five minutes to be appropriate to create a topology that is dense enough for a proper segmentation and model-fitting. Each five minutes, the measured points are assigned to a new topology state cloud to reproduce the volume expansion. In Figure 10, the processing steps performed are illustrated.

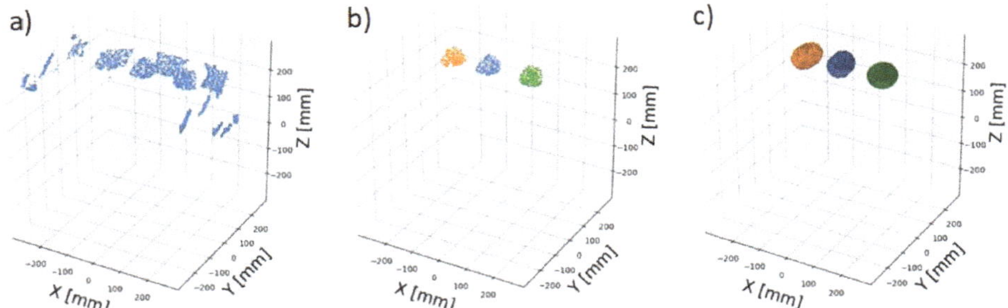

Figure 10. Processing steps: (**a**) Measured topography point cloud of the fermentation chamber inside; (**b**) Segmented dough pieces; (**c**) Model-fitted dough pieces.

Proper segmentation and plausible model-fitting of the points by means of superellipsoids could be seen. The metallic interior surface of the fermentation chamber does not disturb the measurement because the light scatters and bounces in uncontrollable directions. This fact improves the function of the segmentation neural network because of the more precise point cloud structure. The processing time results from the sum of the segmentation and the model fitting time. Currently, the former amounts to less than one second due to the use of an NVIDIA GeForce GTX 1060 GPU. The latter highly depends on the number of points that have to be taken into consideration during the minimization of the objective function. With the current configuration and a number of approximately 100 points per dough piece, a processing time of fewer than three seconds on average was achieved.

To obtain a ground-truth representation of the dough development, we measured the width and height of each dough piece once every five minutes and calculated the volumes assuming an ellipsoidal shape with circular base area by means of:

$$V = \frac{4}{3} \times \pi \times \left(\frac{W}{h}\right)^2 \times h \tag{9}$$

Figure 11 represents our ground truth values of the width, height, the quotient width/height and the volume to be compared with our results.

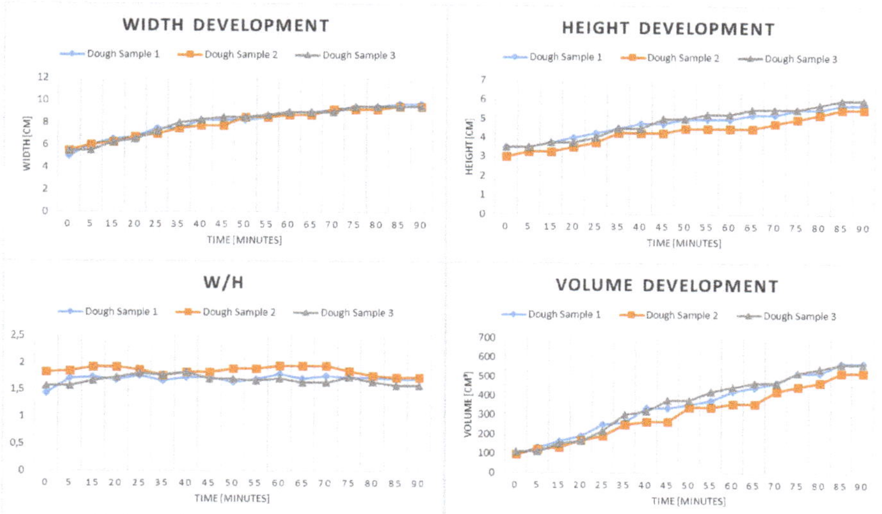

Figure 11. Course of width (**upper left**) and height (upper right) of three dough objects, the quotient W/h (**lower left**) and calculated volume (**lower right**).

The results contradict the assumptions about the determination of the optimal fermentation state made in the introduction of this work. According to a baking expert who was consulted during the experiment, the fermentation would have been stopped after 55 min to obtain an optimal fermentation state. Both the widths and the heights of the dough samples increased approximately linearly also after the designated optimal time and no notable change in the courses, which also caused a nearly linear volume increase. Figure 12 shows the difference between the reference and the calculated volume using our method and the relative deviation with regard to the dough volume.

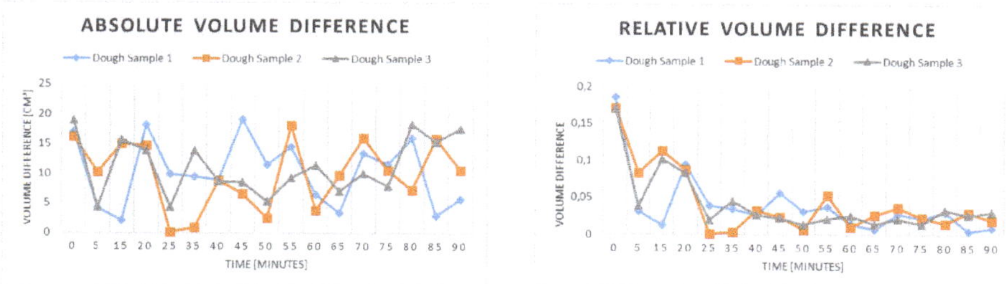

Figure 12. Volume Difference: absolute (**left**) and relative (**right**).

We obtained median volume differences of approximately 10 cm^3 for the three dough pieces during the fermentation period of 90 min, which resembles our measurement of the static test objects. Furthermore, the relative volume difference with regards to the dough volume decreased with increasing fermentation time, which resulted in more accurate volume estimation when the process approaches the optimal state. The method for estimating the volume of watermelons by means of image processing proposed in [33] achieves a deviation of approximately 7.7 % on average. Approximately the same deviation (7.8 %) is stated by the authors in [34] while estimating the volume of kiwi fruits using image processing techniques. Compared to those, our method performs slightly weaker at the beginning with a small object volume, but with increasing fermentation time we obtain more accurate volume estimations because the relative volume error decreases.

The results show that, contrary to the assumptions, the optimal fermentation state cannot be determined only by considering the dough volume gradient. Nevertheless, a pre-defined volume growth factor, which is set by an expert before the start of fermentation, that signifies an optimal state, can be detected properly using our method. In our case, we obtain a factor of approximately three that signifies the desired state.

5. Conclusions and Future Work

In this paper, we proposed a novel method for the continuous monitoring of the volumetric parameters of dough piece during fermentation. We showed that a proper segmentation of a three-dimensional topography and a reliable volume estimation could be achieved by the combination of machine learning and 3D model fitting. Using our proposed method, an essential and promising advancement of the current state of the art has been proposed, which finally enables the possibility of parallel monitoring of multiple dough pieces, instead of single ones, to get a better overview of the complete fermentation interior. In that way, certain planes of fermenting dough pieces can be analyzed individually and independently from each other. We found out that the optimal fermentation state cannot be determined only by considering the volume gradient. It remains to be examined whether other parameters like the surface moisture of the dough can be analyzed to obtain the optimal time automatically. With our method, a predefined desired final volume can be detected automatically, which already can relieve the baking staff. Due to the fact that, using our system, the fermentation chamber does not have to be opened during the fermentation to check the state, a much smoother process without interruptions can be established.

Future work will concentrate on optimizing the model fitting method and minimizing the processing time, which currently takes approximately three seconds on average per object. That is of great interest, since a fermentation chamber contains several dozens of dough pieces whose volumes have to be monitored simultaneously. A difference of one second of computation time for one object applied to 50 dough pieces would lead to a reduction of 50 s altogether, which could make the difference between optimal and over-fermentation. For the acceleration of the calculation, the complete transfer of the process to hardware, like field programmable gate array (FPGA), is conceivable. Another future research task lies in the concurrent detection of different kinds of dough object types like pretzels, donuts, and so on. Therefore, more data has to be captured and our neural network has to be trained to be able to perform the classification of different classes with a high classification rate and accuracy. It should be examined if it is possible to use simulated point clouds for the training, which would lead to an unlimited amount of training data and hence to a much more precise network.

6. Patents

A patent application for the system for the automatic capturing of the fermentation chamber topology and the determination of the volume of dough pieces has been lodged [8].

Author Contributions: Conceptualization, L.A.G. and A.-K.R.; methodology, L.A.G.; software, L.A.G.; validation, L.A.G.; writing—original draft preparation, L.A.G.; writing—review and editing, M.L., A.-K.R. and M.F.; visualization, L.A.G.; supervision, M.L. and M.F.; project administration, L.A.G., M.L., A.-K.R. and M.F.; funding acquisition, M.L., A.-K.R. and M.F.

Funding: This work is part of the research project F.I.T. Fully-Automated Fermentation System, funded by the German Federal Ministry of Economic Affairs and Energy (BMWi, funding code 16KN057228).

Conflicts of Interest: The authors declare no conflict of interest.

References

1. Zentralverband des Deutschen Bäckerhandwerks e. V. Entwicklung Beschäftigte & Betriebe: Beschäftigtenzahlen im Bäckerhandwerk. Available online: https://www.baeckerhandwerk.de/baeckerhandwerk/zahlen-fakten/entwicklung-beschaeftigte-betriebe/ (accessed on 13 December 2018).
2. HuffPost Deutschland. Konkurrenz für McDonald's und Burger King: Das ist der überraschende Spitzenreiter bei Schnell-Restaurants. Available online: https://www.huffingtonpost.de/2015/05/04/schnell-restaurant-spitzenreiter_n_7205306.html (accessed on 15 March 2019).
3. Wing, D.; Scott, A. *The Bread Builders: Hearth Loaves and Masonry Ovens*; Chelsea Green Publishing: White River Junction, VT, USA, 1999; ISBN 1890132055.
4. Ivorra, E.; Amat, S.V.; Sánchez, A.J.; Barat, J.M.; Grau, R. Continuous monitoring of bread dough fermentation using a 3D vision Structured Light technique. *J. Food Eng.* **2014**, *130*, 8–13. [CrossRef]
5. Benz, F. *Backwarenherstellung. Fachkunde für Bäcker, 6*; Dr Schroedel: Hannover, Germany, 1986; ISBN 350791414X.
6. Priyadharshini, K.; Akila, R. A Survey on Computer Vision Technology for Food Quality Evaluation. *Int. J. Innov. Res. Comput. Commun. Eng.* **2016**, *4*, 14860–14865.
7. Lösche, K.; Reichenbach, A. Aerosole & Bio-Additive in der Bäckerei: Innovative Wege zur Verbesserung von Qualität, Haltbarkeit und Energieeffizienz. In *Bäckereitechnologie: Forschung und Innovationen*; f2m Food multimedia Gmbh: Hamburg, Germany, 2015; pp. 50–59.
8. BIBA—Bremer Institut für Produktion und Logistik GmbH, ttz Bremerhaven. *Vorrichtung und Verfahren zur Prozessüberwachung mehrerer Teiglinge in einer Prozesskammer (Pending)*, 2018; 102018124378.2.
9. Elmehdi, H.M.; Page, J.H.; Scanlon, M.G. Monitoring Dough Fermentation Using Acoustic Waves. *Food Bioprod. Process.* **2003**, *81*, 217–223. [CrossRef]
10. Skaf, A.; Nassar, G.; Lefebvre, F.; Nongaillard, B. A new acoustic technique to monitor bread dough during the fermentation phase. *J. Food Eng.* **2009**, *93*, 365–378. [CrossRef]
11. Bajd, F.; Serša, I. Continuous monitoring of dough fermentation and bread baking by magnetic resonance microscopy. *Magn. Reson. Imaging* **2011**, *29*, 434–442. [CrossRef] [PubMed]
12. Pour-Damanab, A.S.; Jafary, A.; Rafiee, S. Monitoring the dynamic density of dough during fermentation using digital imaging method. *J. Food Eng.* **2011**, *107*, 8–13. [CrossRef]
13. Scholz-Reiter, B.; Thamer, H. *Multi-View Sensor Fusion of Synthetic ToF Images for Object Recognition of Universal Logistic Goods*; GfaI: Berlin, Germany, 2011.
14. Qi, C.R.; Su, H.; Mo, K.; Guibas, L.J. Pointnet: Deep learning on point sets for 3d classification and segmentation. In Proceedings of the IEEE Conference on Computer Vision and Pattern Recognition, Honolulu, HI, USA, 21–26 July 2017.
15. He, K.; Gkioxari, G.; Dollar, P.; Girshick, R. Mask R-CNN. In Proceedings of the 2017 IEEE International Conference on Computer Vision, ICCV, Venice, Italy, 22–29 October 2017; IEEE: Piscataway, NJ, USA, 2017; pp. 2980–2988, ISBN 978-1-5386-1032-9.
16. Fang, X.S. The extruded generalized cylinder: A deformable model for object recovery. In Proceedings of the IEEE Conference on Computer Vision and Pattern Recognition, Seattle, WA, USA, 21–23 June 1994; IEEE: Piscataway, NJ, USA, 1994.
17. Biederman, I. Recognition-by-components: A theory of human image understanding. *Psychol. Rev.* **1987**, *94*, 115–147. [CrossRef] [PubMed]
18. Barr, A.H. Superquadrics and angle-preserving transformations. *IEEE Comput. Graph. Appl.* **1981**, *1*, 11–23. [CrossRef]
19. Solina, F.; Bajcsy, R. Recovery of parametric models from range images: The case for superquadrics with global deformations. *IEEE Trans. Pattern Anal. Mach. Intell.* **1990**, *12*, 131–147. [CrossRef]
20. Thamer, H.; Scholz-Reiter, B. 3D-Objekterkennung von heterogenen Stückgütern—Flexible Automatisierung basierend auf 3D-Bildverarbeitung. *Ind. Manag.* **2014**, *31*, 35–38.
21. West, J.; Ventura, D.; Warnick, S. *Spring Research Presentation: A Theoretical Foundation for Inductive Transfer*; Brigham Young University, College of Physical and Mathematical Sciences: Provo, UT, USA, 2007; Volume 1.
22. Lin, T.-Y.; Maire, M.; Belongie, S.; Bourdev, L.; Girshick, R.; Hays, J.; Perona, P.; Ramanan, D.; Zitnick, C.L.; Dollár, P. Microsoft COCO: Common Objects in Context. 2014. Available online: http://arxiv.org/pdf/1405.0312v3 (accessed on 11 October 2019).

23. Perumal, L. Quaternion and its application in rotation using sets of regions. *Int. J. Eng. Technol. Innov.* **2016**, *1*, 35–52.

24. Whaite, P.; Ferrie, F.P. From uncertainty to visual exploration. In Proceedings of the Third International Conference on Computer Vision, Osaka, Japan, 4–7 December 1990; Kak, A.C., Tsuji, S., Eklundh, J.-O., Eds.; IEEE: Piscataway, NJ, USA, 1990; pp. 690–697, ISBN 0-8186-2057-9.

25. Bardinet, E.; Ayache, N.; Cohen, L.D. Fitting of iso-surfaces using superquadrics and free-form deformations. In Proceedings of the IEEE Workshop on Biomedical Image Analysis, Seattle, WA, USA, 24–25 June 1994; IEEE Computer Society Press: Washington, D.C., USA, 1994; pp. 184–193, ISBN 0-8186-5802-9.

26. Duncan, K.; Sarkar, S.; Alqasemi, R.; Dubey, R. Multi-scale superquadric fitting for efficient shape and pose recovery of unknown objects. In Proceedings of the IEEE International Conference on Robotics and Automation (ICRA), Karlsruhe, Germany, 6–10 May 2013; IEEE: Piscataway, NJ, USA, 2013; pp. 4238–4243, ISBN 978-1-4673-5643-5.

27. Biegelbauer, G.; Vincze, M. Efficient 3D Object Detection by Fitting Superquadrics to Range Image Data for Robot's Object Manipulation. In Proceedings of the 2007 IEEE International Conference on Robotics and Automation, Rome, Italy, 10–14 April 2007; IEEE: Piscataway, NJ, USA, 2006; pp. 1086–1091, ISBN 1-4244-0602-1.

28. Wächter, A.; Biegler, L.T. On the implementation of an interior-point filter line-search algorithm for large-scale nonlinear programming. *Math. Program.* **2006**, *106*, 25–57. [CrossRef]

29. Jaklič, A.; Leonardis, A.; Solina, F. Superquadrics and their geometric properties. In *Segmentation and Recovery of Superquadrics*; Springer: Berlin/Heidelberg, Germany, 2000; pp. 13–39.

30. Sinha, N. *Handbook of Food Products Manufacturing, 2 Volume Set*; John Wiley & Sons: Hoboken, NJ, USA, 2007; ISBN 0470113545.

31. Kerker, M. *The Scattering of Light and other Electromagnetic Radiation: Physical Chemistry: A Series of Monographs*; Academic Press: Cambridge, MA, USA, 2013; ISBN 1483191745.

32. Bhattacharya, M.; Hanna, M.A. Viscosity modelling of dough in extrusion. *Int. J. Food Sci. Technol.* **1986**, *21*, 167–174. [CrossRef]

33. Koc, A. Determination of watermelon volume using ellipsoid approximation and image processing. In *Postharvest Biology and Technology*; Elsevier: Amsterdam, The Netherlands, 2007; pp. 366–371.

34. Rashidi, M.; Gholami, M. Determination of kiwifruit volume using ellipsoid approximation and image-processing methods. *Int. J. Agric. Biol.* **2008**, *10*, 375–380, ISSN 1560-8530.

Article

Multi-Sensor Face Registration Based on Global and Local Structures

Wei Li [1],*, Mingli Dong [2],*, Naiguang Lu [2], Xiaoping Lou [2] and Wanyong Zhou [1]

[1] Hebei Engineering Research Center for Assembly and Inspection Robot, North China Institute of Aerospace Engineering, Langfang 065000, China; zhouwanyong@nciae.edu.cn

[2] Key Laboratory of the Ministry of Education for Optoelectronic Measurement Technology and Instrument, Beijing Information Science and Technology University, Beijing 100192, China; nglv2002@sina.com (N.L.); louxiaoping@bistu.edu.cn (X.L.)

* Correspondence: liweikilary@163.com (W.L.); dongml@bistu.edu.cn (M.D.); Tel.: +86-10-8242-6892 (W.L.)

Received: 10 September 2019; Accepted: 28 October 2019; Published: 30 October 2019

Abstract: The work reported in this paper aims at utilizing the global geometrical relationship and local shape feature to register multi-spectral images for fusion-based face recognition. We first propose a multi-spectral face images registration method based on both global and local structures of feature point sets. In order to combine the global geometrical relationship and local shape feature in a new Student's t Mixture probabilistic model framework. On the one hand, we use inner-distance shape context as the local shape descriptors of feature point sets. On the other hand, we formulate the feature point sets registration of the multi-spectral face images as the Student's t Mixture probabilistic model estimation, and local shape descriptors are used to replace the mixing proportions of the prior Student's t Mixture Model. Furthermore, in order to improve the anti-interference performance of face recognition techniques, a guided filtering and gradient preserving image fusion strategy is used to fuse the registered multi-spectral face image. It can make the multi-spectral fusion image hold more apparent details of the visible image and thermal radiation information of the infrared image. Subjective and objective registration experiments are conducted with manual selected landmarks and real multi-spectral face images. The qualitative and quantitative comparisons with the state-of-the-art methods demonstrate the accuracy and robustness of our proposed method in solving the multi-spectral face image registration problem.

Keywords: multi-sensor; face registration; inner-distance; Student's-t Mixtures Model; image fusion

---

1. Introduction

Image fusion can often analyze and extract the complementary information of multi-sensor data, and compose a robust or informative image, which can provide more complex and detailed target scene representation [1–3]. Since the fusion process of multi-sensor images is in accordance with human visual perception, it has an important research significance for object detection and target recognition in the areas of remote sensing image, medical imaging analysis, military target detection and video surveillance [4–6]. E.g., face recognition is one of the major applications of multi-sensor data, such as visible and thermal infrared images. Visible images can provide face texture details with high spatial resolution, while thermal infrared images are not likely to be interfered with by illumination variation or face disguise with high thermal contrast. Therefore, it is beneficial for face recognition to fuse the multi-sensor data, which can combine the advantages of texture detail and thermal radiation information in the multi-spectral face images.

However, image registration is a prerequisite for the success of multi-spectral image fusion, which is an essential and challenging step in the process of image fusion research [7]. Currently, multi-spectral images registration can generally be implemented in two ways: Hardware-based registration and software-based registration. In order to ensure that the multi-spectral images are strictly geometrically aligned, hardware-based registration can be realized based on the coaxial catadioptric optical system through the beam splitter [8]. But the sophisticated and cumbersome imaging sensor equipment for generating co-registered image pairs may not be practical in many face recognition scenarios, due to its high cost and low availability. In contrast, software-based registration can capture dual band images simultaneously and independently by utilizing different off-the-shelf low cost wavebands sensors. Since no additional hardware is needed, it may be more appropriate for practical detection and recognition scenarios compared to the hardware-based registration [9]. In general, software-based registration can be classified into two categories: Area-based and feature-based approaches [10]. Area-based methods attempt to deal directly with the image intensity values to detect, for instance, cross-correlation [11], Fourier transform and mutual information [12,13]. On the contrary, feature-based methods attempt to indirectly extract salient structures or features in the images, such as points of high curvature, corners, strong edges, intersections of lines, structural contours and silhouettes within the images [14–16]. For visible and thermal infrared image pairs usually involve quite different intensity values, and the area-based methods are good usually only on a selected small region of the images. Instead, salient structures with a strong edge could often represent a significant common feature of the heterogeneous images. In view of this, this paper focuses on the feature-based registration methods for the visible and thermal infrared face images.

Feature-based registration methods first extract different feature point sets of salient structures in the multi-spectral images, and then the registration problem can be simplified to determine the correct correspondence and the inherent spatial transformation between two point sets of extracted features. Unlike the face recognition problem [17,18], the multi-sensor face registration problem has not enough common features available to build a convolutional neural network (CNNs) framework to predict the classification model; a popular strategy is to regard the alignment of two point sets as probability density estimation, e.g., the Gaussian Mixture Model (GMM) [19–21] and Student's t Mixture Model (SMM) [22–24]. Because GMM formulations are intensively used, and the heavily tailed SMM is more robust and precise against noise and outliers than the classical GMM, it is a potential research direction to take advantage of SMM for image registration methods [25–27]. In general, both GMM and SMM probabilistic methods are intended to exploit global relationships in the point sets. However, the neighborhood structures among the feature points also contribute to the alignment of two point sets. Thus, another interesting point matching strategy would be aimed first at using local neighborhood structures as the feature descriptor to recover the feature point correspondences, e.g., Shape Context (SC) [28] and Inner-Distance Shape Context (IDSC) [29]. But for multi-spectral face images registration, the local neighborhood structures of feature points are not discriminative enough, because it will inevitably include a number of mismatching points. To address this issue, in this paper we modify the SMM probabilistic framework and take full advantage of the global and local structures during the multi-spectral face images registration process.

Our contribution in this paper includes the following three aspects: Firstly, in order to keep the global and local structures of feature point sets in the registration process, we use IDSC as the local shape descriptors of feature point sets, and treat the confidence of feature matching as the mixing proportion of the finite mixture model, which aims to describe the local shape and global geometrical relationship of feature point sets in a new SMM probabilistic framework. Secondly, in order to make the multi-spectral fusion image hold more apparent details of a visible image and the thermal radiation information of an infrared image, we propose a guided filtering and gradient preserving image fusion strategy for multi-spectral face images fusion. It can improve the robustness of face recognition techniques in varying illumination conditions. Thirdly, we construct a simple multi-spectral imaging system in order to acquire higher resolution multi-spectral face images from visible and thermal infrared cameras simultaneously.

2. The Proposed Registration Method for Visible and Infrared Face Images

For the software-based registration problem, a basic process of infrared and visible images fusion mainly includes the following procedures: Feature extraction and description, feature point sets registration, image transformation and interpolation and finally image fusion [1,30]. As is shown in Figure 1, the first step is to extract robust common salient features that might be preserved in the multi-spectral face images. Considering the fact that visible and infrared images are characterizations of two different modalities, we use edge maps to represent the common salient features of multi-spectral face images. After obtaining the salient features of multi-spectral images, the edge maps can usually be discretized as two feature point sets. Thus, the registration procedure is converted into determining the correct correspondence and estimating the spatial transformation between the feature point sets for multi-spectral face images. Furthermore, the other visible image points can be interpolated using a smoothing Thin-Plate Spline (TPS) [31] based on the pixel coordinates of the thermal infrared image. Finally, the multi-spectral image fusion process takes place in the output image, and the fused image can preserve both the apparent details of the visible image and the thermal radiation information of the infrared image, which will greatly improve the accuracy and efficiency of detection and recognition. In this paper, our attention mainly focuses on the following two aspects: Feature point sets registration and multi-spectral face images fusion.

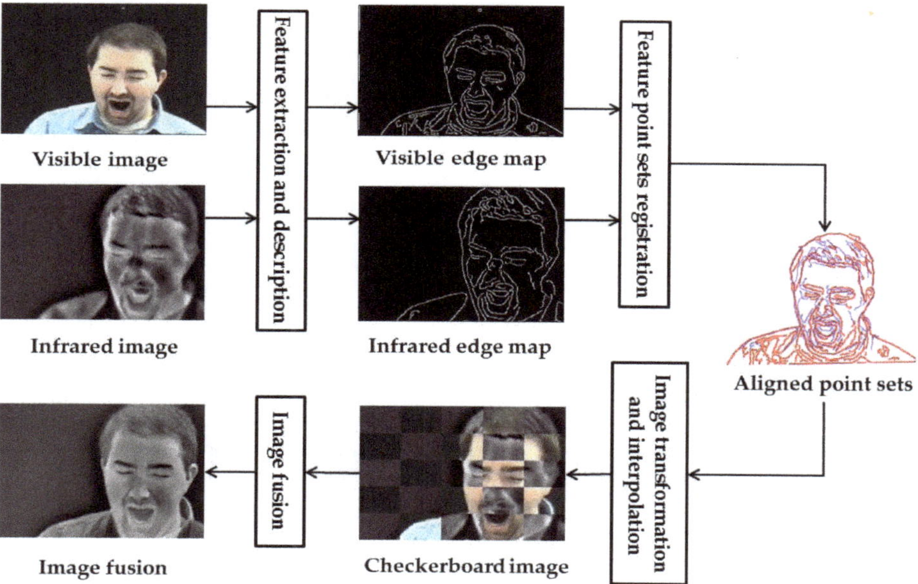

Figure 1. Visible and infrared face images fusion basic process for software-based registration.

2.1. Inner-Distance Shape Context for 2D Feature Descriptors

In order to establish the initial correspondences between the two feature point sets of visible and infrared face images, we first use IDSC as the local feature descriptors, and solve the point sets matching problem through a bipartite graph method [32]. Compared to the SC feature matching method, the IDSC feature matching method uses inner-distance to replace the Euclidean distance, which is robust to articulation shape and part structure. For example, Figure 2 shows the schematic illustration of face silhouette feature descriptors based on IDSC, in the IDSC feature histograms from Figure 2c, the horizontal axis $n_\theta = 12$ denotes the numbers of orientation bins and the vertical axis $n_d = 5$ denotes the numbers of logarithm distance bins.

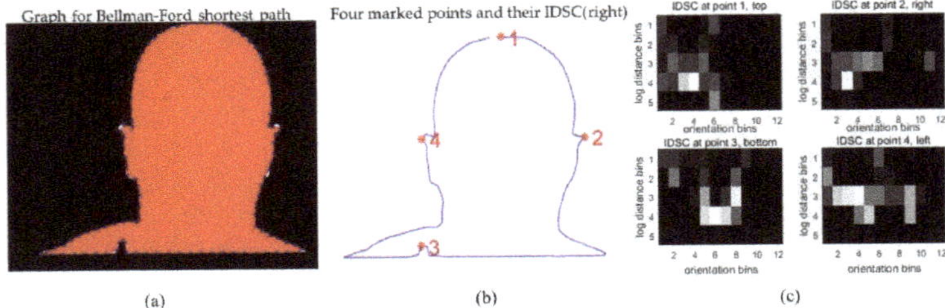

Figure 2. Schematic illustration of face silhouette feature descriptors based on Inner-Distance Shape Context (IDSC). (**a**) Bellman-Ford shortest path graph built using face silhouette landmark points; (**b,c**) Four marked points and their corresponding IDSC feature histograms.

As can be seen from Figure 2c, the IDSC feature histograms for all marked points are quite different, and the excellent discriminative performance demonstrates the inner-distance's ability to capture local structures in the point sets. Consider a point p_i in the point set X and a point q_j in the point set Y. Let $C_{i,j}$ denote the cost of matching p_i to q_j, $h_{X,p_i}(k)$ and $h_{Y,q_j}(k)$ denote the normalized histogram of the relative coordinates of the remaining points, respectively, and K be the number of histogram bins. As the IDSC is represented based on the histogram, it can be measured by the χ^2 test statistic:

$$C_{i,j} = C\left(p_i, q_j\right) = \frac{1}{2}\sum_{k=1}^{K}\frac{\left[h_{X,p_i}(k) - h_{Y,q_j}(k)\right]^2}{h_{X,p_i}(k) + h_{Y,q_j}(k)}, \tag{1}$$

In general, the smaller the matching cost $C_{i,j}$ is, the more similar the local appearance at points p_i and q_j are. Once the cost matrix C for all correspondence points in the point set X and Y is obtained, the correspondences Ω between the two point sets are an instance of an assignment problem, which can be solved by the Hungarian method.

2.2. Student's t Mixtures for Feature Point Set Registration

In order to fully exploit the global structures in the feature point sets, we treat the alignment of two point sets as a probability density estimation of an SMM distribution with the IDSC local feature descriptors. Compared to the GMM distribution, the main advantage of SMM is that Student's t distribution has a heavier tail than the exponentially decaying tail of Gaussian distribution. As is shown in the Figure 3, each component of the SMM has an additional parameter υ called the degrees of freedom, which can change the shape of a Student's t distribution curve, and hence SMM provides a more robust model than the classical GMM.

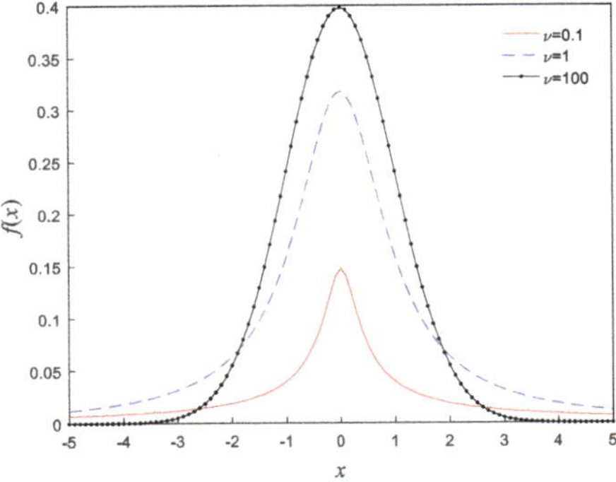

Figure 3. The Student's t distribution for various degrees of freedom.

Given a point set $X_{N \times D} = (x_1, \dots, x_N)^{\mathrm{T}}$, which can be considered as an observation datum, the other point set $Y_{M \times D} = (y_1, \dots, y_M)^{\mathrm{T}}$ is treated as the SMM centroids, where N and M represent the number of points in X and Y, respectively, and D represents the dimension of each point. Then the goal of registration is to align the centroids point set Y to the observation data point set X. Typically, the point sets contain noise and outliers, which can be supposed to be a uniform distribution $1/N$, and the weight of the uniform distribution is denoted as $\eta \in [0, 1]$. Let $w_{ij} \in [0, 1]$ denote a mixing proportion corresponding to the ith component of the SMM for point x_j ($\sum_{i=1}^{M} w_{ij} = 1$), then the mixture probability density function can take the form:

$$f(x|\psi) = \eta \frac{1}{N} + (1 - \eta) \sum_{i=1}^{M} \omega_{ij} f(x_j | y_i, \Sigma_i, v_i), \tag{2}$$

where $\psi = (\psi_1, \psi_2, \dots, \psi_M)$ is mixture parameter set with $\psi_i = (w_{ij}, y_i, \Sigma_i, v_i)$, and $f(x_j | y_i, \Sigma_i, v_i)$ is the Student's t distribution probability density function for the ith component of SMM; we express that:

$$f(x_j | y_i, \Sigma_i, v_i) = \frac{\Gamma\left(\frac{v_i + D}{2}\right) |\Sigma_i|^{-\frac{1}{2}}}{(\pi v_i)^{\frac{1}{2}D} \Gamma\left(\frac{v_i}{2}\right) \left[1 + \frac{d(x_j, y_i; \Sigma_i)}{v_i}\right]^{\frac{v_i + D}{2}}}, \tag{3}$$

where $d(x_j, y_i; \Sigma_i)$ is the Mahalanobis squared distance between observation point x_j and centroid point y_i, while Σ_i, Γ and v represent the covariance matrix, Gamma function and degrees of freedom, respectively.

2.3. Multi-Spectral Face Registration Using Global and Local Structures

In order to estimate the mixture parameter ψ in the mixture probability density function (2), according to [22,23], we first rewrite the function (2) as a complete data logarithm likelihood function $\ln L_C(\psi) = \sum_{j=1}^{N} \ln \sum_{i=1}^{M} w_{ij} f(x_j | y_i, \Sigma_i, v_i)$, and then use the Expectation Maximization (EM) to train the logarithm likelihood function. In general, the EM algorithm can be divided into the following two steps: Expectation step (E-step) and Maximization step (M-step), and the iterative process proceeds by alternating between the E- and M-steps until convergence.

E-step: We first use the current parameter $\psi^{(k)}$ to calculate the posterior probability $\tau_{ij}{}^{(k)}$, and according to the Bayesian Theorem, it can be expressed in terms of the observation data x_j belonging to the ith component of the SMM

$$\tau_{ij}{}^{(k)} = \frac{w_{ij}{}^{(k)} f(x_j | y_i{}^{(k)}, \Sigma_i{}^{(k)}, v_i{}^{(k)})}{f(x_j | \psi^{(k)})}, \tag{4}$$

Next we need to calculate the kth iteration conditional expectation of the logarithm likelihood function $\ln L_C(\psi)$, which can be calculated as follows:

$$Q(\psi | \psi^{(k)}) = Q_1(w | \psi^{(k)}) + Q_2(v | \psi^{(k)}) + Q_3(y, \Sigma | \psi^{(k)}), \tag{5}$$

where Q_1, Q_2 and Q_3 are respectively written as follows:

$$Q_1\left(w|\psi^{(k)}\right) = \sum_{j=1}^{N}\sum_{i=1}^{M} \tau_{ij}^{(k)} \ln w_{ij}$$

$$Q_2\left(v|\psi^{(k)}\right) = \sum_{j=1}^{N}\sum_{i=1}^{M} \tau_{ij}^{(k)}\left(-\ln\Gamma\left(\tfrac{v_i}{2}\right) + \tfrac{v_i}{2}\ln\left(\tfrac{v_i}{2}\right) + \tfrac{v_i}{2}\left\{\sum_{j=1}^{N}\left(\ln u_{ij}^{(k)} - u_{ij}^{(k)}\right) + \gamma\left(\tfrac{v_i^{(k)}+D}{2}\right) - \ln\left(\tfrac{v_i^{(k)}+D}{2}\right)\right\}\right),$$

$$Q_3\left(y,\Sigma_i|\psi^{(k)}\right) = \sum_{j=1}^{N}\sum_{i=1}^{M} \tau_{ij}^{(k)}\left(-\tfrac{D}{2}\ln(2\pi) - \tfrac{1}{2}\ln|\Sigma_i| + \tfrac{D}{2}\ln u_{ij}^{(k)} - \frac{u_{ij}^{(k)}\|x_j-y_i\|^2}{2\Sigma_i}\right)$$

with the gamma in Q_2 that denotes the Digamma function, and the $u_{ij}^{(k)}$ in Q_3 that represents the conditional expectation about additional missing data in the complete-data; it can be written as:

$$u_{ij}^{(k)} = \frac{v_i^{(k)} + D}{v_i^{(k)} + D\left(x_j, y_i^{(k)}; \Sigma_i^{(k)}\right)}, \tag{6}$$

In addition, we use the displacement function ρ to denote the non-rigid transformation in the point set: $X = T(Y, \rho) = Y + \rho(Y)$, and add the regularization term $\varphi(\rho)$ to the Q_3, which can enforce the smoothness of the displacement function in the alignment of two point sets. So this Q_3 term can be rewritten as:

$$\widetilde{Q}_3 = Q_3(\rho, \Sigma_i^{(k+1)}|\psi^{(k)}) + \frac{\lambda}{2}\varphi(\rho), \tag{7}$$

where λ is the regularization parameter and controlling the trade-off of two terms in Equation (7). By reproducing Kernel Hilbert Space and Fourier transformation, the displacement function ρ can be denoted as:

$$\rho(y_j) = \sum_{i=1}^{M} h_i G(y_i, y_j), \quad G(y_i, y_j) = e^{-\frac{1}{2}\|\frac{y_i-y_j}{\beta}\|^2}, \tag{8}$$

where G is a Gaussian kernel matrix, β determines the width of the smoothing Gaussian filter, and the Equation (8) can be conveniently denoted in matrix form as: $\rho(Y) = GH$, where $H_{M\times D} = (h_1, \dots, h_M)$ is a matrix of coefficients. Thus, the Equation (7) can be expanded in the following matrix form:

$$\widetilde{Q}_3 = \sum_{j=1}^{N}\sum_{i=1}^{M} \tau_{ij}^{(k)}\left(-\frac{D}{2}\ln(2\pi) - \frac{1}{2}\ln|\Sigma_i| + \frac{D}{2}\ln u_{ij}^{(k)} - \frac{u_{ij}^{(k)}\|x_j - y_i\|^2}{2\Sigma_i}\right) + \frac{\lambda}{2}tr\left((H^T)^{(k)}GH^{(k)}\right), \tag{9}$$

M-step: On the M-step at the $(k + 1)^{\text{th}}$ iteration of the EM algorithm, considering that Q_1, Q_2 and Q_3 in Equation (5) can be computed independently of each other, thus the maximization of the objective function Q_1, Q_2 and Q_3 with respect to the parameters w, v, Σ, ρ can be operated separately. The mixing proportion for SMM is updated by our consideration of the first term Q_1. Given the feature descriptors in Section 2.1, it can obtain the coarse correspondences Ω between the point sets X and Y. For an observation data x_j, we define ι ($0 \le \iota \le 1$) as a confidence by IDSC feature matching, and then the mixing proportion w_{ij} by incorporating the local structures among neighboring points can be updated by the following rule:

$$w_{ij}^{(k+1)} = \begin{cases} \sum_{j=1}^{N} \frac{\tau_{ij}^{(k)}}{N}, & \exists\left(x_j \to y_i\right) \in \Omega \\ \iota, & \exists\left(x_j \to y_i\right) \in \Omega \end{cases}, \tag{10}$$

If observation data x_j does not have a corresponding point y_i in the label Ω, the mixing proportion $w_{ij}^{(k+1)}$ is given by the average of the posterior probabilities $\tau_{ij}^{(k)}$ of the SMM component membership. If the observation datum x_j corresponds to the centroid point y_i in the label Ω, the $w_{ij}^{(k+1)}$ is given by a constant confidence ι.

Note that in Equation (10) the mixing proportion w_{ij} does not only depend on prior assignment by local structures, but also depends upon the posterior probability of observation data x_j belonging to the ith component of the SMM by global structures. In addition, we obtain degree of freedom v by taking the corresponding derivative of Q_2 to zero, and the updated value $v^{(k+1)}$ is a solution of the following equation:

$$1-\gamma\left(\frac{v_i^{(k+1)}}{2}\right)+\ln\left(\frac{v_i^{(k+1)}}{2}\right)+\frac{\sum\limits_{j=1}^{N}\tau_{ij}^{(k)}\left(\ln u_{ij}^{(k)}-u_{ij}^{(k)}\right)}{\sum\limits_{j=1}^{N}\tau_{ij}^{(k)}}+\gamma\left(\frac{v_i^{(k)}+D}{2}\right)-\ln\left(\frac{v_i^{(k)}+D}{2}\right)=0, \tag{11}$$

Then, to update the estimates of covariance matrices Σ and displacement function ρ in Equation (7), we also need to take the derivative of $\widetilde{Q}_3$, and the update values can be denoted as:

$$H^{(k+1)}=\left(diag\left(P^{(k)}1\right)G+\lambda\Sigma_i^{(k)}I\right)^{-1}\left(P^{(k)}X-diag\left(P^{(k)}1\right)Y\right), P_{ij}^{(k)}=\tau_{ij}^{(k)}u_{ij}^{(k)}$$

$$\Sigma_i^{(k+1)}=\frac{\sum\limits_{j=1}^{N}\tau_{ij}^{(k)}u_{ij}^{(k)}\|x_j-T(y_i,\rho)\|^2}{\sum\limits_{j=1}^{N}\tau_{ij}^{(k+1)}}, \ (i=1,\dots,N) \tag{12}$$

where 1 is a column vector of all ones, and I is an identity matrix, $diag(\bullet)$ denotes a diagonal matrix and $G(i,\bullet)$ denotes the column vector in the kernel matrix G. Furthermore, for our centroid point y_i in the point set Y, the non-rigid transformation $T(Y,\rho)=Y+\rho(Y)$ can be expressed by Equation (8) as follows:

$$T(y_i,\rho)=y_i+G(i,\cdot)H_i^{(k+1)}, \tag{13}$$

At the end, the iterative process alternates between E-steps and M-steps until satisfying the following convergence condition:

$$\left\|\frac{L_C\left(\psi^{(k+1)}\right)-L_C\left(\psi^{(k)}\right)}{L_C\left(\psi^{(k+1)}\right)}\right\|\leq\varepsilon, \tag{14}$$

where ε is a convergence threshold. After a number of repeatability testing, we use the following setting for the parameters $\eta=0.1$, $\iota=0.8$, $\lambda=3$, $\beta=2$, $\varepsilon=10^{-5}$ in our experiment. Furthermore, to register the visible and infrared images accordingly, the image transformation and interpolation is performed on the visible image. Since our multi-spectral face registration method is implemented by using global and local structures, then the registration method is to be named as Face Registration using the Global and Local Structure (FR-GLS) in the rest of this paper.

3. The Proposed Fusion Strategies for Visible and Infrared Face Images

After obtaining the accurate alignment of the visible and infrared images, the second problem is then to solve the multi-spectral images fusion. In order to make sure that the fused image can preserve both the apparent details of the visible image and thermal radiation information of the infrared image, we propose the fusion strategies based on Guided Filtering and Gradient preserving (GF-GP). Firstly, we use two-scale image decomposition and reconstruction with a guided filtering algorithm to get an initial fusion image, and then the output fusion image is optimized by combining the intensity distribution of the infrared image and the intensity variation of the visible image. According to [33], since each source image can be separated into a base layer containing the large-scale details, and the detail layer containing the small-scale details, the two-scale image fusion F_G is given by the guided filtering algorithm as follows:

$$F_G=\overline{B}+\overline{D}, \tag{15}$$

where the fused base layer $\overline{B}$ and the fused detail layer $\overline{D}$ can be denoted as a weighted average cost of the base B_n and detail B_n layers of different source images:

$$\overline{B} = \sum_{n=1}^{N} W_n^B B_n, \quad \overline{D} = \sum_{n=1}^{N} W_n^D D_n, \tag{16}$$

where W_n^B and W_n^D represent the refined weight maps of the base B_n and detail layers D_n via guided image filtering respectively. In order to optimize the initial fusion image F_G, here the visible, infrared and fused images are defined by V, I and F, respectively. Given that the thermal radiation information in the infrared image is typically characterized by the pixel intensity values, the pixel intensity values are quite different between the target and background. Thus, we constrain the fused image F to have the similar pixel intensity distribution with the infrared source image I, and the optimization problem can be formulated as:

$$\delta_1 = \|F - I\|, \tag{17}$$

where $\|\bullet\|$ denotes the l_1 norm. Besides, given that the pixel intensity distribution in the same physical location might be discrepant for infrared and visible face images, they are manifestations of two different phenomena. Note that the detailed information of object edge and texture is mainly characterized by the pixel gradient values in the visible image. Hence, we constrain the fused image F to have the similar pixel gradient values with the visible source image V, the optimization problem can also be formulated as:

$$\delta_2 = \|\nabla F - \nabla V\|, \tag{18}$$

where ∇ denotes the gradient operator. By combining the Equations (17) and (18), the optimization problem of the fusion image F can be formulated as minimizing the following objective function:

$$\delta_F = \delta_1 + \alpha \delta_2 = \|F - I\| + \alpha\|\nabla F - \nabla V\|, \tag{19}$$

where α is a weighting factor, the optimal F could be found by using gradient descent strategy. Furthermore, it is important to note that if the α is small, the fusion image preserves the more thermal radiation information of the infrared image, otherwise, the fusion image preserves more edge and texture information of the visible image. We set the weighting coefficient $\alpha = 5$ in our experiment, because it can achieve good subjective visual quality in most cases.

4. Experimental Results

To demonstrate the robustness and efficiency of our proposed registration method in solving the multi-spectral face image registration problem, subjective and objective evaluation experiments are conducted with the UTK-IRIS standard multi-spectral face database [34] and self-constructed multispectral face data sets. The experiments are performed on a desktop with 3.3 GHz Intel Core™ i5-4590 CPU, 8 GB memory and MATLAB Code.

4.1. Registration on Real Face Images with the UTK-IRIS Database

In this section, the performance of face image registration with our FR-GLS method is evaluated by using the public UTK-IRIS Database. The images in the database have a spatial resolution of 320×240 pixels and the database contains individuals with various poses, facial expressions and illumination variations. We first give an intuitive impression of registration and fusion results, and then we provide a quantitative comparison with three typical methods such as Coherent Point Drift (CPD) [19], Regularized Gaussian Fields (RGF) [20] and SMM [25], based on the specified landmarks of visible and thermal infrared image pairs. The reasons for choosing these three comparison methods rest on the following two considerations. On the one hand, CPD and SMM use both global structures based on finite mixture models to parameterize the transformations. On the other hand, besides the global relationships in the feature point sets, RGF uses local features descriptor such as SC to recover the accurate transformation.

However, our FR-GLS has two major advantages compared to RGF: (i) Unlike the local features descriptor used in RGF, we use a more robust local features descriptor such as IDSC to initialize the correspondences between the two shape features; (ii) In order to preserve the global and local structures of feature point sets during the registration process, a heavy tail distribution in our FR-GLS is used to replace the traditional Gaussian distribution in the RGF method. Furthermore, we use the IDSC descriptor to assign the mixing proportion w_{ij} of the SMM.

4.1.1. Qualitative Evaluation

In order to obtain a qualitative evaluation of the registration results issued by the software-based registration process in Figure 1, several typical frame pairs of four individuals involving different poses and illumination changes are given in Figure 4a–d. Specifically, the original visible images and thermal infrared images are shown in the first and second row, respectively. Then the middle row shows the edge maps extracted by the canny edge detector. We use the red point set to denote the edge map of the thermal infrared images, and the blue point set corresponds to the edge map of the visible images. Subsequently, the visible image is aligned to the thermal infrared image by our FR-GLS method, and the registration performance is demonstrated by the checkerboard in the fourth row, where the aligned visible image is superimposed onto the corresponding thermal infrared image. We can see that the seams between two grids in the checkerboard are natural, and this demonstrates the feasibility and effectiveness of the proposed registration method qualitatively. Finally, the last row represents the fusion results of the visible and thermal IR images by our GF-GP method. We can see that the fused images in the last row contain both the apparent details of the visible image and the thermal radiation information of the infrared image, which can be clearly seen from noses, cheeks, ears, mouths and glasses. Therefore, it will be beneficial for the performance of follow-up face detection and recognition under the case of illumination changes or disguise.

4.1.2. Quantitative Evaluation

In order to give a quantitative comparison of our FR-GLS and other typical methods, such as CPD [19], RGF [20] and SMM [25], we further utilize several visible and infrared image pairs of the four individuals in Figure 4, and manually selected several landmarks and their correspondences (about 20 pairs) in each visible and infrared image pairs as ground truth. For example, the landmarks involve the salient features like the cheeks, eyes, nose, mouth, eyebrows, ears and glasses, which are highly distinguished. Referring to the metric used in [20,21,35], the recall rate is proposed to be a qualitative indicator for evaluating the registration results on all landmarks set pairs of an individual. Here the recall rate, or true positive rate, is defined as the proportion of the true positive correspondences between the landmark pairs to the ground truth correspondences, and a true positive correspondence is counted when the pair falls within a given accuracy threshold of Euclidean distance. Here the Euclidean distance can be defined as 2-norm between a landmark in the aligned feature point set of visible images and the corresponding landmark in the feature point set of infrared images. The recall curves plot the recall rates under different threshold values.

Figure 5 shows the recall rates against different threshold values using the registration methods of CPD, SMM, RGF and FR-GLS, considering four individuals. Each individual contains about 100 hundred-image pairs corresponding to every spectrum. By the definition of recall, we can see that the FR-GLS and RGF methods for the global and local structures are significantly superior to the CPD and SMM methods for the global structures only based on finite mixture models. Moreover, the curves of our FR-GLS method are consistently above the curves of the RGF method for different accuracy thresholds and individuals, which could be attributable to the combination of the robust SMM and IDSC local feature descriptor.

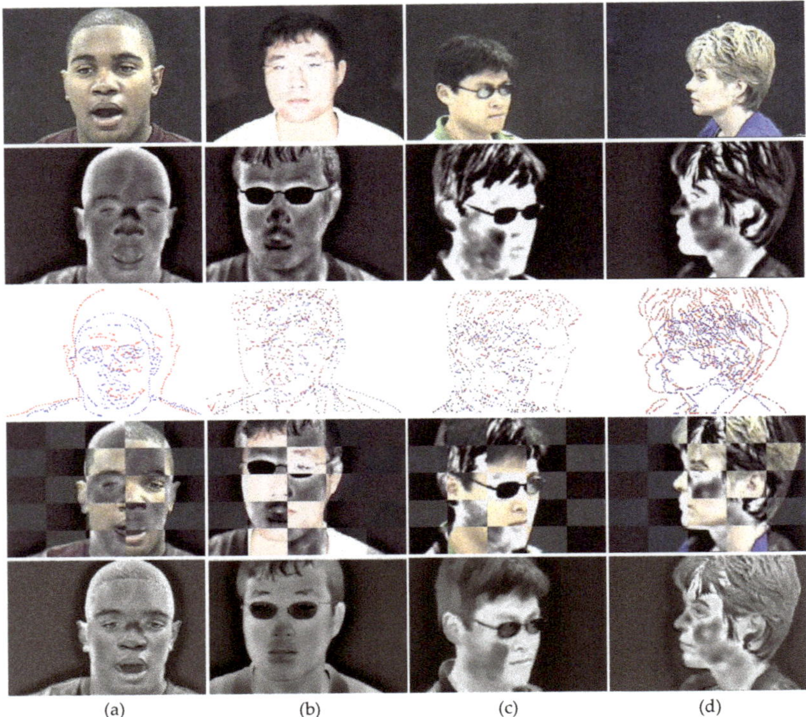

Figure 4. Registration and fusion results of our method on four typical unregistered visible/infrared image pairs in the database of UTK-IRIS. (**a**) Charles; (**b**) Heo; (**c**) Meng; (**d**) Sharon.

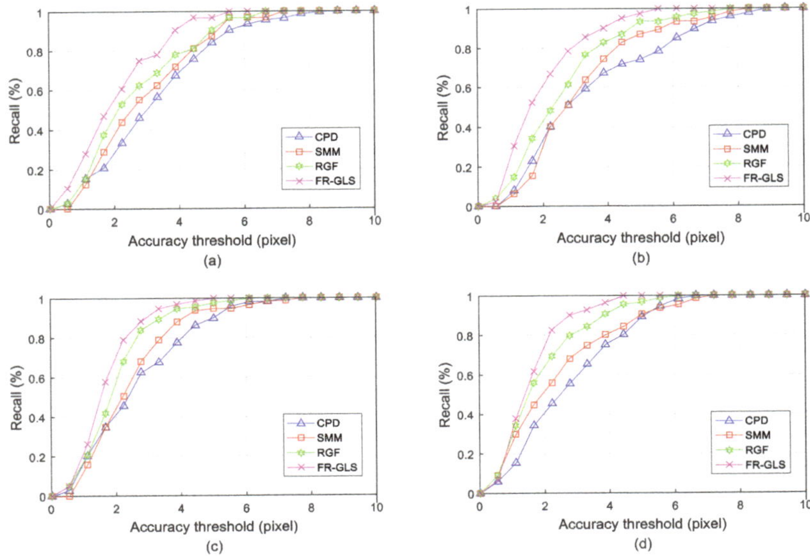

Figure 5. Quantitative comparisons of multispectral face image pairs with four individuals in Figure 4. (**a**) Charles; (**b**) Heo; (**c**) Meng; (**d**) Sharon.

The average registration errors of each individual for different registration methods are summarized in Table 1. We again see that the FR-GLS and RGF methods with the global and local structures significantly outperform the CPD and SMM methods just with finite mixture models.

Moreover, our FR-GLS method can achieve consistently better performance compared to the RGF method. The average registration errors of this RGF method are about 2.40, 2.24, 2.24 and 1.92 pixels on the four individuals, respectively. By contrast, the average registration errors of our FR-GLS method are reduced to about 2.25, 2.06, 1.99 and 1.78 pixels, respectively. In addition, The SMM method with SMM has a slightly better performance than the CPD method with GMM, where the average registration errors are 2.88, 2.40, 2.56 and 2.13 pixels, respectively. Whereas, the average registration errors of the CPD method are up to 3.04, 2.58, 2.73 and 2.35 pixels, respectively. It can quantitatively verify the heavy-tailed property of SMM.

Table 1. The average registration errors comparison of CPD, SMM, RGF and our FR-GLS method on four typical individuals in Figure 4.

Methods	The Average Registration Errors			
	Charles	Heo	Meng	Sharon
CPD [19]	3.04	2.58	2.73	2.35
SMM [25]	2.88	2.40	2.56	2.13
RGF [20]	2.40	2.24	2.24	1.92
FR-GLS	2.25	2.06	1.99	1.78

4.2. Registration on Real Face Images of Self-Built Multi-Spectral Database

In order to demonstrate the performance of our registration in the high-resolution multi-spectral imaging system, we built a four degree of freedom (DOF) pan and tilt head vision platform for multispectral database acquisition, which can be seen in Figure 6.

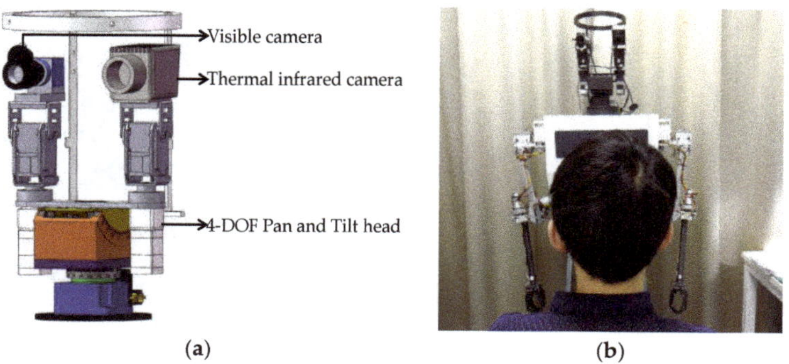

Figure 6. 4-DOF pan and tilt head vision platform for multi-spectral database acquisition: (**a**) Design mechanical model of vision platform; (**b**) Image acquisition scenario.

This 4-DOF pan and tilt head vision platform can realize up-down pitch movement and left-right yaw movement, respectively. Moreover, a Belgium Raven-640-Analog thermal infrared camera with 10 mm short lens is used to acquire a $74° × 59°$ field of view (FOV) infrared image (Figure 6a left), while a DAHENG MER-310-12UC color camera with an Optotune EL-10-30-Ci variable-focus liquid lens is used to acquire a $28° × 22°$ FOV visible image (Figure 6a right). Meanwhile, a visible image is always taken simultaneously with the infrared image to form a multi-spectral image pair, and the visible and infrared images are both compressed into $640 × 480$ pixels. The multi-spectral vision equipment and image acquisition scenario is shown in Figure 6b. The self-built multi-spectral database is composed of 20 different individuals involving different illumination conditions and pose variations.

In addition, for each individual, the sampled images are captured from 13 different angles (tilt head from −90° to +90°) and two illumination conditions (dark and light).

4.2.1. Qualitative Comparison

Referring to the qualitative evaluation in Section 4.1, Figure 7 shows an intuitive impression of registration and fusion results on six typical frame pairs in the self-built multi-spectral database.

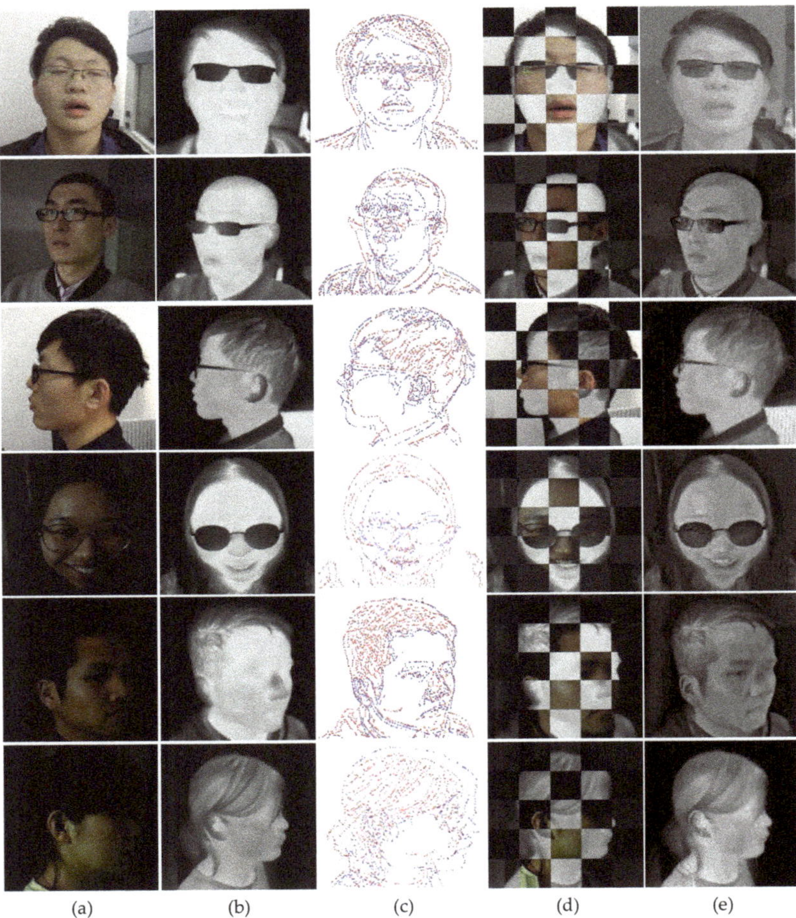

(a)   (b)   (c)   (d)   (e)

Figure 7. Registration and fusion results of our method considering six typical unregistered visible and infrared image pairs in the self-built multi-spectral face datasets. (**a**) Visible image; (**b**) Thermal infrared image; (**c**) Canny edge maps; (**d**) Superimposed checkerboard pattern of visible and infrared images; (**e**) Fusion results.

As shown in the Figure 7 above. Firstly, several original visible images and thermal infrared images of six individuals are given in Figure 7a,b, respectively, and those face images involve different poses and illumination changes. In addition, the image pairs in the bright light are shown in the first three rows, while the image pairs in the dark condition are shown in the last three rows with different poses. Then, Figure 7c shows the discrete point set of edge maps, which are extracted by the canny edge detector. For distinguishing each spectral image, we also use the blue point set to denote the edge map of the visible image, and the red point set corresponds to the edge map of the thermal infrared image. We can see that contour point sets of the face images in the edge map pairs are significantly different for the thermal infrared and visible images. Those discrepancies will result in a lot of outlier and noise for the edge maps alignment, especially in the textured regions of hairs and clothes. Hence, it is hard to match the two edge map pairs accurately.

Furthermore, note that we just focus on the registration results of the face region, the redundant information, such as the edge points of clothes, which is not related to face detection, might be ignored during the matching. Subsequently, the visible image is aligned to the thermal infrared image by our FR-GLS method, and the registration performance is demonstrated by checkerboard images in Figure 7d, where the aligned visible image is superimposed to the corresponding thermal infrared image. In addition to the uninterested regions of hairs and clothes, we can see from Figure 7d that the seams between two grids in the checkerboard images are natural, and this can demonstrate the feasibility and effectiveness of the proposed registration method qualitatively. Finally, for further verification, Figure 7e shows the fusion results of the visible and thermal infrared images by our GF-GP method. Compared to the original thermal infrared images in the second column (Figure 7b), we can see that the fused images in the last column (Figure 7e) seem to be sharpened, which contain both the apparent details of the visible image and the thermal radiation information of the infrared image. Thus, it will be beneficial for follow-up face detection and recognition in case of illumination changes, e.g., the outdoor environment illumination changed from bright to dark.

4.2.2. Quantitative Comparison

To have a quantitative evaluation, we also manually select a set of landmarks in the multi-spectral images (about 40 pairs of landmarks for each spectral image), and then treat the recall on all real face images of an individual as the metric. Figure 8 shows the recall comparison results of the four methods on six individuals, where each individual contains about 52 image pairs.

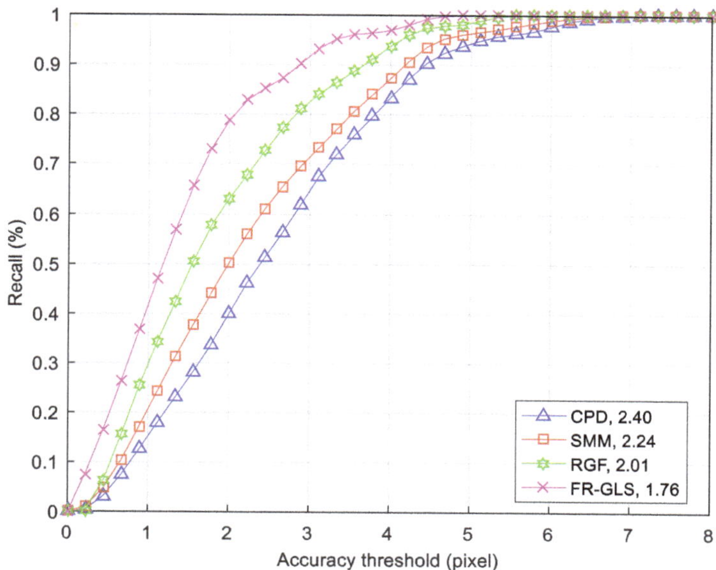

Figure 8. Quantitative comparisons of multi-spectral face image pairs with 6 individuals in Figure 7.

We can see from the Figure 8 above, FR-GLS (mauve fork) and RGF (green star) methods with the global and local structures are significantly superior to the CPD and SMM methods with global structures, and the curve of the Student's t distribution model (red square) is mainly above the curve of the Gaussian distribution model (blue triangle). Furthermore, the curve of our FR-GLS method, combined with SMM and IDSC, is consistently above those of the other three methods. In conclusion, for the six pairs of typical multi-spectral images in Figure 7, the average registration errors of CPD, SMM, RGF and our FR-GLS method are about 2.40, 2.24, 2.01 and 1.76 pixels, respectively.

5. Conclusions

In this paper, we have introduced a novel multi-sensor face images registration method. It uses the IDSC to describe the local feature of a face image, which is more stable than SC, and then the Student's t Mixture probabilistic model is used to estimate the transformation between visible and infrared images by combining global and local structures of feature point sets. In order to verify the correctness of the registration method intuitively, we propose a multi-spectral image fusion strategy based on guided filtering and gradient preserving. Experimental results on a standard real face database and self-built multi-spectral face database demonstrate that the proposed method is able to achieve much more registration accuracy compared to other state-of-the-art registration methods. Therefore, it will be beneficial to improve the reliability of a fusion-based face recognition system. In the future, given that the ultimate goal of registration and fusion is to enhance the recognition rate, a series of more comprehensive and scientific evaluations for multi-sensor face images registration will be conducted through some combination of the latest deep neural network face recognition method.

Author Contributions: Conceptualization, W.L. and W.Z.; Methodology, W.L.; Software, W.Z.; Validation, N.L. and W.Z.; Formal Analysis, W.L.; Investigation, W.L.; Resources, M.D.; Data Curation, W.L.; Writing-Original Draft Preparation, W.L.; Writing-Review & Editing, W.L. and W.Z.; Visualization, W.L.; Supervision, N.L. and M.D.; Project Administration, X.L.; Funding Acquisition, M.D and X.L.

Funding: This research was funded by National Natural Science Foundation of China, grant number No. 51475046, No. 51475047 and the National High-tech R&D Program of China. grant number No. 2015AA042308.

Conflicts of Interest: The authors declare no conflict of interest.

References

1. Ma, J.; Ma, Y.; Li, C. Infrared and visible image fusion methods and applications: A survey. *Inf. Fusion* **2019**, *45*, 153–178. [CrossRef]
2. Ma, J.; Yu, W.; Liang, P. FusionGAN: A generative adversarial network for infrared and visible image fusion. *Inf. Fusion* **2019**, *48*, 11–26. [CrossRef]
3. Huang, Y.; Bi, D.; Wu, D. Infrared and visible image fusion based on different constraints in the non-subsampled shearlet transform domain. *Sensors* **2018**, *18*, 1169. [CrossRef]
4. Ma, J.; Chen, C.; Li, C. Infrared and visible image fusion via gradient transfer and total variation minimization. *Inf. Fusion* **2016**, *31*, 100–109. [CrossRef]
5. Paramanandham, N.; Rajendiran, K. Infrared and visible image fusion using discrete cosine transform and swarm intelligence for surveillance applications. *Infrared Phys. Technol.* **2018**, *88*, 13–22. [CrossRef]
6. Zhu, Z.; Qi, G.; Chai, Y. A geometric dictionary learning based approach for fluorescence spectroscopy image fusion. *Appl. Sci.* **2017**, *7*, 161. [CrossRef]
7. Singh, R.; Vatsa, M.; Noore, A. Integrated multilevel image fusion and match score fusion of visible and infrared face images for robust face recognition. *Pattern Recognit.* **2008**, *41*, 880–893. [CrossRef]
8. Vizgaitis, J.N.; Hastings, A.R. Dual band infrared picture-in-picture systems. *Opt. Eng.* **2013**, *52*, 061306. [CrossRef]
9. Kong, S.G.; Heo, J.; Boughorbel, F. Multiscale fusion of visible and thermal IR images for illumination-invariant face recognition. *Int. J. Comput. Vis.* **2007**, *71*, 215–233. [CrossRef]

10. Oliveira, F.P.M.; Tavares, J.M.R.S. Medical image registration: A review. *Comput. Method Biomech.* **2014**, *17*, 73–93. [CrossRef]

11. Avants, B.B.; Epstein, C.L.; Grossman, M. Symmetric diffeomorphic image registration with cross-correlation: Evaluating automated labeling of elderly and neurodegenerative brain. *Med. Image Anal.* **2008**, *12*, 26–41. [CrossRef]

12. Pan, W.; Qin, K.; Chen, Y. An adaptable-multilayer fractional fourier transform approach for image registration. *IEEE Trans. Pattern Anal.* **2008**, *31*, 400–414. [CrossRef]

13. Zhuang, Y.; Gao, K.; Miu, X. Infrared and visual image registration based on mutual information with a combined particle swarm optimization—Powell search algorithm. *Optik Int. J. Light Electron Opt.* **2016**, *127*, 188–191. [CrossRef]

14. Sotiras, A.; Davatzikos, C.; Paragios, N. Deformable medical image registration: A survey. *IEEE Trans. Med. Imaging* **2013**, *32*, 1153–1190. [CrossRef]

15. Viergever, M.A.; Maintz, J.B.A.; Klein, S. A survey of medical image registration–under review. *Med. Image Anal.* **2016**, *33*, 140–144. [CrossRef]

16. Moghbel, M.; Mashohor, S.; Mahmud, R. Review of liver segmentation and computer assisted detection/diagnosis methods in computed tomography. *Artif. Intell. Rev.* **2018**, *50*, 497–537. [CrossRef]

17. Zhang, Y.; Zhao, D.; Sun, J. Adaptive convolutional neural network and its application in face recognition. *Neural Process. Lett.* **2016**, *43*, 389–399. [CrossRef]

18. Zhang, K.; Zhang, Z.; Li, Z. Joint face detection and alignment using multitask cascaded convolutional networks. *IEEE Signal. Proc. Lett.* **2016**, *23*, 1499–1503. [CrossRef]

19. Myronenko, A.; Song, X. Point set registration: Coherent point drift. *IEEE Trans. Pattern Anal. Mach. Intell.* **2010**, *32*, 2262–2275. [CrossRef]

20. Ma, J.; Zhao, J.; Ma, Y. Non-rigid visible and infrared face registration via regularized Gaussian fields criterion. *Pattern Recognit.* **2015**, *48*, 772–784. [CrossRef]

21. Tian, T.; Mei, X.; Yu, Y. Automatic visible and infrared face registration based on silhouette matching and robust transformation estimation. *Infrared Phys. Technol.* **2015**, *69*, 145–154. [CrossRef]

22. Peel, D.; McLachlan, G.J. Robust mixture modelling using the t distribution. *Stat. Comput.* **2000**, *10*, 339–348. [CrossRef]

23. Gerogiannis, D.; Nikou, C.; Likas, A. Robust Image Registration using Mixtures of t-distributions. In Proceedings of the 2007 IEEE 11th International Conference on Computer Vision, Rio de Janeiro, Brazil, 14–21 October 2007; pp. 1–8.

24. Gerogiannis, D.; Nikou, C.; Likas, A. The mixtures of Student's t-distributions as a robust framework for rigid registration. *Image Vis. Comput.* **2009**, *27*, 1285–1294. [CrossRef]

25. Zhou, Z.; Zheng, J.; Dai, Y. Robust non-rigid point set registration using student's-t mixture mode. *PloS ONE* **2014**, *9*, e91381.

26. Zhou, Z.; Tu, J.; Geng, C. Accurate and robust non-rigid point set registration using student's t mixture model with prior probability modeling. *Sci. Rep. UK* **2018**, *8*, 8742. [CrossRef]

27. Maiseli, B.; Gu, Y.; Gao, H. Recent developments and trends in point set registration methods. *J. Vis. Commun. Image Represent.* **2017**, *46*, 95–106. [CrossRef]

28. Belongie, S.; Malik, J.; Puzicha, J. Shape matching and object recognition using shape contexts. *IEEE Trans. Pattern Anal. Mach. Intell.* **2002**, *24*, 509–522. [CrossRef]

29. Ling, H.; Jacobs, D.W. Shape classification using the inner-distance. *IEEE Trans. Pattern Anal. Mach. Intell.* **2007**, *29*, 286–299. [CrossRef]

30. Li, H.; Wu, X.J. DenseFuse: A Fusion Approach to Infrared and Visible Images. *IEEE Trans. Image Process.* **2018**, *28*, 2614–2623. [CrossRef]

31. Chui, H.; Rangarajan, A. A new point matching algorithm for non-rigid registration. *Comput. Vis. Image Underst.* **2003**, *89*, 114–141. [CrossRef]

32. Riesen, K.; Bunke, H. Approximate graph edit distance computation by means of bipartite graph matching. *Image Vis. Comput.* **2009**, *27*, 950–959. [CrossRef]

33. Li, S.; Kang, X.; Hu, J. Image fusion with guided filtering. *IEEE Trans. Image Process.* **2013**, *22*, 2864–2875. [PubMed]

34. UTK-IRIS Database. Available online: http://www.cse.ohio-state.edu/otcbvs-bench/ (accessed on 10 September 2015).
35. Dwith, C.; Ghassemi, P.; Pfefer, T. Free-form deformation approach for registration of visible and infrared facial images in fever screening. *Sensors* **2018**, *18*, 125. [CrossRef] [PubMed]

 *applied*
sciences

Article

Multifocus Image Fusion Using a Sparse and Low-Rank Matrix Decomposition for Aviator's Night Vision Goggle

Bo-Lin Jian [1], Wen-Lin Chu [2], Yu-Chung Li [3] and Her-Terng Yau [1,*]

[1] Department of Electrical Engineering, National Chin-Yi University of Technology, Taichung 41170, Taiwan; BoLin@ncut.edu.tw

[2] Department of Mechanical Engineering, National Chin-Yi University of Technology, Taichung 44170, Taiwan; wlchu@ncut.edu.tw

[3] Department of Mechanical Engineering, National Cheng Kung University, Tainan 70101, Taiwan; a238966777@gmail.com

* Correspondence: pan1012@ms52.hinet.net or htyau@ncut.edu.tw; Tel.: +886-423924505#7229

Received: 20 February 2020; Accepted: 17 March 2020; Published: 23 March 2020

Abstract: This study proposed the concept of sparse and low-rank matrix decomposition to address the need for aviator's night vision goggles (NVG) automated inspection processes when inspecting equipment availability. First, the automation requirements include machinery and motor-driven focus knob of NVGs and image capture using cameras to achieve autofocus. Traditionally, passive autofocus involves first computing of sharpness of each frame and then use of a search algorithm to quickly find the sharpest focus. In this study, the concept of sparse and low-rank matrix decomposition was adopted to achieve autofocus calculation and image fusion. Image fusion can solve the multifocus problem caused by mechanism errors. Experimental results showed that the sharpest image frame and its nearby frame can be image-fused to resolve minor errors possibly arising from the image-capture mechanism. In this study, seven samples and 12 image-fusing indicators were employed to verify the image fusion based on variance calculated in a discrete cosine transform domain without consistency verification, with consistency verification, structure-aware image fusion, and the proposed image fusion method. Experimental results showed that the proposed method was superior to other methods and compared the autofocus put forth in this paper and the normalized gray-level variance sharpness results in the documents to verify accuracy.

Keywords: autofocus; night vision goggles; image fusion; sparse and low-rank matrix decomposition

---

1. Introduction

Night vision goggles (NVG) equipment can be used as nighttime visual aids by helicopter pilots in low-light environments. In particular, the NVG availability situation will directly affect the safety of nighttime aerial reconnaissance missions. Therefore, highly equipment availability should be maintained through regularly maintaining inspections and verifications. At present day, the aviator's NVG model AN/AVS-6 (V) 1 and AN/AVS-6 (V) 2 still rely on lots of manpower to perform their calibration. While processing, NVGs need to be placed on a test bench in order to commence manual focus adjustment operations by observing the image through the eyepiece. After the focus adjustment operation is completed, to make sure that the equipment is in line with calibration standards is confirmed only by human eyes. Since automatic image detection technology is sophisticated and widely applied, the researcher aims to reduce the staff's education time and to achieve the goal of proper equipment used by means of automatic image detection [1].

After gaining an insight into the performance and limitations of NVG, the autofocus operating process can be more accurately developed. Referring to the document of basic structure [2], the image quality of NVG relies on the electromagnetic spectrum signals detected by the enlarged image intensifier. The electro-optic system of the image intensifier is an important component. This component significantly affects resolution and light amplification. However, this component is subject to damage under strong light or high-humidity environments, and the general architectural diagram of the image intensifier is as shown in Figure 1 [2]. As the image intensifier will affect aviator's safety, image intensifier detection has become a standardized process. The current aviator's nighttime NVG test bench (TS-3895A/UV) [3] can provide the nighttime low-light environment required for NVG calibration. However, the test bench itself is unable to automatically adjust the NVG focal length. In addition to the drawback of needing to observe NVG eyepiece images by human eye before manually adjusting nighttime NVG focal length, human factors may lead to inaccurate test results. Therefore, this project intends to use a direct current (DC) servo driver to promote the focal knob of NVG to achieve the purpose of adjusting focus and acquiring quantitative value of rotation angle. For the configuration and design, refer to the document [1]. At present the autofocusing methods can be divided into active autofocusing and passive autofocusing [4]. Active autofocusing involves installing external infrared or other tools to measure distance between camera lens and target. Passive autofocusing, on the other hand, involves calculating sharpness information of a single image obtained from the camera. After calculating the sharpness of multiple images, the sharpness curve is acquired. The peak value of the sharpness curve is the best focal distance. Since this case proposes to adjust focus via image information of NVG, the passive autofocusing method was adopted. The key to the application of this method lies in whether effective sharpness points can be calculated through image information. Light luminance is the key affecting the passive autofocusing system. In previous studies, many types of sharpness computing methods were compared [5] to determine merits and drawbacks, which were applied in NVG's autofocusing [1]. In passive autofocusing, regardless of sharpness computing method, the subsequent image intensifier display on the screen undergoes defect testing, all of which are independent processes. Jian and Peng proposed autofocusing process for NVGs [1], which uses gradient-based variable step search and variation of normalized gray-level as the main method for accomplishing autofocus. Wang et al. [6] suggested the application of a robust principal component analysis method in multifocus image fusion. Additionally, an increasing number of related themes have undergone research [7], and low-rank matrix and sparse matrix themes aroused the study interest. Therefore, further development and application in NGV autofocusing and image fusion to aid in identifying NVG equipment availability were explored.

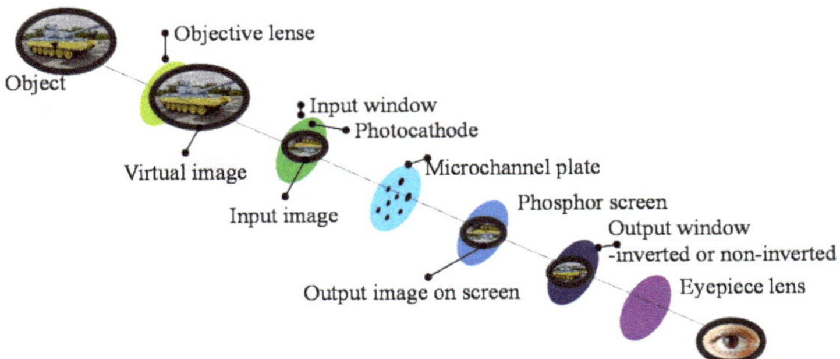

Figure 1. Structure of image intensifier [2].

The configuration of the mechanism comprises an NVG testing autofocusing system that includes a platform, motor, mechanism, and camera, as shown in Figure 2, as well as the multifocus problem caused by the inaccuracy between lens and NVG, as shown in Figure 3. This study adopted the image fusion method to resolve multifocus problems. Targeting how to correctly fuse images to ensure results presenting better information representativeness compared to any single input image is also an important topic in image fusion [7]. So far, a large quantity of image fusion techniques have been proposed. Among them, wavelet transport-based image fusion is a popular subject of research [8,9] because it can maintain precision of spectrum while increasing and improving accuracy of the space. When using wavelet decomposition, if only a few decomposition stages are used, the fused image's accuracy of space will be poorer. On the contrary, if too many decomposition stages are used, spatial similarity between the fused image and the original will be poorer [10]. Among those fusion methods, structure-aware image fusion [11] and image fusion in the discrete cosine transform (DCT) domain [12,13] are quite classic methods and widely used in various fields [14,15]. The following discusses wavelet-based image fusion in recent years. Vanmali et al. [16] proposed a quantitative measure using structural dissimilarity to measure the ringing artifacts. Ganasala and Prasad [17] especially focused on poor contrast and high-computational complexity issues of fusion outcomes. Seal and Panigrahy [18] focused on translation-invariant à trous wavelet transform and fractal dimension using a differential box counting method. Hassan et al. [19] implemented image fusion methods that are combined with wavelet transform and the learning ability of artificial neural networks. In recent years, deep learning networks have also been used to execute image fusion [20–22]. In general, deep learning networks' fusion quality depend on the sample characteristics at the time of data training. Image fusion based on low-rank matrix and sparse matrix characteristics has been a popular topic in recent years. Maqsood and Javed [23] proposed a multimodal image fusion scheme, which was based on two-scale image decomposition and sparse representation. This technology mainly uses the edge information of the sparse matrix for fusion. Ma et al. [24] proposed a multifocus image fusion method, mainly established in one fusion rule of sparse coefficients, which is based on the optimum theory and solved by the orthogonal matching pursuit method. Wang and Bai [25] proposed a novel strategy on the low frequency fusion assisted through sparse representation. Wang [26] proposed a novel fusion method based on sparse representation and non-subsampled contourlet transform, and used some indicators to prove the fusion result was excellent. Fu et al. [27] proposed a multifocus image fusion method through distributed compressed sensing (DCS). This method is mainly considered the high-frequency images' information. The final result was using visual and quantitative metric evaluations to analyze the results of the fusion. Among all the methods for decomposing data into low-rank matrix and sparse matrix, the most classic is robust principal component analysis (RPCA). There have been quite extensive expansion and application of RPCA, where RPCA via the principal component pursuit (PCP) method has been used to reduce the amount of calculation, with numerous extensions and expansions [28], including stable principal component pursuit (SPCP) [28], quantization based principal component pursuit (QPCP) [29], block based principal component pursuit (BPCP) [30], and local principal component pursuit (LPCP) [31]. Additionally, other methods for solving low-rank matrix and sparse matrix also include the subspace tracking series method [32], matrix completion series method [33], and nonnegative matrix factorization series method [34]. Of the discussions on these various methods, so far, studies have provided different pros to decompose matrices [28,35,36]. This study attempts to fuse the images of different focal distances by decomposing low-rank matrix and sparse matrix, not only taking into consideration decomposition and recombination of a single image [37,38] but also considering simultaneously decomposing and fusing more than two images [6,7] and even expanding to multiple images. Among those studies to date, there has not yet been a correct image fusion rating standard, and different fields result in different conclusions. Nevertheless, the rating standard currently provides evidential fusion results and field applicability related studies [39–41]. Thus, the indicators provided in the study by Liu [42] et al. were adopted to carry out fusion rating. The program for fusion rating standard used is provided by the website below: https://github.com/zhengliu6699/imageFusionMetrics, which discusses feasibility of applying deep semi- nonnegative matrix factorization (NMF) model [34] method in autofocusing and image fusion.

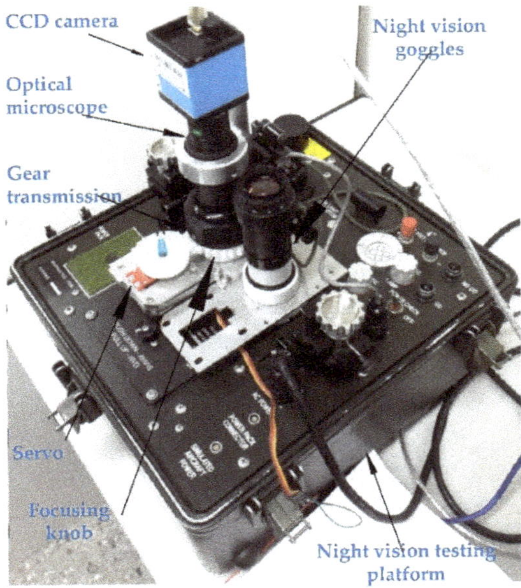

Figure 2. System installation [1].

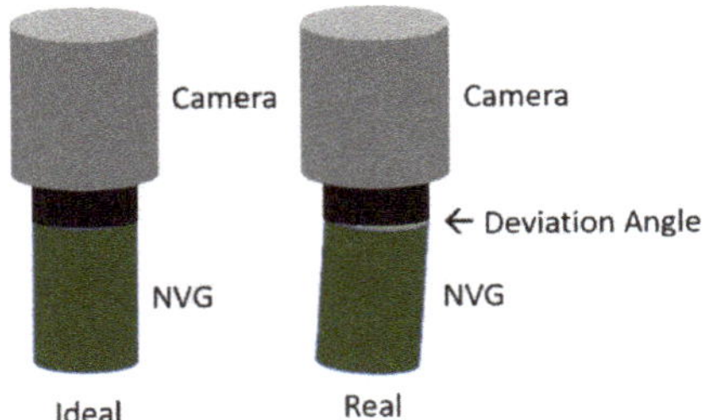

Figure 3. Situation of a minor error between lens and night vision goggles (NVG).

2. Materials and Methods

In order to examine the pros and cons of the NVG image autofocusing and fusion method proposed, the processing method was as shown in Figure 4. The process mainly consisted of several blocks, including low-rank and sparse matrix, image fusion, and autofocus. The image samples, proposing method, matrix decomposition process, and fusion method used were included in low-rank and sparse matrix and image fusion blocks. The explanation for this part is found in the description of tested images section and image fusion using low-rank and sparse matrix section. Autofocus block explains the use of sparse matrix information to complete the sharpness computing and to obtain the best focal image.

2.1. Description of Tested Images

In order to carry out various fusion method ratings, aircraft, clock, disk, lab, leopard, and toy images commonly used in research were used here to test the qualities of fusion methods. The images for testing were as shown in Figure 5a–l. In addition, the images for NVG testing of fusion results were as shown in Figure 5m,n. Through the aviaiton nighttime NVG testing bench (TS-3895A/UV) with NVG, the DC servo driver-driven focal knob was able to collect NVG testing images at the rotation angles ranging between 1 to 110 degrees. In order to simplify and facilitate the description of subsequent algorithms, the images collected were converted from colored ones into the gray-level, 110 images in total for autofocusing the algorithm used. To compare the quality of the traditional methods with the method in this paper, the same image sources as those of Jian and Peng [1] were used.

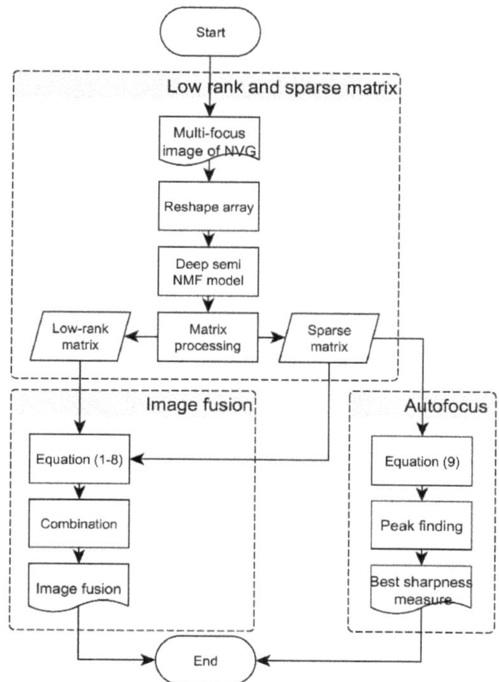

Figure 4. Autofocusing and fusion algorithm handling method.

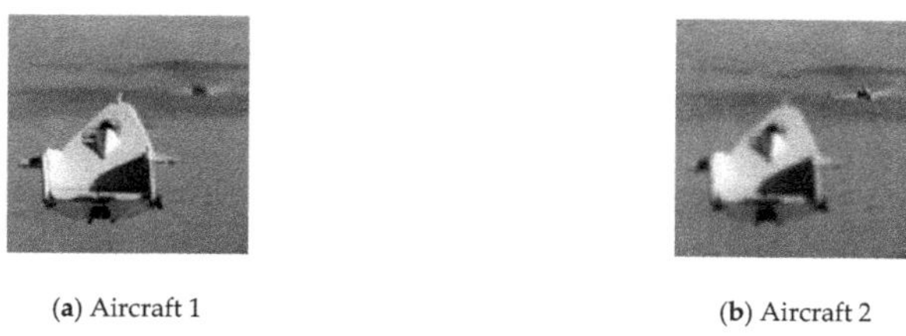

(**a**) Aircraft 1   (**b**) Aircraft 2

Figure 5. *Cont.*

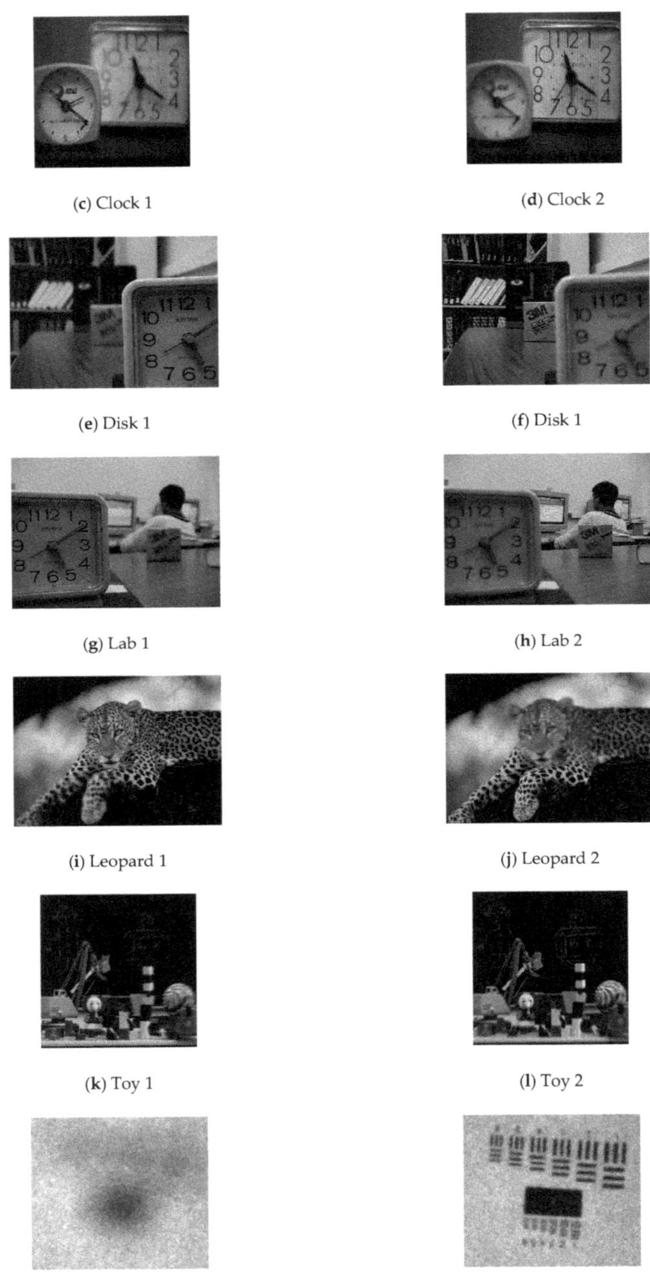

(**c**) Clock 1

(**d**) Clock 2

(**e**) Disk 1

(**f**) Disk 1

(**g**) Lab 1

(**h**) Lab 2

(**i**) Leopard 1

(**j**) Leopard 2

(**k**) Toy 1

(**l**) Toy 2

(**m**) Motor rotation angle of 60 (incorrect focal distance)   (**n**) Motor rotation angle of 96 (correct focal distance)

Figure 5. Autofocusing and fusion algorithm handling method.

2.2. Image Fusion Using Low-Rank and Sparse Matrix

According to the fusion algorithm processing flow in Figure 4, the image fusion algorithm was verified. In the process, the diagram of a matrix structure of the source image for multifocus image of NVG was as shown in Figure 6. In particular, **I** is a two-dimenisonal image matrix, t is the total number of images, the image height is n, and the width is m.

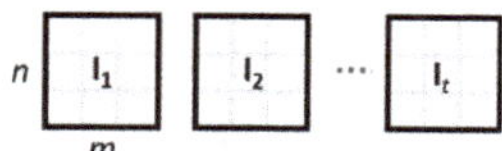

Figure 6. Diagram of matrix structure of the source image.

First, single source images were converted into one-dmeiensional vectors. Data arrangement in this step was as shown in Figure 7, where $\mathbf{I}^R$ was a one-dimensional vector. Each one-dimensional vector $\mathbf{I}_t^R$ was sequenced from top to bottom according to frame sequence. After combination, it was named **D** matrix (data matrix).

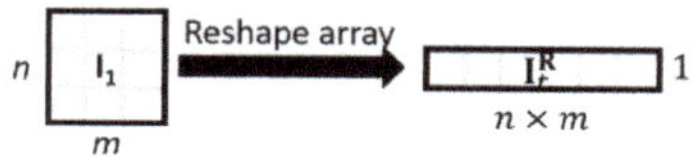

Figure 7. Diagram of the image matrix converted into one-dimensional vectors.

Then, the deep semi-NMF model method proposed by Trigeorgis, Bousmalis, Zafeiriou, and Schuller [34] was used here to obtain the low-dimensional representation. The equation is as shown in Equation (1):

$$\mathbf{D} \approx \mathbf{Z} \times \mathbf{H}. \tag{1}$$

In particular, **D** is data matrix, **Z** is loadings matrix, and **H** is features matrix. This study adopted the method with low-dimesional characteristics to obtain **A** and **E** matrices. **A** is low-rank matrix and **E** is sparse matrix:

$$\mathbf{A} = \mathbf{Z} \times \mathbf{H}, \tag{2}$$

$$\mathbf{E} = \mathbf{D} - \mathbf{A}. \tag{3}$$

In **A** matrix, the relationship bewteen respective row vectors and images was as shown in Equation (4):

$$\mathbf{A} = \begin{bmatrix} {}^A\mathbf{I}_1^R \\ {}^A\mathbf{I}_2^R \\ \vdots \\ {}^A\mathbf{I}_t^R \end{bmatrix}, \; {}^A\mathbf{I}_1^R : \text{the size is 1 by L} \tag{4}$$

In particular, the length of L is $n \times m$. In **A** matrix, the respective row vectors are, ${}^A\mathbf{I}_1^R$, ${}^A\mathbf{I}_2^R, \ldots, {}^A\mathbf{I}_t^R$, respectively. In this paper, **D** matrix formed by respective images was decomposed into **A** and **E** matrices. Therefore, these two charactersitics were targeted for processing. In particular, **A** process involved reshaping the row vectors from ${}^A\mathbf{I}_1^R$ through ${}^A\mathbf{I}_t^R$ to the two-dimensional images of ${}^A\mathbf{I}_1 - {}^A\mathbf{I}_t$ and obtaining the mean value. The result was called ${}^A\mathbf{I}_{best}$, and the process was as shown in Equation (5):

$$ {}^A\mathbf{I}_{best} = \left({}^A\mathbf{I}_1 + {}^A\mathbf{I}_2 + \ldots + {}^A\mathbf{I}_t \right)/t \tag{5}$$

The first image corresponding to sparse matrix ${}^E\mathbf{I}_1$ and mask offset manipulation was as shown in Figure 8.

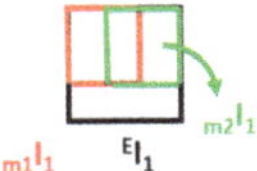

Figure 8. Diagram of the corresponding image in sparse matrix and mask offset.

The corresponding image in sparse matrix was as shown in Equations (6) and (7):

$$OptInd_z^x = \operatorname*{argmax}_{x \in [1,\ t]} \left(Var\left(_{m_z}I_x\right)\right),\ Z = 1,\ 2,\ \dots, len \tag{6}$$

$$^{E}\mathbf{I_{best}} = {}^{E}\mathbf{I}(OptIndex_z^x) \tag{7}$$

In particular, Var in Equation (6) is the computed variance, and *len* is the total number of images undergoing mask offset. The best label obtained according to Equation (6) was used to acquire the $^{E}\mathbf{I_{best}}$ image in Equation (7), through which the best edge information was retained. Finally, the image obtained in Equation (5) was added to the images in $^{A}\mathbf{I_{best}}$ and Equation (7) to get the best fusion image $\mathbf{I_{best}}$, as shown in Equation (8):

$$\mathbf{I_{best}} = {}^{A}\mathbf{I_{best}} + {}^{E}\mathbf{I_{best}} \tag{8}$$

2.3. Autofocus Using Sparse Matrix

In addition to image fusion, autofocusing using sparse matrix method process was also put forward in this study. The sparse feature of sparse matrix was mainly used to test the focus stripe correlation generated from the testing bench. Since low-rank matrix had the main components of focus stripe, it had the same signficiance pointed out in Equation (3) where sparse matrix was the orignal image subtracted by low-rank matrix. Hence, the correpsonding frame information in sparse matrix can be used as a reference for sharpness. This concept was applied in the acutal practice. First, 110 images of different focal distances were compiled into **D** matrix accoridng to the arrangement diagram in Figures 6 and 7. As shown in Equations (1) through (4), they were decomposed into low-rank matrix **A** and sparse matrix **E**. Then, $^{E}I_i^R$, $i = 1,2,3 \dots t$ corrsponding to different focal distances in E matrix were directly used to calulate the images corresponding to single $^{E}I_i^R$, to tally the results. An image frame corresponding to the lowest value was the sharpest frame. In particular, the lowest point in Figure 4 was the frame with the best sharpness. The calculation can be simplified into Equation (9).

$$
\begin{aligned}
FP \ &= \ \operatorname*{argmin}_{i \in [1,\ t]} \left(\sum_{k=1}^{N} {}^{E}I_i^R(k) \right) \\
&= \ \operatorname*{argmin}_{i \in [1,\ t]} \left(\sum_{x=1}^{m} \sum_{y=1}^{n} {}^{E}I_i(x,y) \right)
\end{aligned}
\tag{9}
$$

where *FP* is focus position and $N = m \times n$ is image size.

3. Experiment Results and Discussion

3.1. Image Fusion Results

According to the image fusion method propsed in Figure 4 and the based-on variance calculated in discrete cosine transform domain without consistency verification (DctVar) and based-on variance calculated in discrete cosine transform domain with consistency verification (DctVarCv) [12,13] and structure-aware image fusion (SAIF) [11] methods, comparison and verification were carried out. This study was to evaluate the quality of image fusion using relevant indicators compiled by Liu et al. [42]. The original images verified were as shown in the description of the tested images section. The fusion results of the respective images were as shown from Figures 9–14. Among which, (a) is the mehtod proposed in this study, (b) is the fused image result of DctVar method, (c) is the fused image result of DctVarCv method, and (d) is the fused image result of SAIF method. The fusion indicator results of vairous images were as shown from Tables 1–6. Among these four fusion methods, gray background's fusion quality indicator pointed out the best results for the 12 indicators, which in seuqence were: Q_{MI}, Q_{TE}, Q_{NCIE}, Q_G, Q_M, Q_{SF}, Q_P, Q_S, Q_C, Q_Y, Q_{CV}, Q_{CB}. The higher the value, the better the fusion quality.

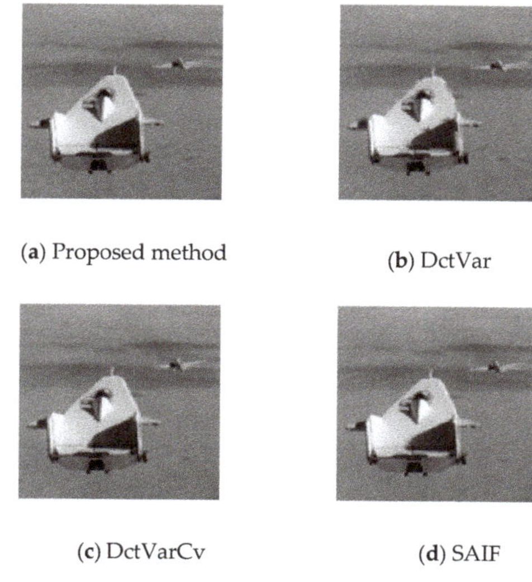

(**a**) Proposed method   (**b**) DctVar

(**c**) DctVarCv   (**d**) SAIF

Figure 9. The fusion result of the aircraft image.

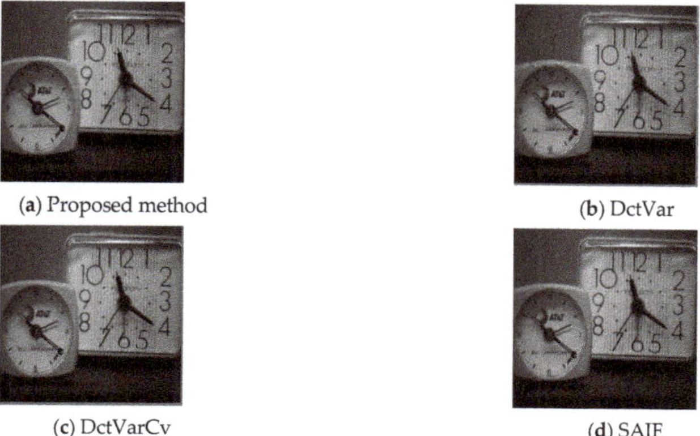

(**a**) Proposed method   (**b**) DctVar

(**c**) DctVarCv   (**d**) SAIF

Figure 10. The fusion result of the clock image.

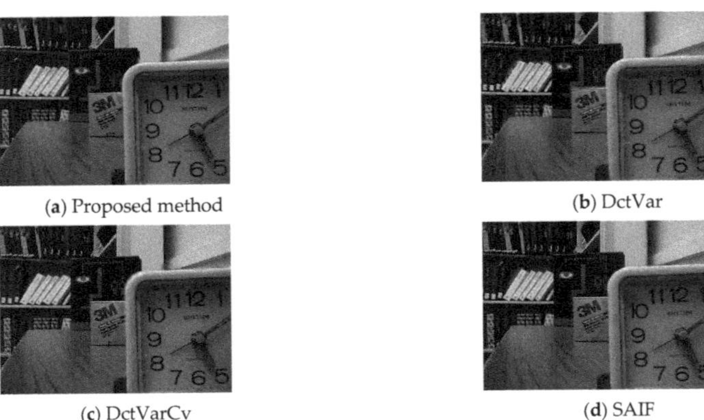

(**a**) Proposed method   (**b**) DctVar

(**c**) DctVarCv   (**d**) SAIF

Figure 11. The fusion result of the clock image.

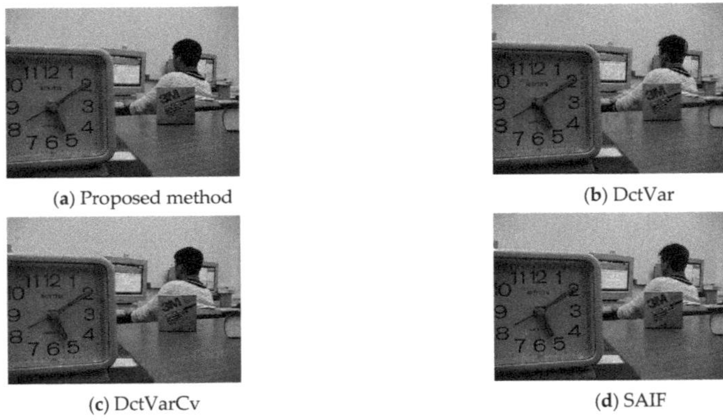

(**a**) Proposed method   (**b**) DctVar

(**c**) DctVarCv   (**d**) SAIF

Figure 12. The fusion result of the lab image.

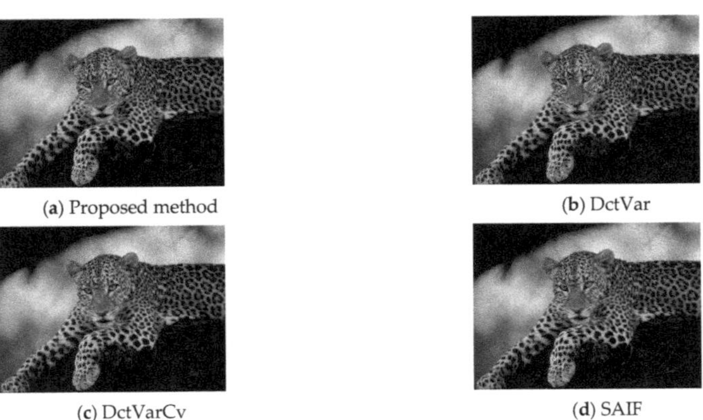

(**a**) Proposed method   (**b**) DctVar

(**c**) DctVarCv   (**d**) SAIF

Figure 13. The fusion result of the leopard image.

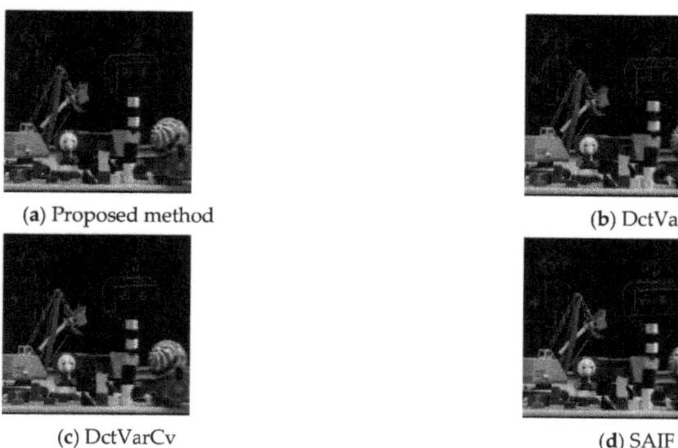

(**a**) Proposed method   (**b**) DctVar

(**c**) DctVarCv   (**d**) SAIF

Figure 14. The fusion result of the toy image.

Table 1. List of aircraft fusion results and fusion quality indicators.

Aircraft	Proposed Method	DctVar	DctVarCv	SAIF	Optimum
Q_{MI}	1.37	1.32	1.36	1.31	This study (4) > DctVarCv (3) > DctVar (2) > SAIF (1)
Q_{TE}	0.442	0.434	0.441	0.441	This study (4) > DctVarCv (3) > SAIF (2) > DctVar (1)
Q_{NCIE}	0.847	0.842	0.845	0.842	This study (4) > DctVarCv (3) > DctVar (2) > SAIF (1)
Q_G	0.672	0.648	0.674	0.662	DctVarCv (4) > This study (3) > SAIF (2) > DctVar (1)
Q_M	2.312	2.253	2.311	2.082	This study (4) > DctVarCv (3) > DctVar (2) > SAIF (1)
Q_{SF}	-0.059	-0.091	-0.060	-0.068	This study (4) > DctVarCv (3) > SAIF (2) > DctVar (1)
Q_P	0.79	0.7	0.80	0.78	DctVarCv (4) > This study (3) > SAIF (2) > DctVar (1)
Q_S	0.948	0.948	0.948	0.953	SAIF (4) > DctVar (3) > This study (2) > DctVarCv (1)
Q_C	0.89	0.83	0.88	0.88	This study (4) > DctVarCv (3) > SAIF (2) > DctVar (1)
Q_Y	0.974	0.931	0.977	0.965	DctVarCv (4) > This study (3) > SAIF (2) > DctVar (1)
Q_{CV}	9	18	9	9	DctVar (4) > This study (3) > DctVarCv (2) > SAIF (1)
Q_{CB}	0.7597	0.709	0.76	0.745	DctVarCv (4) > This study (3) > SAIF (2) > DctVar (1)
Total score	41	20	37	22	

Optimum rule: Index headed the table with maximum points. Normalized mutual information (Q_{MI}); Fusion metric based on Tsallis entropy (Q_{TE}); Nonlinear correlation information entropy (Q_{NCIE}); Gradient-based fusion performance (Q_G); Image fusion metric based on a multiscale scheme (Q_M); Image fusion metric based on spatial frequency (Q_{SF}); Image fusion metric based on phase congruency (Q_P); Piella's metric (Q_S); Cvejie's metric (Q_C); Yang's metric (Q_Y); Chen–Varshney metric (Q_{CV}); Chen–Blum metric (Q_{CB}).

Table 2. List of clock fusion results and fusion quality indicators.

Clock	Proposed Method	DctVar	DctVarCv	SAIF	Optimum
Q_{MI}	1.21	1.18	1.19	1.14	This study (4) > DctVarCv (3) > DctVar (2) > SAIF (1)
Q_{TE}	0.415	0.406	0.41	0.411	This study (4) > SAIF (3) > DctVarCv (2) > DctVar (1)
Q_{NCIE}	0.8447	0.8424	0.8441	0.8398	This study (4) > DctVarCv (3) > DctVar (2) > SAIF (1)
Q_G	0.682	0.662	0.68	0.676	This study (4) > DctVarCv (3) > SAIF (2) > DctVar (1)
Q_M	2.56	2.58	2.6	2.35	DctVarCv (4) > DctVar (3) > This study (2) > SAIF (1)
Q_{SF}	-0.04	0.17	0.16	-0.06	DctVar (4) > DctVarCv (3) > This study (2) > SAIF (1)
Q_P	0.804	0.629	0.739	0.803	This study (4) > SAIF (3) > DctVarCv (2) > DctVar (1)
Q_S	0.946	0.926	0.933	0.956	SAIF (4) > This study (3) > DctVarCv (2) > DctVar (1)
Q_C	0.798	0.756	0.77	0.801	SAIF (4) > This study (3) > DctVarCv (2) > DctVar (1)
Q_Y	0.98	0.9	0.96	0.96	This study (4) > SAIF (3) > DctVarCv (2) > DctVar (1)
Q_{CV}	13	104	98	12	DctVar (4) > DctVarCv (3) > This study (2) > SAIF (1)
Q_{CB}	0.77	0.65	0.72	0.75	This study (4) > SAIF (3) > DctVarCv (2) > DctVar (1)
Total score	40	22	31	27	

Optimum rule: Index headed the table with maximum points. Normalized mutual information (Q_{MI}); Fusion metric based on Tsallis entropy (Q_{TE}); Nonlinear correlation information entropy (Q_{NCIE}); Gradient-based fusion performance (Q_G); Image fusion metric based on a multiscale scheme (Q_M); Image fusion metric based on spatial frequency (Q_{SF}); Image fusion metric based on phase congruency (Q_P); Piella's metric (Q_S); Cvejie's metric (Q_C); Yang's metric (Q_Y); Chen–Varshney metric (Q_{CV}); Chen–Blum metric (Q_{CB}).

Table 3. List of disk fusion results and fusion quality indicators.

Disk	Proposed Method	DctVar	DctVarCv	SAIF	Optimum
Q_{MI}	1.12	1.11	1.15	1	DctVarCv (4) > This study (3) > DctVar (2) > SAIF (1)
Q_{TE}	0.384	0.372	0.387	0.373	DctVarCv (4) > This study (3) > SAIF (2) > DctVar (1)
Q_{NCIE}	0.836	0.837	0.84	0.831	DctVarCv (4) > DctVar (3) > This study (2) > SAIF (1)
Q_G	0.68	0.7	0.69	0.68	DctVar (4) > DctVarCv (3) > SAIF (2) > This study (1)
Q_M	2.3	2.8	2.7	2.2	DctVar (4) > DctVarCv (3) > This study (2) > SAIF (1)
Q_{SF}	-0.04	-0.01	-0.04	-0.04	DctVar (4) > DctVarCv (3) > This study (2) > SAIF (1)
Q_P	0.777	0.666	0.795	0.797	SAIF (4) > DctVarCv (3) > This study (2) > DctVar (1)
Q_S	0.92	0.92	0.92	0.93	SAIF (4) > DctVarCv (3) > DctVar (2) > This study (1)
Q_C	0.769	0.746	0.756	0.766	This study (4) > SAIF (3) > DctVarCv (2) > DctVar (1)
Q_Y	0.983	0.919	0.989	0.956	DctVarCv (4) > This study (3) > SAIF (2) > DctVar (1)
Q_{CV}	13	142	27	17	DctVar (4) > DctVarCv (3) > SAIF (2) > This study (1)
Q_{CB}	0.76	0.68	0.78	0.73	DctVarCv (4) > This study (3) > SAIF (2) > DctVar (1)
Total score	27	28	40	25	

Optimum rule: Index headed the table with maximum points. Normalized mutual information (Q_{MI}); Fusion metric based on Tsallis entropy (Q_{TE}); Nonlinear correlation information entropy (Q_{NCIE}); Gradient-based fusion performance (Q_G); Image fusion metric based on a multiscale scheme (Q_M); Image fusion metric based on spatial frequency (Q_{SF}); Image fusion metric based on phase congruency (Q_P); Piella's metric (Q_S); Cvejie's metric (Q_C); Yang's metric (Q_Y); Chen–Varshney metric (Q_{CV}); Chen–Blum metric (Q_{CB}).

Table 4. List of leopard fusion results and fusion quality indicators.

Leopard	Proposed Method	DctVar	DctVarCv	SAIF	Optimum
Q_{MI}	1.4509	1.4708	1.471	1.4631	DctVarCv (4) > DctVar (3) > SAIF (2) > This study (1)
Q_{TE}	0.4598	0.4524	0.4539	0.4601	SAIF (4) > This study (3) > DctVarCv (2) > DctVar (1)
Q_{NCIE}	0.8677	0.8695	0.8692	0.8688	DctVar (4) > DctVarCv (3) > SAIF (2) > This study (1)
Q_G	0.856	0.857	0.857	0.859	SAIF (4) > DctVarCv (3) > DctVar (2) > This study (1)
Q_M	2.476	2.7	2.695	2.667	DctVar (4) > DctVarCv (3) > SAIF (2) > This study (1)
Q_{SF}	-0.0133	-0.0116	-0.0118	-0.0113	SAIF (4) > DctVar (3) > DctVarCv (2) > This study (1)
Q_P	0.947	0.947	0.949	0.952	SAIF (4) > DctVarCv (3) > This study (2) > DctVar (1)
Q_S	0.9734	0.9737	0.9737	0.9742	SAIF (4) > DctVar (3) > DctVarCv (2) > This study (1)
Q_C	0.945	0.944	0.945	0.946	SAIF (4) > This study (3) > DctVarCv (2) > DctVar (1)
Q_Y	0.9923	0.9904	0.9921	0.9929	SAIF (4) > This study (3) > DctVarCv (2) > DctVar (1)
Q_{CV}	13.2	13.8	13.4	12.7	DctVar (4) > DctVarCv (3) > This study (2) > SAIF (1)
Q_{CB}	0.836	0.872	0.874	0.844	DctVarCv (4) > DctVar (3) > SAIF (2) > This study (1)
Total score	20	30	33	37	

Optimum rule: Index headed the table with maximum points. Normalized mutual information (Q_{MI}); Fusion metric based on Tsallis entropy (Q_{TE}); Nonlinear correlation information entropy (Q_{NCIE}); Gradient-based fusion performance (Q_G); Image fusion metric based on a multiscale scheme (Q_M); Image fusion metric based on spatial frequency (Q_{SF}); Image fusion metric based on phase congruency (Q_P); Piella's metric (Q_S); Cvejie's metric (Q_C); Yang's metric (Q_Y); Chen–Varshney metric (Q_{CV}); Chen–Blum metric (Q_{CB}).

3.2. Image Fusion Results of the Discussion

In the air craft image fusion result, seven indicators point out that the method in this study derived the best fusion result. From the subjective human eye observation, it was deemed that the fusion result using the DctVar method was clearly the poorest, while the other results were more approximate. In the clock image fusion result, the seven indicdators showed the study derived the best fusion result, while the subjective human eye observation deemed the DctVar and DctVarCv results to be the poorest. The study and the SAIF methods were equally matched in terms of the details. In the disk image fusion result, the five indicators showed the DctVarCv method derived the best fusion result, while the subejctive human eye observation deemed the DctVar and DctVarCv reuslts to be the poorest. The square effect clearly existed in the images. On the other hand, the SAIF method derived the best fusion result. There was a considerable difference between the indicator reuslts and the human eye congition. It was speculated that the square effect exerted less influence on the ratings of the indicdators. In the lab image fusion results, the five indicators showed the DctVarCv derived the best fusion results. The subejctive human eye observation deemed DctVar and SAIF results to be the poorest, with suqare effect and halo lines in the head region. The DctVarCv and the study derived the best fusion results. In the leopard image fusion results, the seven indicadtors showed SAIF derived the best fusion result, but the respecitve indidactors were basically quite approximate. The subjecitve eye observation deemed the respective methods derived the same results. In the toy image fusion results, the six indicators sshowed DctVarCv derived the best fusion result. The subejctive eye observation deemed the study and SAIF derived the best fusion result, while the square effect existed in the DctVar and DctVarCv in the details. Therefore, it was speculated that the indicators were unable to determine the influence of the square effect.

Table 5. List of lab fusion results and fusion quality indicators.

Lab	Proposed Method	DctVar	DctVarCv	SAIF	Optimum
Q_{MI}	1.26	1.22	1.27	1.18	DctVarCv (4) > This study (3) > DctVar (2) > SAIF (1)
Q_{TE}	0.43	0.417	0.427	0.418	This study (4) > DctVarCv (3) > SAIF (2) > DctVar (1)
Q_{NCIE}	0.843	0.841	0.844	0.839	DctVarCv (4) > This study (3) > DctVar (2) > SAIF (1)
Q_G	0.73	0.74	0.73	0.72	DctVar (4) > DctVarCv (3) > This study (2) > SAIF (1)
Q_M	2.34	2.7	2.695	2.398	DctVar (4) > DctVarCv (3) > SAIF (2) > This study (1)
Q_{SF}	-0.03	-0.01	-0.03	-0.03	DctVar (4) > DctVarCv (3) > This study (2) > SAIF (1)
Q_P	0.784	0.674	0.799	0.795	DctVarCv (4) > SAIF (3) > This study (2) > DctVar (1)
Q_S	0.95	0.948	0.951	0.956	SAIF (4) > DctVarCv (3) > This study (2) > DctVar (1)
Q_C	0.802	0.786	0.798	0.791	This study (4) > DctVarCv (3) > SAIF (2) > DctVar (1)
Q_Y	0.98	0.92	1	0.95	DctVarCv (4) > This study (3) > SAIF (2) > DctVar (1)
Q_{CV}	5	18	5	8	DctVar (4) > SAIF (3) > This study (2) > DctVarCv (1)
Q_{CB}	0.72	0.65	0.76	0.72	DctVarCv (4) > This study (3) > SAIF (2) > DctVar (1)
Total score	31	26	39	24	

Optimum rule: Index headed the table with maximum points. Normalized mutual information (Q_{MI}); Fusion metric based on Tsallis entropy (Q_{TE}); Nonlinear correlation information entropy (Q_{NCIE}); Gradient-based fusion performance (Q_G); Image fusion metric based on a multiscale scheme (Q_M); Image fusion metric based on spatial frequency (Q_{SF}); Image fusion metric based on phase congruency (Q_P); Piella's metric (Q_S); Cvejie's metric (Q_C); Yang's metric (Q_Y); Chen–Varshney metric (Q_{CV}); Chen–Blum metric (Q_{CB}).

Table 6. List of toy fusion results and fusion quality indicators.

Toy	Proposed Method	DctVar	DctVarCv	SAIF	Optimum
Q_{MI}	1.17	1.16	1.2	1.06	DctVarCv (4) > This study (3) > DctVar (2) > SAIF (1)
Q_{TE}	0.431	0.419	0.43	0.436	SAIF (4) > This study (3) > DctVarCv (2) > DctVar (1)
Q_{NCIE}	0.836	0.836	0.837	0.831	DctVarCv (4) > This study (3) > DctVar (2) > SAIF (1)
Q_G	0.63	0.62	0.65	0.63	DctVarCv (4) > This study (3) > SAIF (2) > DctVar (1)
Q_M	1.5	2.1	2	1.7	DctVar (4) > DctVarCv (3) > SAIF (2) > This study (1)
Q_{SF}	-0.11	-0.08	-0.11	-0.11	DctVar (4) > DctVarCv (3) > SAIF (2) > This study (1)
Q_P	0.771	0.695	0.824	0.821	DctVarCv (4) > SAIF (3) > This study (2) > DctVar (1)
Q_S	0.934	0.931	0.936	0.948	SAIF (4) > DctVarCv (3) > This study (2) > DctVar (1)
Q_C	0.8	0.756	0.823	0.816	DctVarCv (4) > SAIF (3) > This study (2) > DctVar (1)
Q_Y	0.94	0.86	0.98	0.95	DctVarCv (4) > SAIF (3) > This study (2) > DctVar (1)
Q_{CV}	32	35	31	29	DctVar (4) > This study (3) > DctVarCv (2) > SAIF (1)
Q_{CB}	0.73	0.66	0.77	0.76	DctVarCv (4) > SAIF (3) > This study (2) > DctVar (1)
Total score	27	23	41	29	

Optimum rule: Index headed the table with maximum points. Normalized mutual information (Q_{MI}); Fusion metric based on Tsallis entropy (Q_{TE}); Nonlinear correlation information entropy (Q_{NCIE}); Gradient-based fusion performance (Q_G); Image fusion metric based on a multiscale scheme (Q_M); Image fusion metric based on spatial frequency (Q_{SF}); Image fusion metric based on phase congruency (Q_P); Piella's metric (Q_S); Cvejie's metric (Q_C); Yang's metric (Q_Y); Chen–Varshney metric (Q_{CV}); Chen–Blum metric (Q_{CB}).

Overall, the respective methods showed advantageousness. Under the premise that the square effect was not considered, the DctVar and DctVarCv results under various indicator rations produced advantageous results. Moreover, under subjective observations, the image details were also sound. Without taking into account the halo effect, SAIF had the best details. Compared to other methods, the study was almost unaffected by the square effect and the halo effect, while DctVar and DctVarCv showed strong square effect. Under subjective observations, it already severely affected the fusion results. In the subjective rating of details, the study was the same as SAIF. However, in terms of indicator ratings, the study received 18 best ratings, which was superior to SAIF with 15 best ratings. Further, the study was not affected by the halo effect; thus the relatively more stable fusion result.

The NVG images in Figure 5m,n were used to test fusion quality. The fusion results of DctVar, DctVarCv, and SAIF methods were as shown in Figure 15a–d. Results showed that the image results using the DctVar method presented many squares that were completely unusable. Hence, the study was significantly superior to the DctVar method. In the surrounding of the focus stripe of DctVarCv, there was a large square, while a circular halo rose in the center for the SAIF. Therefore, the fusion test results of NVG images pointed out the study was also superior to DctVarCv and SAIF.

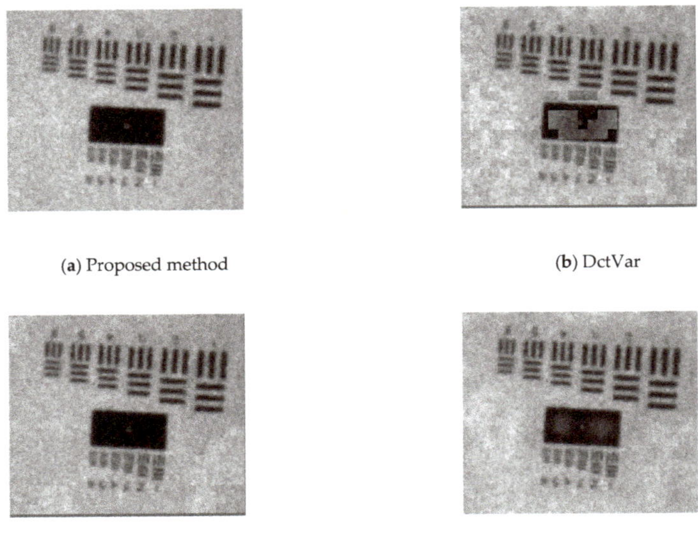

(a) Proposed method	(b) DctVar
(c) DctVarCv	(d) SAIF

Figure 15. The fusion result of the NVG image (60 and 96 degrees).

In addition to the night vision goggle image explored above, this study also observed night vision goggle images fusion with different focal lengths. The result were as shown in Figures 16–18. Intuitively, the fusion results showed that the method proposed in this study has still some incomplete treatments in the peripheral edge area. Otherwise, compared to our methods, it was more advanced than other methods and the discussion was also consistent with the previous paragraph.

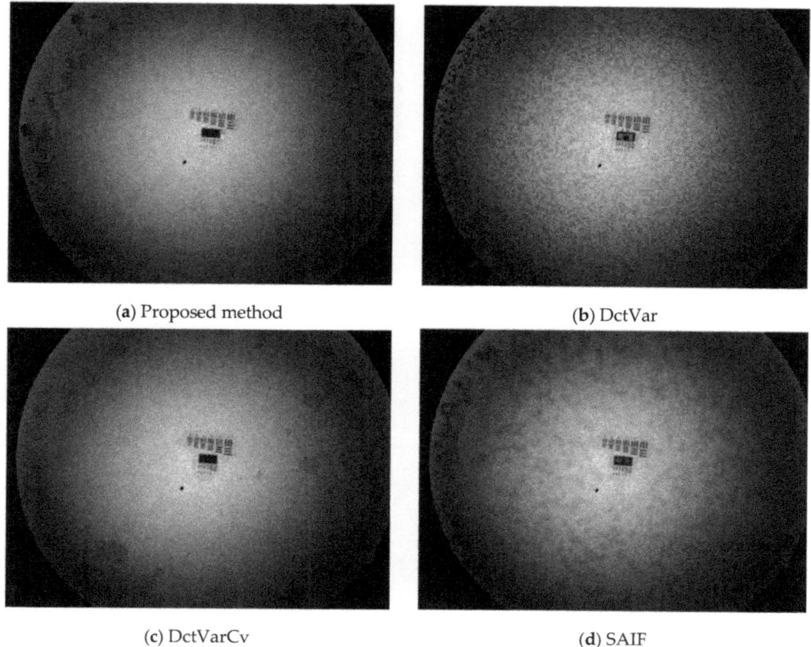

(**a**) Proposed method

(**b**) DctVar

(**c**) DctVarCv

(**d**) SAIF

Figure 16. The fusion result of the NVG image (61 and 96 degrees).

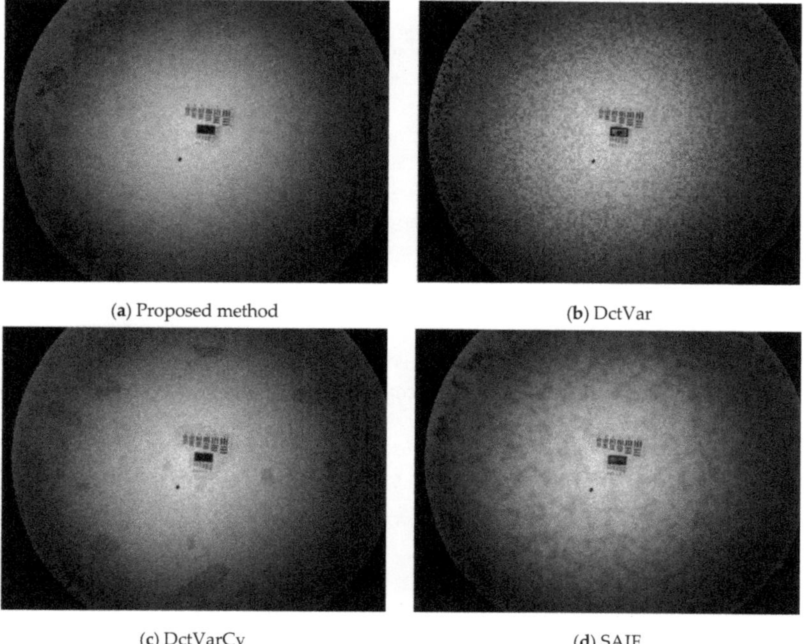

(**a**) Proposed method

(**b**) DctVar

(**c**) DctVarCv

(**d**) SAIF

Figure 17. The fusion result of the NVG image (62 and 96 degrees).

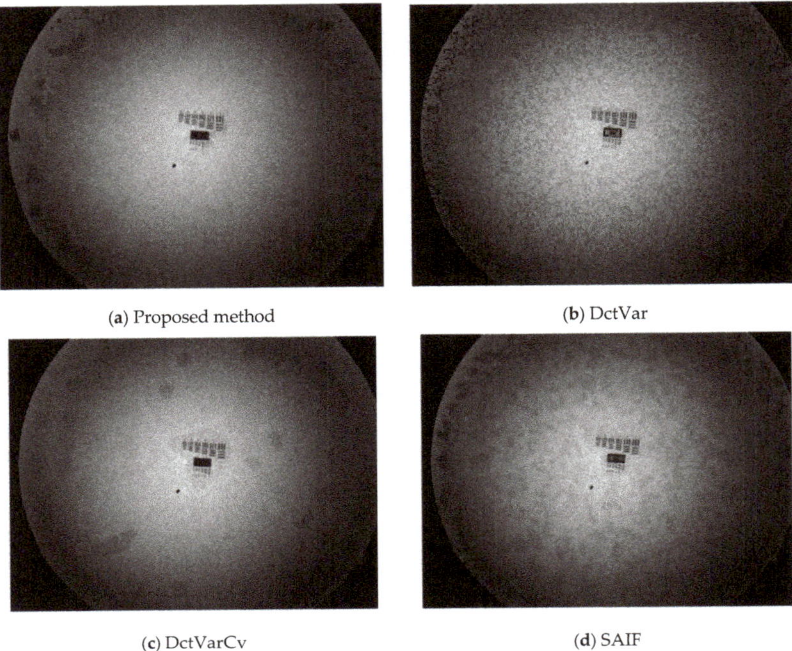

(**a**) Proposed method   (**b**) DctVar

(**c**) DctVarCv   (**d**) SAIF

Figure 18. The fusion result of the NVG image (63 and 96 degrees).

3.3. Autofocus Results

The image sources are the NVG images introduced in the description of the tested images section. According to the process introduced in the autofocus using sparse matrix section, an experiment was performed and Equation (9) was carried out for computing. The statistical result was as shown in in Figure 19a. The lowest point was the frame with the best sharpness. In the example, the 96th frame was the sharpest frame. The image in this frame is Figure 19c. In order to compare the accuracy of the sharpness in this study, Figure 19b was the same source image. The normalized gray-level variance sharpness method proved that the autofocusing application of NVG was effective. According to the calculation result using the normalized gray-level variance sharpness method, the 96th frame was also the one with the best sharpness. Figure 19d is the image in the vicinity of the sharpest point. Compared to Figure 19c, the 90th frame showed obvious differences, proving the robustness of the method in this study and the normalized gray-level variance sharpness method. The method in this study featured the advantages of simple and easy-to-understand computing. In Equation (9), only the sum of the sparse matrices corresponding to the respective images needed to be computed, and the least value was sought as the best sharpness point. This part also validated the concept that "the corresponding frame information in the sparse matrix serves as a reference for sharpness" proposed in the autofocus using sparse matrix section.

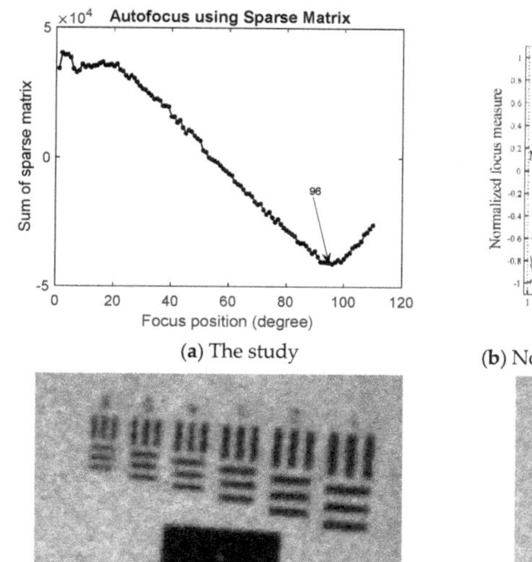

(a) The study

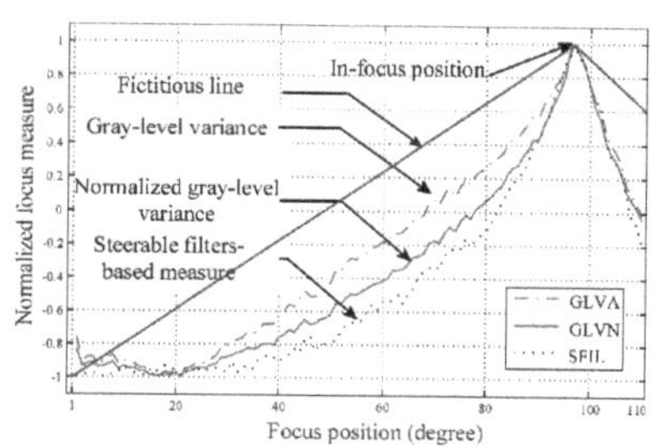

(b) Normalized gray-level variance sharpness [1]

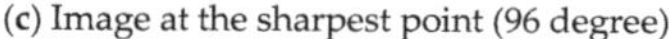

(c) Image at the sharpest point (96 degree)

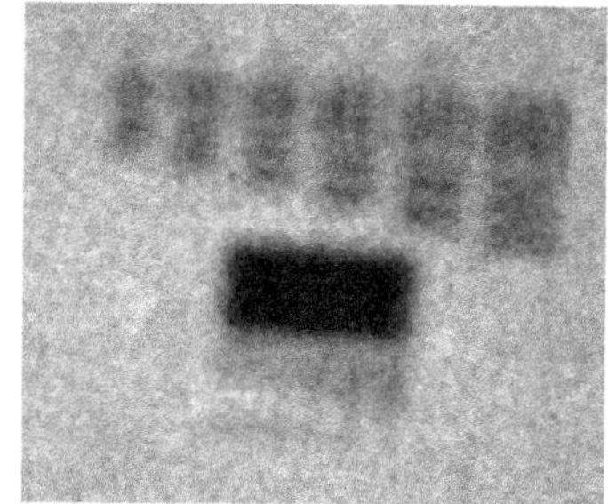

(d) Image nearby the sharpest point (90 degree)

Figure 19. Result comparison of this study and normalized gray-level variance sharpness method under correct focal distances.

4. Conclusions

The decomposition process of the deep semi-NMF model was employed in this study to obtain the sparse and low-rank matrix information, based on which information the autofocusing requirements were completed. Additionally, the simple calculation can be completed using the sparse matrix. Experimental results also proved that under NVG images, the autofocusing method and the traditional normalized gray-level variance sharpness method derived the same calculation results, both deriving the sharpest image frame. Furthermore, in solving the multifocus problem arising from mechanism errors, taking into account the 12 image fusion indicators and the square effect and halo, overall experimental results proved that the method in this study was superior to the other three methods in terms of image testing. On top of it, 18 best ratings were obtained under the image fusion indicator rations. Finally, the autofocusing and image fusion algorithm put forth in this study possessed substantive value in terms of enhancing automated testing equipment process.

Author Contributions: Conceptualization, B.-L.J. and H.-T.Y.; methodology, B.-L.J.; software, W.-L.C.; validation, Y.-C.L.; formal analysis, B.-L.J.; investigation, B.-L.J.; resources, H.-T.Y.; data curation, W.-L.C.; writing—original draft preparation, B.-L.J. and H.-T.Y.; writing—review and editing, W.-L.C.; visualization, Y.-C.L.; supervision, H.-T.Y.; project administration, B.-L.J.; funding acquisition, B.-L.J. and H.-T.Y. All authors have read and agreed to the published version of the manuscript.

Funding: This research was funded by Ministry of Science and Technology of the Republic of China, Taiwan, grant numbers 107-2218-E-167-004 and 108-2218-E-167-005.

Acknowledgments: This work was supported in part by the Ministry of Science and Technology of the Republic of China, Taiwan, under Contract MOST 107-2218-E-167-004 and 108-2218-E-167-005.

Conflicts of Interest: The authors declare no conflict of interest.

References

1. Jian, B.L.; Peng, C.C. Development of an automatic testing platform for aviator's night vision goggle honeycomb defect inspection. *Sensors (Basel)* **2017**, *17*, 1403. [CrossRef] [PubMed]
2. Sabatini, R.; Richardson, M.A.; Cantiello, M.; Toscano, M.; Fiorini, P.; Zammit-Mangion, D.; Gardi, A. Experimental flight testing of night vision imaging systems in military fighter aircraft. *J. Test. Eval.* **2014**, *42*, 1–16. [CrossRef]
3. Chrzanowski, K. Review of night vision metrology. *Opto-Electron. Rev.* **2015**, *23*, 149–164. [CrossRef]
4. Jang, J.; Yoo, Y.; Kim, J.; Paik, J. Sensor-based auto-focusing system using multi-scale feature extraction and phase correlation matching. *Sensors (Basel)* **2015**, *15*, 5747–5762. [CrossRef] [PubMed]
5. Pertuz, S.; Puig, D.; Garcia, M.A. Analysis of focus measure operators for shape-from-focus. *Pattern Recognit.* **2013**, *46*, 1415–1432. [CrossRef]
6. Wan, T.; Zhu, C.C.; Qin, Z.C. Multifocus image fusion based on robust principal component analysis. *Pattern Recognit. Lett.* **2013**, *34*, 1001–1008. [CrossRef]
7. Zhang, Q.; Liu, Y.; Blum, R.S.; Han, J.G.; Tao, D.C. Sparse representation based multi-sensor image fusion for multi-focus and multi-modality images: A review. *Inf. Fusion* **2018**, *40*, 57–75. [CrossRef]
8. Singh, R.; Khare, A. Fusion of multimodal medical images using daubechies complex wavelet transform-a multiresolution approach. *Inf. Fusion* **2014**, *19*, 49–60. [CrossRef]
9. Liu, Y.; Liu, S.P.; Wang, Z.F. A general framework for image fusion based on multi-scale transform and sparse representation. *Inf. Fusion* **2015**, *24*, 147–164. [CrossRef]
10. Zhang, Q.; Ma, Z.K.; Wang, L. Multimodality image fusion by using both phase and magnitude information. *Pattern Recognit. Lett.* **2013**, *34*, 185–193. [CrossRef]
11. Li, W.; Xie, Y.G.; Zhou, H.L.; Han, Y.; Zhan, K. Structure-aware image fusion. *Optik* **2018**, *172*, 1–11. [CrossRef]
12. Haghighat, M.B.A.; Aghagolzadeh, A.; Seyedarabi, N. Multi-focus image fusion for visual sensor networks in dct domain. *Comput. Electr. Eng.* **2011**, *37*, 789–797. [CrossRef]
13. Haghighat, M.B.A.; Aghagolzadeh, A.; Seyedarabi, H. Real-time fusion of multi-focus images for visual sensor networks. In Proceedings of the 2010 6th Iranian Conference on Machine Vision and Image Processing, Isfahan, Iran, 27–28 October 2010; pp. 1–6.
14. Dogra, A.; Goyal, B.; Agrawal, S. From multi-scale decomposition to non-multi-scale decomposition methods: A comprehensive survey of image fusion techniques and its applications. *IEEE Access* **2017**, *5*, 16040–16067. [CrossRef]
15. Paramanandham, N.; Rajendiran, K. Infrared and visible image fusion using discrete cosine transform and swarm intelligence for surveillance applications. *Infrared Phys. Technol.* **2018**, *88*, 13–22. [CrossRef]
16. Vanmali, A.V.; Kataria, T.; Kelkar, S.G.; Gadre, V.M. Ringing artifacts in wavelet based image fusion: Analysis, measurement and remedies. *Inf. Fusion* **2020**, *56*, 39–69. [CrossRef]
17. Ganasala, P.; Prasad, A.D. Medical image fusion based on laws of texture energy measures in stationary wavelet transform domain. *Int. J. Imaging Syst. Technol.* **2019**, 1–14. [CrossRef]
18. Seal, A.; Panigrahy, C. Human authentication based on fusion of thermal and visible face images. *Multimed. Tools Appl.* **2019**, *78*, 30373–30395. [CrossRef]
19. Hassan, M.; Murtza, I.; Khan, M.A.Z.; Tahir, S.F.; Fahad, L.G. Neuro-wavelet based intelligent medical image fusion. *Int. J. Imaging Syst. Technol.* **2019**, *29*, 633–644. [CrossRef]
20. Liu, Y.; Chen, X.; Peng, H.; Wang, Z.F. Multi-focus image fusion with a deep convolutional neural network. *Inf. Fusion* **2017**, *36*, 191–207. [CrossRef]
21. Liu, X.Y.; Liu, Q.J.; Wang, Y.H. Remote sensing image fusion based on two-stream fusion network. *Inf. Fusion* **2020**, *55*, 1–15. [CrossRef]
22. Lin, S.Z.; Han, Z.; Li, D.W.; Zeng, J.C.; Yang, X.L.; Liu, X.W.; Liu, F. Integrating model-and data-driven methods for synchronous adaptive multi-band image fusion. *Inf. Fusion* **2020**, *54*, 145–160. [CrossRef]
23. Maqsood, S.; Javed, U. Multi-modal medical image fusion based on two-scale image decomposition and sparse representation. *Biomed. Signal Process. Control* **2020**, *57*, 101810. [CrossRef]
24. Ma, X.L.; Hu, S.H.; Liu, S.Q.; Fang, J.; Xu, S.W. Multi-focus image fusion based on joint sparse representation and optimum theory. *Signal Process.-Image Commun.* **2019**, *78*, 125–134. [CrossRef]
25. Wang, Z.; Bai, X. High frequency assisted fusion for infrared and visible images through sparse representation. *Infrared Phys. Technol.* **2019**, *98*, 212–222. [CrossRef]

26. Wang, K. Rock particle image fusion based on sparse representation and non-subsampled contourlet transform. *Optik* **2019**, *178*, 513–523. [CrossRef]
27. Fu, G.-P.; Hong, S.-H.; Li, F.-L.; Wang, L. A novel multi-focus image fusion method based on distributed compressed sensing. *J. Vis. Commun. Image Represent.* **2020**, *67*, 102760. [CrossRef]
28. Bouwmans, T.; Zahzah, E.H. Robust pca via principal component pursuit: A review for a comparative evaluation in video surveillance. *Comput. Vis. Image Underst.* **2014**, *122*, 22–34. [CrossRef]
29. Yan, Z.B.; Chen, C.Y.; Yao, Y.; Huang, C.C. Robust multivariate statistical process monitoring via stable principal component pursuit. *Ind. Eng. Chem. Res.* **2016**, *55*, 4011–4021. [CrossRef]
30. Tang, G.; Nehorai, A. Robust principal component analysis based on low-rank and block-sparse matrix decomposition. In Proceedings of the Information Sciences and Systems (CISS), 2011 45th Annual Conference, Baltimore, MD, USA, 23 March 2011; pp. 1–5.
31. Wohlberg, B.; Chartrand, R.; Theiler, J. Local principal component analysis for nonlinear datasets. In Proceedings of the International Conference on Acoustics, Speech, and Signal Processing (ICASSP) 2012, Kyoto, Japan, 25–30 March 2012.
32. Narayanamurthy, P.; Vaswani, N. Provable dynamic robust pca or robust subspace tracking. *IEEE Trans. Inf. Theory* **2018**, *64*, 1547–1577.
33. Kang, Z.; Peng, C.; Cheng, Q. Top-n recommender system via matrix completion. In Proceedings of the Thirtieth AAAI Conference on Artificial Intelligence, Phoenix, AZ, USA, 12–17 February 2016; pp. 179–185.
34. Trigeorgis, G.; Bousmalis, K.; Zafeiriou, S.; Schuller, B. A deep semi-nmf model for learning hidden representations. In Proceedings of the International Conference on Machine Learning, Bejing, China, 22–24 June 2014; pp. 1692–1700.
35. Vaswani, N.; Narayanamurthy, P. Static and dynamic robust pca and matrix completion: A review. *Proc. IEEE* **2018**, *106*, 1359–1379. [CrossRef]
36. Bouwmans, T.; Sobral, A.; Javed, S.; Jung, S.K.; Zahzah, E.-H. Decomposition into low-rank plus additive matrices for background/foreground separation: A review for a comparative evaluation with a large-scale dataset. *Comput. Sci. Rev.* **2017**, *23*, 1–71. [CrossRef]
37. Liu, X.H.; Chen, Z.B.; Qin, M.Z. Infrared and visible image fusion using guided filter and convolutional sparse representation. *Opt. Precis. Eng.* **2018**, *26*, 1242–1253.
38. Li, H.; Wu, X.-J. Multi-focus noisy image fusion using low-rank representation. *arXiv* **2018**, arXiv:1804.09325.
39. El-Hoseny, H.M.; Abd El-Rahman, W.; El-Rabaie, E.M.; Abd El-Samie, F.E.; Faragallah, O.S. An efficient dt-cwt medical image fusion system based on modified central force optimization and histogram matching. *Infrared Phys. Technol.* **2018**, *94*, 223–231. [CrossRef]
40. Liu, Z.; Blasch, E.; Bhatnagar, G.; John, V.; Wu, W.; Blum, R.S. Fusing synergistic information from multi-sensor images: An overview from implementation to performance assessment. *Inf. Fusion* **2018**, *42*, 127–145. [CrossRef]
41. Somvanshi, S.S.; Kunwar, P.; Tomar, S.; Singh, M. Comparative statistical analysis of the quality of image enhancement techniques. *Int. J. Image Data Fusion* **2017**, *9*, 131–151. [CrossRef]
42. Liu, Z.; Blasch, E.; Xue, Z.; Zhao, J.; Laganiere, R.; Wu, W. Objective assessment of multiresolution image fusion algorithms for context enhancement in night vision: A comparative study. *IEEE Trans. Pattern Anal. Mach. Intell.* **2012**, *34*, 94–109. [CrossRef]

 *applied sciences*

Article

Error Resilience for Block Compressed Sensing with Multiple-Channel Transmission

Hsiang-Cheh Huang [1], Po-Liang Chen [1] and Feng-Cheng Chang [2,*]

[1] Department of Electrical Engineering, National University of Kaohsiung, Kaohsiung City 81148, Taiwan; hchuang@nuk.edu.tw (H.-C.H.); hch.nuk@gmail.com (P.-L.C.)

[2] Department of Innovative Information and Technology, Tamkang University, New Taipei City 25137, Taiwan

* Correspondence: 135170@mail.tku.edu.tw; Tel.: +886-3987-3088

Received: 22 November 2019; Accepted: 19 December 2019; Published: 24 December 2019

Abstract: Compressed sensing is well known for its superior compression performance, in existing schemes, in lossy compression. Conventional research aims to reach a larger compression ratio at the encoder, with acceptable quality reconstructed images at the decoder. This implies looking for compression performance with error-free transmission between the encoder and the decoder. Besides looking at compression performance, we applied block compressed sensing to digital images for robust transmission. For transmission over lossy channels, error propagation or data loss can be expected, and protection mechanisms for compressed sensing signals are required for guaranteed quality of the reconstructed images. We propose transmitting compressed sensing signals over multiple independent channels for robust transmission. By introducing correlations with multiple-description coding, which is an effective means for error resilient coding, errors induced in the lossy channels can effectively be alleviated. Simulation results presented the applicability and superiority of performance, depicting the effectiveness of protection of compressed sensing signals.

Keywords: block compressed sensing; error resilience; reconstruction

---

1. Introduction

Data compression has long been an important topic in the field of signal processing. With the broad use of smartphone and tablet cameras, vast amounts of multimedia content, mostly images, have accumulated. Thus, how a mechanism for efficiently performing data compression on the multimedia contents is urgent. There have been successful and popular standards for image compression including the well-known JPEG, which employs discrete cosine transform (DCT), and JPEG2000, which applies discrete wavelet transform (DWT), to still images. With the evolution of new techniques, advancements in data compression can also be expected, and compressed sensing techniques present new and novel concepts.

Compressed sensing (CS) is one recently developed technique of lossy data compression research and applications [1–4]. In addition to international standards in multimedia compression, such as JPEG [5] or JPEG2000 [6], it would be constructive to explore new innovative compression techniques. The major research looking at JPEG, JPEG2000, and CS focuses on compression performance by balancing the amount of compressed data and the reconstructed quality. In CS, this requires a sampling rate that is far less than the Nyquist rate and has the ability to reconstruct the original signal of lossy compression. The primary goal in compressed sensing research is the compression capability. Therefore, a method for effectively decoding the extremely small amount of compressed signals and comparing this to counterparts in JPEG or JPEG2000 would be of great interest and is currently a major challenge [7–9]. In this paper, in addition to looking at compression performance, we consider practical scenarios of robust transmission of compressed sensing signals using block compressed sensing (BCS) [10–12]. We transmitted compressed signals over lossy channels to observe the effect of packet losses. To reduce quality degradation of robust transmission [13–15], we employed the transmission of BCS signals over multiple independent channels. Multiple description coding (MDC) [16,17] was employed to protect and reconstruct the BCS signals. To better explore the performance of BCS, we used adaptive sampling [18,19]. This way, the enhanced quality of the reconstructed image can be observed for error controlled transmission. Therefore, the novelty of this paper is the error resilient transmission of BCS with MDC. We observed enhanced performance using our algorithm.

This paper is organized as follows. In Section 2 we briefly describe the fundamentals and mathematical representations of BCS. In Section 3 we present the proposed method for error resilient transmission of compressed sensing signals over multiple independent and lossy channels. The simulation results are demonstrated in Section 4. Here we point out the vulnerability of compressively sensed signals transmitted over a single channel and how our algorithm for multiple-channel transmission for grey-level and color images reduced image quality degradation. Finally, we address the conclusion of this paper in Section 5.

2. Fundamental Concepts of Block Compressed Sensing

The field of compressed sensing aims to look for new sampling schemes that go against conventional sampling theorems or the well-known Nyquist-Shannon theorem. With compressed sensing a rate much smaller than twice the maximal bandwidth can be achieved to meet perfect reconstruction recovery.

Compressed sensing, based on the representations in [1,2], is composed of the *sparsity* principle, and the *incoherence* principle. Based on the concepts of compressed sensing depicted in Figure 1, we divided the original image $\mathbf{X}$ with the size of $M \times N$ into a set of small blocks $\mathbf{X}_k$. Each block $\mathbf{X}_k$ has the size $B \times B$, and the subscript k denoted the index of the block corresponding to the original image, $1 \leq k \leq \frac{M}{B} \times \frac{M}{B}$. With the partition of $\mathbf{X}$ into $\mathbf{X}_k$, we performed the below operations block by block to turn compressed sensing into BCS [10].

- The *sparsity* principle implies the information rate in data compression. In BCS this was expected to use a much smaller sampling rate than conventionally required, and it can be represented via Ψ, $\Psi \in C^{B^2 \times B^2}$, where C denotes the complex number in the $B^2 \times B^2$ matrix. Ψ was the basis to reach sparsity with a k-sparse coefficient vector $\mathbf{X}_k$, $\mathbf{X}_k \in C^{B^2 \times 1}$, with the condition that

$$\mathbf{f}_k = \Psi \mathbf{X}_k \tag{1}$$

  where $\mathbf{f}_k$ denotes the reconstruction corresponding to the original signal, $\mathbf{X}_k$.
- The *incoherence* principle extends the duality between time and frequency. The measurement basis Φ, $\Phi \in C^{m \times B^2}$, which acts like noiselet, was employed to sense the signal $\mathbf{f}_k$, with the condition that

$$\mathbf{Y}_k = \Phi \mathbf{f}_k \tag{2}$$

  where $\mathbf{Y}_k$ denotes the measurement vector, as depicted in Figure 1a. We noted that Equation (2) was an underdetermined system.

For the reconstruction from BCS signals at the decoder, several methods can be employed. Considering Equations (1) and (2), by minimizing the L1-norm of $\mathbf{X}_k$, i.e., $\min\|\mathbf{X}_k\|_1$ subject to $\mathbf{Y}_k = \Phi \Psi \mathbf{X}_k$, compressed sensing guarantees perfect recovery with a probability close to 1.0. Both the 'equality constraint' and the 'inequality quadratic constraint' are widely employed conditions for minimization [20,21]. The equality constraint means that

$$\min\|\mathbf{X}_k\|_1 \text{ subject to } \mathbf{Y}_k = \Phi \Psi \mathbf{X}_k \tag{3}$$

where, $\|\cdot\|_1$ denotes the L1-norm. The inequality quadratic constraint implies that

$$\|\min\mathbf{X}_k\|_1 \text{ subject to } \|\Phi \Psi \mathbf{X}_k - \mathbf{Y}_k\|_2 \leq \varepsilon \tag{4}$$

where, $\|\cdot\|_2$ denotes the L2-norm. Thus, the inequality quadratic constraint leaves some tolerance for minimization.

When an image $\mathbf{X}$ is represented by a BCS scheme, it focuses on the local characteristics of the image. $\mathbf{X}$ was divided into a set of blocks, $\mathbf{X}_k$. Therefore, it might be inefficient to assign the same number of measurement dimension to each sampled vector corresponding to the different image block. Due to the local characteristics, one block in the image had significantly different sparsity than the other. With adaptive sampling in BCS, the entropy of a block may be used to evaluate the included information. It was expected to have better reconstruction quality with error-free transmission from adaptive sampling. Regarding adaptive sampling (AS) [18], the normalized DCT coefficient c'_k can be calculated by

$$c'_k = \frac{c_k - c_{k,\min}}{c_{k,\max} - c_{k,\min}} \tag{5}$$

where $c_{k,\max}$ and $c_{k,\min}$ denote the maximal and minimal DCT coefficients in block $\mathbf{X}_k$. Then, the entropy in $\mathbf{X}_k$ can be calculated by

$$H_k = -\int_0^1 p(c'_k) \cdot \log p(c'_k) dc'_k \tag{6}$$

With the aid of adaptive sampling, better performances can be observed for error-free transmission. We will explore the use of adaptive sampling for lossy compression of BCS signals.

3. Proposed Algorithm

For the effective delivery of compressed sensing signals, and considering the robust transmission depicted in [14], we employed the use of transmission over multiple mutually independent lossy channels. Figure 1 describes the block diagram of our system.

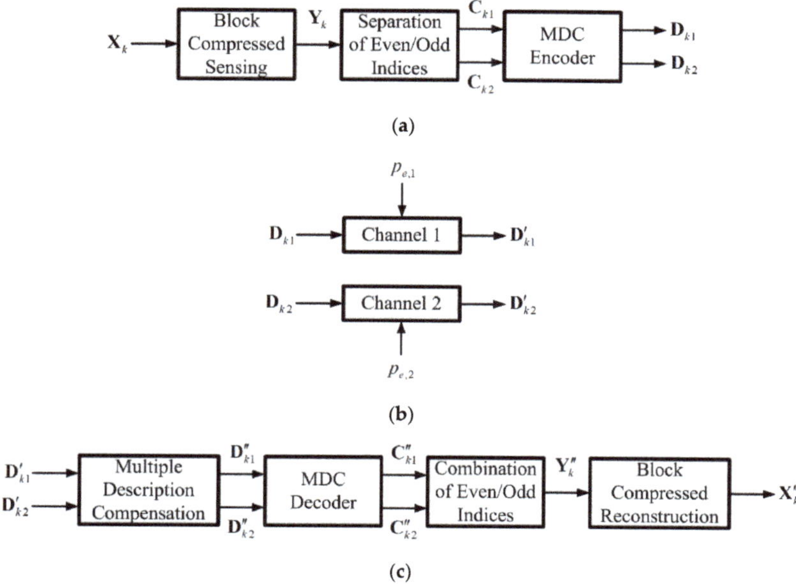

Figure 1. Block diagrams for transmission with block compressed sensing (BCS) and multiple description coding (MDC): (**a**) BCS encoder and protection with MDC; (**b**) Lossy transmission over two channels with $p_{e,1}$ and $p_{e,2}$; (**c**) MDC and BCS decoder for protection and reconstruction.

In Figure 1a, the input image $\mathbf{X}$ is divided into $\mathbf{X}_k$. It is then compressed with BCS and the compressed sensing signal is denoted by $\mathbf{Y}_k$. With the notations described in Section 2, we denoted $\mathbf{Y}_k = \{Y_{k,1}, Y_{k,2}, \cdots, Y_{k,m}\}$ and set m to an even number for application with multiple description coding. To enable easy separation of BCS coefficients, we chose the odd numbered indices to form $\mathbf{C}_{k1} = \{Y_{k,1}, Y_{k,3}, \cdots, Y_{k,m-1}\}$, and the even numbered ones to form $\mathbf{C}_{k2} = \{Y_{k,2}, Y_{k,4}, \cdots, Y_{k,m}\}$. For ease of representation, we rearranged the notations to be

$$\mathbf{C}_{k1} = \left\{Y_{k,1}, Y_{k,3}, \cdots, Y_{k,m-1}\right\} = \left\{C_{k1,1}, C_{k1,2}, \cdots, C_{k1,\frac{m}{2}}\right\} \tag{7}$$

and

$$\mathbf{C}_{k2} = \left\{Y_{k,2}, Y_{k,4}, \cdots, Y_{k,m}\right\} = \left\{C_{k2,1}, C_{k2,2}, \cdots, C_{k2,\frac{m}{2}}\right\} \tag{8}$$

After that, we employed multiple description transform coding (MDTC) [15] in MDC to form the elements in the two descriptions of $\mathbf{D}_{k1}$ and $\mathbf{D}_{k2}$ in Equation (9):

$$\begin{bmatrix} D_{k1,i} \\ D_{k2,i} \end{bmatrix} = \begin{bmatrix} r_2 \cos\theta_2 & -r_2 \sin\theta_2 \\ -r_1 \cos\theta_1 & r_1 \sin\theta_1 \end{bmatrix} \begin{bmatrix} C_{k1,i} \\ C_{k2,i} \end{bmatrix}, \tag{9}$$

where $i = 1, 2, \cdots, \frac{m}{2}$. The 2×2 matrix in Equation (9) has the condition that $r_1 r_2 \sin(\theta_1 - \theta_2) = 1$, leading to the determinant of one. The resulting elements in Equation (9) form the two descriptions $\mathbf{D}_{k1} = \left\{D_{k1,1}, D_{k1,2}, \cdots, D_{k1,\frac{m}{2}}\right\}$ and $\mathbf{D}_{k2} = \left\{D_{k2,1}, D_{k2,2}, \cdots, D_{k2,\frac{m}{2}}\right\}$.

Next, descriptions $\mathbf{D}_{k1}$ and $\mathbf{D}_{k2}$ were transmitted over two independent lossy channels, with the loss probability of $p_{e,1}$ and $p_{e,1}$ for Channels 1 and 2, as depicted in Figure 1b. The received descriptions may become $\mathbf{D}'_{k1} = \left\{D'_{k1,1'}, D'_{k1,2'}, \cdots, D'_{k1,\frac{m}{2}}\right\}$ and $\mathbf{D}'_{k2} = \left\{D'_{k2,1'}, D'_{k2,2'}, \cdots, D'_{k2,\frac{m}{2}}\right\}$ due to the possibility of induced errors. Note that $\mathbf{D}'_{k1}$ and $\mathbf{D}'_{k2}$ may not be identical to their counterparts $\mathbf{D}_{k1}$ and $\mathbf{D}_{k2}$, respectively.

At the decoder, as shown in Figure 1c by employing [16], compensation should be applied between received descriptions. Compensated descriptions $\mathbf{D}''_{k1} = \left\{D''_{k1,1'}, D''_{k1,2'}, \cdots, D''_{k1,\frac{m}{2}}\right\}$ and $\mathbf{D}''_{k2} = \left\{D''_{k2,1'}, D''_{k2,2'}, \cdots, D''_{k2,\frac{m}{2}}\right\}$ was calculated first. The elements in the descriptions $D''_{k1,i}$ and $D''_{k2,i'}$ $i = 1, 2, \cdots, \frac{m}{2}$ were compensated as follows:

$$D''_{k1,i} = \left(\frac{r_1 \cos\theta_1 \cos\theta_2 \sigma_1^2 + r_1 \sin\theta_1 \sin\theta_2 \sigma_2^2}{-\cos\theta_2 \sigma_1^2 + \sin\theta_2 \sigma_2^2} \right) D'_{k1,i'} \tag{10}$$

$$D''_{k2,i} = \left(\frac{r_2 \cos\theta_1 \cos\theta_2 \sigma_1^2 + r_2 \sin\theta_1 \sin\theta_2 \sigma_2^2}{\cos\theta_2 \sigma_1^2 - \sin\theta_2 \sigma_2^2} \right) D'_{k1,i}. \tag{11}$$

By taking the inverse operations of Equation (9), compensated BCS coefficients were obtained in Equation (12):

$$\begin{bmatrix} C''_{k1,i} \\ C''_{k2,i} \end{bmatrix} = \begin{bmatrix} r_2 \cos\theta_2 & -r_2 \sin\theta_2 \\ -r_1 \cos\theta_1 & r_1 \sin\theta_1 \end{bmatrix}^{-1} \begin{bmatrix} D''_{k1,i} \\ D''_{k2,i} \end{bmatrix}. \tag{12}$$

By gathering all the elements together, $\mathbf{C}''_{k1}$ and $\mathbf{C}''_{k2}$ were formed from the compensated descriptions in Equation (12). Here we denote

$$\mathbf{C}''_{k1} = \left\{C''_{k1,1'}, C''_{k1,2'}, \cdots, C''_{k1,\frac{m}{2}}\right\} = \left\{Y''_{k,1'}, Y''_{k,3'}, \cdots, Y''_{k,m-1}\right\} \tag{13}$$

and

$$\mathbf{C}''_{k2} = \left\{C''_{k2,1'}, C''_{k2,2'}, \cdots, C''_{k2,\frac{m}{2}}\right\} = \left\{Y''_{k,2'}, Y''_{k,4'}, \cdots, Y''_{k,m}\right\} \tag{14}$$

After the combination of the even and odd indexed components in Equations (13) and (14) and Figure 1c, we obtained $\mathbf{Y}_k''$. Finally, using BCS we reconstructed block $\mathbf{X}_k''$. After completing the reconstruction of all the blocks, we composed the reconstructed image $\mathbf{X}''$.

4. Simulation Results

In our simulations we provided three sets of experiments based on three test images. The first was cameraman grey-level test image with a size of 256×256. The second was the ducks color image, taken by the authors, with a size of 1024×1024. The third was the Pasadena-houses color image with a size of 1760×1168 [22]. These three images were employed in the experiments.

Here we start the first set of experiments with the test image cameraman. In Figure 2, we present the use of entropy for adaptive sampling. Considering the practical implementation of adaptive sampling, we applied quantization to the entropy values with a step size of 0.1. Figure 2a presents the original test image, cameraman. The relationship between the number of blocks and entropy is depicted in Figure 2b. Smaller entropy values imply smoother blocks.

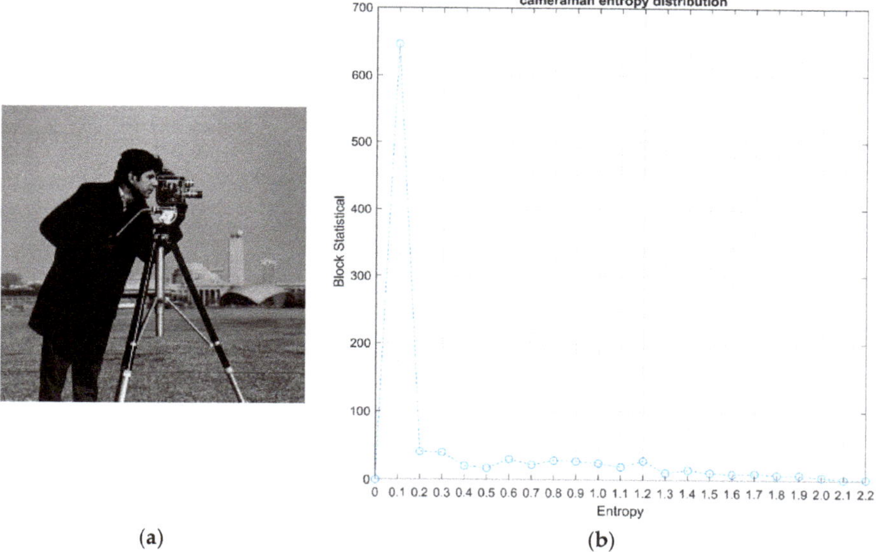

(a)   (b)

Figure 2. The test image cameraman: (**a**) original image with a size of 256×256; and (**b**) the entropy distribution.

Figure 3 presents the error-free transmission of BCS for cameraman. Considering the practical applications, we chose $B = 8$ in BCS in Equation (1). The measurement rate was set to $\frac{12}{64} = 0.1875$, meaning that $m = 12$ compressed sensing coefficients were selected per block on the average. The reconstructed images were assessed with the peak signal-to-noise ratio (PSNR) and the structural similarity (SSIM) [23]. Figure 3a shows only the reconstruction for equality constraint (EQ). Because adaptive sampling was not applied, it is implied that every block was reconstructed with 12 BCS coefficients. Figure 3b presents the reconstruction for equality constraint (EQ) with adaptive sampling (AS) [10]. Here we use the abbreviation EQ + AS for the results in Figure 3b. Because adaptive sampling was applied, the number of BCS coefficients varied from one block to another. Blocks with larger entropies were designated a higher number of BCS coefficients, and vice versa. The average number of BCS coefficients for all the blocks was 12. Via adaptive sampling, we easily observed enhanced performances. Figure 3c,d led to the results for quadratic constraint (QC) and quadratic constraint with adaptive sampling (QC + AS), respectively. Again, with adaptive sampling, better performances were observed. We also compared the results of Figure 3a,c. With the quadratic constraint in Equation (4) it performed slightly better than its counterpart, equality constraint, in Equation (3). Similar phenomena can also be found by comparing Figure 3b,d.

(a) EQ
PSNR: 19.14 dB, SSIM: 0.61

(b) EQ+AS
PSNR: 24.24 dB, SSIM: 0.76

(c) QC
PSNR: 19.20 dB, SSIM: 0.62

(d) QC+AS
PSNR: 24.35 dB, SSIM: 0.77

Figure 3. Reconstruction of BCS coefficients: (**a**) equality constraint only (EQ); (**b**) equality constraint and adaptive sampling (EQ + AS); (**c**) quadratic constraint only (QC); and (**d**) quadratic constraint and adaptive sampling (QC + AS).

In Figure 4 we present the lossy transmission for block compressed sensing with adaptive sampling. In Figure 4a,b, when the BCS coefficients were transmitted over the two independent and lossy channels in Figure 1b, we set $p_{e,1} = p_{e,2} = 0.1$. Here we employed the equality constraint (EQ) for reconstruction. Constructing the received BCS coefficients directly led to the result in Figure 4a, meaning that some protection may be required. In Figure 4b, we applied multiple description coding from Equation (9) for protection. The parameters were chosen to be $\theta_1 = \frac{\pi}{3}$, $\theta_2 = \frac{-\pi}{4}$, and $r_2 = 3$, and this led to $r_1 = \frac{1}{r_2 \sin(\theta_1 - \theta_2)} = 0.3451$, from calculations using Equation (9). With the compensation techniques of MDC from Equation (9), we observed that the reconstructed quality was greatly improved. In addition, for the evaluation of reconstruction under severely lost channels we set $p_{e,1} = p_{e,2} = 0.5$, with the results shown in Figure 4c,d. Severe degradation was easily visible in Figure 4c, as was the improvement of reconstructed quality after protection in Figure 4d.

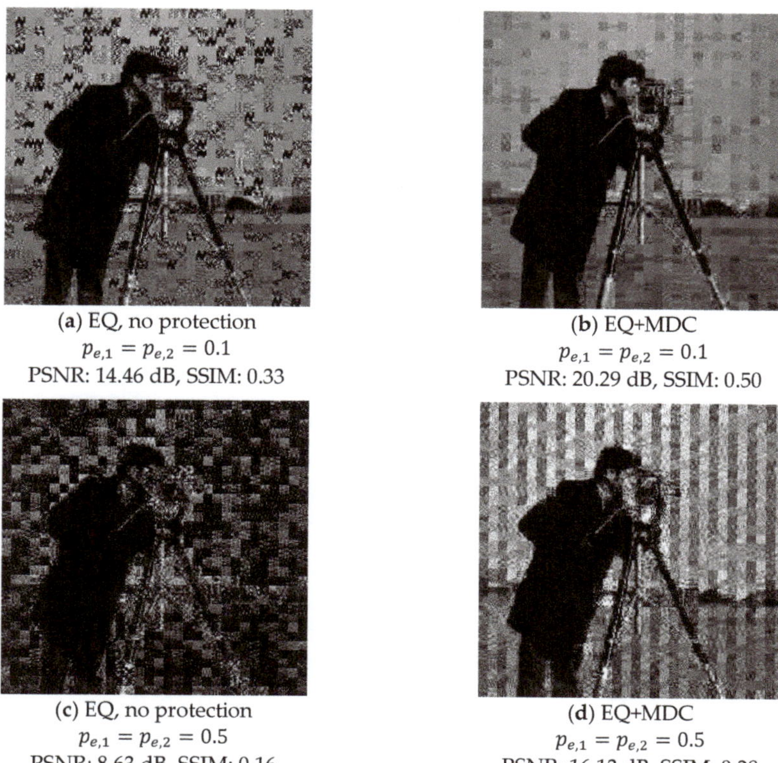

(a) EQ, no protection
$p_{e,1} = p_{e,2} = 0.1$
PSNR: 14.46 dB, SSIM: 0.33

(b) EQ+MDC
$p_{e,1} = p_{e,2} = 0.1$
PSNR: 20.29 dB, SSIM: 0.50

(c) EQ, no protection
$p_{e,1} = p_{e,2} = 0.5$
PSNR: 8.63 dB, SSIM: 0.16

(d) EQ+MDC
$p_{e,1} = p_{e,2} = 0.5$
PSNR: 16.13 dB, SSIM: 0.28

Figure 4. Reconstruction of BCS coefficients with equality constraint over lossy channels.

In Figure 5 we applied adaptive sampling and quadratic constraint with different selections of the value ε in Equation (4) for experiments. We chose $\varepsilon = 250$ for Figure 5a,c,e, and $\varepsilon = 320$ for Figure 5b,d,f. In Figure 5a,b, we applied error-free transmission, or $p_{e,1} = p_{e,2} = 0.0$, over the two independent and lossy channels. We observed that Figure 5a, or the one with $\varepsilon = 250$, performed slightly better. In Figure 5c,d, we set $p_{e,1} = 0.1$ and $p_{e,2} = 0.0$, meaning that Channel 1 was lossy, and Channel 2 was error-free. Here we noticed that Figure 5d, or the one with $\varepsilon = 320$, had better PSNR and SSIM values. We also noted that even though the PSNR value in Figure 5d was larger than that of Figure 5c, the SSIM value was not. SSIM considers the local characteristics of the images, and PSNR takes the error of the whole image into account. Thus, there may sometimes be mismatches between the two measures. Finally, in Figure 5e,f, we set $p_{e,1} = p_{e,2} = 0.1$. Due to data loss in both channels, reconstructions in Figure 5e,f depict inferior results to their counterparts in Figure 5c,d. Here, Figure 5f, or the one with $\varepsilon = 320$, performed better than Figure 5e. Based the performances depicted in Figure 5, we may conclude that with the careful selection of the tolerance value or, in this case, $\varepsilon = 320$, a better quality reconstructed image can be acquired.

(a) QC, $\varepsilon = 250$
$p_{e,1} = p_{e,2} = 0.0$
PSNR: 25.66 dB, SSIM: 0.77

(b) QC, $\varepsilon = 320$
$p_{e,1} = p_{e,2} = 0.0$
PSNR: 25.52 dB, SSIM: 0.75

(c) QC+MDC, $\varepsilon = 250$
$p_{e,1} = 0.1$, $p_{e,2} = 0.0$
PSNR: 23.08 dB, SSIM: 0.63

(d) QC+MDC, $\varepsilon = 320$
$p_{e,1} = 0.1$, $p_{e,2} = 0.0$
PSNR: 23.42 dB, SSIM: 0.58

(e) QC+MDC, $\varepsilon = 250$
$p_{e,1} = p_{e,2} = 0.1$
PSNR: 14.68 dB, SSIM: 0.31

(f) QC+MDC, $\varepsilon = 320$
$p_{e,1} = p_{e,2} = 0.1$
PSNR: 15.97 dB, SSIM: 0.36

Figure 5. Reconstruction of BCS coefficients with quadratic constraint (QC) over lossy channels with protection.

In Figure 6 we present the evaluations of the reconstructed image quality over a range of lossy probabilities between 0 and 0.5, with a measurement rate of 0.1. We observed that the increase of loss tended toward inferior quality reconstructed images, as was expected. In addition, protection with multiple description coding alleviated the effect caused by data loss, which led to better quality reconstructed images. Finally, we found that with the careful selection of the ε value, quadratic constraint performed slightly better than equality constraint in terms of reconstructed image quality.

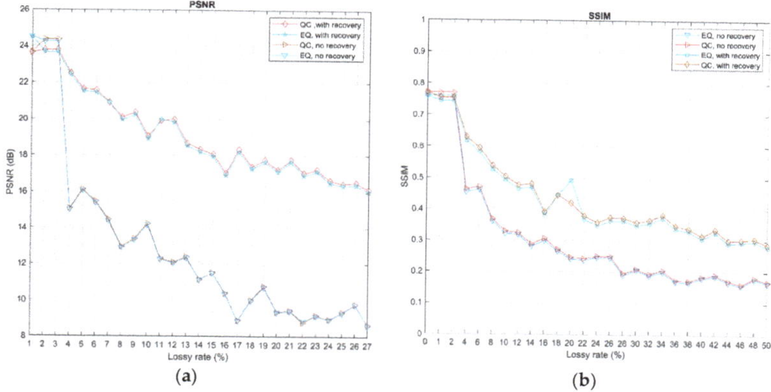

(a)   (b)

Figure 6. Reconstructions for EQ and QC constraints for cameraman with different lossy probabilities:
(**a**) Peak Signal-to-Noise (PSNR) curves and (**b**) Structural Similarity (SSIM) curves.

In Figures 7–11 we demonstrate the second set of experiments for the color image ducks with a size of 1024×1024. The color image is composed of three color planes, namely, red, green, and blue. Figure 7a depicts the color image ducks, and Figure 7b–d shows the entropies in the three color planes for adaptive sampling (AS).

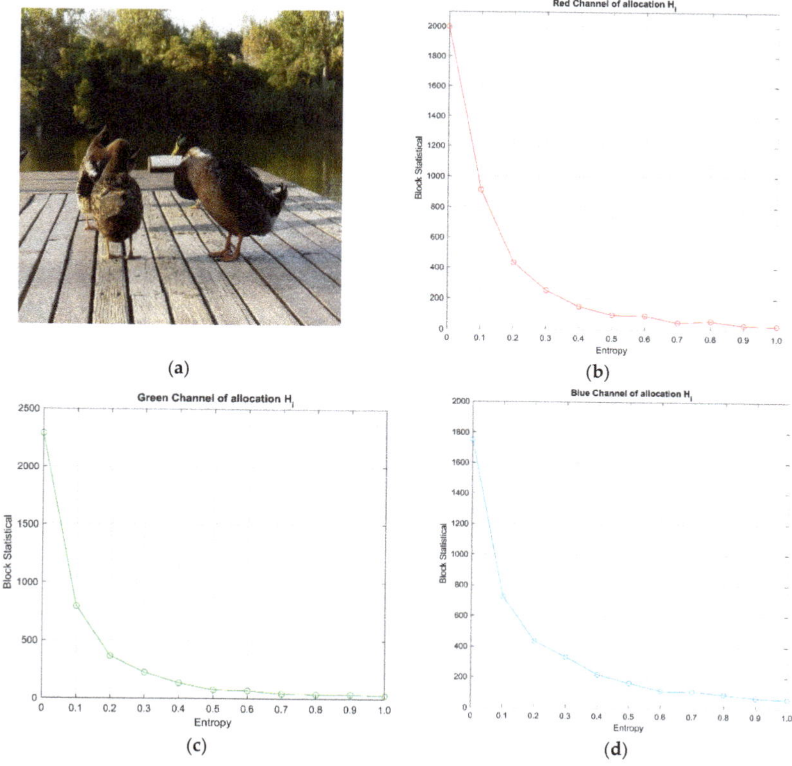

(a)   (b)

(c)   (d)

Figure 7. Entropy distributions for adaptive sampling in ducks: (**a**) Original image; (**b**) The entropy in the red plane; (**c**) The entropy in the green plane; (**d**) The entropy in the blue plane.

Figure 8 presents the error-free transmission of BCS for ducks. The measurement rate was set to $\frac{12}{64} = 0.1875$ for each color plane, and reconstructed images can be assessed with the PSNR and SSIM measures. Note that the abbreviations in the captions of Figure 8 are identical to their counterparts in Figure 3. Figure 8a,b show only the reconstruction for equality constraint (EQ) and quadratic constraint (QC), respectively. The PSNR and SSIM values are also shown. Figure 8c,d present the reconstruction for EQ + AS and QC + AS, respectively.

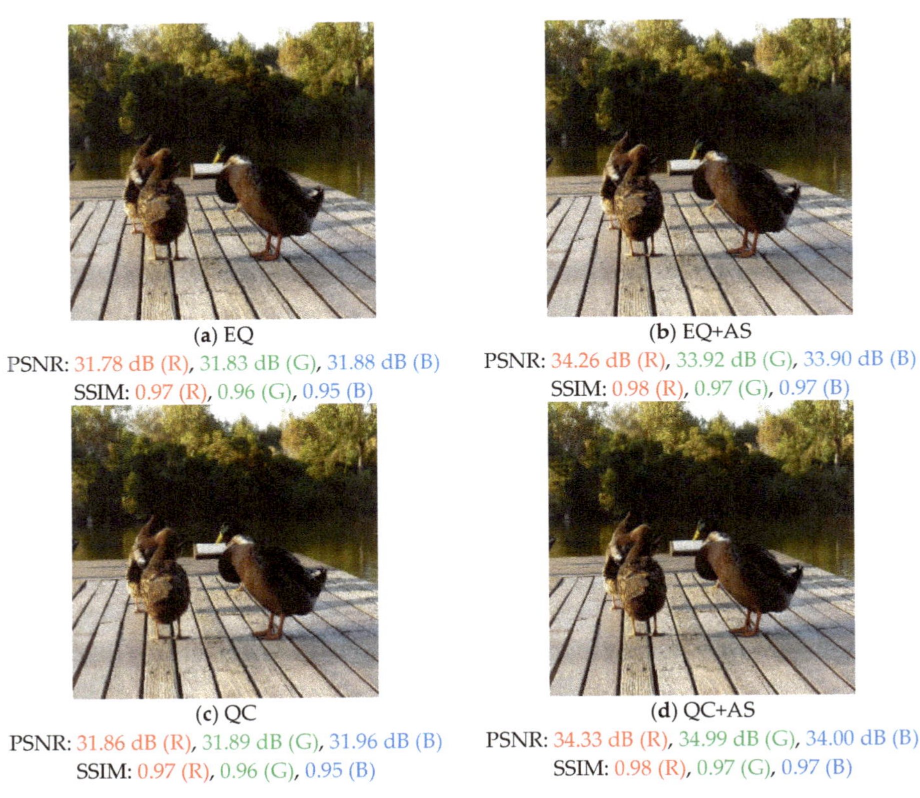

(**a**) EQ
PSNR: 31.78 dB (R), 31.83 dB (G), 31.88 dB (B)
SSIM: 0.97 (R), 0.96 (G), 0.95 (B)

(**b**) EQ+AS
PSNR: 34.26 dB (R), 33.92 dB (G), 33.90 dB (B)
SSIM: 0.98 (R), 0.97 (G), 0.97 (B)

(**c**) QC
PSNR: 31.86 dB (R), 31.89 dB (G), 31.96 dB (B)
SSIM: 0.97 (R), 0.96 (G), 0.95 (B)

(**d**) QC+AS
PSNR: 34.33 dB (R), 34.99 dB (G), 34.00 dB (B)
SSIM: 0.98 (R), 0.97 (G), 0.97 (B)

Figure 8. Reconstruction of BCS coefficients for the three color planes: (**a**) EQ only; (**b**) EQ + AS; (**c**) QC only; and (**d**) QC + AS.

With adaptive sampling, better PSNR and SSIM performance can be observed. In addition, the performance with QC was slightly better than that with Equation (9). This observation for the color image of ducks in Figure 8 is similar to that for the grey-level image of cameraman in Figure 3.

In Figure 9, we apply the multiple description coding of Equation (9) to protect the BCS coefficients with EQ or QC constraints. Regarding the demonstrations of performance of the proposed method, for the BCS coefficients in the red and green planes we set $p_{e,1} = p_{e,2} = 0.2$, and the BCS coefficients in the blue plane were treated as error-free transmissions or $p_{e,1} = p_{e,2} = 0.0$. In Figure 9a,c, because the BCS coefficients could be lost during transmission, the degradation of the reconstructed image was affected whether or not the EQ or QC constraints were applied. In Figure 9b,d, due to high correlation between color planes and protection with multiple description coding, the reconstructed image quality was significantly improved.

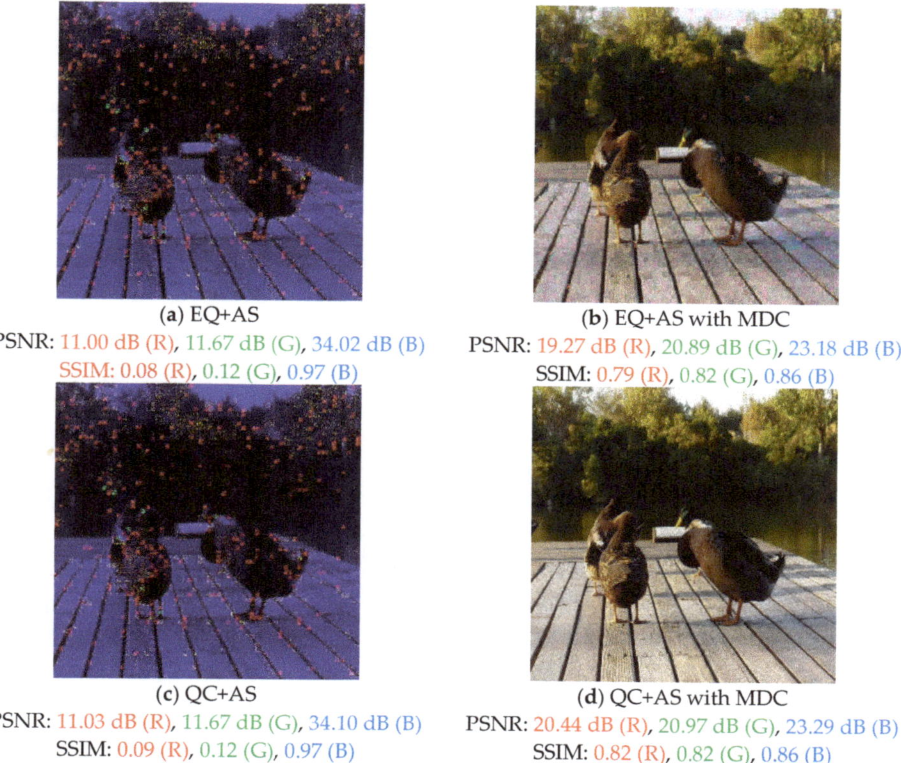

(**a**) EQ+AS
PSNR: 11.00 dB (R), 11.67 dB (G), 34.02 dB (B)
SSIM: 0.08 (R), 0.12 (G), 0.97 (B)

(**b**) EQ+AS with MDC
PSNR: 19.27 dB (R), 20.89 dB (G), 23.18 dB (B)
SSIM: 0.79 (R), 0.82 (G), 0.86 (B)

(**c**) QC+AS
PSNR: 11.03 dB (R), 11.67 dB (G), 34.10 dB (B)
SSIM: 0.09 (R), 0.12 (G), 0.97 (B)

(**d**) QC+AS with MDC
PSNR: 20.44 dB (R), 20.97 dB (G), 23.29 dB (B)
SSIM: 0.82 (R), 0.82 (G), 0.86 (B)

Figure 9. Demonstration of lossy transmission in the red and green planes: (**a**) EQ + AS; (**b**) EQ + AS with MDC; (**c**) QC + AS; and (**d**) QC + AS with MDC.

In Figure 10, we provide three sets of angles θ_1 and θ_2 in MDC with $r_1 = r_2 = 1$ from Equation (9). First, we chose $\theta_1 = \frac{\pi}{6}$ and $\theta_2 = \frac{-\pi}{6}$, which led to opposite signs between angles. Second, we choose $\theta_1 = \frac{\pi}{3}$ and $\theta_2 = \frac{-\pi}{6}$, meaning that the two angles were orthogonal. Third, by combining the first two selections, we chose $\theta_1 = \frac{\pi}{4}$ and $\theta_1 = \frac{-\pi}{4}$. For Figure 10a,c,e, reconstruction was based on the equality constraint, while for Figure 10b,d,f, reconstruction was based on the quadratic constraint. Comparing the three selections, the orthogonal angles may lead to better performance to combat channel errors in reconstruction.

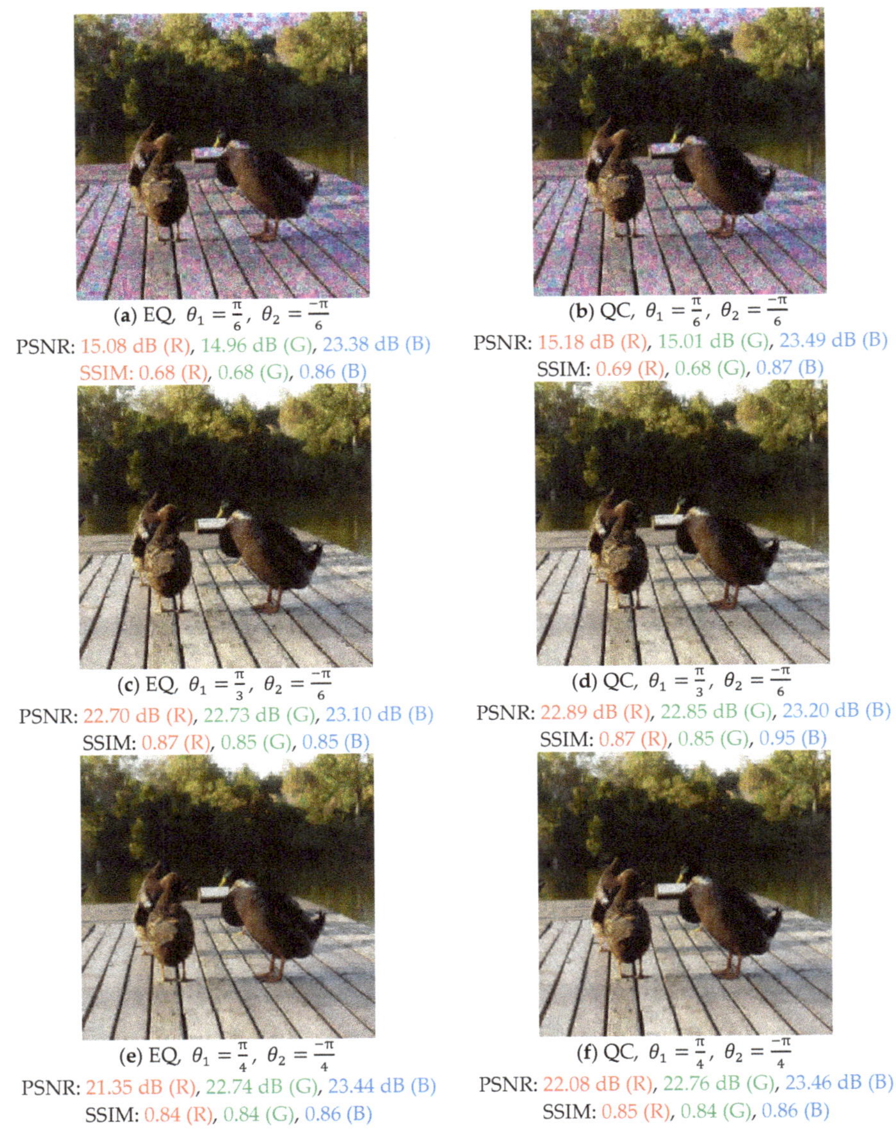

(a) EQ, $\theta_1 = \frac{\pi}{6}$, $\theta_2 = \frac{-\pi}{6}$
PSNR: 15.08 dB (R), 14.96 dB (G), 23.38 dB (B)
SSIM: 0.68 (R), 0.68 (G), 0.86 (B)

(b) QC, $\theta_1 = \frac{\pi}{6}$, $\theta_2 = \frac{-\pi}{6}$
PSNR: 15.18 dB (R), 15.01 dB (G), 23.49 dB (B)
SSIM: 0.69 (R), 0.68 (G), 0.87 (B)

(c) EQ, $\theta_1 = \frac{\pi}{3}$, $\theta_2 = \frac{-\pi}{6}$
PSNR: 22.70 dB (R), 22.73 dB (G), 23.10 dB (B)
SSIM: 0.87 (R), 0.85 (G), 0.85 (B)

(d) QC, $\theta_1 = \frac{\pi}{3}$, $\theta_2 = \frac{-\pi}{6}$
PSNR: 22.89 dB (R), 22.85 dB (G), 23.20 dB (B)
SSIM: 0.87 (R), 0.85 (G), 0.95 (B)

(e) EQ, $\theta_1 = \frac{\pi}{4}$, $\theta_2 = \frac{-\pi}{4}$
PSNR: 21.35 dB (R), 22.74 dB (G), 23.44 dB (B)
SSIM: 0.84 (R), 0.84 (G), 0.86 (B)

(f) QC, $\theta_1 = \frac{\pi}{4}$, $\theta_2 = \frac{-\pi}{4}$
PSNR: 22.08 dB (R), 22.76 dB (G), 23.46 dB (B)
SSIM: 0.85 (R), 0.84 (G), 0.86 (B)

Figure 10. Demonstration of angle selection in multiple description coding with equality and quadratic constraints. BCS coefficients experienced a lossy rate of $p_{e,1} = p_{e,2} = 0.1$ in three planes.

In Figure 11, we display the evaluations of reconstructed image qualities over the range of lossy probabilities between the range of 0 and 0.5, with a measurement rate of 0.1. We observed that with increased loss rates the inferior quality of the reconstructed images could be monitored. In addition, protection with multiple description coding, or the three curves at the upper portion, alleviated the effect caused by data loss, which led to better quality reconstructed images. Finally, we found that with the careful selection of the ε value from Equation (4), quadratic constraint performed slightly better than the equality constraint in reconstructed image quality.

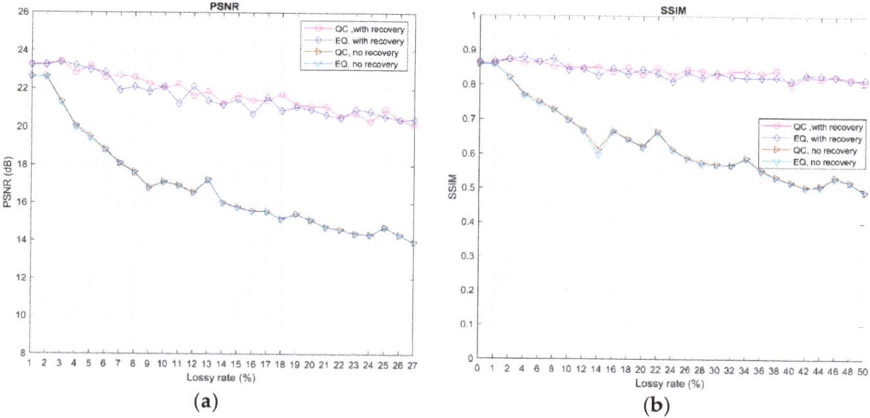

Figure 11. Reconstructions for EQ and QC constraints for ducks with different lossy probabilities: (**a**) PSNR curves and (**b**) SSIM curves.

In Figures 12–16 we demonstrated the results of the third set of experiments for the color image Pasadena-houses with the size of 1760 × 1168. Unlike the test image cameraman in Figure 3a and ducks in Figure 7a, which are square-shared images, Figure 12a was a rectangular-shaped image. The color image was composed of three color planes, namely red, green, and blue. Figure 12a depicts the color image Pasadena-houses, and Figure 12b–d displays the entropy distributions of the three color planes for adaptive sampling (AS).

Figure 13 presents the error-free transmission of BCS for Pasadena-houses. The measurement rate was set to $\frac{12}{64} = 0.1875$ for each color plane whether AS was applied or not. The reconstructed images were assessed with PSNR and SSIM. Figure 13a,b show the reconstruction for equality constraint (EQ) and quadratic constraint (QC), respectively, and the PSNR and SSIM values are also shown. Figure 13c,d present the reconstruction for EQ + AS and QC + AS, respectively. With adaptive sampling, better PSNR and SSIM performances were observed due to the local characteristics in entropy distribution. Again, QC performed slightly better than EQ. This observation fit for the color images Pasadena-houses in Figure 12, ducks in Figure 8, and the grey-level image cameraman in Figure 3.

Figure 12. *Cont.*

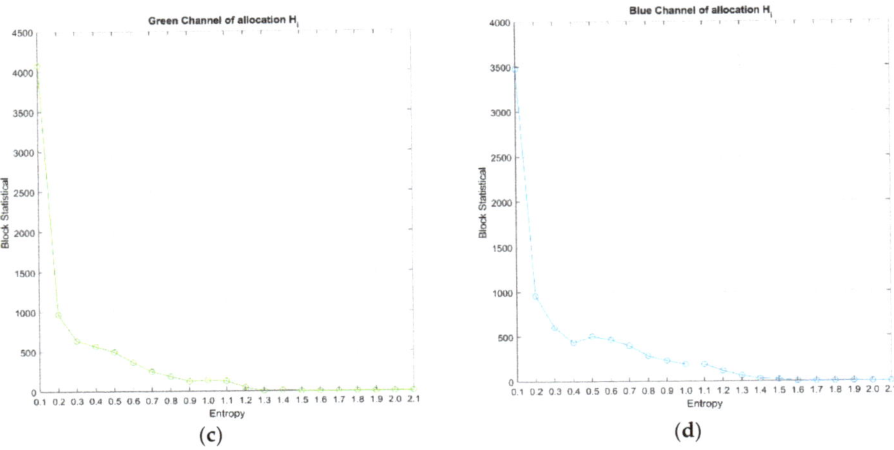

Figure 12. Entropy distributions for adaptive sampling in Pasadena-houses: (**a**) original image; (**b**) entropy in the red plane; (**c**) entropy in the green plane; and (**d**) entropy in the blue plane.

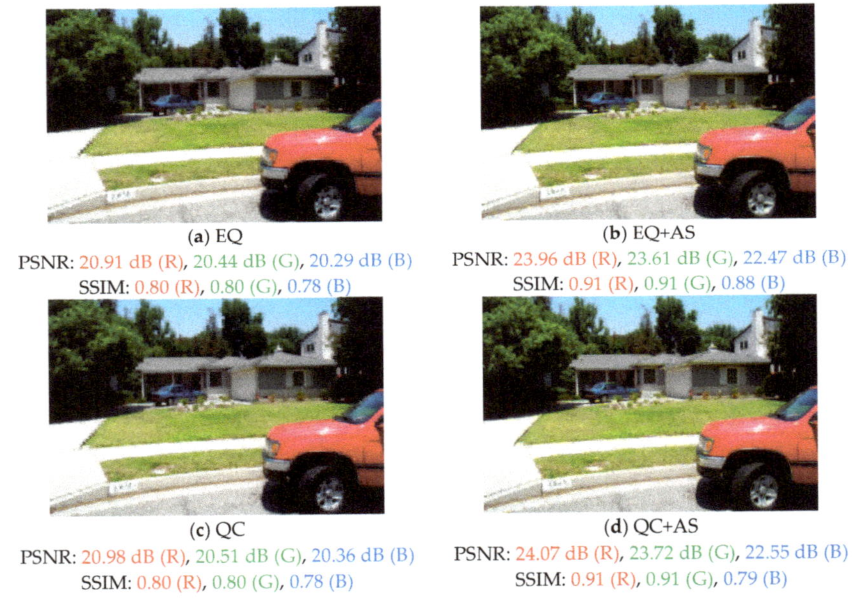

Figure 13. Reconstruction of BCS coefficients for the three color planes: (**a**) EQ only; (**b**) EQ + AS; (**c**) QC only; and (**d**) QC + AS.

In Figure 14 we applied multiple description coding from Equation (9) to protect the BCS coefficients with EQ or QC constraints. By following the scenarios in Figure 9 for the BCS coefficients in the red and green planes, we set $p_{e,1} = p_{e,2} = 0.2$. The BCS coefficients in the blue plane were treated as error-free transmissions, or $p_{e,1} = p_{e,2} = 0.0$. In Figure 14a,c, because the BCS coefficients could be lost during transmission, the degradation of the reconstructed image was affected whether or not the EQ or QC constraints were applied. In Figure 14b,d, due to the high correlation between color planes and the protection with multiple description coding, the reconstructed image quality was significantly improved.

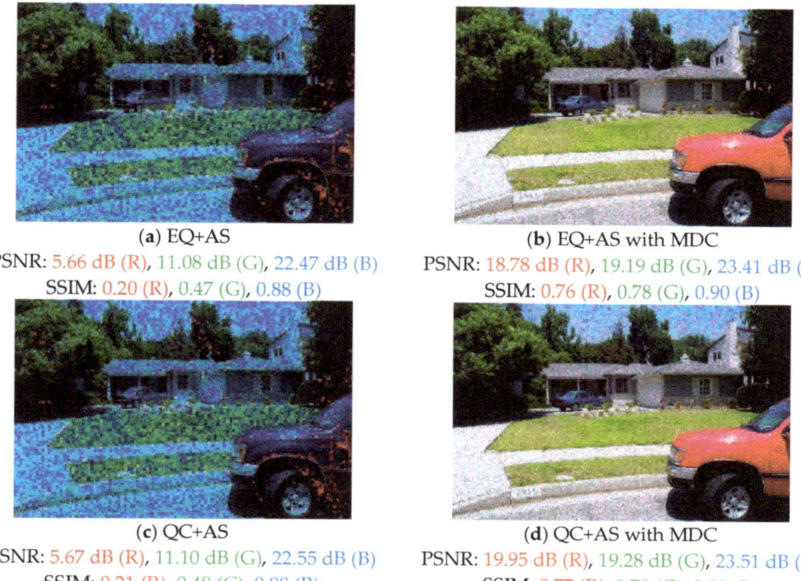

(a) EQ+AS
PSNR: 5.66 dB (R), 11.08 dB (G), 22.47 dB (B)
SSIM: 0.20 (R), 0.47 (G), 0.88 (B)

(b) EQ+AS with MDC
PSNR: 18.78 dB (R), 19.19 dB (G), 23.41 dB (B)
SSIM: 0.76 (R), 0.78 (G), 0.90 (B)

(c) QC+AS
PSNR: 5.67 dB (R), 11.10 dB (G), 22.55 dB (B)
SSIM: 0.21 (R), 0.48 (G), 0.88 (B)

(d) QC+AS with MDC
PSNR: 19.95 dB (R), 19.28 dB (G), 23.51 dB (B)
SSIM: 0.77 (R), 0.78 (G), 0.90 (B)

Figure 14. Demonstration of lossy transmission in the red and green planes: (**a**) EQ + AS; (**b**) EQ + AS with MDC; (**c**) QC + AS; and (**d**) QC + AS with MDC.

In Figure 15, by following the same parameter settings used in Figure 10, we provided three sets of selections of angles θ_1 and θ_2 in MDC with $r_1 = r_2 = 1$ in Equation (9). First, we chose $\theta_1 = \frac{\pi}{6}$ and $\theta_2 = \frac{-\pi}{6}$, which led to opposite signs between angles. Second, we chose $\theta_1 = \frac{\pi}{3}$ and $\theta_2 = \frac{-\pi}{6}$, meaning that the two angles were orthogonal. Third, by combining the first two selections, we chose $\theta_1 = \frac{\pi}{4}$ and $\theta_2 = \frac{-\pi}{4}$. For Figure 15a,c,e, reconstruction was based on the equality constraint, while for Figure 15b,d,f, reconstruction was based on the quadratic constraint. Comparing the three selections, angles that were orthogonal may have led to better performance combatting channel errors in reconstruction.

In Figure 16 we demonstrated the evaluations of reconstructed image quality over a range of lossy probabilities between 0 and 0.5, with the measurement rate of 0.1. We observed that, with increased loss rates, inferior quality reconstructed images were seen whether or not the protection was deployed. In addition, protection with multiple description coding, or the three curves at the upper portion in Figure 16a,b, alleviated the effect of data loss, which led to better quality reconstructed images.

(a) EQ, $\theta_1 = \frac{\pi}{6}$, $\theta_2 = \frac{-\pi}{6}$
PSNR: 12.37 dB (R), 12.04 dB (G), 23.04 dB (B)
SSIM: 0.51 (R), 0.53 (G), 0.89 (B)

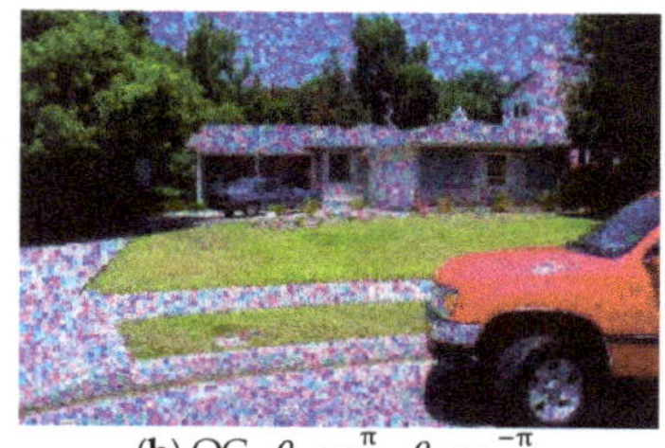

(b) QC, $\theta_1 = \frac{\pi}{6}$, $\theta_2 = \frac{-\pi}{6}$
PSNR: 12.42 dB (R), 12.06 dB (G), 23.14 dB (B)
SSIM: 0.52 (R), 0.53 (G), 0.89 (B)

Figure 15. *Cont.*

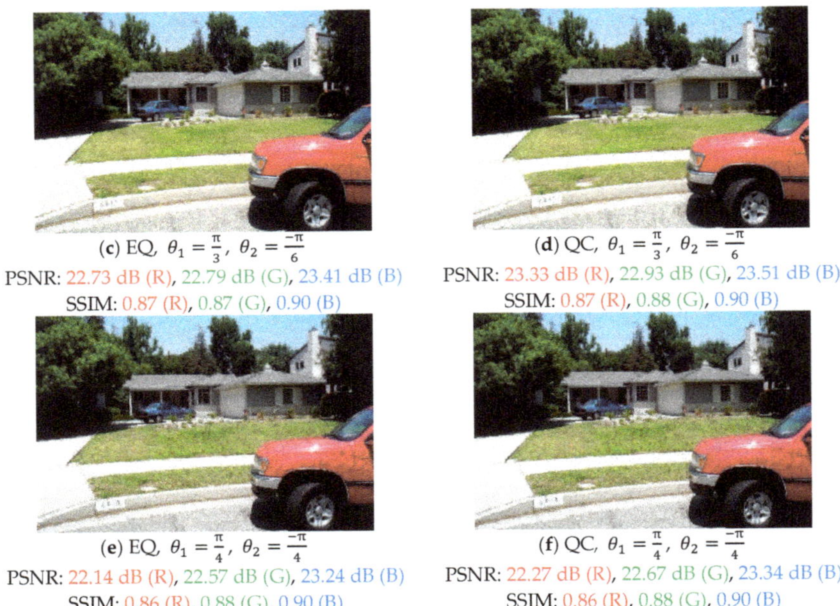

(**c**) EQ, $\theta_1 = \frac{\pi}{3}$, $\theta_2 = \frac{-\pi}{6}$
PSNR: 22.73 dB (R), 22.79 dB (G), 23.41 dB (B)
SSIM: 0.87 (R), 0.87 (G), 0.90 (B)

(**d**) QC, $\theta_1 = \frac{\pi}{3}$, $\theta_2 = \frac{-\pi}{6}$
PSNR: 23.33 dB (R), 22.93 dB (G), 23.51 dB (B)
SSIM: 0.87 (R), 0.88 (G), 0.90 (B)

(**e**) EQ, $\theta_1 = \frac{\pi}{4}$, $\theta_2 = \frac{-\pi}{4}$
PSNR: 22.14 dB (R), 22.57 dB (G), 23.24 dB (B)
SSIM: 0.86 (R), 0.88 (G), 0.90 (B)

(**f**) QC, $\theta_1 = \frac{\pi}{4}$, $\theta_2 = \frac{-\pi}{4}$
PSNR: 22.27 dB (R), 22.67 dB (G), 23.34 dB (B)
SSIM: 0.86 (R), 0.88 (G), 0.90 (B)

Figure 15. Demonstration of angle selection in multiple description coding with equality and quadratic constraints. BCS coefficients experience lossy rate of $p_{e,1} = p_{e,2} = 0.1$ in three planes.

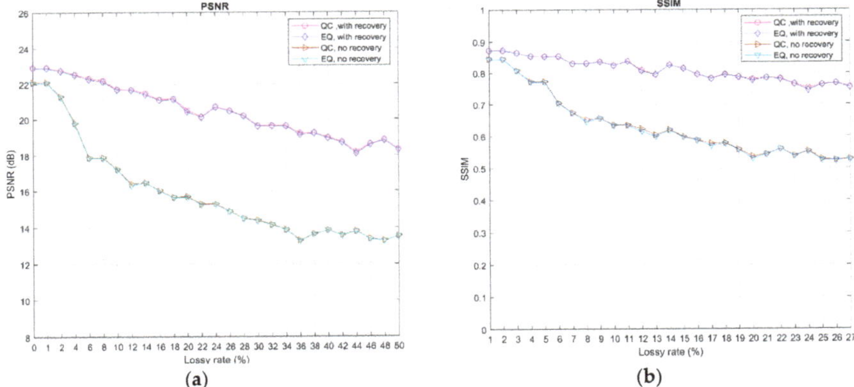

Figure 16. Reconstructions for EQ and QC constraints for Pasadena-houses with different lossy probabilities: (**a**) PSNR curves and (**b**) SSIM curves.

From the results of the three sets of experiments, which corresponded to the three test images presented above, cameraman, ducks, and Pasadena-houses, protection of BCS with multiple description coding pointed to the applicability for transmitting over lossy channels. MDC worked together with EQ or QC reconstruction methods for BCS signals. Compensation between received descriptions with MDC helped reduce the effects of lossy channels. The alleviation of reconstructed image quality with error resilient coding is coveted.

5. Conclusions

In this paper, we presented the error resilient transmission scheme for block compressed sensing. For error-free transmission, we found that adaptive sampling enhanced the reconstructed image quality under both equality and the quadratic constraints. For lossy transmission, we observed the vulnerability of compressively sensed information for transmission over lossy channels, and noted the need to provide protection schemes to alleviate the effects of data loss. We proposed our algorithm to transmit compressed information over multiple independent and lossy channels, and to work with multiple description coding for protection and reconstruction. For grey-level images, multiple description coding demonstrated effective protection. For color images, high correlations between the color planes can further aid better quality of reconstruction. The simulation results presented enhanced performance with multiple description coding and adaptive sampling. A wide range of lossy probabilities were simulated to verify the effectiveness of multiple description coding for protecting block compressed sensing. In future, we intend to look for other effective means and the ways to choose the parameter values to ensure error resilient transmission for compressed sensing of images.

Author Contributions: Conceptualization, H.-C.H.; formal analysis, P.-L.C. and F.-C.C.; investigation, P.-L.C.; methodology, H.-C.H. and F.-C.C.; software, P.-L.C.; writing–original draft, H.-C.H.; writing–review and editing, H.-C.H. and F.-C.C. All authors have read and agreed to the published version of the manuscript.

Funding: This research was funded by the Ministry of Science and Technology of Taiwan, R.O.C., grant number MOST 107-2221-E-390-018-MY2.

Conflicts of Interest: The authors declare no conflict of interest.

References

1. Candes, E.J.; Wakin, M.B. An introduction to compressive sampling. *IEEE Signal Process. Mag.* **2008**, *25*, 21–30. [CrossRef]
2. Romberg, J. Imaging via compressive sampling. *IEEE Signal Process. Mag.* **2008**, *25*, 14–20. [CrossRef]
3. Arildsen, T.; Larsen, T. Compressed sensing with linear correlation between signal and measurement noise. *Signal Process.* **2014**, *98*, 275–283. [CrossRef]
4. Rani, M.; Dhok, S.B.; Deshmukh, R.B. A systematic review of compressive sensing: Concepts, implementations and applications. *IEEE Access* **2018**, *6*, 4875–4894. [CrossRef]
5. Pennebaker, W.B.; Mitchell, J.L. *JPEG: Still Image Data Compression Standard*; Kluwer Academic Publishers: Norwell, MA, USA, 1992.
6. Taubman, D.; Marcellin, M. *JPEG2000: Image Compression Fundamentals, Standards and Practice*; Kluwer Academic Publishers: Norwell, MA, USA, 2001.
7. l_1-Magic: Recovery of Sparse Signal Via Convex Programming. Available online: https://statweb.stanford.edu/~candes/l1magic/ (accessed on 21 November 2019).
8. Candes, E.J.; Tao, T. Decoding by linear programming. *IEEE Trans. Inf. Theory* **2005**, *51*, 1–19. [CrossRef]
9. Blumensath, T. Compressed sensing with nonlinear observations and related nonlinear optimization problems. *IEEE Trans. Inf. Theory* **2013**, *59*, 3466–3474. [CrossRef]
10. Gao, Z.; Xiong, C.; Ding, L.; Zhou, C. Image representation using block compressive sensing for compression application. *J. Vis. Commun. Image Represent.* **2013**, *24*, 885–894. [CrossRef]
11. Bigot, J.; Boyer, C.; Weiss, P. An analysis of block sampling strategies in compressed sensing. *IEEE Trans. Inf. Theory* **2016**, *62*, 2125–2139. [CrossRef]
12. Wang, J.; Zhang, J.; Wang, W.; Yang, C. A perturbation analysis of nonconvex block-sparse compressed sensing. *Commun. Nonlinear Sci. Numer. Simul.* **2015**, *29*, 416–426. [CrossRef]
13. Deng, C.; Lin, W.; Lee, B.S.; Lau, C.T. Robust image coding based upon compressive sensing. *IEEE Trans. Multimed.* **2012**, *14*, 278–290. [CrossRef]
14. Lu, C.S.; Chen, H.W. Compressive image sensing for fast recovery from limited samples: A variation on compressive sensing. *Inf. Sci.* **2015**, *325*, 33–47. [CrossRef]
15. Dai, L.; Wang, Z.; Yang, Z. Compressive sensing based time domain synchronous OFDM transmission for vehicular communications. *IEEE J. Sel. Areas Commun.* **2013**, *31*, 460–469.

16. Wang, Y.; Orchard, M.T.; Vaishampayan, V.; Reibman, A.R. Multiple description coding using pairwise correlating transforms. *IEEE Trans. Image Process.* **2001**, *10*, 351–366. [CrossRef] [PubMed]

17. Hemami, S.S. Reconstruction-optimized lapped orthogonal transforms for robust image transmission. *IEEE Trans. Circuits Syst. Video Technol.* **1996**, *6*, 168–181. [CrossRef]

18. Zhu, S.; Zeng, B.; Gabbouj, M. Adaptive sampling for compressed sensing based image compression. *J. Vis. Commun. Image Represent.* **2015**, *30*, 94–105. [CrossRef]

19. Malloy, M.L.; Nowak, R.D. Near-optimal adaptive compressed sensing. *IEEE Trans. Inf. Theory* **2014**, *60*, 4001–4012. [CrossRef]

20. Bora, A.; Jalal, A.; Price, E.; Dimakis, A.G. Compressed sensing using generative models. In Proceedings of the 34th International Conference on Machine Learning, Sydney, Australia, 6–11 August 2017; pp. 537–546.

21. Metzler, C.A.; Maleki, A.; Baraniuk, R.G. From denoising to compressed sensing. *IEEE Trans. Inf. Theory* **2016**, *62*, 5117–5144. [CrossRef]

22. Computational Image Archive. Available online: http://www.vision.caltech.edu/html-files/archive.html (accessed on 21 November 2019).

23. Wang, Z.; Bovik, A.C.; Sheikh, H.R.; Simoncelli, E.P. Image quality assessment: From error visibility to structural similarity. *IEEE Trans. Image Process.* **2004**, *13*, 600–612. [CrossRef] [PubMed]

Article

Image Completion with Hybrid Interpolation in Tensor Representation

Rafał Zdunek * and Tomasz Sadowski

Department of Electronics, Wroclaw University of Science and Technology, Wybrzeze Wyspianskiego 27, 50-370 Wroclaw, Poland; tomasz.sadowski@pwr.edu.pl
* Correspondence: rafal.zdunek@pwr.edu.pl; Tel.: +48-71-320-3215

Received: 10 December 2019; Accepted: 20 January 2020; Published: 22 January 2020

Abstract: The issue of image completion has been developed considerably over the last two decades, and many computational strategies have been proposed to fill-in missing regions in an incomplete image. When the incomplete image contains many small-sized irregular missing areas, a good alternative seems to be the matrix or tensor decomposition algorithms that yield low-rank approximations. However, this approach uses heuristic rank adaptation techniques, especially for images with many details. To tackle the obstacles of low-rank completion methods, we propose to model the incomplete images with overlapping blocks of Tucker decomposition representations where the factor matrices are determined by a hybrid version of the Gaussian radial basis function and polynomial interpolation. The experiments, carried out for various image completion and resolution up-scaling problems, demonstrate that our approach considerably outperforms the baseline and state-of-the-art low-rank completion methods.

Keywords: image completion; tensor decomposition models; image interpolation; image up-scaling

---

1. Introduction

Image completion aims to synthesize missing regions in an incomplete image or a video sequence with the content-aware information captured from accessible or unperturbed regions. The missing regions may result from removing unwanted sub-areas, occlusion, or unobserved or considerably damaged random pixels.

Image completion is one of the fundamental research topics in the area of computer vision and graphics technology, motivated by widespread applications in various applied sciences [1]. It is used for the restoration of old photographs, paintings, and films by removing scratches, dust spots, occlusions, or other user-marked objects, such as annotations, subtitles, stamps, logos, etc. In telecommunications, image completion techniques may fix error concealment problems or recover missing data-blocks lost during transmission or video compression [2,3]. The recent works also emphasize their usefulness in remote sensing imaging to remove occlusions, such as clouds and "dead" pixels [4,5].

The topic of image completion has been extensively studied for at least two decades, resulting in the development of several computational approaches to address this problem. The pioneering work in fully automated digital image inpainting dates back to 2000 when Bertalmio et al. [6] introduced a milestone algorithm that was able to automatically fill in missing regions from their neighboring information without providing user-defined sample images. This algorithm is based on partial differential equations (PDEs) that determine isophotes (brightness-level lines) along which the information on the surrounding structure is propagated upwards. It is a fully automatic algorithm, but it inpaints efficiently only in narrow lines, does not recover textures, and produces blurred results. Another approach to image inpainting is to synthesize the texture. This concept was introduced by Efros and Leung [7] in 1999, but their algorithm requires a sample texture image to recover the texture in the missing region. Bertalmio et al. [8] also developed a hybrid version of both inpainting techniques, in which an incomplete image is decomposed into its texture and structure components, one reconstructed by texture synthesis, and another by PDE-based image inpainting. When a region to be completed is relatively large, the exemplar-based image synthesis algorithms [9–11] or their hybrid versions [12] seem to be more appropriate. The hybrid strategies have been studied in many other research papers [13–16], and currently, image completion based on simultaneous fill-in of texture and structure is a fundamental approach, especially for recovering large missing regions. Various neural network architectures, e.g., convolutional neural network and generative adversarial network, can also perform hybrid image completion [17–20]. A survey of image completion methods can be found in [1,5,21,22].

The above-mentioned image completion methods are efficient even for very large missing regions; however, they cannot be applied for incomplete images with many uniformly distributed missing pixels (e.g., about 90 %). No texture information can be learned from any subregions of such an image. There is also no continuous boundary of missing regions, and hence the neighboring information cannot be propagated towards the centers of missing regions. In such cases, different methods must be applied. Assuming that an incomplete image has a low-rank structure and satisfies the incoherence condition [23], usually represented by clusters of similar patches, the image completion boils down to a rank minimization problem. Since it is an NP-hard problem, many computational strategies have been proposed to approximate it by a convex relaxation, usually involving matrix or tensor decomposition methods. One of them assumes a convex approximation of a rank minimization problem with a nuclear-norm minimization problem, which can be solved easily using singular value decomposition (SVD) [24–26]. Due to the orthogonality condition, low-rank representations yielded by SVD contain negative values, which is not profitable for representing a large class of images. Other low-rank models (not necessarily restricted to SVD) can be used to tackle this problem [27].

One of the commonly used methods for extracting low-rank part-based representations from nonnegative matrices is nonnegative matrix factorization (NMF) [28]. It has already found many relevant applications in image analysis, and can also be used for solving image completion problems [29,30]. In this approach, the missing regions are sequentially updated with an NMF-based low-rank approximation of an observed image, which resembles the phenomena of propagating the neighboring information towards the missing regions in the PDE-based inpainting methods. However, due to the non-convexity of alternating optimization, NMF is sensitive to its initialization, especially for insufficiently sparse data. The ambiguity effects can be considerably relaxed if tensor decomposition models are used [31]. Moreover, multi-linear decomposition models prevent cross-modal interactions, which is particularly useful for image representations. Such a low-rank image completion methodology is mostly based on the concept of tensor completion issues that have been extensively studied [32–36]. There are many tensor decomposition models that are used for image completion tasks, including the fundamental ones, such as CANDECOMP/PARAFAC (CP) [37,38] and Tucker decomposition [39–41], as well as tensor networks, such as tensor ring [42], tensor train [43,44], hierarchical Tucker decomposition [45], and other tensor decomposition models [46].

Tensor completion methods are intrinsically addressed for processing color images (3D multi-way arrays), but they can also be applied to gray-scale images using various tensorization or scanning operators, e.g., the ket augmentation [47]. They are also very flexible in incorporating penalty terms or constraints; however, their efficiency strongly depends on the image to be completed. In practice, a low-rank representation is always a certain approximation of the underlying image, controlled by the rank of a tensor decomposition model. The problem of rank selection is regarded in terms of a trade-off between under- and over-fitting, and many approaches exist to tackle it. For example, Yokota et al. [38] proposed to increase the rank with recursive updates. In the early recursive steps, a low-rank structure is recovered, and then it is gradually updated to a higher-rank structure that contains more details. The rank can also be controlled by the decreasing rank procedure [46] or using the singular value thresholding strategy [43]. Thus, one of the drawbacks of this kind of image completion method is the problem of selecting the optimal rank of decomposition, and it is usually resolved by heuristic procedures. Moreover, the tensor decomposition-based image-completion methods usually involve a high computational cost if all factor matrices or core tensors are updated in each iterative step.

To relax the problem with the right rank selection while keeping computational costs at a very low level, we propose a very simple alternative approach to low-rank image completion. Our strategy is based on the Tucker decomposition model in which the full-rank factor matrices are previously estimated with a hybrid connection of two interpolation methods. Since the factor matrices are precomputed, only the core tensor must be estimated. Despite the full-rank assumption, it involves a relatively low computational cost because the core tensor is sparse with non-zero entries determined by the available pixels. Motivated by several works [48–51] on the use of various interpolation methods for solving image completion problems, other recent works [52–56] on image processing aspects, and the concept of tensor product spline surfaces [57], we show the relationship of the Tucker decomposition with factorizable radial basis function (RBF) interpolation and use it to compute the factor matrices. RBF interpolation is a mesh-free method, which is very profitable for recovering irregularly distributed missing pixels, but it may incorrectly approximate linear structure. Hence, we combine it with low-degree polynomial interpolation, and both interpolation methods can be expressed by the Tucker decomposition model. Adopting the idea of PDE-based inpainting, we propose to compute the interpolants using only restricted surrounding subareas, instead of all accessible pixels. Hence, the whole image is divided into overlapping blocks, and the Tucker decomposition model is applied to each block separately. As many interpolation methods do not approximate the boundary entries well, the overlapping is necessary to avoid discontinuity effects. The proposed methodology is applied to various image completion problems, where the incomplete images are obtained from true images by removing many random pixels or many small holes. One of the experiments is performed for resolution up-scaling, where a low-resolution image is up-scaled to high resolution using the proposed algorithm. All the experiments demonstrate that the proposed method considerably outperforms the baseline and state-of-the-art low-rank image-completion methods in terms of the performance, and it is much faster than the methods based on tensor decompositions.

The remainder of this paper is organized as follows. Section 2 reviews some preliminary knowledge about fundamental tensor operations, the Tucker decomposition model, low-rank tensor completion, and RBF interpolation. The proposed algorithm is described in Section 3. The numerical experiments performed on various image completion problems are given and discussed in Section 4. The last section contains the conclusions.

2. Preliminary

Mathematical notations and preliminaries of tensor decomposition models are adopted from [31]. Tensors are multi-way arrays, denoted by calligraphic letters (e.g., $\mathcal{Y}$). Let $\mathcal{Y} \in \mathbb{R}^{I_1 \times \dots \times I_N}$ be the N-way array. The elements of $\mathcal{Y}$ are denoted as $y_{i_1,\dots,i_N}$, where $\forall n : 1 \leq i_n \leq I_n, n = 1,\dots,N$. Boldface uppercase letters (e.g., $\mathbf{Y}$) denote matrices; lowercase boldface ones stand for vectors (e.g., $\mathbf{y}$); non-bold letters are scalars. The vector $\mathbf{y}_j$ contains the j-th column of $\mathbf{Y}$. The symbol $|| \cdot ||_F$ denotes the Frobenius norm of a matrix; $|| \cdot ||$ denotes the 2-nd norm. The sets of real numbers and natural numbers are represented by $\mathbb{R}$ and $\mathcal{N}$, respectively. The symbols $\lfloor x \rfloor$ and $\lceil x \rceil$ stand for the floor and ceiling functions of x, respectively.

Let $\mathcal{M} = [m_{i_1,\dots,i_N}] \in \mathbb{R}^{I_1 \times \dots \times I_N}$ be the N-th way observed tensor with missing entries, and $\Omega = [\omega_{i_1,\dots,i_N}] \in \mathbb{R}^{I_1 \times \dots \times I_N}$ be a binary tensor that indicates the locations of available entries in $\mathcal{M}$. If $m_{i_1,\dots,i_N}$ is observed, then $\omega_{i_1,\dots,i_N} = 1$; otherwise, $\omega_{i_1,\dots,i_N} = 0$. The locations of positive entries in $\bar{\Omega} = 1 - \Omega$ correspond to missing entries in $\mathcal{M}$. The number of observed entries is equal to $|\Omega| = \{\sharp(i_1,\dots,i_N) : \omega_{i_1,\dots,i_N} > 0\}$.

Definition 1. *Let*

$$\hat{\mathcal{Y}} = [\hat{y}_{i_1,\dots,i_N}], \quad \text{where} \quad \hat{y}_{i_1,\dots,i_N} = \begin{cases} m_{i_1,\dots,i_N} & \text{if} & \omega_{i_1,\dots,i_N} = 1, \\ 0 & \text{otherwise} \end{cases} \tag{1}$$

be a zero-filled incomplete tensor.

Definition 2. *Let*

$$\mathcal{Y}(\Omega) = [\bar{y}_{i_1,\dots,i_N}], \quad \text{where} \quad \bar{y}_{i_1,\dots,i_N} = \begin{cases} m_{i_1,\dots,i_N} & \text{if} & \omega_{i_1,\dots,i_N} = 1, \\ \varnothing & \text{otherwise} \end{cases} \tag{2}$$

be a subtensor of $\mathcal{M}$ that contains only the entries pointed to by Ω.

Thus: $|\mathcal{Y}(\Omega)| = |\Omega|$. Let $\omega = \mathrm{vec}(\Omega) \in \mathbb{R}^{\prod_{n=1}^{N} I_n}$ be a vectorized version of Ω.

2.1. Image Completion with Tucker Decomposition

For the N-th order tensor $\mathcal{Y} \in \mathbb{R}^{I_1 \times I_2 \times \dots \times I_N}$, the Tucker decomposition [58] with the ranks $J = (J_1, J_2, \dots, J_N)$ can be formulated as

$$\mathcal{Y} = \mathcal{G} \times_1 \boldsymbol{U}^{(1)} \times_2 \boldsymbol{U}^{(2)} \times_3 \dots \times_N \boldsymbol{U}^{(N)}, \tag{3}$$

where $\mathcal{G} = [g_{j_1,\dots,j_N}] \in \mathbb{R}^{J_1 \times J_2 \times \dots \times J_N}$ for $J_n \leq I_n$ is the core tensor, and $\boldsymbol{U}^{(n)} = [\boldsymbol{u}_1^{(n)}, \dots, \boldsymbol{u}_{J_n}^{(n)}] = [u_{i_n,j_n}] \in \mathbb{R}^{I_n \times J_n}$ for $n = 1, \dots, N$ is the factor matrix capturing the features across the n-th mode of $\mathcal{Y}$. The operator $\times_n$ stands for the standard tensor-matrix contraction along the n-th mode, which is defined as

$$\left[\mathcal{G} \times_n \boldsymbol{U}^{(n)} \right]_{j_1,\dots,j_{n-1},i_n,j_{n+1},\dots,j_N} = \sum_{j_n=1}^{J_n} g_{j_1,\dots,j_N} u_{i_n,j_n}^{(n)}. \tag{4}$$

The optimization problem for low-rank image completion with the Tucker decomposition can be formulated as follows:

$$\min_{\mathcal{Z},\mathcal{G},\mathbf{U}^{(1)},\dots,\mathbf{U}^{(N)}} \frac{1}{2}||\mathcal{Z} - \mathcal{Y}||_F^2 + \Phi(\mathcal{G}, \{\boldsymbol{U}^{(n)}\}), \quad \text{s.t.} \quad \mathcal{Z}_\Omega = \mathcal{M}_\Omega, \ \mathcal{Z}_\Omega \geq 0,$$

$$\forall n, j_n : ||\boldsymbol{u}_{j_n}^{(n)}|| = 1, \ j_n = 1, \dots, J_n, \ n = 1, \dots N. \tag{5}$$

where $\mathcal{Y}$ is given by model (3), and $\Phi(\cdot)$ is a penalty function that imposes the desired constraints onto the core tensor $\mathcal{G}$ and the factor matrices $\{\boldsymbol{U}^{(n)}\}$. The projection $\mathcal{Z}_\Omega = \mathcal{M}_\Omega$ means that $z_{i_1,\dots,i_N}$ is replaced with $m_{i_1,\dots,i_N}$ if $\omega_{i_1,\dots,i_N} = 1$; otherwise, no changes. Assuming $\forall n : J_n < I_n$, the tensor $\mathcal{Y}$ in (3) has a low Tucker rank.

Problem (5) can be solved by performing iterative updates with the Tucker decomposition in each step. The Tucker decomposition can be computed in many ways, depending on the constraints imposed on the estimated factors. If the nonnegativity constraints are used (as specified), any nonnegative least square (NNLS) solver can be applied in the alternating optimization scheme. Neglecting the computational cost of using an NNLS solver and the cost of computing the core $\mathcal{G}$ in (3), the total computational complexity for approximating the solution to (5) in K iterations can be roughly estimated as $\mathcal{O}(K \sum_{n=1}^{N} J_n \prod_{p=1}^{N} I_p)$.

2.2. RBF Interpolation

The RBF interpolation [59] is a commonly used mesh-free method for approximating unstructured and high-dimensional data with high-order interpolants. Given the set of I distinct data points (x_i, y_i)

for $i = 1, \ldots, I$, $\forall i : x_i \in \mathbb{R}^P$, $y_i \in \mathbb{R}$, the aim is to find the approximating function $y(x)$, which is referred to as the interpolant, satisfying the condition: $\forall i : y(x_i) = y_i$. For the RBFs $\psi : [0, \infty) \to \mathbb{R}$, the interpolant takes the form

$$y(x) = \sum_{j=1}^{J} w_j \psi\left(||x - x_j||\right), \tag{6}$$

where $\{w_i\}$ are real-value weighting coefficients. For I data points, we have

$$y_i = \sum_{j=1}^{J} w_j \psi\left(||x_i - x_j||\right), \quad \text{where} \quad i = 1, \ldots, I \tag{7}$$

The weighting coefficients $\{w_i\}$ can be computed from the system of linear equations:

$$y = \Psi w, \tag{8}$$

where $y = [y_1, \ldots, y_I]^T \in \mathbb{R}^I$, $\Psi = [\psi\left(||x_i - x_j||\right)] \in \mathbb{R}^{I \times J}$, and $w = [w_1, \ldots, w_J]^T \in \mathbb{R}^J$. If $I \geq J$ and Ψ is a full-rank matrix, system (8) can be solved with any linear least-square solver.

3. Proposed Algorithm

For interpolation of N-way incomplete tensor $\mathcal{M}$, the points x_i and x_j in (7) are expressed by index values: $x_i = [i_1, i_2, \ldots, i_N] \in \mathbb{R}^N$ and $x_j = [j_1, j_2, \ldots, j_N] \in \mathbb{R}^N$. The distance measure in (6) can also be regarded in a wider sense, and hence the norm l_2 can be replaced with the distance function $D(x_i, x_j)$. For any distance function, the following conditions are satisfied: $D(x, x) = 0$, $\forall x \neq y : D(x, y) > 0$, $D(x, y) = D(y, x)$, and $D(x, z) \leq D(x, y) + D(y, z)$. In the N-dimensional space, any data point $y_{i_1, \ldots, i_N}$ can be modelled with the interpolant

$$\begin{aligned}
y_{i_1, i_2, \ldots, i_N} &= \sum_{j=1}^{J} w_j \psi\left(-\frac{D(x_i, x_j)}{\tau}\right) \\
&= \sum_{j_1=1}^{J_1} \sum_{j_2=1}^{J_2} \cdots \sum_{j_N=1}^{J_N} w_{j_1, j_2, \ldots, j_N} \psi\left(-\frac{D([i_1, i_2, \ldots, i_N], [j_1, j_2, \ldots, j_N])}{\tau}\right),
\end{aligned} \tag{9}$$

where $\tau > 0$ is a scaling factor.

Let $D([i_1, i_2, \ldots, i_N], [j_1, j_2, \ldots, j_N])$ be an additively separable distance function, i.e.,

$$D([i_1, i_2, \ldots, i_N], [j_1, j_2, \ldots, j_N]) = \sum_{n=1}^{N} d^{(n)}(i_n, j_n), \tag{10}$$

where $d^{(n)}(i_n, j_n)$ is a one-variable function that expresses the distance metrics for the variables (i_n, j_n) in the n-th mode. We also assume that the RBF ψ is multiplicatively separable. That is

$$\psi\left(\sum_{n=1}^{N} x_n\right) = \prod_{n=1}^{N} \psi^{(n)}(x_n). \tag{11}$$

Considering separability conditions (10), (11), model (9) can be reformulated as follows:

$$\begin{aligned}
y_{i_1, \ldots, i_N} &= \sum_{j_1=1}^{J_1} \cdots \sum_{j_N=1}^{J_N} w_{j_1, \ldots, j_N} \psi\left(-\frac{\sum_{n=1}^{N} d^{(n)}(i_n, j_n)}{\tau}\right) \\
&= \sum_{j_1=1}^{J_1} \cdots \sum_{j_N=1}^{J_N} w_{j_1, \ldots, j_N} \prod_{n=1}^{N} \psi^{(n)}\left(-\frac{d^{(n)}(i_n, j_n)}{\tau}\right) = \sum_{j_1=1}^{J_1} \cdots \sum_{j_N=1}^{J_N} w_{j_1, \ldots, j_N} \prod_{n=1}^{N} f_{i_n, j_n}^{(n)} \\
&= \sum_{j_N=1}^{J_N} \cdots \left(\sum_{j_1=1}^{J_1} w_{j_1, \ldots, j_N} f_{i_1, j_1}^{(1)}\right) f_{i_N, j_N}^{(N)},
\end{aligned} \tag{12}$$

where $f_{i_n, j_n}^{(n)} = \psi^{(n)}\left(-\frac{d^{(n)}(i_n, j_n)}{\tau}\right)$.

Following the standard tensor-matrix contraction rule given in (4), formula (12) can be presented in the following form:

$$\mathcal{Y} = \mathcal{W} \times_1 F^{(1)} \times_2 \ldots \times_N F^{(N)},\tag{13}$$

where $\mathcal{Y} = [y_{i_1,\ldots,i_N}] \in \mathbb{R}^{I_1 \times \ldots \times I_N}$, $\mathcal{W} = [w_{j_1,\ldots,j_N}] \in \mathbb{R}^{J_1 \times \ldots \times J_N}$, and $\forall n : F^{(n)} = [f_{i_n,j_n}^{(n)}] \in \mathbb{R}^{I_n \times J_n}$.

Remark 1. *The model given in (13) has a form identical to model (3), where $\mathcal{W}$ is the core tensor, and $\{F^{(n)}\}$ are factor matrices. Hence, the RBF interpolation model in N-dimensional space, in which separability conditions (10), (11) are satisfied, boils down to the Tucker decomposition model, where the factor matrices $\{F^{(n)}\}$ are previously determined by rational functions.*

If $d^{(n)}(i_n, j_n) = i_n^{j_n - 1}$ is expressed by an exponential function, $\tau = 1$, and $\psi(\xi) = -\xi$, then $f_{i_n,j_n}^{(n)} = \psi^{(n)}\left(-\frac{d^{(n)}(i_n,j_n)}{\tau}\right) = i_n^{j_n - 1}$, and model (12) takes the form of the standard multivariate polynomial regression:

$$y_{i_1,\ldots,i_N} = \sum_{j_1=1}^{J_1} \cdots \sum_{j_N=1}^{J_N} w_{j_1,\ldots,j_N} \prod_{n=1}^{N} (i_n)^{j_n - 1},\tag{14}$$

where numbers $(J_1, \ldots, J_N)$ determine the degrees of the polynomial with respect to each mode, i.e., $\forall n : J_n = D_n + 1$, where D_n is a degree of the polynomial along the n-th mode. Thus the multivariate polynomial regression can be also presented in the form of the Tucker decomposition model.

The RBF ψ can take various forms, such as Gaussian, polyharmonic splines, multiquadrics, inverse multiquadrics, and inverse quadratics [59]. The Gaussian RBF (GRBF), expressed by $\psi(\xi) = \exp\{\xi\}$, is commonly used for interpolation; however, it cannot be used for constructing an interpolant with polynomial precision. For example, the GRBF does not approximate a linear function $y(x)$ well. Hence, the polynomial regression in (14) is more suitable for linear or slowly varying functions, but it often yields underestimated interpolants if the polynomial has too low of a degree. An increase in the degree leads to an ill-conditioned regression problem and unbiased least squares estimates. To tackle this problem, both interpolation approaches can be combined, which leads to the following interpolation model:

$$y_{i_1,\ldots,i_N} = \sum_{j_1=1}^{J_1} \cdots \sum_{j_N=1}^{J_N} w_{j_1,\ldots,j_N} \prod_{n=1}^{N} \exp\left(-\frac{d^{(n)}(i_n,j_n)}{\tau}\right) + \sum_{r_1=1}^{R_1} \cdots \sum_{r_N=1}^{R_N} c_{r_1,\ldots,r_N} \prod_{n=1}^{N} (i_n)^{r_n - 1}.\tag{15}$$

Model (15) can be equivalently presented in the tensor-matrix contraction form:

$$\mathcal{Y} = \mathcal{W} \times_1 F^{(1)} \times_2 \ldots \times_N F^{(N)} + \mathcal{C} \times_1 P^{(1)} \times_2 \ldots \times_N P^{(N)},\tag{16}$$

where $\mathcal{C} = [c_{r_1,\ldots,r_N}] \in \mathbb{R}^{R_1 \times \ldots \times R_N}$ is the core tensor of the polynomial regression model, and $\forall n : P^{(n)} = [p_{i_n,r_n}^{(n)}] = [1, i_n, i_n^2, \ldots, i_n^{R_n - 1}] \in \mathbb{R}^{I_n \times R_n}$ is the respective factor matrix with $p_{i_n,r_n}^{(n)} = i_n^{r_n - 1}$. System (16) has $\prod_{n=1}^{N} J_n + \prod_{n=1}^{N} R_n$ variables, and only $\prod_{n=1}^{N} I_n$ equations, resulting in ambiguity of its solution under the assumption that $\forall n : J_n = I_n$. To relax this problem, a side-condition is imposed:

$$\mathcal{W} \times_1 P^{(1)T} \times_2 \ldots \times_N P^{(N)T} = 0.\tag{17}$$

Vectorizing models (16), (17) and performing the straightforward mathematical operations, we get

$$\begin{bmatrix} y \\ 0 \end{bmatrix} = \begin{bmatrix} F & P \\ P^T & 0 \end{bmatrix} \begin{bmatrix} w \\ c \end{bmatrix},\tag{18}$$

where $y = \text{vec}(\mathcal{Y}) \in \mathbb{R}^{\prod_{n=1}^{N} I_n}$, $F = F^{(N)} \otimes \ldots \otimes F^{(1)} \in \mathbb{R}^{\prod_{n=1}^{N} I_n \times \prod_{n=1}^{N} J_n}$, $P = P^{(N)} \otimes \ldots \otimes P^{(1)} \in \mathbb{R}^{\prod_{n=1}^{N} I_n \times \prod_{n=1}^{N} R_n}$, $w = \text{vec}(\mathcal{W}) \in \mathbb{R}^{\prod_{n=1}^{N} J_n}$, and $c = \text{vec}(\mathcal{C}) \in \mathbb{R}^{\prod_{n=1}^{N} R_n}$. The symbol $\otimes$ denotes the Kronecker product, and $\text{vec}(\cdot)$ means the vectorization operator.

To apply model (16) for image completion, let $\hat{\mathcal{Y}}$ be obtained according to Definition 1. System (18) can be applied to model non-zero entries in $\hat{\mathcal{Y}}$, using the following transformations: $\hat{y} = \text{vec}(\hat{\mathcal{Y}}) = S\text{vec}(\mathcal{Y})$, $\hat{F} = SFS$, $\hat{P} = SP$, and $\hat{w} = Sw$, where $S = \text{diag}\{\text{vec}(\Omega)\} \in \mathbb{R}^{\prod_{p=1}^{N} I_p \times \prod_{p=1}^{N} I_p}$ is a diagonal matrix with binary values on the main diagonal. The matrices $\hat{F}$ and $\hat{P}$ have $|\bar{\Omega}|$ zero-value rows, and $\hat{F}$ also has the same number of zero-value columns. After removing the zero-value rows and columns from the transformed system, the observed entries in $\mathcal{M}$ can be expressed by the model

$$\begin{bmatrix} y(\omega) \\ 0 \end{bmatrix} = \begin{bmatrix} F(\omega, \omega) & P(\omega, :) \\ P(\omega, :)^T & 0 \end{bmatrix} \begin{bmatrix} w(\omega) \\ c \end{bmatrix}, \tag{19}$$

where $y(\omega) = \text{vec}(\mathcal{Y}(\Omega)) \in \mathbb{R}^{|\Omega|}$ is a vectorized version of observed entries in $\mathcal{M}$, $F(\omega, \omega) \in \mathbb{R}^{|\Omega| \times |\Omega|}$ is a submatrix of $\hat{F}$ by selecting non-zero rows and columns, $P(\omega, :) \in \mathbb{R}^{|\Omega| \times \prod_{n=1}^{N} R_n}$ is obtained from $\hat{P}$ by removing all zero-value rows, and $w(\omega) = \text{vec}(\mathcal{W}(\Omega)) \in \mathbb{R}^{|\Omega|}$, taking into account rule (2). Note that $|\Omega| << \prod_{n=1}^{N} I_n$, if the number of missing entries in $\mathcal{M}$ is relatively large.

The matrix $F(\omega, \omega)$ is positive-definite because it is generated by a GRBF. The matrix $P(\omega, :)$ might be ill-conditioned if the polynomial functions of higher degrees are used, but the second term in (16) is used to better approximate linear relationships, and hence there is no need to use higher degrees. In this approach, $\forall n : R_n = 3$, which leads to second-degree polynomials. The system matrix in (19) is therefore symmetric and positive-definite, and any least-square (LS) solver can be used to compute the vectors $w(\omega)$ and c from (19), given $y(\omega)$.

System (18) can also be used to compute the missing entries in $\mathcal{M}$. Having the estimates for $w(\omega)$ and c, the missing entries can be obtained by solving the following system of linear equations:

$$\begin{bmatrix} \bar{y} \\ 0 \end{bmatrix} = \begin{bmatrix} F(\bar{\omega}, \omega) & P(\bar{\omega}, :) \\ P(\omega, :)^T & 0 \end{bmatrix} \begin{bmatrix} w(\omega) \\ c \end{bmatrix}, \tag{20}$$

where $F(\bar{\omega}, \omega) \in \mathbb{R}^{|\Omega| \times |\Omega|}$ is obtained from F by removing the rows and selecting the columns that are indexed by Ω, and the vector $\bar{y} = y(\bar{\omega})$ contains the estimates for the missing entries in $\mathcal{M}$. Hence, the completed image is expressed by

$$\mathcal{Y} = [y_{i_1, \ldots, i_N}], \quad \text{where} \quad y_{i_1, \ldots, i_N} = \begin{cases} m_{i_1, \ldots, i_N} & \text{if} \quad \omega_{i_1, \ldots, i_N} = 1, \\ \bar{y}_{i_1, \ldots, i_N} & \text{otherwise} \end{cases} \tag{21}$$

The system matrix in (19) has the order of $|\Omega| + \prod_{n=1}^{N} R_n$. Applying Gaussian elimination to (19), the computational complexity amounts to $\mathcal{O}\left((|\Omega| + \prod_{n=1}^{N} R_n)^3\right)$, and it is relatively large if the number of missing entries in $\mathcal{M}$ is small. It is therefore necessary to reduce the computational complexity for the LS problem associated with (19). To tackle this issue, let input tensor $\hat{\mathcal{Y}} = \left[\hat{\mathcal{Y}}^{(s_1, \ldots, s_N)}\right] \in \mathbb{R}^{I_1 \times \ldots \times I_N}$ be partitioned into small overlapping subtensors $\{\hat{\mathcal{Y}}^{(s_1, \ldots, s_N)}\}$, where $\forall n : s_n = 1, \ldots, S_n$ and S_n is the number of partitions of $\hat{\mathcal{Y}}$ along its n-th mode. The total number of subtensors is equal to $S_Y = \prod_{n=1}^{N} S_n$. Each subtensor can be expressed by $\hat{\mathcal{Y}}^{(s_1, \ldots, s_N)} = \left[y_{\Gamma(s_1), \ldots, \Gamma(s_N)}\right] \in \mathbb{R}^{L_1 \times \ldots \times L_N}$, where $\forall n : L_n = \lceil \frac{I_n}{S_n} \rceil \leq I_n$. The set $\Gamma(s_n)$ contains the indices of the entries in $\hat{\mathcal{Y}}$ which belong to $\hat{\mathcal{Y}}^{(s_1, \ldots, s_N)}$ along its n-th mode, and it can be expressed by

$$\forall n : \Gamma(s_n) = \begin{cases} \{\gamma(s_n) + 1, \ldots, \gamma(s_n) + L_n\} \in \mathcal{N}^{L_n} & \text{for} \quad s_n < S_n \\ \{\gamma(S_n), \ldots, I_n\} & \text{for} \quad s_n = S_n \end{cases} \tag{22}$$

where $\gamma(s_n) = \lfloor (s_n - 1)(L_n - \eta_n) \rfloor$. The parameter $\eta_n = \lfloor \frac{\theta_n L_n}{100} \rfloor$ determines the number of overlapping pixels along the n-th mode, and $\theta_n \in [0, 99]$ expresses the percentage of the overlap along the n-th mode.

The proposed methodology for image completion should be applied separately to each subtensor. The missing pixels are completed using a very limited volume of the input tensor but mostly restricted to their nearest neighborhood. It is thus a strategy that resembles PDE-based image inpainting, but it is much more flexible for highly dissipated known pixels and allows us to reduce the computational cost dramatically. Moreover, the factor matrices $\{F^{(n)}\}$ and $\{P^{(n)}\}$ in (16) are expressed by radial functions, and hence can be precomputed before using the subtensor partitioning procedure. In RBF-based interpolation methods, the samples or pixels that are close to the boundary of the region of interest are usually not well approximated. However, due to overlapping, boundary perturbation effects in our approach are considerably relaxed.

The pseudo-code of the proposed image completion algorithm is presented in Algorithm 1.

Algorithm 1: Tensorial Interpolation for Image Completion (TI-IC)

Input : $\mathcal{M} \in \mathbb{R}^{I_1 \times \dots \times I_N}$ – input tensor, $\Omega = [\omega_{i_1,\dots,i_N}] \in \mathbb{R}^{I_1 \times \dots \times I_N}$ - a binary tensor of indicators for the missing entries, $d(\cdot, \cdot)$ – distance metrics, τ - scaling factor, $R = \{R_1, \dots, R_N\}$ – degrees of the interpolation polynomials, $\{S_1, \dots, S_N\}$ – partitioning rates, $\{\theta_1, \dots, \theta_N\}$ – rates of overlapping (in percentage).

Output: $\mathcal{Y} \in \mathbb{R}^{I_1 \times \dots \times I_N}$ – completed tensor

1 **for** $n = 1, \dots, N$ **do**
2 $\quad$ Compute factor matrices $F^{(n)}$ using metrics d, and $P^{(n)}$ using set R.
3 $F = F^{(N)} \otimes \dots \otimes F^{(1)}, P = P^{(N)} \otimes \dots \otimes P^{(1)},$
4 **foreach** $(s_1, \dots, s_N)$ **do**
5 $\quad$ Create $\hat{\mathcal{Y}}$ from $\mathcal{M}$ using (1), and select $\hat{\mathcal{Y}}^{(s_1,\dots,s_N)} = \left[y_{\Gamma(s_1),\dots,\Gamma(s_N)} \right] \in \mathbb{R}^{L_1 \times \dots \times L_N}$ from $\hat{\mathcal{Y}}$
6 $\quad$ and $\Omega^{(s_1,\dots,s_N)} = \left[\omega_{\Gamma(s_1),\dots,\Gamma(s_N)} \right] \in \mathbb{R}^{L_1 \times \dots \times L_N}$ from Ω, using rule (22),
7 $\quad$ Compute $\omega^{(s_1,\dots,s_N)} = \mathrm{vec}(\Omega^{(s_1,\dots,s_N)})$ and $\bar{\omega}^{(s_1,\dots,s_N)} = 1 - \omega^{(s_1,\dots,s_N)}$,
$\quad$ and $y(\omega^{(s_1,\dots,s_N)}) = \mathrm{vec}(\hat{\mathcal{Y}}^{(s_1,\dots,s_N)})$,
8 $\quad$ Select $F(\omega^{(s_1,\dots,s_N)}, \omega^{(s_1,\dots,s_N)}) \in \mathbb{R}^{|\Omega^{(s_1,\dots,s_N)}| \times |\Omega^{(s_1,\dots,s_N)}|}$ and
$\quad$ $F(\bar{\omega}^{(s_1,\dots,s_N)}, \omega^{(s_1,\dots,s_N)}) \in \mathbb{R}^{|\bar{\Omega}^{(s_1,\dots,s_N)}| \times |\Omega^{(s_1,\dots,s_N)}|}$ from F,
$\quad$ $P(\omega^{(s_1,\dots,s_N)}, :) \in \mathbb{R}^{|\Omega^{(s_1,\dots,s_N)}| \times \prod_{n=1}^{N} R_n}$ from P,
9 $\quad$ Compute $w(\omega^{(s_1,\dots,s_N)})$ and c solving system (19),
10 $\quad$ Given $w(\omega^{(s_1,\dots,s_N)})$ and c, compute $\bar{y}^{(s_1,\dots,s_N)}$ from (20), and construct $\mathcal{Y}^{(s_1,\dots,s_N)}$ using (21).
11 $\mathcal{Y} = \left[\mathcal{Y}^{(s_1,\dots,s_N)} \right] \in \mathbb{R}^{I_1 \times \dots \times I_N}.$

Remark 2. *The computational complexity for calculating matrices F and P in Algorithm 1 amounts to $\mathcal{O}\left(\prod_{n=1}^{N} I_n(J_n + R_n) \right)$. Assuming Gaussian elimination is used for solving system (19), we have $\mathcal{O}\left(S_Y(|\Omega^{(s_1,\dots,s_N)}| + \prod_{n=1}^{N} R_n)^3 \right)$ for S_Y subtensors, and $\mathcal{O}(S_Y(|\bar{\Omega}^{(s_1,\dots,s_N)}| + \prod_{n=1}^{N} R_n)(|\Omega^{(s_1,\dots,s_N)}| + \prod_{n=1}^{N} R_n))$ for computing $\bar{y}^{(s_1,\dots,s_N)}$ from (20). Let $|\Omega| = \xi \prod_{n=1}^{N} I_n$ and $\forall (s_1, \dots, s_N) : |\Omega^{(s_1,\dots,s_N)}| = \xi \prod_{n=1}^{N} L_n$, where $0 \leq \xi \leq 1$. Neglecting matrix P in (19) because it is much smaller than F, the computational complexity for solving system (19) can be roughly estimated as $\mathcal{O}\left(S_Y(|\Omega^{(s_1,\dots,s_N)}|)^3 \right) = \mathcal{O}\left(S_Y \xi^3 (\prod_{n=1}^{N} L_n)^3 \right) \approx \mathcal{O}\left(\xi^3 \prod_{n=1}^{N} \frac{I_n^3}{S_n^2} \right)$. Note that the computational complexity for solving the same system without partitioning tensors $\mathcal{Y}$ into subtensors and under the same assumption can be roughly estimated as $\mathcal{O}\left(|\Omega|^3 \right) = \mathcal{O}\left(\xi^3 \prod_{n=1}^{N} I_n^3 \right)$. Hence, the partitioning strategy decreases the computational complexity S_Y^2 times with respect to the non-partitioned system, and if $\forall n : J_n = I_n$, the complexity for precomputing matrix F might predominate.*

4. Results

This section presents an experimental study that was carried out to demonstrate the performance of the proposed algorithms. The tests were performed for a few image completion problems using the following RGB images: Barbara, Lena, Peppers, and Monarch, which are presented in Figure 1. All of them have a resolution of 512×512 pixels.

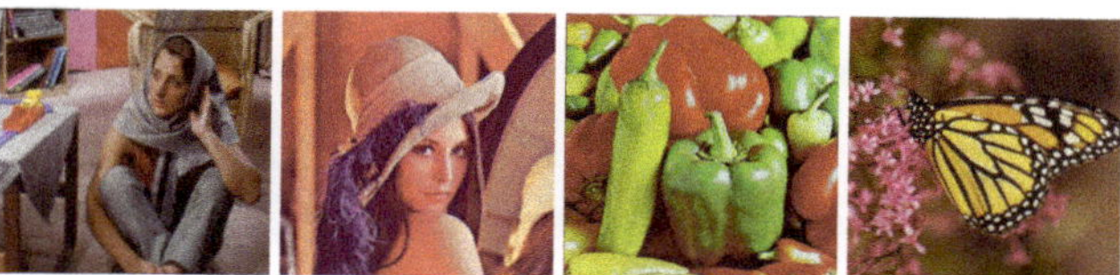

Figure 1. Original images: Barbara, Lena, Peppers, and Monarch (from left to right).

4.1. Setup

The incomplete images were obtained by removing some entries from tensors representing the original images. The following test cases were analyzed:

- **A**: 90% uniformly distributed random missing tensor fibers in its third mode (color), which corresponds to 90% missing pixels ("dead pixels"),
- **B**: 95% uniformly distributed random missing tensor entries ("disturbed pixels"),
- **C**: 200 uniformly distributed random missing circles—created in the same way as in the first case, but the disturbances are circles with a random radius not exceeding 10 pixels,
- **D**: resolution up-scaling—an original image was down-sampled twice by removing the pixels according to a regular grid mask with edges equal to 1 pixel. The aim was to recover the missing pixels on the edges.

We compared the proposed methods with the following: FAN (filtering by adaptive normalization) and EFAN (efficient filtering by adaptive normalization) [60], SmPC-QV (smooth PARAFAC tensor completion with quadratic variation) [38], LRTV (low-rank total-variation) [61], TMAC-inc (low-rank tensor completion by parallel matrix factorization with the rank-increasing) [62], C-SALSA (constrained split augmented Lagrangian shrinkage algorithm) [63], fALS (filtered alternating least-squares) [64], and KA-TT (ket augmentation tensor train) [65]. FAN and EFAN are based on adaptive Gaussian low-pass filtration. SmPC-QV performs low-rank tensor completion with smoothness-penalized CP decomposition and gradually increasing CP rank. LRTV accomplishes low-rank matrix completion using total variation regularization. TMAC-inc also belongs to a family of low-rank tensor completion, and in this approach, an incomplete tensor is unfolded with respect to all modes, and the resulting matrices are completed by applying low-rank matrix factorizations together with the adaptive rank-adjusting strategy. C-SALSA performs image completion using the variable splitting approach to solve an LS image reconstruction problem with a strong nonsmooth regularizer. fALS and KA-TT combine low-pass filtration with standard Tucker decomposition and tensor train models, respectively. The proposed algorithm is referred to as Tensorial Interpolation for Image Completion (TI-IC), and it is presented in Algorithm 1. It combines two strategies: RBF-interpolation with an exponential function, and multivariate polynomial regression. To emphasize the importance of both terms in the model (16), we also present the results obtained separately for each of them, using the same partitioning strategy in each case. The TI-IC algorithm with only the exponential term is referred to as TI-IC(Exp). When only the polynomial regression is used, TI-IC will be denoted as TI-IC(Poly).

TI-IC is flexible with respect to the choice of the distance function $d^{(n)}(\cdot, \cdot)$, degrees of the interpolation polynomials, partitioning, and overlapping rates. Since $\{i_n, j_n\}$ lie on a line, $d^{(n)}(i_n, j_n) = |i_n - j_n|$ seems to be the best choice. Factor matrices $\{P^{(n)}\}$ were determined by quadratic polynomials, hence $\forall n : R_n = 3$. Higher-order polynomials result in ill-conditioning of the system matrix in (19) and do not noticeably improve the performance. The partitioning and overlapping rates were set experimentally to $[S_1, S_2, S_3] = [32, 32, 1]$ and $[\theta_1, \theta_2, \theta_3] = [33.33, 33.33, 0]$. As the resolution of $\mathcal{M}$ is 512×512, the overlapping amounts to 5 pixels across the first and second mode for each subtensor $\hat{\mathcal{Y}} \in \mathbb{R}^{16 \times 16 \times 3}$. For larger subtensors, computational time increased considerably, and we did not observe a noticeable improvement in the quality of recovered images. For smaller subtensors, the performance decreased. The scaling factor in the exponential RBFs was also determined experimentally, and to compute $F^{(n)}$ we set $\tau = 3$ for TI-IC, and $\tau = 5$ for TI-IC(Exp).

In the iterative algorithms, the maximum number of iterations was set to 1000, and the threshold for the residual error was equal to 10^{-12}. The maximum rank was limited to 50.

The algorithms were implemented in MATLAB 2016a and run on the distributed cluster server in the Wroclaw Centre for Networking and Supercomputing (WCSS) (https://www.wcss.pl/en/) using PLGRID (http://www.plgrid.pl/en) queues and parallel workers. The resources were limited to 10 cores (ncpus) and 32 GB RAM (mem). The workers can be employed to run each algorithm for various initializations in parallel, or they can be used to process subtensors $\hat{\mathcal{Y}}^{(s_1,\dots,s_N)}$ in Algorithm 1 in parallel, in such a way that each subtensor is processed by one CPU core. The block partitioning procedure was implemented with the `blockproc` function in MATLAB 2016a, which has an option to run the computation across the available workers. We analyzed both options, i.e., when it was enabled and when it was disabled.

4.2. Image Completion

The recovered images were validated quantitatively using the signal-to-interference ratio (SIR) measure [31], defined as $\mathrm{SIR} = 20 \log_{10} \frac{\|\mathcal{M}\|_F}{\|\mathcal{M}-\mathcal{Y}\|_F}$. The SIR values were averaged over the colormaps. The speeds of the algorithms were compared by measuring the averaged runtime of each test.

Figure 2 illustrates the incomplete image (top left) used in Test A and the results obtained with the algorithms: FAN, EFAN, SmPC-QV, LRTV, C-SALSA, TMac-inc, fALS, KA-TT, TI-IC(Exp), TI-IC(Poly), and TI-IC. The images reconstructed in tests B, C, and D with the same algorithms are depicted in Figures 3–5, respectively. Due to the random initialization of some baseline algorithms, all the tests were repeated 100 times, and the SIR samples are presented in Figure 6 in the form of box-plots, separately for each test. The mean runtime of the evaluated algorithms and the corresponding standard deviations for each test case are listed in Table 1. The algorithms run on a parallel pool of MATLAB workers are denoted with an asterisk.

Figure 2. Test A (90% randomly missing pixels) for the image "Barbara".

Figure 3. Test B (95% randomly missing entries in the incomplete tensor) for the image "Lena".

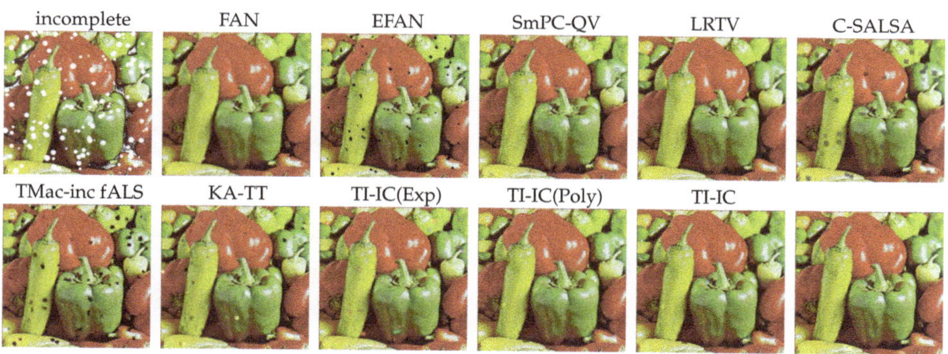

Figure 4. Test C (200 missing circles of maximum 10-pixel radius) for the image "Peppers".

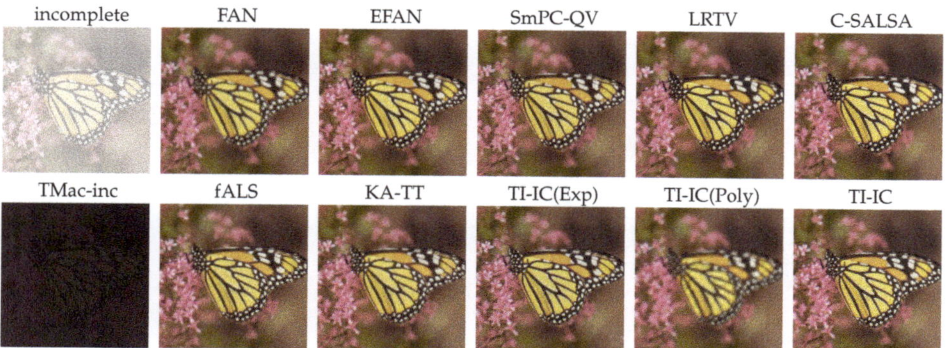

Figure 5. Test D: resolution up-scaling for the image "Monarch".

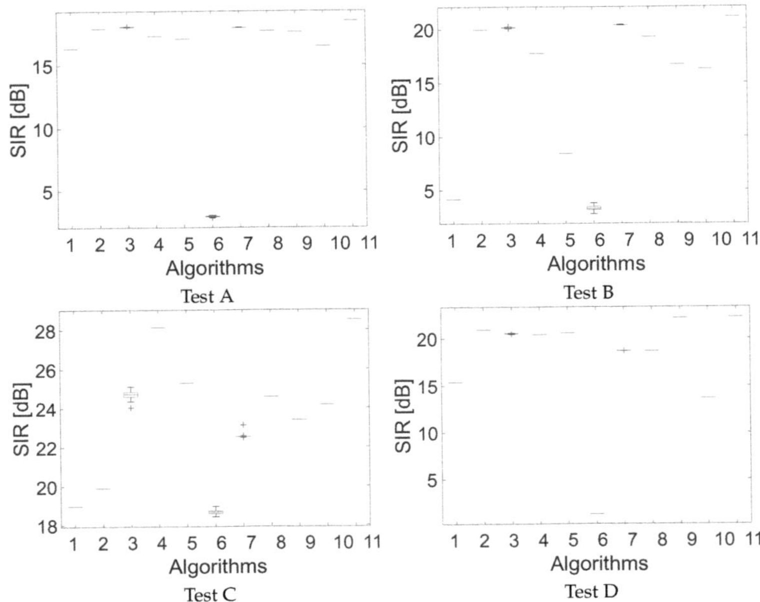

Figure 6. Box-plots of signal-to-interference ratio (SIR) performance for the tests (**A**–**D**) with the algorithms 1 = FAN, 2 = EFAN, 3 = SmPC-QV, 4 = LRTV, 5 = C-SALSA, 6 = TMac-inc, 7 = fALS, 8 = KA-TT, 9 = TI-IC(Exp), 10 = TI-IC(Poly), 11 = TI-IC.

Table 1. Mean runtime (in seconds) of the algorithms and the corresponding standard deviations for each test case. An asterisk denotes the use of parallel processing with a parallel pool of workers in MATLAB.

Algorithm/Test	Test A	Test B	Test C	Test D
FAN	0.25 ± 0.08	0.25 ± 0.06	0.24 ± 0.04	0.2 ± 0.04
EFAN	0.06 ± 0.01	0.06 ± 0.01	0.15 ± 0.01	0.07 ± 0.01
SmPC-QV	534.67 ± 91.82	579.12 ± 109.36	416.68 ± 82.97	507.74 ± 101.19
LRTV	876.64 ± 89.25	943.6 ± 78.45	966.72 ± 87.01	977.86 ± 117.85
C-SALSA	73.87 ± 17.08	86.47 ± 22.52	66.33 ± 18.84	354.58 ± 91.99
TMac-inc	122.37 ± 21.75	120.8 ± 40.98	440.05 ± 54.35	193.22 ± 23.29
fALS	545.39 ± 63.67	478.98 ± 47.47	879.41 ± 74.83	206.23 ± 74.93
KA-TT	433.8 ± 81.18	379.54 ± 63.55	392.59 ± 54.09	419.3 ± 70.94
TI-IC(Exp)	3.77 ± 1.03	1.39 ± 0.17	133.14 ± 10.92	15.08 ± 1.97
TI-IC(Exp) *	1.7 ± 0.12	1.14 ± 0.04	15.39 ± 0.56	2.81 ± 0.1
TI-IC(Poly)	0.57 ± 0.11	0.13 ± 0.02	1.73 ± 0.11	0.71 ± 0.1
TI-IC(Poly) *	0.49 ± 0.03	0.36 ± 0.01	0.58 ± 0.02	0.49 ± 0.01
TI-IC	6.01 ± 1.49	2.91 ± 0.78	406.75 ± 51.51	24.59 ± 4.77
TI-IC *	1.96 ± 0.05	1.19 ± 0.04	40.98 ± 1.79	4.15 ± 0.18

4.3. Discussion

The experiments were carried out for typical but challenging image completion problems. In test A, we knew only 10% of the pixels in the "Barbara" image, and the aim was to recover the 90%

missing pixels. The results illustrated in Figures 2 and 6A show that good quality reconstructions were obtained when the EFAN, SmPC-QV, fALS, KA-TT, TI-IC(Exp), and TI-IC algorithms were used, but the image recovered with TI-IC had the highest SIR score. TI-IC was also quite fast in this test (see Table 1). It lost in the runtime category only to FAN and EFAN, but its parallel version TI-IC *) was more than 250 times faster than SmPC-QV. The latter performs the CP decomposition, but the difference in computational speed comes from the fact that in our method, the factor matrices are precomputed, and only the core tensor in the Tucker decomposition is estimated using the data. The results obtained in Test B are presented in Figures 3 and 6B. They confirm the conclusions drawn from Test A, but it should be noted that TI-IC strengthened its leading position in the SIR performance. Moreover, its runtime was shorter than that in the previous test because only 5% of the entries were known, and hence the system matrix in (19) was smaller. Test C compared algorithms for the completion of many small-scale missing regions (holes), distributed across the image. The results presented in Figures 4 and 6C show that EFAN and SmPC-QV failed to provide satisfactory reconstructions in this test, but TI-IC outperformed the other algorithms considerably. Obviously, a lower number of missing pixels in the image to be completed results in a noticeable increase in the runtime, but it was still below the runtime of low-rank tensor completion methods, such as SmPC-QV, TMac-inc, fALS, and LRTV. In Test D, only 50 % of pixels were unknown, but not all the tested algorithms handled this case well. In this test, TI-IC also yielded the best reconstruction (see Figures 5 and 6D) but only slightly better than that obtained with TI-IC(exp). Hence, the low-degree polynomial regression in Test D does not affect the result considerably, and due to the computation time, it can be neglected. In other tests, both approaches (RBF interpolation and polynomial regression), combined appropriately, were essential to yielding high-quality results.

5. Conclusions

In this study, we showed the relationship between the models of RBF interpolation and Tucker decomposition (Remark 1). We combined the exponential RBF interpolation and polynomial regression in one model and experimentally demonstrated that such a hybrid method achieved the highest SIR scores in all the tests. The proposed algorithm (TI-IC) can be applied to a wide spectrum of image-completion problems. The incomplete images can contain many single missing entries or missing pixels distributed across the image, a large number of small-scale regions (holes), or regularly shaped missing regions, such as in resolution up-scaling problems. The TI-IC algorithm is also computationally efficient. It provides reconstructions of the highest quality and in a much shorter time than the tested low-rank tensor image-completion methods. Its runtime depends on the number of missing entries in an input tensor, and it is shorter if more entries are unknown. The computational complexity of the proposed method can be controlled by the block partitioning strategy, as proven in Remark 2. Assuming the overlapping in this partitioning strategy, we avoided visible disturbances around boundary entries of the blocks, which is an intrinsic effect of RBF interpolation methods. Furthermore, due to the use of RBFs, the overlapping blocks can be processed in parallel computer architectures, and our experiments demonstrated that the use of a parallel pool of workers in MATLAB considerably shortened the runtime of the proposed algorithm.

Summing up, the proposed algorithm outperforms all the tested image completion methods for a wide spectrum of tests. Its computational runtime is also satisfactory and considerably shorter than that for the low-rank tensor decompositions. The proposed algorithm can also be efficiently implemented on parallel computer architectures.

Author Contributions: Conceptualization, R.Z.; Methodology, R.Z.; Software, R.Z. and T.S.; Investigation, R.Z. and T.S.; Validation, R.Z. and T.S.; Visualization, T.S.; Writing—original draft, R.Z.; Writing—review and editing, R.Z.; Funding acquisition, R.Z. All authors have read and agreed to the published version of the manuscript.

Funding: This research was supported by grant 2015/17/B/ST6/01865, funded by the National Science Center in Poland. Calculations were performed at the Wroclaw Centre for Networking and Supercomputing under grant no. 127.

Conflicts of Interest: The authors declare no conflict of interest.

References

1. Zarif, S.; Faye, I.; Rohaya, D. Image Completion: Survey and Comparative Study. *Int. J. Pattern Recognit. Artif. Intell.* **2015**, *29*, 1554001. [CrossRef]
2. Hu, W.; Tao, D.; Zhang, W.; Xie, Y.; Yang, Y. The Twist Tensor Nuclear Norm for Video Completion. *IEEE Trans. Neural Netw. Learn. Syst.* **2017**, *28*, 2961–2973. [CrossRef] [PubMed]
3. Ebdelli, M.; Meur, O.L.; Guillemot, C. Video Inpainting with Short-Term Windows: Application to Object Removal and Error Concealment. *IEEE Trans. Image Process.* **2015**, *24*, 3034–3047. [CrossRef] [PubMed]
4. He, W.; Yokoya, N.; Yuan, L.; Zhao, Q. Remote Sensing Image Reconstruction Using Tensor Ring Completion and Total Variation. *IEEE Trans. Geosci. Remote Sens.* **2019**, *57*, 8998–9009. [CrossRef]
5. Lakshmanan, V.; Gomathi, R. A Survey on Image Completion Techniques in Remote Sensing Images. In Proceedings of the 4th International Conference on Signal Processing, Communication and Networking (ICSCN), Piscataway, NJ, USA, 16–18 March 2017; pp. 1–6.
6. Bertalmio, M.; Sapiro, G.; Caselles, V.; Ballester, C. Image Inpainting. In *Proceedings of the 27th Annual Conference on Computer Graphics and Interactive Techniques, New Orleans, LA, USA, 23–28 July 2000*; ACM Press: New York, NY, USA, 2000; pp. 417–424.
7. Efros, A.A.; Leung, T.K. Texture Synthesis by Non-Parametric Sampling. In Proceedings of the International Conference on Computer Vision (ICCV), Kerkyra, Corfu, Greece, 20–25 September 1999; Volume 2, pp. 1033–1038.
8. Bertalmio, M.; Vese, L.; Sapiro, G.; Osher, S. Simultaneous Structure and Texture Image Inpainting. *Trans. Img. Proc.* **2003**, *12*, 882–889. [CrossRef]
9. Criminisi, A.; Perez, P.; Toyama, K. Object Removal by Exemplar-based Inpainting. In Proceedings of the IEEE Computer Vision and Pattern Recognition (CVPR), Madison, WI, USA, 16–22 June 2003.
10. Sun, J.; Yuan, L.; Jia, J.; Shum, H.Y. Image Completion with Structure Propagation. *ACM Trans. Graph.* **2005**, *24*, 861–868. [CrossRef]
11. Darabi, S.; Shechtman, E.; Barnes, C.; Goldman, D.B.; Sen, P. Image Melding: Combining Inconsistent Images Using Patch-based Synthesis. *ACM Trans. Graph.* **2012**, *31*, 82:1–82:10. [CrossRef]
12. Buyssens, P.; Daisy, M.; Tschumperle, D.; Lezoray, O. Exemplar-Based Inpainting: Technical Review and New Heuristics for Better Geometric Reconstructions. *IEEE Trans. Image Process.* **2015**, *24*, 1809–1824. [CrossRef]
13. Hesabi, S.; Jamzad, M.; Mahdavi-Amiri, N. Structure and Texture Image Inpainting. In Procedings of the International Conference on Signal and Image Processing, Chennai, India, 15–17 December 2010; pp. 119–124.
14. Jia, J.; Tang, C.-K. Image Repairing: Robust Image Synthesis by Adaptive ND Tensor Voting. In Proceedings of the IEEE Computer Society Conference on Computer Vision and Pattern Recognition, Madison, WI, USA, 18–20 June 2003; Volume 1, p. 7762318.
15. Drori, I.; Cohen-Or, D.; Yeshurun, H. Fragment-based Image Completion. *ACM Trans. Graph.* **2003**, *22*, 303–312. [CrossRef]
16. Shao, X.; Liu, Z.; Li, H. An Image Inpainting Approach Based On the Poisson Equation. In Proceedings of the Second International Conference on Document Image Analysis for Libraries (DIAL'06), Lyon, France, 27–28 April 2006.
17. Iizuka, S.; Simo-Serra, E.; Ishikawa, H. Globally and Locally Consistent Image Completion. *ACM Trans. Graph.* **2017**, *36*, 107:1–107:14. [CrossRef]
18. Pathak, D.; Krähenbühl, P.; Donahue, J.; Darrell, T.; Efros, A. Context Encoders: Feature Learning by Inpainting. In Proceedings of the IEEE Conference on Computer Vision and Pattern Recognition (CVPR), Las Vegas, NV, USA, 27–30 June 2016.
19. Guo, J.; Liu, Y. Image completion using structure and texture GAN network. *Neurocomputing* **2019**, *360*, 75–84. [CrossRef]
20. Zhao, D.; Guo, B.; Yan, Y. Parallel Image Completion with Edge and Color Map. *Appl. Sci.* **2019**, *9*, 3856. [CrossRef]
21. Liu, C.; Peng, Q.; Xun, W. Recent Development in Image Completion Techniques. In Proceedings of the IEEE International Conference on Computer Science and Automation Engineering, Shanghai, China, 10–12 June 2011; Volume 4, pp. 756–760.

22. Atapour-Abarghouei, A.; Breckon, T.P. A comparative review of plausible hole filling strategies in the context of scene depth image completion. *Comput. Graph.* **2018**, *72*, 39–58. [CrossRef]

23. Candès, E.J.; Recht, B. Exact Matrix Completion via Convex Optimization. *Found. Comput. Math.* **2009**, *9*, 717. [CrossRef]

24. Cai, J.F.; Candès, E.J.; Shen, Z. A Singular Value Thresholding Algorithm for Matrix Completion. *SIAM J. Optim.* **2010**, *20*, 1956–1982. [CrossRef]

25. Zhang, D.; Hu, Y.; Ye, J.; Li, X.; He, X. Matrix completion by Truncated Nuclear Norm Regularization. In Proceedings of the IEEE Conference on Computer Vision and Pattern Recognition, Providence, RI, USA, 16–21 June 2012; pp. 2192–2199.

26. Zhang, M.; Desrosiers, C. Image Completion with Global Structure and Weighted Nuclear Norm Regularization. 2017 International Joint Conference on Neural Networks (IJCNN), Anchorage, AK, USA, 14–19 May 2017; pp. 4187–4193.

27. Wen, Z.; Yin, W.; Zhang, Y. Solving a low-rank factorization model for matrix completion by a nonlinear successive over-relaxation algorithm. *Math. Program. Comput.* **2012**, *4*, 333–361. [CrossRef]

28. Lee, D.D.; Seung, H.S. Learning the parts of objects by non-negative matrix factorization. *Nature* **1999**, *401*, 788–791. [CrossRef]

29. Wang, Y.; Zhang, Y. Image Inpainting via Weighted Sparse Non-Negative Matrix Factorization. In Proceedings of the 18th IEEE International Conference on Image Processing, Brussels, Belgium, 11–14 September 2011; pp. 3409–3412.

30. Sadowski, T.; Zdunek, R. Image Completion with Smooth Nonnegative Matrix Factorization. In Proceedings of the 17th International Conference on Artificial Intelligence and Soft Computing ICAISC, Zakopane, Poland, 3–7 June 2018; Part II, pp. 62–72.

31. Cichocki, A.; Zdunek, R.; Phan, A.H.; Amari, S.I. *Nonnegative Matrix and Tensor Factorizations: Applications to Exploratory Multi-way Data Analysis and Blind Source Separation*; Wiley and Sons: Chichester, UK, 2009.

32. Tomasi, G.; Bro, R. PARAFAC and missing values. *Chemom. Intell. Lab. Syst.* **2005**, *75*, 163–180. [CrossRef]

33. Acar, E.; Dunlavy, D.M.; Kolda, T.G.; Mørup, M. Scalable Tensor Factorizations with Missing Data. In Proceedings of the SIAM International Conference on Data Mining, Columbus, OH, USA, 29 April–1 May 2010; pp. 701–712.

34. Liu, J.; Musialski, P.; Wonka, P.; Ye, J. Tensor Completion for Estimating Missing Values in Visual Data. *IEEE Trans. Pattern Anal. Mach. Intell.* **2013**, *35*, 208–220. [CrossRef]

35. Song, Q.; Ge, H.; Caverlee, J.; Hu, X. Tensor Completion Algorithms in Big Data Analytics. *ACM Trans. Knowl. Discov. Data* **2019**, *13*, 6:1–6:48. [CrossRef]

36. Gao, B.; He, Y.; Lok Woo, W.; Yun Tian, G.; Liu, J.; Hu, Y. Multidimensional Tensor-Based Inductive Thermography with Multiple Physical Fields for Offshore Wind Turbine Gear Inspection. *IEEE Trans. Ind. Electron.* **2016**, *63*, 6305–6315. [CrossRef]

37. Zhao, Q.; Zhang, L.; Cichocki, A. Bayesian CP Factorization of Incomplete Tensors with Automatic Rank Determination. *IEEE Trans. Pattern Anal. Mach. Intell.* **2015**, *37*, 1751–1763. [CrossRef] [PubMed]

38. Yokota, T.; Zhao, Q.; Cichocki, A. Smooth PARAFAC Decomposition for Tensor Completion. *IEEE Trans. Signal Process.* **2016**, *64*, 5423–5436. [CrossRef]

39. Gui, L.; Zhao, Q.; Cao, J. Brain Image Completion by Bayesian Tensor Decomposition. In Proceedings of the 22nd International Conference on Digital Signal Processing (DSP), London, UK, 23–25 August 2017; pp. 1–4.

40. Geng, X.; Smith-Miles, K. Facial Age Estimation by Multilinear Subspace Analysis. In Proceedings of the 2009 IEEE International Conference on Acoustics, Speech and Signal Processing, ICASSP '09, Taipei, Taiwan, 19–24 April 2009; pp. 865–868.

41. Chen, B.; Sun, T.; Zhou, Z.; Zeng, Y.; Cao, L. Nonnegative Tensor Completion via Low-Rank Tucker Decomposition: Model and Algorithm. *IEEE Access* **2019**, *7*, 95903–95914. [CrossRef]

42. Wang, W.; Aggarwal, V.; Aeron, S. Efficient Low Rank Tensor Ring Completion. In Proceedings of the IEEE International Conference on Computer Vision (ICCV), Venice, Italy, 22–29 October 2017; pp. 5698–5706.

43. Bengua, J.A.; Phien, H.N.; Tuan, H.D.; Do, M.N. Efficient Tensor Completion for Color Image and Video Recovery: Low-Rank Tensor Train. *IEEE Trans. Image Process.* **2017**, *26*, 2466–2479. [CrossRef] [PubMed]

44. Ko, C.Y.; Batselier, K.; Yu, W.; Wong, N. Fast and Accurate Tensor Completion with Tensor Trains: A System Identification Approach. *arXiv* **2018**, arXiv:1804.06128.

45. Silva, C.D.; Herrmann, F.J. Hierarchical Tucker Tensor Optimization—Applications to Tensor Completion. In Proceedings of the SAMPTA, Bremen, Germany, 1–5 July 2013.

46. Zhou, P.; Lu, C.; Lin, Z.; Zhang, C. Tensor Factorization for Low-Rank Tensor Completion. *IEEE Trans. Image Process.* **2018**, *27*, 1152–1163. [CrossRef]

47. Latorre, J.I. Image Compression and Entanglement. *arXiv* **2005**, arXiv:quant-ph/0510031.

48. Huo, X.; Tan, J.; He, L.; Hu, M. An automatic video scratch removal based on Thiele type continued fraction. *Multimed. Tools Appl.* **2014**, *71*, 451–467. [CrossRef]

49. Karaca, E.; Tunga, M.A. Interpolation-based image inpainting in color images using high dimensional model representation. In Proceedings of the 24th European Signal Processing Conference (EUSIPCO), Budapest, Hungary, 29 August–2 September 2016; pp. 2425–2429.

50. Sapkal, M.S.; Kadbe, P.K.; Deokate, B.H. Image inpainting by Kriging interpolation technique for mask removal. 2016 International Conference on Electrical, Electronics, and Optimization Techniques (ICEEOT), Palanchur, India, 3–5 March 2016; pp. 310–313.

51. He, L.; Xing, Y.; Xia, K.; Tan, J. An Adaptive Image Inpainting Method Based on Continued Fractions Interpolation. *Discret. Dyn. Nat. Soc.* **2018**, *2018*. [CrossRef]

52. Guariglia, E. Primality, Fractality, and Image Analysis. *Entropy* **2019**, *21*, 304. [CrossRef]

53. Guariglia, E. Harmonic Sierpinski Gasket and Applications. *Entropy* **2018**, *20*, 714. [CrossRef]

54. Gao, B.; Lu, P.; Woo, W.L.; Tian, G.Y.; Zhu, Y.; Johnston, M. Variational Bayesian Subgroup Adaptive Sparse Component Extraction for Diagnostic Imaging System. *IEEE Trans. Ind. Electron.* **2018**, *65*, 8142–8152. [CrossRef]

55. Frongillo, M.; Riccio, G.; Gennarelli, G. Plane wave diffraction by co-planar adjacent blocks. In Proceedings of the Loughborough Antennas Propagation Conference (LAPC), Loughborough, UK, 14–15 November 2016; pp. 1–4.

56. Ghassemian, H. A review of remote sensing image fusion methods. *Inf. Fusion* **2016**, *32*, 75–89. [CrossRef]

57. Prautzsch, H.; Boehm, W.; Paluszny, M. Tensor Product Surfaces. In *Bézier and B-Spline Techniques*; Springer: Berlin/Heidelberg, Germany, 2002; pp. 125–140.

58. Tucker, L.R. The Extension of Factor Analysis to Three-Dimensional Matrices. In *Contributions to Mathematical Psychology*; Gulliksen, H., Frederiksen, N., Eds.; Holt, Rinehart and Winston: New York, NY, USA, 1964; pp. 110–127.

59. Buhmann, M.D. *Radial Basis Functions—Theory and Implementations, Volume 12*; *Cambridge Monographs on Applied and Computational Mathematics*; Cambridge University Press: Cambridge, UK, 2009.

60. Achanta, R.; Arvanitopoulos, N.; Susstrunk, S. Extreme Image Completion. In Proceedings of the IEEE International Conference on Acoustics, Speech and Signal Processing (ICASSP), New Orleans, LA, USA, 5–9 March 2017; pp. 1333–1337.

61. Yokota, T.; Hontani, H. Simultaneous Visual Data Completion and Denoising Based on Tensor Rank and Total Variation Minimization and Its Primal-Dual Splitting Algorithm. In Proceedings of the IEEE Conference on Computer Vision and Pattern Recognition (CVPR), Honolulu, HI, USA, 21–26 July 2017; pp. 3843–3851.

62. Xu, Y.; Hao, R.; Yin, W.; Su, Z. Parallel matrix factorization for low-rank tensor completion. *Inverse Probl. Imaging* **2015**, *9*, 601–624. [CrossRef]

63. Afonso, M.V.; Bioucas-Dias, J.M.; Figueiredo, M.A.T. Fast Image Recovery Using Variable Splitting and Constrained Optimization. *IEEE Trans. Image Process.* **2010**, *19*, 2345–2356. [CrossRef]

64. Sadowski, T.; Zdunek, R. Image Completion with Filtered Alternating Least Squares Tucker Decomposition. In Proceedings of the IEEE SPA Conference: Algorithms, Architectures, Arrangements, and Applications, Poznan, Poland, 18–20 September 2019; pp. 241–245.

65. Zdunek, R.; Fonal, K.; Sadowski, T. Image Completion with Filtered Low-Rank Tensor Train Approximations. In Proceedings of the 15th International Work-Conference on Artificial Neural Networks, IWANN, Gran Canaria, Spain, 12–14 June 2019; pp. 235–245.

 *applied sciences*

Article

A Correction Method for Heat Wave Distortion in Digital Image Correlation Measurements Based on Background-Oriented Schlieren

Chang Ma, Zhoumo Zeng *, Hui Zhang and Xiaobo Rui

State Key Laboratory of Precision Measurement Technology and Instrument, Tianjin University, Tianjin 300072, China
* Correspondence: zhmzeng@tju.edu.cn; Tel.: +86-22-2740-2366

Received: 19 August 2019; Accepted: 11 September 2019; Published: 13 September 2019

Abstract: Digital image correlation (DIC) is a kind of displacement and strain measurement technique. It can realize non-contact and full-field measurement and is widely used in the testing and research of mechanical properties of materials at high temperatures. However, many factors affect measurement accuracy. As the high temperature environment is complex, the impact of heat waves on DIC is the most significant factor. In order to correct the disturbance in DIC measurement caused by heat waves, this paper proposes a method based on the background-oriented schlieren (BOS) technique. The spot pattern on the surface of a specimen in digital image correlation can be used as the background in the background-oriented schlieren technique. The BOS technique can measure the distortion information of the images caused by heat flow field. The specimen images taken through the heat waves can be corrected using the distortion information. Besides, the characteristics of distortions due to heat waves are also studied in this paper. The experiment results verify that the proposed method can effectively eliminate heat wave disturbances in DIC measurements.

Keywords: digital image correlation; high-temperature measurement; heat waves; thermal disturbance; background-oriented schlieren

1. Introduction

The mechanical properties of materials are significant in the utilization of materials. Aerospace, engine, petrochemical, and other areas are developing rapidly. In these fields, materials must work stably at high temperatures [1–3]. Therefore, it is vital to study the mechanical properties of materials at high temperatures. At present, a variety of measurement methods for material deformation in a high-temperature environment have been developed, among which the digital image correlation (DIC) method [4–6] has attracted more and more research interest. The digital image correlation method is a non-contact optical measurement technique which measures the displacement and strain caused by deformation of the material. Due to its advantages of simple operation, wide application range, and full-field measurement [7–10], digital image correlation has become one of the most active optical measurement methods and is widely used in scientific research and engineering practice.

However, the digital image correlation measurement is based on the images of the specimen. The quality of the images dramatically affects the measurement accuracy of the DIC measurement. In high-temperature environments, heat waves caused by heat sources can distort the images. How to eliminate the influence of thermal disturbance on measurement accuracy is a problem that needs to be solved when using the DIC method to measure objects at high temperatures. Many experts and scholars have studied this problem and made many efforts to eliminate the influence of heat waves on DIC measurement results by improving experimental devices and algorithms.

One type of effort mainly focuses on hardware improvement. Novak et al. [11] added an "air knife" in their experimental devices. The "air knife" is positioned to blow across the sample surface, and its role is to minimize the thermal turbulence and thoroughly mix air in the lens of sight of the imaging systems, thereby reducing apparent distortions caused by heat waves. Jenner et al. [12] designed a unique system for high-temperature measurement, including a camera, loading mechanism, heating furnace, etc. to measure the strain of high-strength steel at high temperatures. L. Chen et al. [13] used an air controller to mix the air between the furnace window and lens. Bao et al. [14] used a color speckle and color camera to separate the displacement due to heat flow disturbance, which improved the measurement accuracy of DIC. This method can realize real-time correction but requires making a specific spot pattern on the specimen surface. Pan et al. [15] conducted experiments in a vacuum wind tunnel and adopted violet illumination and filters to eliminate thermal radiation and thermal disturbances. No significant disturbance of the heat waves was observed in the vacuum. The influence of heat waves can be reduced by improving the experimental system, but the corresponding system will become more complex, and the costs will increase. Besides, it is difficult to eliminate the influence of heat waves by only improving the system hardware [16]. Other solutions have to be found to solve the problem.

Another type of effort aiming at the improvement of algorithms has been carried out. There have been different ideas on algorithms to remove the effect of heat waves. The first idea is to simulate the heat wave environment through numerical simulation. Zhang et al. [17] used numerical simulation models to correct the effects of heat waves. They proposed using numerical simulation to obtain a heat flow field model, combining the ray-tracing principle to analyze the image distortion caused by the heat flow field and correcting the high-temperature deformation measurement results. The results proved that this method is feasible if the experimental parameters are accurately controlled so that the simulation model is consistent with the actual situation. However, actual environments iare complex and variable, and it is difficult to predict the distortion caused by heat waves accurately. Another algorithm idea is to regard the effect of heat waves on the image as a kind of noise and process the image using a denoise algorithm to achieve better DIC measurement results. Song et al. [16] proposed a high-temperature strain measurement method by combining the digital image correlation method and the Improved Random Sample Consensus (IRANSAC) smoothing algorithm. The IRANSAC algorithm reduces the noise from the airflow disturbance. This method smooths the noisy displacement field and reduces the noise level. Due to the heat waves on images being close to Gaussian noise, Y. Hu et al. [18] used a classical algorithm, the inverse filtering method, to remove the Gaussian noise and to process the images disturbed by the heat waves. The results proved that the method is useful to some degree. In some cases, the denoise algorithm can remove the noise caused by heat waves. However, with DIC as a measurement method, using the denoise algorithm may cause new errors, which affect the measurement results of DIC. A third idea is to research the characteristics of the disturbance caused by heat waves on imaging and according to the characteristics, correct the images. Hao et al. [19] extended the principal component analysis (PCA) method to extract the disturbance characteristics and then corrected the calculation results of DIC. Since the distortion caused by heat waves is random, Su et al. [20] proposed the grayscale-average technique, which corrects the distortion by using multiple measurements to average. The results proved that the signal to noise ratio of the processed images was significantly improved. However, in this method, it was assumed that a point on the image, under the influence of the heat waves, is oscillating around the real value. However, after research, it was found that a point on the image, through the heat waves, oscillates around an offset value. After the average processing, most of the distortion was corrected, but there was still a little distortion. Moreover, this small distortion was not negligible in DIC, either.

In this paper, a method based on the background-oriented schlieren (BOS) technique is proposed to correct the heat wave distortion based on the study of characteristics of distortions due to heat waves. The background-oriented schlieren technique [21] can realize the measurement of refractive index field information that causes the light refraction. By recording the image of the background spot pattern in the presence or absence of heat wave distortion, the distortion displacement field caused by the heat waves can be obtained. According to the obtained distortion displacement field, the DIC measurement results are the corrected. The remainder of this paper is organized as follows: In Section 2, the characteristics of distortions due to heat waves are analyzed, and the flow of the proposed method is introduced in detail. The experiment system is shown in Section 3. Section 4 is the experiment results and discussion. The error level of DIC measurement with or without the influence of the heat waves is analyzed, and experiments show the characteristics of heat waves on imaging distortions. Also, experiments confirmed the effectiveness of the proposed method. The conclusion is in Section 5.

2. Theoretical Background

2.1. The Principle of the Influence of Heat Waves on DIC Measurement

The measurement of the DIC method is based on the images of the measured object taken by the camera. If heat waves exist between the camera and the measured object, the heat waves may distort the images taken by the camera, affecting the measurement accuracy of DIC. The way the heat waves cause image distortion can be divided into two aspects. One aspect is the temperature difference between the area affected by the heat waves and the other areas. This temperature difference causes a heterogeneous refractive index field, which causes the light to refract. The other aspect is the hot air flow caused by heat waves.

First of all, the refraction of light caused by the temperature difference is analyzed, regardless of the fluidity of the hot air.

Figure 1 is a schematic diagram of a simplified test system with a region of heterogeneous refractive index between the object to be measured and the imaging system. The heterogeneous refractive index region is caused by heat waves.

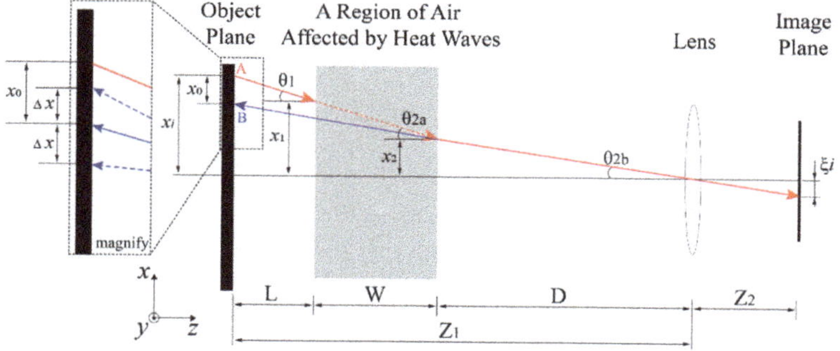

Figure 1. Schematic diagram of simplified experimental setup.

The path of light through a heterogeneous refractive index field is governed by Fermat's principle. According to Fermat's principle, light travels along the shortest path of the optical distance. In general, the path a ray of light follows is governed by the set of differential equations [22,23]:

$$\frac{d^2x}{dz^2} = \left[1 + \left(\frac{dx}{dz}\right)^2 + \left(\frac{dy}{dz}\right)^2\right]\left[\frac{1}{n}\frac{\partial n}{\partial x} - \frac{dx}{dz}\frac{1}{n}\frac{\partial n}{\partial z}\right] \tag{1a}$$

$$\frac{d^2y}{dz^2} = \left[1 + \left(\frac{dx}{dz}\right)^2 + \left(\frac{dy}{dz}\right)^2\right]\left[\frac{1}{n}\frac{\partial n}{\partial y} - \frac{dy}{dz}\frac{1}{n}\frac{\partial n}{\partial z}\right] \tag{1b}$$

where n is the refractive index of air and the z-axis is aligned with the optical axis of the imaging system. In order to simplify the analysis process, only the paraxial ray is discussed, and the angle between the ray and the z-axis is a small angle. Thus, $\frac{dx}{dz} \ll 1$ and $\frac{dy}{dz} \ll 1$. Furthermore, it is assumed that the refractive index changes in the same magnitude in all three spatial directions. With these simplifying assumptions, the light ray enters the heterogeneous refractive index field at the same location it would have passed through if the refractive index field had been homogeneous, that is, $\theta_{2a} = \theta_1$. However, when the light ray leaves the heterogeneous refractive index field, it will propagate at a new angle θ_{2b}.

The refractive index of air can be expressed as $n(x, y, z) = n_0 + n\prime(x, y, z)$ [24], n_0 is the refractive index at homogenous room temperature. $n\prime(x, y, z)$ represents the variation of the refractive index from the base value. Because of $n\prime \ll n_0$, there is $\frac{1}{n} \approx \frac{1}{n_0}$. Under the above assumptions, the propagation equation of light can be simplified as

$$\frac{d^2x}{dz^2} = \frac{1}{n_0} \frac{\partial n\prime}{\partial x} \tag{2a}$$

$$\frac{d^2y}{dz^2} = \frac{1}{n_0} \frac{\partial n\prime}{\partial y} \tag{2b}$$

The analysis is now focused on a light ray in the x–z plane as shown in Figure 1. Because of $\frac{dx}{dz} = \tan(\theta)$, according to the Equation (2a), there is

$$\tan(\theta_{2b}) = \tan(\theta_1) - \frac{1}{n_0} \int_{-\frac{W}{2}}^{\frac{W}{2}} \left(\frac{\partial n\prime}{\partial x}\right) dz \tag{3}$$

Due to the existence of the heat waves, the light is refracted. In Figure 1, the light emitted from the point A on the object plane appears at the point B in the perspective of the image plane. According to the geometric relationship, the distortion x_0 can be expressed as

$$x_0 = [\tan(\theta_1)(L + W)] - [\tan(\theta_{2b})(L + W)] = (L + W)\frac{1}{n_0} \int_{-\frac{W}{2}}^{\frac{W}{2}} \left(\frac{\partial n\prime}{\partial x}\right) dz \tag{4}$$

According to Equation (4), if such an assumption is made, the air region affected by heat waves is stable and does not flow, and only the temperature is different from the other region, then $n\prime$ and W are constant. The distortion x_0 is also a constant.

However, hot air flows and continuously changes at random. If the heat waves caused by the heat source is stable and a point on the object plane is observed on the image plane after passing through the region affected by the heat waves, there will occur a main distortion x_0. However, due to the fluidity of the heat waves, as shown in Figure 1, this point will randomly oscillate around x_0, and based on the main distortion x_0, a random distortion Δx occurs. This phenomenon will be verified in the experiment. The correction method proposed in this paper is to use the BOS technique to correct the main distortion and use the time-averaging method to correct the random distortion. In this way, the DIC measurement results with the influence of heat waves removed can be obtained.

2.2. Principle of Background-Oriented Schlieren

The background-oriented schlieren (BOS) technique was proposed by Meier [25]. The BOS technique combines particle image velocimetry (PIV) technology for flow field velocity measurements with traditional schlieren technology. It can measure the refractive index field using the offset of particles in the background pattern [26]. Since its introduction, the BOS technique has attracted the attention of many scholars and is still continuously developed [21]. At present, the BOS technique is mainly used in the field of density measurements of a fluid field, fluid field visualization, temperature measurements, aero-optical wavefront measurements, and optical transfer function measurements [25,27,28]. The spot pattern in the digital image correlation method can be used as the background pattern in the BOS technique. The BOS technique can be divided into two steps. The first step is to extract the distortion displacement field information by using the high precision PIV algorithm. The second step is to use the distortion displacement field information to construct the remapping function to complete the correction of the image. The technical principle of the BOS technique is shown in Figure 2. First, the image of the background pattern without flow field interference is taken as the reference image. Then, in the presence of flow field interference, the background pattern is imaged again as the measurement image. Next, the displacement of the corresponding particles in the two images is extracted by the PIV algorithm to obtain the refraction information of the light. Then the remapping function is constructed by using the obtained disturbance information to complete the correction of the image with distortion. The extraction of the displacement of the particles on the background pattern is the key to the background-oriented schlieren technique.

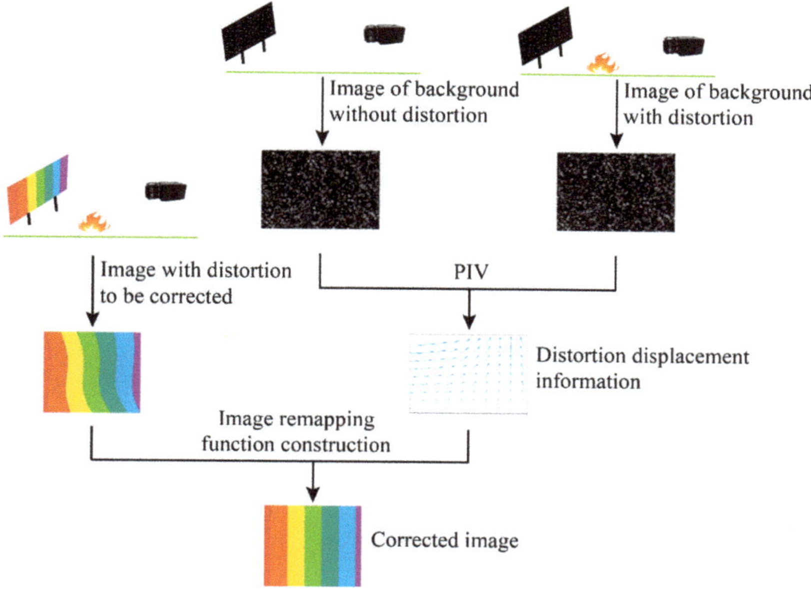

Figure 2. Schematic diagram of optical distortion correlation method based on the BOS technique.

2.3. Correlation Algorithm Flow

When using DIC to measure the displacement and strain of the object, it is necessary to spray spots on the surface of the object. The sprayed spots can be used as the background pattern in the BOS technique. The two methods can be combined. The flow chart of the proposed correction method is shown in Figure 3.

At first, after spraying the spots, when there is no heat source, the object to be measured is photographed to get the reference image. Next, several images of the spot pattern through the heat waves are taken as the background images with disturbance information. The PIV calculation between the background images and the reference image is performed to get disturbance displacement maps. These disturbance displacement maps are averaged over time to get the displacement distribution diagram of the main distortion.

Second, the measured object is loaded and then images of the object through heat waves are taken. These images using the displacement distribution diagram of the main distortion are remapped to eliminate the main distortion in the images.

Third, the images whose main distortion have been removed are averaged to eliminate random distortion. After that, the image with the heat wave disturbance removed is obtained. Then, performing the DIC measurement using this image will give the displacement and strain fields of the specimen, which have corrected the heat wave disturbance.

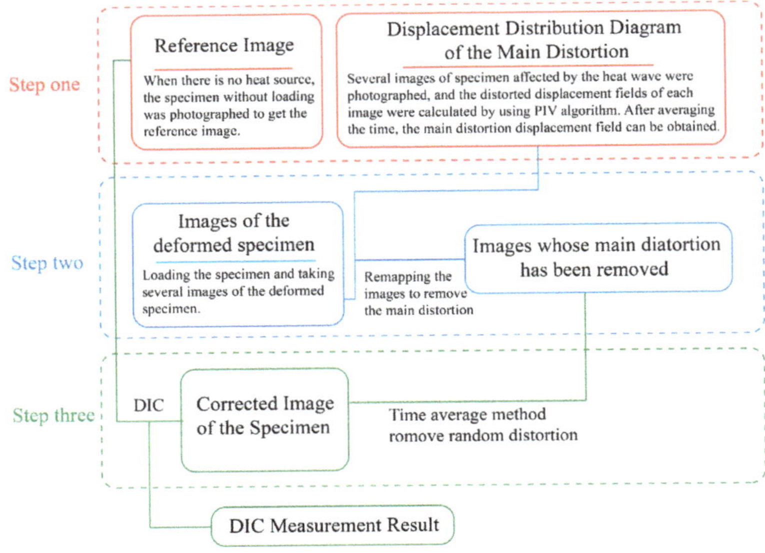

Figure 3. Flow chart of the correction method.

3. Experimental System

In-plane displacement measurement experiments of a disc were carried out to verify that the proposed method can improve the accuracy of DIC measurements.

The schematic diagram of the experimental system is shown in Figure 4. The system consists of a camera, a lens, a hot plate, and a metal disc with black spots on the surface.

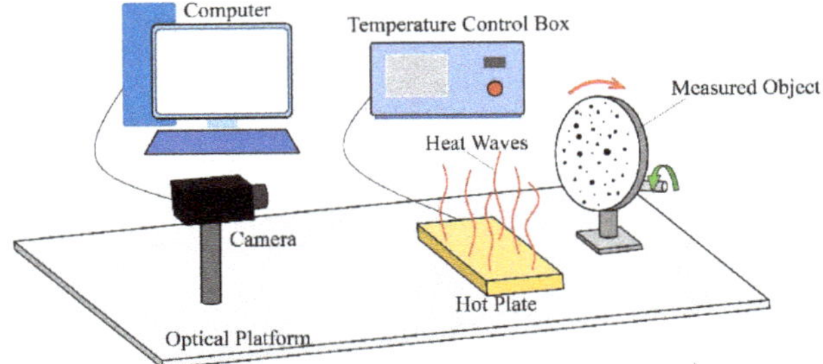

Figure 4. Schematic diagram of the experimental system.

There is one camera to do the 2D DIC measurements. The camera is perpendicular to the surface of the measured object. The camera model is POINT GREY 60S6M (FLIR System, Inc., Portland, OR, USA). The resolution of the camera is 13 fps at 2736 × 2192 pixels. The lens used is a Schneider lens with a focal length of 29.3 mm and a F-number of 2. The measured object is a metal disc, as shown in Figure 5, and the surface is sprayed with the black spots that can be used for DIC measurement. The spots pattern can also be used for the BOS technique. There is a micrometer screw bar on the backside of the disc. The screw bar can rotate the disc to load the in-plane displacement to the disc. The resolution of the rotation is 0.05 degrees. The purpose of the experiment was to measure the in-plane displacement loaded to the disc accurately. During the experiment, the fluorescent light on the ceiling was the only source of illumination, and the laboratory was closed with no additional airflow interference. A hot plate was placed between the measured object and the camera, closer to the measured object. The top of the hot plate (heated portion) had dimensions of 200×100 mm in plane, and the height of the hot plate was 25 mm. The temperature of the hot plate can be accurately controlled by the control box up to 350 °C. Heat waves are formed in the air region above the hot plate, and the specimen is imaged through the heat waves. The acquired images were processed using MATLAB (The MathWorks, Inc., Natick, MA, USA). DIC measurement was performed with the open-source 2D DIC software Ncorr [29]. The experimental device diagram is shown in Figure 6.

Figure 5. Measured object.

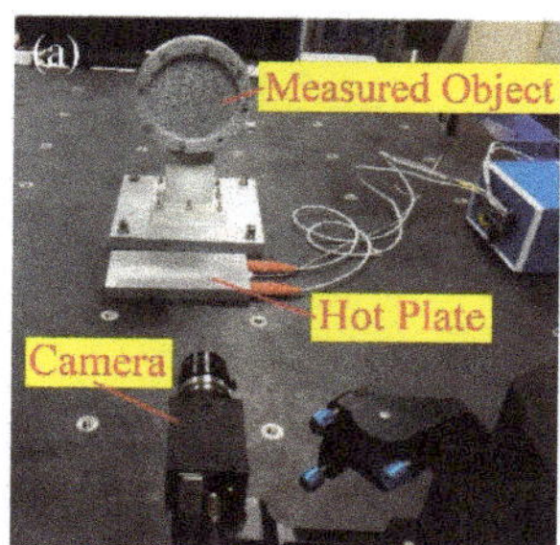

Figure 6. Experimental device diagram: (**a**) The camera is facing vertically to the measured object; (**b**) Temperature control box; (**c**) Hot plate in front of the measured object.

4. Experiments and Results

4.1. Baseline Noise of the Experimental Setup

The baseline noise of the experimental setup in the test conditions of the laboratory without the introduction of a heat source was measured first. The laboratory was in a closed environment with no airflow disturbance. The room temperature was approximately 25 °C, and the hot plate was not activated. When the laboratory environment was stable, the images of the specimen were recorded at 1 Hz over 5 min, and a total of 300 images were obtained. The middle 100 images were selected. The first image of the 100 images was taken as the reference image, and the remaining 99 images were correlated to the reference image to complete the DIC measurement. The displacement and strain fields of the 99 images were obtained. The Region of Interest (ROI) in DIC measurement was set to 452×452 pixels, as shown in Figure 7. The pixel equivalent was 0.109 mm/pixel. The subset radius was 14 pixels and the subset spacing was 1 pixel. An example of the calculation result is shown in Figure 8.

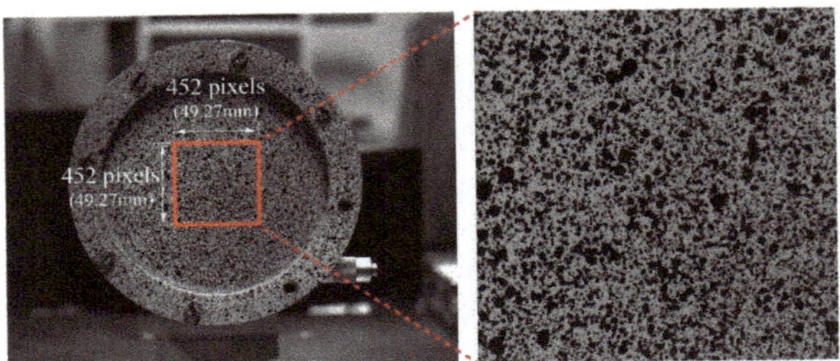

Figure 7. Region of Interest (ROI) in DIC measurement.

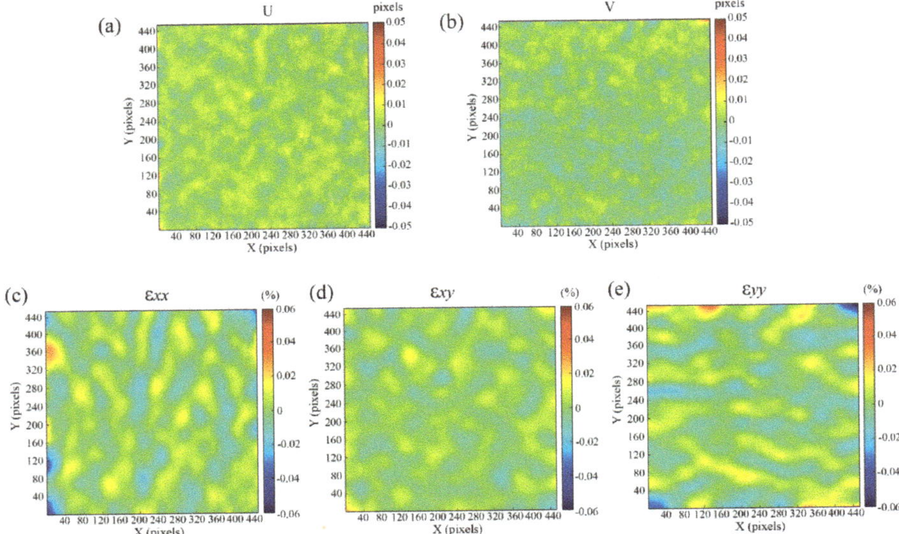

Figure 8. Representative contour plots of the displacements and strains computed in the DIC measurement without the introduction of a heat source: (**a**) Displacement in U direction; (**b**) displacement in V direction; (**c**) strain ε_{xx}; (**d**) strain ε_{xy}; (**e**) strain ε_{yy}.

In order to quantify the errors of the displacement field and strain field, two standard deviations, the spatial standard deviation and the temporal standard deviation, were calculated according to Reference [30]. The spatial standard deviation is calculated by calculating the standard deviation of each displacement field and strain field data to quantify the spatial variation of the displacement and strain. Then, the spatial standard deviation is averaged over 99 images. The temporal standard deviation is calculated by calculating the standard deviation of the 99 displacement fields or the strain field corresponding subsets to quantify the temporal variation. Then, in the ROI region, the standard deviations of all the subsets are spatially averaged. Table 1 shows the calculated spatial standard deviation and temporal standard deviation. According to References [31–34], for a DIC system with extremely well-controlled experimental noise sources, the error does not exceed 0.001 pixels, and the DIC system in a general laboratory environment does not exceed 0.01 pixels. From Table 1, it can be seen that the displacement standard deviations, both spatial and temporal, did not exceed 0.005 pixels, which is generally accepted by the DIC community as being a reasonable noise floor for typical experiments.

Table 1. Baseline noise floor for the experimental setup, without the purposeful introduction of a heat source, quantified by spatial and temporal standard deviations (STD) of the data.

Component	Spatial STD	Temporal STD
U *(pixels)*	0.0037	0.0023
V *(pixels)*	0.0040	0.0024
ε_{xx} *(%)*	0.0077	0.0040
ε_{xy} *(%)*	0.0047	0.0027
ε_{yy} *(%)*	0.0076	0.0039

4.2. Characteristics of Distortions due to Heat Waves

Next, an experiment was conducted to investigate how heat waves affect DIC measurements. First, an image of the specimen was taken without the influence of a heat source as a reference image. The hot plate was turned on and the temperature set to 300 °C by the controller. After the temperature of the hot plate was stabilized at 300 °C, 200 images of the specimen were recorded at 1 Hz. The 200 images taken through heat waves and the reference image were used for DIC measurement. The result of the DIC measurement of the image taken at 100 s is shown in Figure 9.

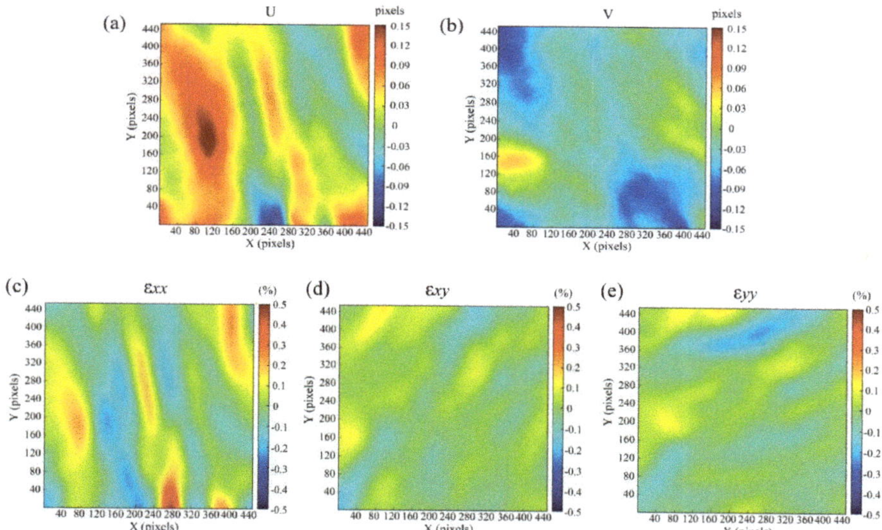

Figure 9. Representative contour plots of the displacements and strains computed in DIC measurement affected by heat waves: (**a**) Displacement in U direction; (**b**) displacement in V direction; (**c**) strain ε_{xx}; (**d**) strain ε_{xy}; (**e**) strain ε_{yy}.

Comparing Figures 8 and 9, the effect of heat waves on the DIC measurement results is evident. The spatial and temporal standard deviations of the displacement fields and strain fields under the influence of heat waves are shown in Table 2. It can be seen from Table 2 that the spatial standard deviation and temporal standard deviation are significantly increased due to the effect of heat waves. The spatial standard deviation was increased by nearly ten times, and the standard deviation of the displacement fields was more than 0.05 pixels, which is unacceptable.

Table 2. Mean error of displacements and strains caused by imaging through heat waves, quantified by spatial and temporal standard deviations (STD) of the data.

Component	Spatial STD	Temporal STD
U (pixels)	0.0516	0.0108
V (pixels)	0.0365	0.0069
ε_{xx} (%)	0.1100	0.0336
ε_{xy} (%)	0.0490	0.0139
ε_{yy} (%)	0.0590	0.0149

Figure 10 shows the variation of the displacement of the ROI central subset over time. It can be seen that the swing of the curves of the displacement in both directions of U and V are not surrounding zero, but there is an offset, the Mu and Mv, as shown in Figure 10. The offsets are the main distortion mentioned above, and the swing around the main distortion is random distortion.

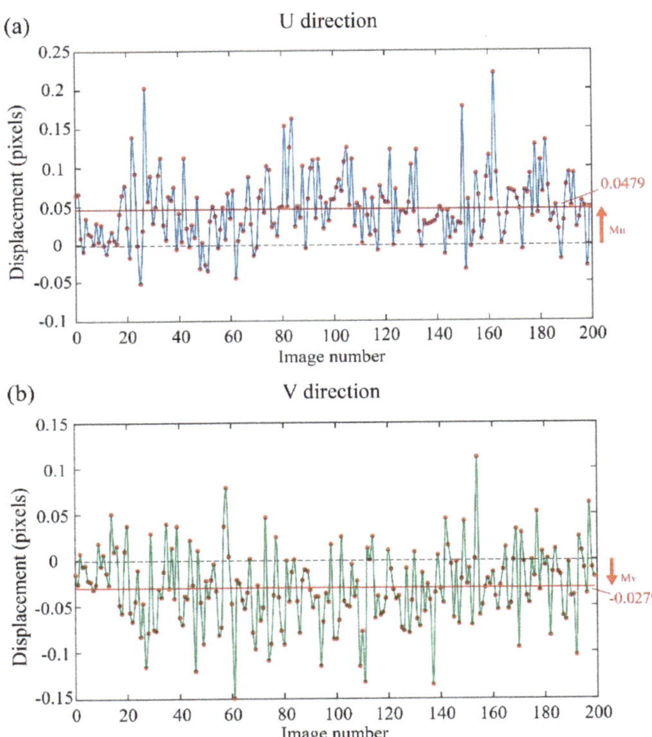

Figure 10. Displacement of the center subset as a function of time: (**a**) U direction; (**b**) V direction.

Next, the situation in which the temperature of the heat source changes drastically was analyzed. The heat source temperature was adjusted to 100, 150, 200, 250, 300 degrees Celsius, respectively. At each temperature, 200 unloaded sample images were taken; the calculated main distortion is shown in Figure 11. As the temperature increases, the main distortion tends to increase, but at different temperatures, this offset is different. Therefore, in the case where the temperature of the heat source changes drastically, the method of correcting the main distortion based on the BOS technique is no longer applicable. But the time-average method can also be used to improve the accuracy of the measurement.

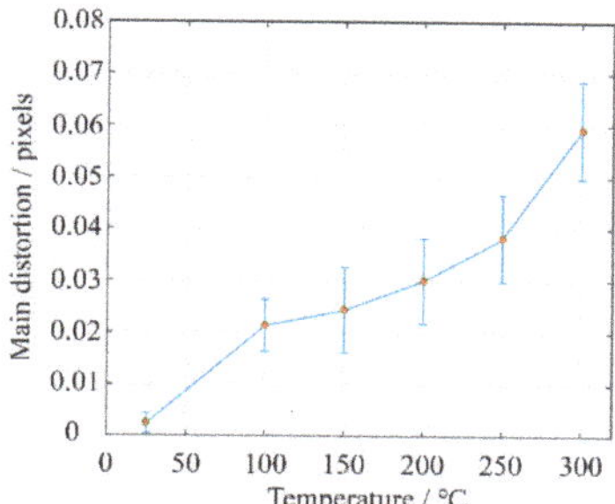

Figure 11. The average main distortion at each temperature.

4.3. Influence of Heat Waves on DIC Measurement Results

Then, the screw bar behind the specimen was rotated and the in-plane displacement to the disc was loaded. DIC was used to measure the in-plane displacement of the disc. The measurement results with and without the heat waves were compared. First, in the absence of the heat source, the micrometer screw bar behind the specimen was rotated to make a slight rotation of the disc. The angle of the rotation was 0.2 degrees. The image before the rotation is the reference image, and the rotated specimen image is correlated with the reference image. The measured displacement field is shown in Figure 12.

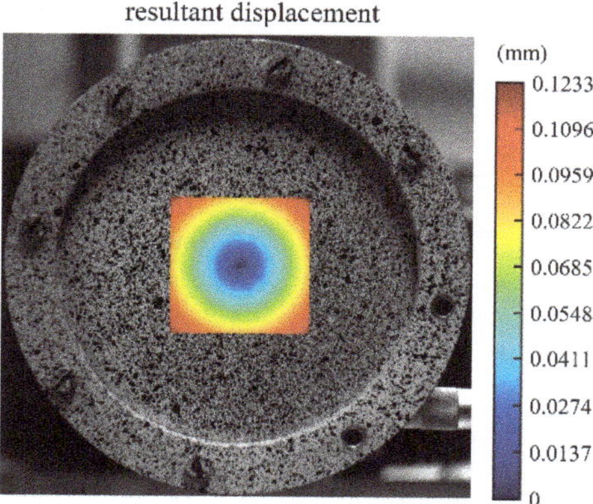

Figure 12. Displacement field measured without the influence of heat waves.

The components of the displacement field in the U and V directions are shown in Figure 13. The displacement fields are expressed in pixel values to facilitate the precision analysis.

Next, the image taken without the heat source and rotation was used as the reference image. The specimen was kept still, the heat source turned on, and the temperature of the hot plate adjusted to 300 °C. After the environment was stable, the micrometer screw bar was rotated to rotate the disc, and the angle of the rotation was still 0.2 degrees. Then, 100 images at a frequency of 1 Hz were taken, the result of the DIC measurement of the images taken at the 50th second is shown in Figure 14. The components of the displacement field in the U and V directions are shown in Figure 15. It can be seen that under the influence of heat waves, the measurement results of the DIC were severely distorted.

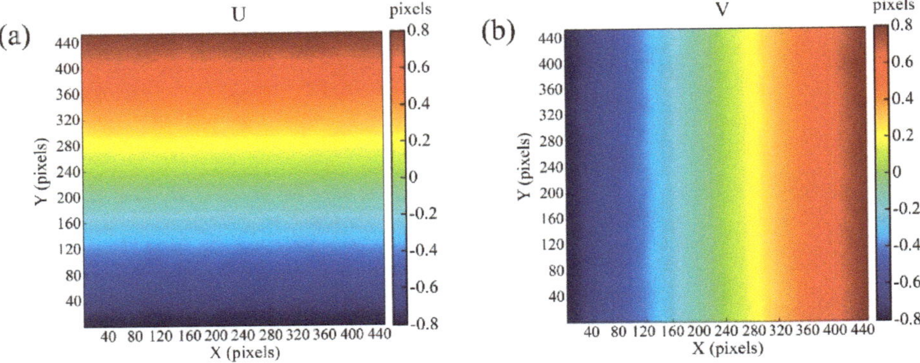

Figure 13. Displacement fields obtained from the DIC measurement without the influence of heat waves: (**a**) Component in U direction; (**b**) component in V direction.

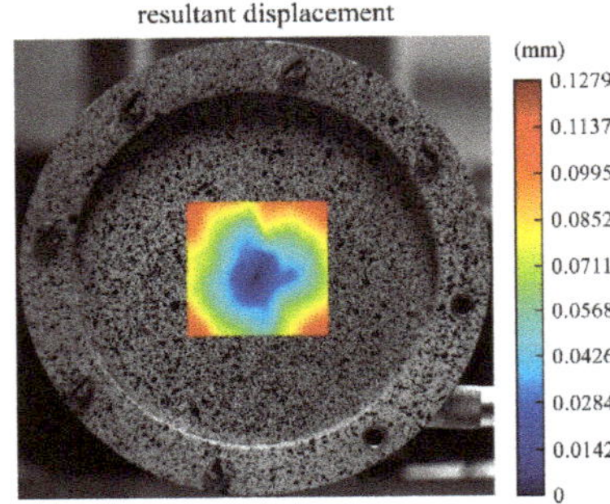

Figure 14. Displacement field measured with the influence of heat waves.

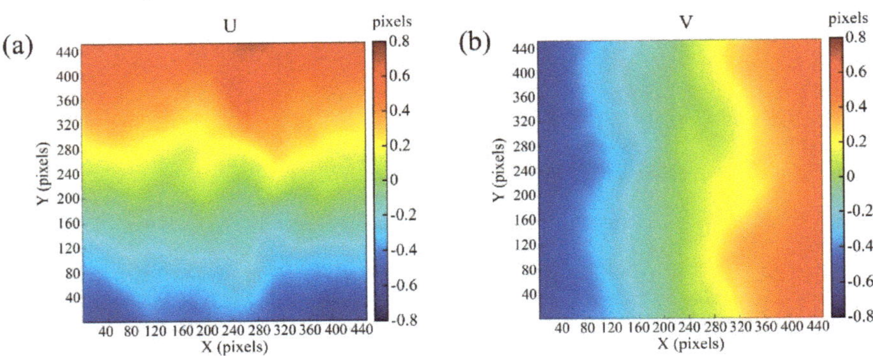

Figure 15. Displacement fields obtained from DIC measurement with the influence of heat waves: (**a**) Component in U direction; (**b**) component in V direction.

The 100 displacement fields were time-averaged to remove random disturbances, and the result is shown in Figure 16. It can be seen that the result improved, but it is still not ideal due to the main disturbance not having been removed.

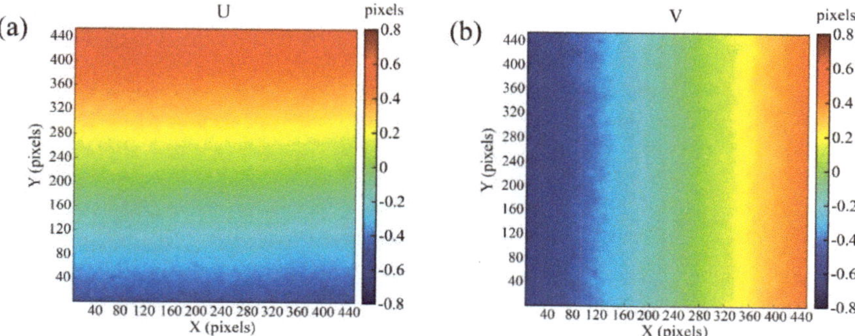

Figure 16. Displacement field after time average: (**a**) Component in U direction; (**b**) Component in V direction.

4.4. Verification of the Correction Algorithm

In order to verify the proposed method, the following experiment was performed. First, in the case where the hot plate was not turned on, and the specimen was not loaded in displacement, an image was taken as the reference image of the DIC and BOS technique. Next, the hot plate was turned on, the temperature of the hot plate adjusted to 300 °C, and 200 images taken after the environment was stable. Using the particle image velocimetry (PIV) algorithm in the BOS technique, the distortion vector field of each image was obtained by calculating the amount of distortion of every particle in the 200 images compared with the corresponding particle in the reference image. By time-averaging 200 distortion vector fields, the vector field of the main distortion was obtained, as shown in Figure 17.

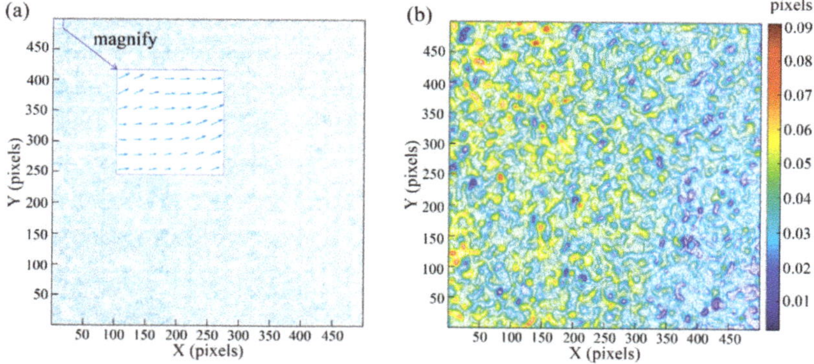

Figure 17. Main distortion measured by the BOS technique: (**a**) Displacement vector distribution (the box represents local amplification results); (**b**) resultant displacement vector distribution.

The micrometer screw was gently turned to rotate the disc by 0.2 degrees. The in-plane displacement to the disc was loaded, and then 100 images of the specimen were taken through the heat waves. The main distortion displacement field was utilized to remap the 100 images to remove the main distortion on them. Then, using the time-average method, 100 images with the main distortion removed were averaged to remove the random distortion. The image with the heat wave disturbance removed was finally obtained, as shown in Figure 18. The corrected image was correlated to the reference image to complete the DIC measurement. The measurement result is shown in Figure 19, and the components of the displacement field in the U and V directions are shown in Figure 20. It can be seen that the influence of the heat waves on the DIC measurement result was corrected, as compared with Figures 14 and 15.

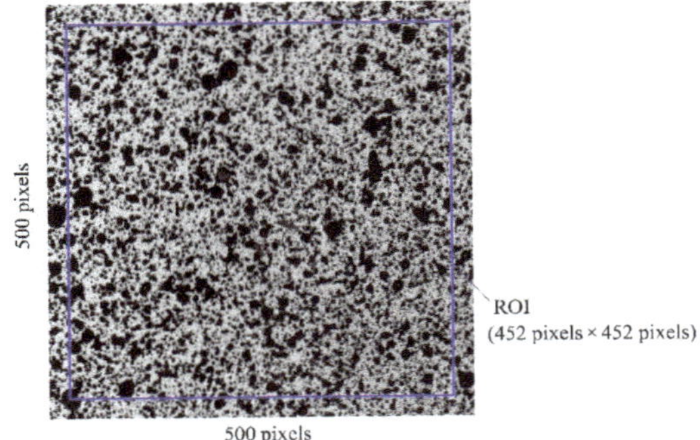

Figure 18. Corrected image obtained by the proposed method (The blue box is the region where the DIC calculation is performed).

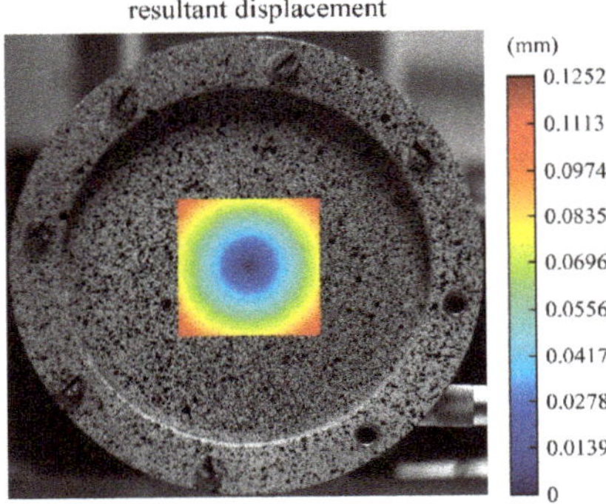

Figure 19. Displacement field after correction.

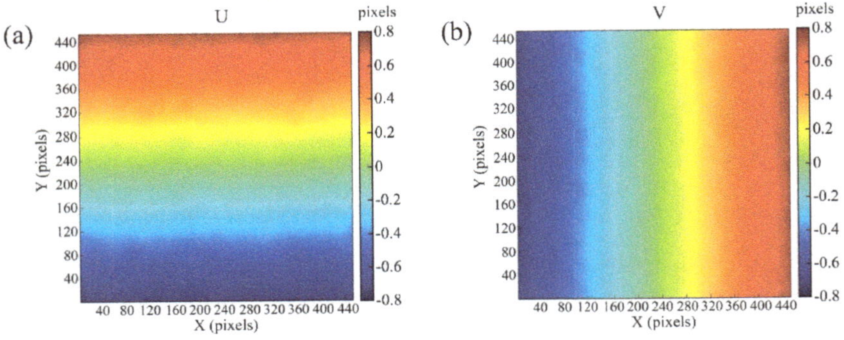

Figure 20. Displacement fields obtained from DIC measurement using the corrected image: (**a**) Component in U direction; (**b**) component in V direction.

In order to further prove that the proposed method can improve the measurement accuracy of DIC, a plot of displacement in U direction vs. Y ($X = 226$) and a plot of displacement in V direction vs. X ($Y = 226$) are shown in Figure 21.

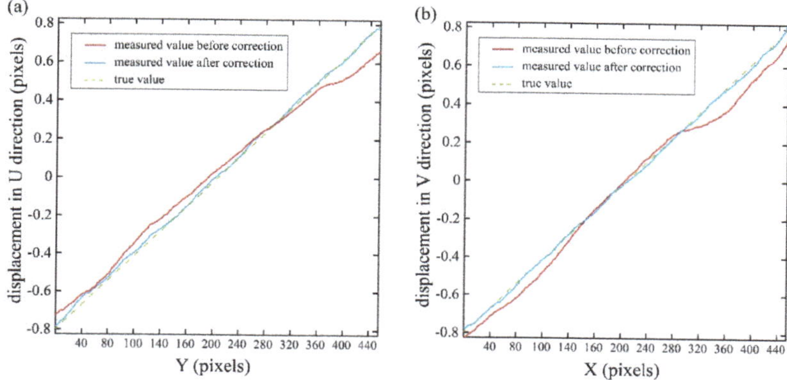

Figure 21. Plots of displacement in U vs. Y and in V vs. X: (**a**) Displacement component in U direction vs. Y (X = 226); (**b**) displacement component in V direction vs. X (Y = 226).

The root means square error (RMSE) is used to evaluate the displacement measurement results before and after the correction. The RMSE reflects the degree to which the measured value deviates from the real value, and can reflect the accuracy of the measurement. The smaller the root mean square error, the higher the measurement accuracy. The calculation formula of the RMSE is shown in Equation (5) [35].

$$RMSE = \sqrt{\frac{\sum_{i=1}^{n}\left(X_i - X_{ture,i}\right)^2}{n}} \tag{5}$$

where X_i is the measured value, the $X_{ture,i}$ is the real value and the n is the number of measurements. The *RMSE* of the measurement of the displacement component in the U direction before correction was 0.0732 pixels, while it was 0.0126 pixels after correction. The *RMSE* of the measurement of the displacement component in V direction before correction was 0.0711 pixels, while it was 0.0102 pixels after correction.

5. Conclusions

This paper proposes a correction method based on the background-oriented schlieren technique. The method can correct the distortion caused by heat waves to digital image correlation measurements. Through theoretical analysis and experiments, the characteristics of the distortion due to heat waves on the images were researched. The effectiveness of the proposed method was verified by experiments. Through the research of this paper, the following conclusions are drawn:

1. The distortion on the images caused by heat waves can be divided into the main distortion and a random distortion. In the experiments performed in this paper, the main distortion reached 0.05 pixels, and the most significant swing amplitude of the random distortion reached 0.2 pixels. The effect of this distortion on the measurement results of digital image correlation is not negligible.
2. Spot patterns used in digital image correlation measurements can also be used in the background-oriented schlieren technique. The background schlieren method can be used to obtain the vector displacement fields of the main distortion caused by heat waves.
3. The main distortion vector obtained by the background-oriented schlieren technique remap the deformed images to eliminate the main distortion. Then, the time-average method should be used to eliminate the random distortion. The experimental results showed that the proposed correction method can effectively remove the disturbance of heat waves and obtain high precision DIC measurement results.

Although the proposed method can effectively improve the measurement accuracy of DIC, it also has a certain limitation. This method is especially suitable for cases where the temperature of the heat source is stable. When the temperature changes drastically, the correction is less than ideal.

Author Contributions: Conceptualization, C.M.; Data curation, Z.Z.; Formal analysis, X.R.; Funding acquisition, H.Z.; Investigation, C.M.; Methodology, C.M.; Project administration, Z.Z.; Resources, Z.Z.; Software, X.R.; Supervision, Z.Z.; Validation, H.Z.; Visualization, C.M.; Writing—original draft, C.M.; Writing—review & editing, H.Z. and X.R.; All authors reviewed the manuscript.

Funding: This research was funded by the National Key R&D Program of China (No. 2016YFF0101802) and Natural Science Foundation of Tianjin (No. 17JCYBJC19300).

Conflicts of Interest: The authors declare no conflict of interest.

References

1. Zettergren, M.D.; Semeter, J.L.; Dahlgren, H. Dynamics of density cavities generated by frictional heating: Formation, distortion, and instability. *Geophys. Res. Lett.* **2015**, *42*, 10120–10125. [CrossRef]

2. Santos, A.C. High temperature modal analysis of a non-uniformly heated rectangular plate: Experiments and simulations. *J. Sound. Vib.* **2019**, *443*, 397–410. [CrossRef]

3. Devesse, W.; De Baere, D.; Guillaume, P. High Resolution Temperature Measurement of Liquid Stainless Steel Using Hyperspectral Imaging. *Sensors* **2017**, *1*, 91. [CrossRef] [PubMed]

4. Pan, B. Digital image correlation for surface deformation measurement: History developments, recent advances and future goals. *Meas. Sci. Technol.* **2018**, *29*, 082001. [CrossRef]

5. Yoneyama, S. Basic principle of digital image correlation for in-plane displacement and strain measurement. *Adv. Compos. Mater.* **2016**, *25*, 105–123. [CrossRef]

6. Hu, Y.J.; Jiang, C.; Liu, W.; Yu, Q.Q.; Zhou, Y.L. Degradation of the In-plane Shear Modulus of Structural BFRP Laminates Due to High Temperature. *Sensors* **2018**, *10*, 3361. [CrossRef]

7. Moazzami, M.; Ayatollahi, M.R.; Chamani, H.R. Determination of higher order stress terms in cracked Brazilian disc specimen under mode I loading using digital image correlation technique. *Opt. Lasers. Eng.* **2018**, *107*, 344–352. [CrossRef]

8. Rahmatabadi, D.; Shahmirzaloo, A.; Hashemi, R. Using digital image correlation for characterizing the elastic and plastic parameters of ultrafine-grained Al 1050 strips fabricated via accumulative roll bonding process. *Mater. Res. Express* **2019**, *6*, 086542. [CrossRef]

9. Sutton, M.A.; Hild, F. Recent advances and perspectives in digital image correlation. *Exp. Mech.* **2015**, *55*, 1–8. [CrossRef]

10. Smrkic, M.F.; Koscak, J.; Damjanovic, D. Application of 2D digital image correlation for displacement and crack width measurement on RC elements. *Gradevinar* **2018**, *70*, 771–781.

11. Novak, M.D.; Zok, F.W. High-temperature materials testing with full-field strain measurement: Experimental design and practice. *Rev. Sci. Instrum.* **2011**, *82*, 115101. [CrossRef] [PubMed]

12. Jenner, F.; Walter, F.W. Application of high-speed video extensometry for high-temperature tensile characterization of boron heat-treated steels. *J. Strain. Anal. Eng. Des.* **2014**, *49*, 378–387. [CrossRef]

13. Chen, L.; Wang, Y.H.; Dan, X.Z. Experimental research of digital image correlation system in high temperature test. In Proceedings of the 7th International Symposium on Precision Mechanical Measurements, Xiamen, China, 7–12 August 2015.

14. Bao, S.; Wang, Y.H.; Liu, L. An error elimination method for high-temperature digital image correlation using color speckle and camera. *Opt. Lasers. Eng.* **2019**, *116*, 47–54. [CrossRef]

15. Dong, Y.; Pan, B. In-situ 3D shape and recession measurements of ablative materials in an arc-heated wind tunnel by stereo-digital image correlation. *Opt. Lasers. Eng.* **2019**, *116*, 75–81. [CrossRef]

16. Song, J.; Yang, J.; Liu, F. High temperature strain measurement method by combining digital image correlation of laser speckle and improved RANSAC smoothing algorithm. *Opt. Lasers. Eng.* **2018**, *111*, 8–18. [CrossRef]

17. Zhang, M.; Miao, H.; Xiong, C. On the correction method for distortion caused by heat flow in high temperature deformation measurement. *J. Exp. Mech.* **2017**, *32*, 718–724.

18. Hu, Y.; Bao, S.; Dan, X. Improvement of high-temperature deformation measurement accuracy based on image restoration method. *Meas. Sci. Technol.* **2018**, *29*, 094002. [CrossRef]

19. Hao, W.F.; Zhu, J.G.; Zhu, Q. Displacement field denoising for high-temperature digital image correlation using principal component analysis. *Mech. Adv. Mater. Struc.* **2016**, *24*, 830–839. [CrossRef]

20. Su, Y.Q.; Yao, X.F.; Wang, S. Improvement on measurement accuracy of high-temperature DIC by grayscale-average technique. *Opt. Lasers. Eng.* **2015**, *75*, 10–16. [CrossRef]

21. Raffel, M. Background-oriented schlieren (BOS) techniques. *Exp. Fluids.* **2015**, *56*, 60. [CrossRef]

22. Born, M.; Wolf, E.; Bhatia, A.B. *Principles of Optics: Electromagnetic Theory of Propagation, Interference and Diffraction of Light*, 7rd ed.; Cambridge University Press: Cambridge, UK, 2016; pp. 99–117.

23. Iizuka, K. *Engineering Optics*, 3rd ed.; Springer: Berlin, Germany, 2008; pp. 235–290.

24. Ciddor, P.E. Refractive index of air: New equations for the visible and near infrared. *Appl. Optics.* **1996**, *35*, 1566–1573. [CrossRef] [PubMed]

25. Meier, G.E.A. Computerized background-oriented schlieren. *Exp. Fluids.* **2002**, *33*, 181–187. [CrossRef]

26. Verso, L.; Liberzon, A. Background oriented schlieren in a density stratified fluid. *Rev. Sci. Instrum.* **2015**, *86*, 103705. [CrossRef] [PubMed]

27. Mizukaki, T.; Wakabayashi, K.; Matsumura, T. Background-oriented schlieren with natural background for quantitative visualization of open-air explosions. *Shock Waves* **2014**, *24*, 69–78. [CrossRef]

28. Tanda, G.; Foss, A.M.; Misale, M. Heat transfer measurements in water using a schlieren technique. *Int. J. Heat. Mass. Tran.* **2014**, *71*, 451–458. [CrossRef]

29. Blaber, J.; Adair, B.; Antoniou, A. Ncorr: Open-source 2D digital image correlation software. *Exp. Mech.* **2015**, *55*, 1105–1122. [CrossRef]

30. Jones, E.M.C.; Reu, P.L.; Antoniou, A. Distortion of digital image correlation (DIC) displacements and strains from heat waves. *Exp. Mech.* **2018**, *58*, 1133–1156. [CrossRef]

31. Sutton, M.A.; Orteu, J.J. *Image Correlation for Shape, Motion and Deformation Measurement: Basic Concepts, Theory and Applications*; Springer: New York, NY, USA, 2009.

32. Schreier, H.W.; Braasch, J.R.; Sutton, M.A. Systematic errors in digital image correlation caused by intensity interpolation. *Opt. Eng.* **2000**, *39*, 2915–2921. [CrossRef]

33. Baldi, A.; Bertolino, F. Experimental analysis of the errors due to polynomial interpolation in digital image correlation. *Strain* **2015**, *51*, 248–263. [CrossRef]

34. Reu, P.L. Introduction to digital image correlation: Best practices and applications. *Exp. Tech.* **2012**, *36*, 3–4. [CrossRef]

35. Chai, T.; Draxler, R.R. Root mean square error (RMSE) or mean absolute error (MAE)?—Arguments against avoiding RMSE in the literature. *Geosci. Model. Dev.* **2014**, *7*, 1247–1250. [CrossRef]

Article

An Effective Optimization Method for Machine Learning Based on ADAM

Dokkyun Yi [1], Jaehyun Ahn [2] and Sangmin Ji [2,*]

[1] Division of Creative Integrated General Studies, Daegu University College, Kyungsan 38453, Korea; dkyi@daegu.ac.kr

[2] Department of Mathematics, College of Natural Sciences, Chungnam National University, Daejeon 34134, Korea; jhahn@cnu.ac.kr

* Correspondence: smji@cnu.ac.kr

Received: 11 December 2019; Accepted: 25 January 2020; Published: 5 February 2020

Abstract: A machine is taught by finding the minimum value of the cost function which is induced by learning data. Unfortunately, as the amount of learning increases, the non-liner activation function in the artificial neural network (ANN), the complexity of the artificial intelligence structures, and the cost function's non-convex complexity all increase. We know that a non-convex function has local minimums, and that the first derivative of the cost function is zero at a local minimum. Therefore, the methods based on a gradient descent optimization do not undergo further change when they fall to a local minimum because they are based on the first derivative of the cost function. This paper introduces a novel optimization method to make machine learning more efficient. In other words, we construct an effective optimization method for non-convex cost function. The proposed method solves the problem of falling into a local minimum by adding the cost function in the parameter update rule of the ADAM method. We prove the convergence of the sequences generated from the proposed method and the superiority of the proposed method by numerical comparison with gradient descent (GD, ADAM, and AdaMax).

Keywords: numerical optimization; ADAM; machine learning; stochastic gradient methods

---

1. Introduction

Machine learning is a field of computer science that gives computer systems the ability to learn with data, without being explicitly programmed. For machines to learn data, a machine learns how to minimize it by introducing a cost function. The cost function is mostly made up of the difference between the true value and the value calculated from the Artificial Neural Network (ANN) [1–7]. Therefore, the cost function varies with the amount of training data, non-linear activation function in ANN, and the structure of ANN. These changes generate both a singularity and local minimum within the cost function. We know that all the differentiable functions at this point (singularity or local minimum) have a first derivative value of zero.

The gradient-based optimization (gradient descent optimization) is widely used to find the minimum value of the cost function [8–15]. Gradient descent (GD) is a method that was first introduced and uses a fixed-point method to make the first derivative of the cost function zero. This method works somewhat well; however, it causes many difficulties in complex ANNs. To overcome this difficulty, a method called ADAM (Adaptive moment estimation) [15] was introduced to add the momentum method and to control the distance of each step. In other words, this method uses the sum of the gradient values multiplied by the weights calculated in the past, which is the idea of momentum. This is called the 'first momentum', and the sum of squares of the gradient is calculated in the same way. This is called the 'second momentum', and the ratio of the first and the second momentum values is calculated and the minimum value is searched for according to the ratio. More detailed information can be found in [16–22]. This method is still widely used and works well in most areas. There is a more advanced method called AdaMax (Adaptive moment estimation with Maximum). The AdaMax method uses the maximum value of the calculation method of the second momentum part in ADAM. This provides a more stable method. Various other methods exist. We are particularly interested in GD, ADAM, and AdaMax, because GD was the first to be introduced, ADAM is still widely used, and AdaMax is a modified method of ADAM. It has been empirically observed that these algorithms fail to converge toward an optimal solution [23]. We numerically analyze the most basic GD method, the most widely used ADAM method, and the modified AdaMax method. As a result of these numerical interpretations, we introduce the proposed method. The existing methods based on gradient descent operate by changing the parameter so that the first derivative of the cost function becomes zero, which results in finding the minimum value of the cost function. For this method to be established, it is assumed that the cost function has a convex property [24,25]. However, the first derivative of the cost function is also zero at a local minimum. Therefore, this existing method may converge to the local minimum in the cost function where the local minimum exists. If the value of the cost function at the local minimum is not as small as is desired, the value of the parameter should change. To solve this problem, we solve the optimization problem by using the first derivative of the cost function and the cost function itself. In this way, if the value of the cost function is non-zero, the parameter changes even if the first derivative of the cost function becomes zero. This is the reason for adding the cost function. Here, we also use the idea of the ADAM method by using these data to add to the parameter changes, and also demonstrate the convergence of the created sequence.

In summary, our research question is why neural networks are often poorly trained by known optimization methods. Our research goal is to find a new optimization method which resolve this phenomenon. For this, first we prove the convergence of the new method. Next, we use the simplest cost function to visualize the movements of our method and basic methods near a local minimum. In addition, then, we compare performances of our method and ADAM on practical datasets such as MNIST and CIFAR10.

This paper is organized as follows. In Section 2, the definition of the cost function and the cause of the non-convex cost function are explained. In Section 3, we briefly describe the known Gradient-Descent-based algorithms. In particular, we introduce the first GD (Gradient Descent) method, the most recently used ADAM method, and finally, the improved ADAM method. In Section 4, we explain the proposed method, the convergence of the proposed method, and other conditions. In Section 5, we present several numerical comparisons between the proposed method and the discussed methods. The first case is the numerical comparison of a one variable non-convex function. We then perform numerical comparisons of non-convex functions in a two-dimensional space. The third case is a numerical comparison between four methods in two-dimensional region separation. Finally, we test with MNIST (Modified National Institute of Standards and Technology) and CIFAR10 (The Canadian Institute for Advanced Research)—the most basic examples of image analysis. MNIST classifies 10 types of grayscale images, as seen in Figure 1, and this experiment shows that our method is also efficient at analyzing images (https://en.wikipedia.org/wiki/MNIST_database). CIFAR10 (The Canadian Institute for Advanced Research) is a dataset that classifies 10 types like MNIST (Modified National Institute of Standards and Technology), but CIFAR10 has RGB images. Therefore, CIFAR10 requires more computation than MNIST. Through numerical comparisons, we confirm that the proposed method is more efficient than the existing methods described in this paper. In Section 6, we present the conclusions and future work.

Figure 1. Examples of MNIST dataset.

2. Cost Function

In this section, we explain basic machine learning among the various modified machine learning algorithms. For convenience, machine learning refers to basic machine learning. To understand the working principle of machine learning, we try to understand the principle from the structure with one neuron. Let x be input data and $H(x)$ be output data, which is obtained by

$$H(x) = \sigma(wx + b),$$

where w is weight, b is bias, and σ is a sigmoid function (it is universally called an activation function, and various functions can be used). Therefore, the result of the function H is a value between 0 and 1. Non-linearity is added to the cost function by using the non-linear activation function. For machine learning, let LS be the set of learning data and let $l > 2$ be the number of the size of LS. In other words, when the first departure point of learning data is 1, the learning dataset is $LS = \{(x_1, y_1), (x_2, y_2), ..., (x_l, y_l)\}$, where x_s is a real number and y_s is a value between 0 and 1. From LS, we can define a cost function as follows:

$$C(w, b) = \frac{1}{l} \sum_{s=1}^{l} (y_s - H(x_s))^2.$$

The machine learning is completed by finding w and b, which satisfies the minimum value of the cost function. Unfortunately, there are several local minimum values of the cost function, because the cost function is not convex. Furthermore, the deepening of the structure of an ANN means that the activation function performs the synthesis many times. This results in an increase in the number of local minimums of the cost function. More complete interpretations and analysis are currently being made in more detail and these will be reported soon.

3. Learning Methods

In this section, we provide a brief description of the well-known optimization methods—GD, ADAM, and AdaMax.

3.1. Gradient Descent Method

GD is the most basic method and the first introduced. In GD, a fixed-point iteration method is introduced with the first derivative of the cost function. A parameter is changed in each iteration as follows.

$$w_{i+1} = w_i - \eta \frac{\partial C(w_i)}{\partial w},$$

The pseudocode version of this method is as follows in Algorithm 1.

Algorithm 1: Pseudocode of Gradient Descent Method.

η : Learning rate
$C(w)$: Cost function with parameters w
w_0 : Initial parameter vector
$i \leftarrow 0$ (Initialize time step)
while w not converged **do**
  $i \leftarrow i + 1$
  $w_{i+1} \leftarrow w_i - \eta \dfrac{\partial C}{\partial w}(w_i)$
end while
return w_i (Resulting parameters)

As in the above formula, if the gradient is zero, the parameter does not change and does not continue to the local minimum.

3.2. ADAM Method

The ADAM method is the most widely used method based on the GD method and the momentum method, and additionally, a variation of the interval. The first momentum is obtained by

$$m_i = \beta_1 m_{i-1} + (1 - \beta_1)\frac{\partial C}{\partial w}.$$

The second momentum is obtained by

$$v_i = \beta_2 v_{i-1} + (1 - \beta_2)\left(\frac{\partial C}{\partial w}\right)^2.$$

$$w_{i+1} = w_i - \eta \frac{\hat{m}_i}{\sqrt{\hat{v}_i} + \epsilon},$$

where $\hat{m}_i = m_i/(1 - \beta_1)$ and $\hat{v}_i = v_i/(1 - \beta_2)$. The pseudocode version of this method is as follows in Algorithm 2.

Algorithm 2: Pseudocode of ADAM Method.

η : Learning rate
$\beta_1, \beta_2 \in [0, 1)$: Exponential decay rates for the moment estimates
$C(w)$: Cost function with parameters w
w_0 : Initial parameter vector
$m_0 \leftarrow 0$
$v_0 \leftarrow 0$
$i \leftarrow 0$ (Initialize timestep)
while w not converged **do**
  $i \leftarrow i + 1$
  $m_i \leftarrow \beta_1 \cdot m_{i-1} + (1 - \beta_1) \cdot \dfrac{\partial C}{\partial w}(w_i)$
  $v_i \leftarrow \beta_2 \cdot v_{i-1} + (1 - \beta_2) \cdot \dfrac{\partial C}{\partial w}(w_i)^2$
  $\hat{m}_i \leftarrow m_i/(1 - \beta_1^i)$
  $\hat{v}_i \leftarrow v_i/(1 - \beta_2^i)$
  $w_{i+1} \leftarrow w_i - \eta \cdot \hat{m}_i/(\sqrt{\hat{v}_i} + \epsilon)$
end while
return w_i (Resulting parameters)

The ADAM method is a first-order method. Thus, it has low time complexity. As the parameter changes repeat, the learning rate becomes smaller due to the influence of $\hat{m}_i / \sqrt{\hat{v}_i + \epsilon}$ and varies slowly around the global minimum.

3.3. AdaMax Method

The AdaMax method is based on the ADAM method and uses the maximum value of the calculation of the second momentum.

$$u_i = max \left(\beta_2 \cdot u_{i-1}, \left| \frac{\partial C}{\partial w}(w_i) \right| \right).$$

$$w_{i+1} = w_i - \eta \frac{\hat{m}_i}{u_i},$$

The pseudocode version of this method is as follows in Algorithm 3.

Algorithm 3: Pseudocode of AdaMax.

η : Learning rate
$\beta_1, \beta_2 \in [0, 1)$: Exponential decay rates for the moment estimates
$C(w)$: Cost function with parameters w
w_0 : Initial parameter vector
$m_0 \leftarrow 0$
$u_0 \leftarrow 0$
$i \leftarrow 0$ (Initialize time step)
while w not converged **do**
  $i \leftarrow i+1$
  $m_i \leftarrow \beta_1 \cdot m_{i-1} + (1 - \beta_1) \cdot \frac{\partial C}{\partial w}(w_i)$
  $u_i \leftarrow max \left(\beta_2 \cdot u_{i-1}, \left| \frac{\partial C}{\partial w}(w_i) \right| \right)$
  $w_{i+1} \leftarrow w_i - (\eta / (1 - \beta_1^i)) \cdot m_i / u_i$
end while
return w_i (Resulting parameters)

---

4. The Proposed Method

The main idea starts with a fixed-point iteration method, the condition of the cost function $(0 \leq C(w, b))$, and the condition of the first derivative of the cost function. We define an auxiliary function H such as

$$H(w) = \lambda C(w) + \frac{\partial C(w)}{\partial w},$$

where λ is determined to be positive or negative according to the initial sign of $\partial C(w) / \partial w$. We make w change if the value of $C(w)$ is large even if it falls to a local minimum using $H(w)$.

Optimization

The iteration method is

$$w_{i+1} = w_i - \eta \frac{\hat{m}_i}{\sqrt{\hat{v}_i + \epsilon}}, \tag{1}$$

where

$$m_i = \beta_1 m_{i-1} + (1 - \beta_1) H(w_i),$$
$$v_i = \beta_2 v_{i-1} + (1 - \beta_2) (H(w_i))^2, \tag{2}$$

$\hat{m}_i = m_i / (1 - \beta_1)$, and $\hat{v}_i = v_i / (1 - \beta_2)$.

Theorem 1. *The iteration method in* (1) *is a way to satisfy convergence.*

Proof. Equation (2) can be altered by

$$
\begin{aligned}
m_i &= \beta_1^i m_0 + (1 - \beta_1) \sum_{k=1}^{i} \beta_1^{k-1} H(w_{i-k+1}) \\
&= (1 - \beta_1) \sum_{k=1}^{i} \beta_1^{k-1} H(w_{i-k+1})
\end{aligned}
$$

and

$$
\begin{aligned}
v_i &= \beta_2^i v_0 + (1 - \beta_2) \sum_{k=1}^{i} \beta_2^{k-1} \left(H(w_{i-k+1}) \right)^2 \\
&= (1 - \beta_2) \sum_{k=1}^{i} \beta_2^{k-1} \left(H(w_{i-k+1}) \right)^2
\end{aligned}
$$

under $m^0 = 0$ and $v^0 = 0$. Therefore, Equation (1) can be altered by

$$
\begin{aligned}
w_{i+1} &= w_i - \eta \frac{\sqrt{1-\beta_2}}{1-\beta_1} \times \frac{(1-\beta_1)\sum_{k=1}^{i} \beta_1^{k-1} H(w_{i-k+1})}{\sqrt{(1-\beta_2)\sum_{k=1}^{i} \beta_2^{k-1}(H(w_{i-k+1}))^2} + \epsilon} \\
&= w_i - \eta \frac{\sum_{k=1}^{i} \beta_1^{k-1} H(w_{i-k+1})}{\sqrt{\sum_{k=1}^{i} \beta_2^{k-1}(H(w_{i-k+1}))^2} + \epsilon} \\
&= w_i - \eta \frac{S_i}{\sqrt{SS_i} + \epsilon},
\end{aligned} \tag{3}
$$

where

$$
S_i = \sum_{k=1}^{i} \beta_1^{k-1} H(w_{i-k+1}) \quad \text{and} \quad SS_i = \sum_{k=1}^{i} \beta_2^{k-1} \left(H(w_{i-k+1}) \right)^2 .
$$

Here, ϵ is introduced to exclude the case of dividing by 0, and it is 0 unless it is divided by 0. As a result of a simple calculation, the following relation can be obtained:

$$
(S_i)^2 \leq \left(\frac{1 - \left(\frac{\beta_1^2}{\beta_2} \right)^i}{1 - \left(\frac{\beta_1^2}{\beta_2} \right)} \right) SS_i.
$$

For more information on the relation between S_i and SS_i, we explain the calculation process at Corollary 1. From Equation (3), we have

$$
w_{i+1} = w_i - \eta \frac{S_i}{(SS_i + \epsilon)^{1/2}}
$$

Using the relation between S_i and SS_i,

$$
\implies \quad |w_{i+1} - w_i| \leq \eta \left(\frac{1 - \left(\frac{\beta_1^2}{\beta_2} \right)^i}{1 - \left(\frac{\beta_1^2}{\beta_2} \right)} \right)
$$

Here, since ϵ is to prevent division by zero, it can be considered to be zero in the calculation. After a sufficiently large number τ, a sufficiently small value η, and using the Taylor's theorem, we can obtain

$$
\begin{aligned}
H(w_{i+1}) &\approx H(w_i) + \frac{\partial H}{\partial w}(w_{i+1} - w_i) \\
H(w_{i+1}) &\approx H(w_i) + \eta \frac{\partial H}{\partial w}\left(\frac{1 - \left(\frac{\beta_1^2}{\beta_2}\right)^i}{1 - \left(\frac{\beta_1^2}{\beta_2}\right)}\right) \\
H(w_{i+1}) &\approx H(w_i) + \xi,
\end{aligned}
\tag{4}
$$

where

$$
\xi = \eta \frac{\partial H}{\partial w}\left(\frac{1 - \left(\frac{\beta_1^2}{\beta_2}\right)^i}{1 - \left(\frac{\beta_1^2}{\beta_2}\right)}\right)
$$

and is negligibly small. This is possible because we set the η value to a very small value.

Looking at the relationship between S_i and S_{i-1}, and using (4),

$$
\begin{aligned}
S_i &= H(w_i) + \beta_1 S_{i-1} \\
\left(S_i - \frac{H(w_i)}{1 - \beta_1}\right) &= \beta_1\left(S_{i-1} - \frac{H(w_{i-1})}{1 - \beta_1}\right) - \frac{\beta_1}{1 - \beta_1}\xi. \\
\left(S_i - \frac{H(w_i)}{1 - \beta_1} + \frac{\beta_1}{(1 - \beta_1)^2}\xi\right) &= \beta_1\left(S_{i-1} - \frac{H(w_{i-1})}{1 - \beta_1} + \frac{\beta_1}{(1 - \beta_1)^2}\xi\right).
\end{aligned}
$$

We have

$$
\left(S_i - \frac{H(w_i)}{1 - \beta_1} + \frac{\beta_1}{(1 - \beta_1)^2}\xi\right) = \beta_1^i\left(S_0 - \frac{H(w_0)}{1 - \beta_1} + \frac{\beta_1}{(1 - \beta_1)^2}\xi\right).
$$

Therefore,

$$
S_i = \frac{H(w_i)}{1 - \beta_1} - \frac{\beta_1}{(1 - \beta_1)^2}\xi + \beta_1^i\left(S_0 - \frac{H(w_0)}{1 - \beta_1} + \frac{\beta_1}{(1 - \beta_1)^2}\xi\right)
$$

If the initial condition S_0 is defined as a fairly large value, the following equation can be obtained:

$$
\begin{aligned}
\frac{S_i}{S_{i-1}} &= \frac{\frac{\frac{H(w_i)}{1 - \beta_1} - \gamma_1}{\gamma_2} + \beta_1^i}{\frac{\frac{H(w_{i-1})}{1 - \beta_1} - \gamma_1}{\gamma_2} + \beta_1^{i-1}} \\
&\approx \beta_1.
\end{aligned}
$$

where

$$
\begin{aligned}
\gamma_1 &= \frac{\beta_1}{(1 - \beta_1)^2}\xi, \\
\gamma_2 &= S_0 - \frac{H(w_0)}{1 - \beta_1} + \frac{\beta_1}{(1 - \beta_1)^2}\xi.
\end{aligned}
$$

Through a similar process, we have

$$SS_i = (H(w_i))^2 + \beta_2 SS_{i-1}$$

$$= \beta_2 \left(SS_{i-1} - \frac{(H(w_{i-1}))^2}{1-\beta_1} - \delta \right),$$

where δ is $2\zeta\left(H(w_{i-1}) + \zeta/2\right)/(1-\beta_2)$ and can remain constant because of a sufficiently small ζ value.

$$SS_i = \frac{(H(w_i))^2}{1-\beta_1} + \delta + \beta_1^i \left(SS_0 - \frac{(H(w_0))^2}{1-\beta_1} - \delta \right).$$

If the initial condition SS_0 is defined as a fairly large value, the following equation can be obtained:

$$\frac{SS_i}{SS_{i-1}} = \frac{\dfrac{(H(w_i))^2 + \delta}{\gamma_3} + \beta_2^i}{\dfrac{(H(w_{i-1}))^2 + \delta}{\gamma_3} + \beta_2^{i-1}}$$

$$\approx \beta_2$$

where

$$\gamma_3 = (1-\beta_1) SS_0 - (H(w_0))^2 - \delta$$

Assuming that the initial values S_0 and SS_0 are sufficiently large, Equation (3) can be changed as follows:

$$\left| \frac{w^{i+1} - w^i}{w^i - w^{i-1}} \right| = \left| \frac{\dfrac{S_i}{S_{i-1}}}{\left(\dfrac{SS_i}{SS_{i-1}}\right)^{1/2}} \right| = \frac{\beta_1}{\beta_2^{1/2}} < 1.$$

For w^i to converge (*a Cauchy sequence*), the following condition should be satisfied:

$$\left| \frac{w^{i+1} - w^i}{w^i - w^{i-1}} \right| \leq \gamma \leq 1.$$

To satisfy this condition, the larger the value of β_2 is, the better. However, if the value of β_2 is larger than 1, the value of v^i becomes negative and we should compute the complex value. As a result, β_2 is preferably as close to 1 as possible. Conversely, the smaller the value of β_1 is, the better. However, if the value of β_1 is small, the convergence of w^i is fast and the change of w^i is small, so the convergence value that we want cannot be achieved. As a result, β_1 is also preferably as close to 1 as possible. Therefore, it is better to decide in the range of $\beta_1^2 \leq \beta_2$. Generally, β_1 is 0.9 and β_2 is 0.999. After computing, we have $|w^{i+1} - w^i| \leq \gamma^{i-\tau} |w^{\tau+1} - w^\tau|$. As the iteration continues, the value of $\gamma^{i-\tau}$ converges to zero. Therefore, after a sufficiently large number (greater than τ) w^{i+1} and w^i are equal. $\square$

Corollary 1. *The relationship between S_i and SS_i is*

$$(S_i)^2 \leq \left(\frac{1 - \left(\frac{\beta_1^2}{\beta_2} \right)^i}{1 - \left(\frac{\beta_1^2}{\beta_2} \right)} \right) SS_i.$$

Proof.

$$
\begin{aligned}
S_i &= \sum_{k=1}^{i} \beta_1^{k-1} H(w_{i-k+1}) \\
&= H(w_i) + \beta_1 H(w_{i-1}) + \cdots + \beta_1^{i-1} H(w_1) \\
&= H(w_i) + \frac{\beta_1}{(\beta_2)^{1/2}} (\beta_2)^{1/2} H(w_{i-1}) \\
&\quad + \cdots + \frac{\beta_1^{i-1}}{\left(\beta_2^{i-1} \right)^{1/2}} \left(\beta_2^{i-1} \right)^{1/2} H(w_1)
\end{aligned}
$$

Using the general Cauchy–Schwarz inequality, we have

$$
\begin{aligned}
S_i^2 &\leq \left\{ 1 + \left(\frac{\beta_1}{\beta_2^{1/2}} \right)^2 + \cdots + \left(\frac{\beta_1^{i-1}}{\beta_2^{(i-1)/2}} \right)^2 \right\} \\
&\quad \times \left\{ (H(w_i))^2 + \beta_2 \left(H(w_{i-1}) \right)^2 + \cdots + \beta_2^{i-1} \left(H(w_1) \right)^2 \right\} \\
&\leq \left(\frac{1 - \left(\frac{\beta_1^2}{\beta_2} \right)^i}{1 - \left(\frac{\beta_1^2}{\beta_2} \right)} \right) SS_i.
\end{aligned}
$$

$\square$

Theorem 2. *The limit of w^i satisfies that $\lim\limits_{i \to \infty} H(w^i) = 0$.*

Proof. When the limit of w^i is w^*, using (5) and continuity of H, the following equation is obtained as

$$
\begin{aligned}
w^* &= w^* - \eta \frac{\sum_{k=\tau}^{*} \beta_1^{k-1} H(w^{i-k+1})}{\sqrt{\sum_{k=\tau}^{*} \beta_2^{k-1} \left(H(w^{i-k+1}) \right)^2} + \epsilon} \\
0 &= \frac{\sum_{k=\tau}^{*} \beta_1^{k-1} H(w^{i-k+1})}{\sqrt{\sum_{k=\tau}^{*} \beta_2^{k-1} \left(H(w^{i-k+1}) \right)^2} + \epsilon}.
\end{aligned}
$$

The effect of ϵ is to avoid making the denominator zero, so the denominator part of the above equation is not zero. We can get $\sum_{k=\tau}^{*} \beta_1^{k-1} H(w^{i-k+1}) = 0$. Since $\beta_1 < 1$, $\beta_1^2 < \beta_2$, and w^i converges to

w^* assuming that $*$ is a fairly large number, w^* and w^{*-1} are close, and β^κ can be regarded as 0 after κ, where κ is an appropriate large integer. Therefore, we can get

$$
\begin{aligned}
0 &= H(w^*) + \beta_1 H(w^{*-1}) + \beta_1^2 H(w^{*-2}) + \dots + \beta_1^\kappa H(w^{*-\kappa}) \\
&\approx H(w^*)\left(1 + \beta_1 + \beta_1^2 + \dots + \beta_1^\kappa\right) \\
&= H(w^*)\frac{1 - \beta_1^{\kappa+1}}{1 - \beta_1}.
\end{aligned}
$$

The extreme value w^* of the sequence w^i becomes a variable value that brings the cost function H close to zero. $\square$

5. Numerical Tests

In these numerical tests, we perform several experiments to show the novelty of the proposed method. The GD method, the ADAM method, the AdaMax method, and the proposed method are compared according to each experiment. Please note that some of the techniques such as batch, epoch, and drop out are not included. β_1 and β_2 used in ADAM and AdaMax are fixed as 0.9 and 0.999, respectively, and ϵ is used as 10^{-8}. These values are the default of β_1, β_2, and ϵ.

5.1. One Variable Non-Convex Function Test

Since the cost function has a non-convex property, we test each method with a simple non-convex function in this experiment. The cost function is $C(w) = (w+5)(w+3)(w-1)(w-10)/800 + 3$ and has the global minimum at $w \approx 7.1047$. The starting point, w_0, is initialized at -9 and the iteration number is 100. The reason this cost function is divided by 800 is that if you do not divide it, the degree of this function is 4, so the value of the function becomes too big and it is too far away from the real problem.

Figure 2 shows the change in the cost function ($C(w)$) according to the change in w.

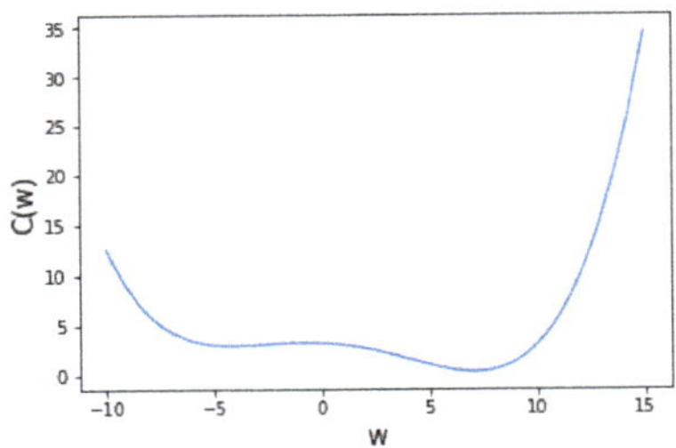

Figure 2. Cost function with one local minimum.

Figure 3 shows the iterations of four methods over $C(w)$ with $w_0 = -9$. In this experiment, GD, ADAM, and AdaMax fall into a local minimum and it is confirmed that there is no motion. On the other hand, the proposed method is confirmed to settle at the global minimum beyond the local minimum. Although the global minimum is near 7, the other methods stayed near -4.

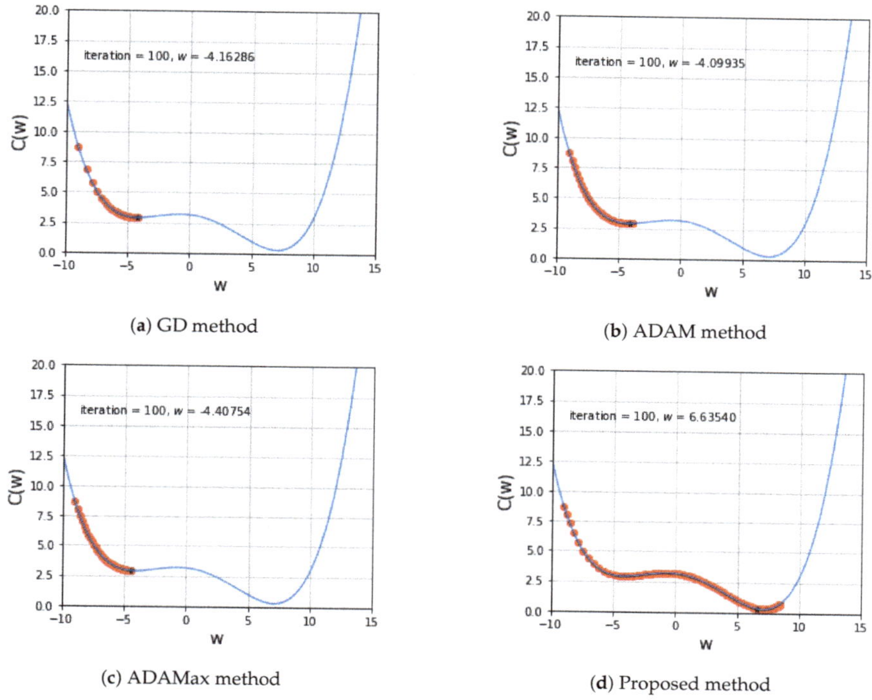

(a) GD method

(b) ADAM method

(c) ADAMax method

(d) Proposed method

Figure 3. Compared to the other scheme, the learning rate is 0.2 and the β_1, β_2, λ used in the proposed method are 0.95, 0.9999, -10^{-2}.

5.2. Two Variables Non-Convex Function Test

In this section, we experiment with three two-variable cost functions. The first and second experiments are to find the global minimum of the Beale function and the Styblinski–Tang function, respectively, and the third experiment is to test whether the proposed method works effectively at the saddle point.

5.2.1. Beale function

The Beale function is defined by $C(w_1, w_2) = (1.5 - w_1 + w_1 w_2)^2 + (2.25 - w_1 + w_1 w_2^2)^2 + (2.625 - w_1 + w_1 w_2^3)^2$ and has the global minimum at (3, 0.5). Figure 4 shows the Beale function.

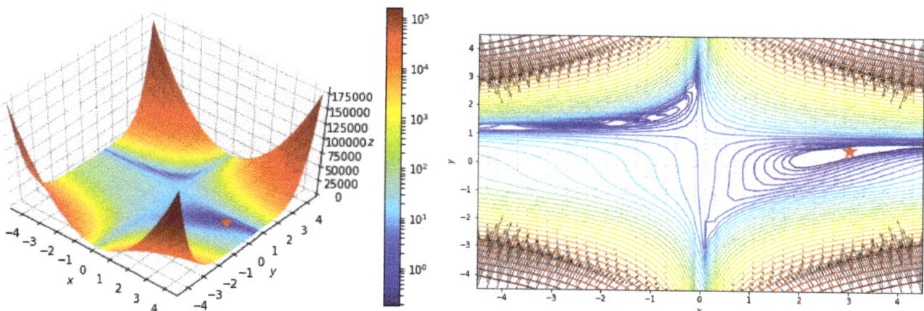

Figure 4. Beale Function.

Figure 5 shows the results of each method. GD's learning late was set to 10^{-4}, which is different from other methods because only GD has a very large gradient and weight divergence. We confirm that all methods converge well because this function is convex around the given starting point.

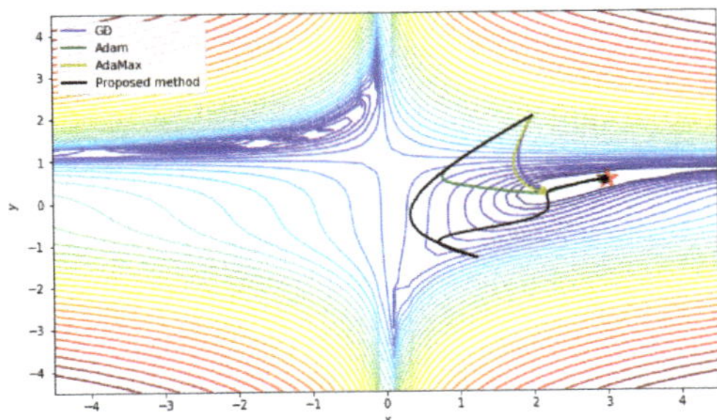

Figure 5. Result of the Beale function with an initial point of (2, 2). $W_0 = (2, 2)$ and the iteration number are 1000. The learning rate is 0.1 and the β_1, β_2, and λ used in proposed method are 0.9, 0.999, and -0.5, respectively. GD's learning late was set to 10^{-4}.

Table 1 shows the weight change of each method according to the change of iteration. As you can see from this table, the proposed method shows the best performance.

Table 1. Change of weights of each method.

Iteration/Method	GD	ADAM	ADAMax	Proposed Method
0	[2.0, 2.0]	[2.0, 2.0]	[2.0, 2.0]	[2.0, 2.0]
100	[1.81, 0.84]	[1.53, 0.24]	[1.75, 0.69]	[2.11, −0.33]
200	[1.85, 0.63]	[2.20, 0.18]	[1.80, 0.53]	[2.50, 0.32]
300	[1.89, 0.52]	[2.39, 0.28]	[1.85, 0.42]	[2.72, 0.41]
400	[1.92, 0.45]	[2.53, 0.34]	[1.90, 0.35]	[2.81, 0.44]
500	[1.95, 0.39]	[2.64, 0.39]	[1.94, 0.30]	[2.87, 0.46]
600	[1.98, 0.35]	[2.73, 0.42]	[1.99, 0.26]	[2.91, 0.47]
700	[2.00, 0.32]	[2.79, 0.44]	[2.03, 0.23]	[2.93, 0.48]
800	[2.03, 0.30]	[2.84, 0.45]	[2.06, 0.22]	[2.95, 0.48]
900	[2.05, 0.28]	[2.88, 0.46]	[2.10, 0.21]	[2.96, 0.49]
1000	[2.06, 0.27]	[2.91, 0.47]	[2.13, 0.21]	[2.97, 0.49]

To see the results from another starting point, we experiment with the same cost function using a different starting point $(-4, 4)$, hyperparameter $\lambda = -0.2$ and the iteration number = 50,000.

Figure 6 shows the results of each method. In this experiment, we confirm that GD, ADAM, and AdaMax fall into a local minimum and stop there, whereas the proposed method reaches the global minimum effectively.

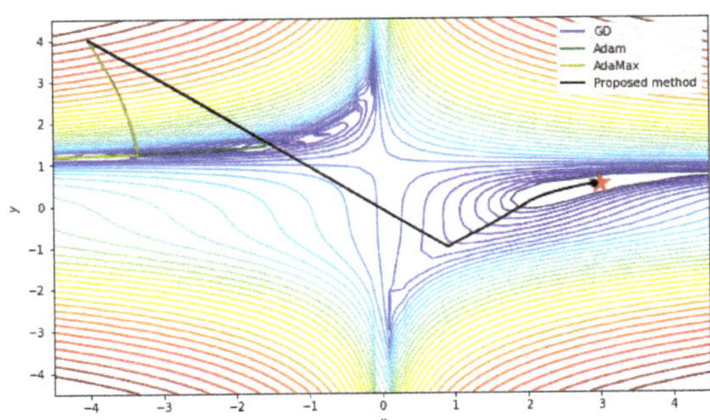

Figure 6. Result of Beale function with initial point $(-4, 4)$.

5.2.2. Styblinski–Tang function

The Styblinski–Tang function is defined by

$$C(w_1, w_2) = \left((w_1^4 - 16w_1^2 + 5w_1) + (w_2^4 - 16w_2^2 + 5w_2)^2 \right) /2 + 80$$

and has the global minimum at $(-2.903534, -2.903534)$.

Figure 7 shows the Styblinski–Tang function. In this experiment, we present a result with the starting point $W_0 = (6, 0)$. Please note that a local minimum point $\approx (2.7468, -2.9035)$ is located between W_0 and the global minimum point.

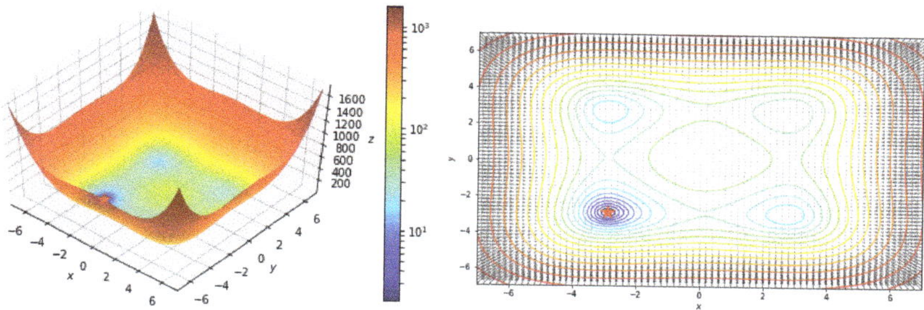

Figure 7. Styblinski–Tang Function.

Figure 8 shows the results of each method. Only the Proposed method find the global minimum, and other methods could not avoid a local minimum.

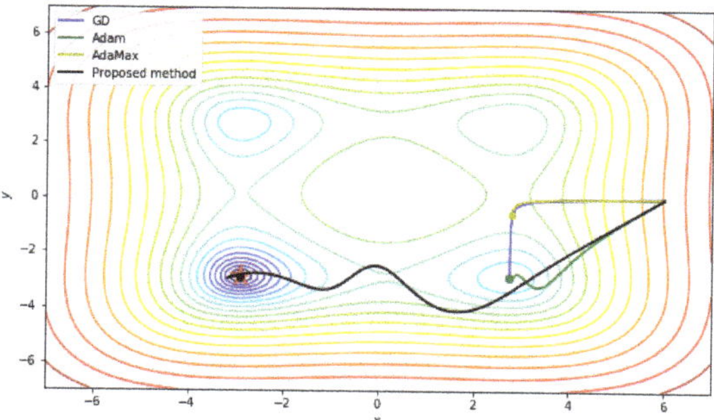

Figure 8. Result of Styblinski–Tang function with initial point $(6, 0)$. The learning rate is 0.1 and the β_1, β_2, λ used in Proposed method are 0.95, 0.9999, 0.2 and GD's learning late was set to 10^{-3}.

5.2.3. Function with a Saddle Point

The cost function $C(w_1, w_2) = w_2^2 - w_1^2 + 2$ is shown in Figure 9. A Hessian matrix of this cost function is $\begin{pmatrix} -2 & 0 \\ 0 & 2 \end{pmatrix}$, so this cost function has a saddle point at $(0, 0)$.

In this experiment, we present results based on two different starting points. The first starting point is $W_0 = (0.001, 0.2)$.

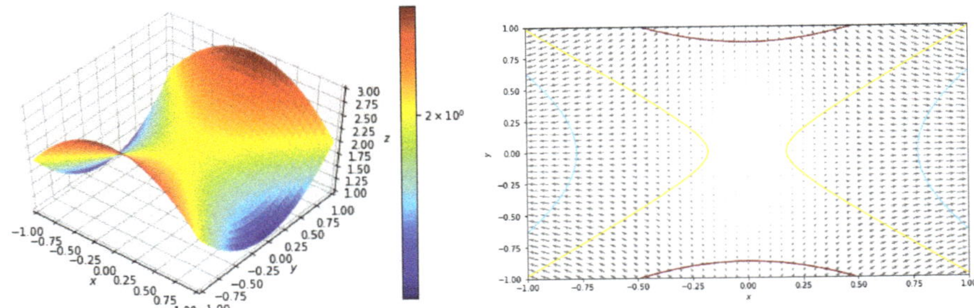

Figure 9. Cost function $C(w_1, w_2) = w_2^2 - w_1^2 + 2$.

Figure 10 and Table 2 shows the results of each method. In this experiment, we see that the proposed method also changes the parameters more rapidly than the other three methods near the saddle point.

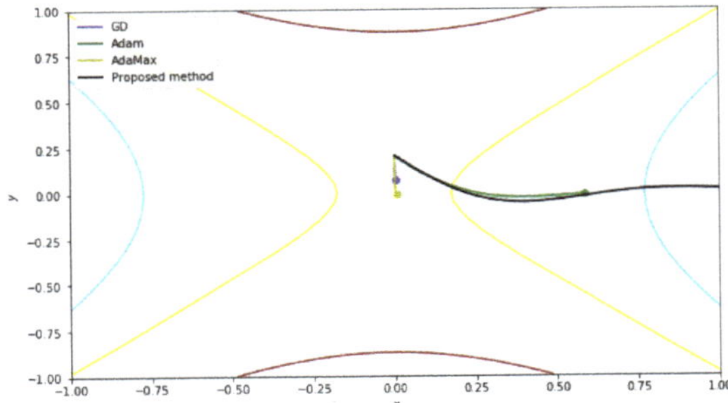

Figure 10. The results near the saddle point with an initial point of (0.001, 2). The iteration number is 50, the learning rate is 10^{-2}, and the β_1, β_2, and λ used in the proposed method are 0.95, 0.9999, and -5×10^{-3}, respectively.

Table 2. Change values of each method.

B	GD	ADAM	ADAMax	Proposed Method
0	$[0.001, 0.2]$	$[0.001, 0.2]$	$[0.001, 0.2]$	$[0.001, 0.2]$
10	$[0.00121899, 0.16341456]$	$[0.0943284, 0.10245869]$	$[0.00157076, 0.11185508]$	$[0.16787479, 0.03591625]$
20	$[0.00148595, 0.13352159]$	$[0.20284632, 0.02224525]$	$[0.00235688, 0.05096114]$	$[0.38849311, -0.05289886]$
30	$[0.00181136, 0.10909686]$	$[0.32453918, -0.02053808]$	$[0.00347925, 0.0150328]$	$[0.61448606, -0.01491036]$
40	$[0.00220804, 0.08914008]$	$[0.45704978, -0.0233784]$	$[0.00511317, -0.00198256]$	$[0.83303301, 0.02029141]$
50	$[0.00269159, 0.07283394]$	$[0.59806857, -0.0075764]$	$[0.007516, -0.00722015]$	$[1.04047722, 0.01258689]$

Then, we performed an experiment with the same cost function, but with another starting point (0, 0.01), and with an iteration number of 100. The hyper parameters in this experiment are the same as above. The only thing that was changed was the starting point.

Figure 11 shows the result of this experiment with an initial point of (0, 0.01). In this experiment, since one of the coordinates of the initial value is 0, the other methods do not work.

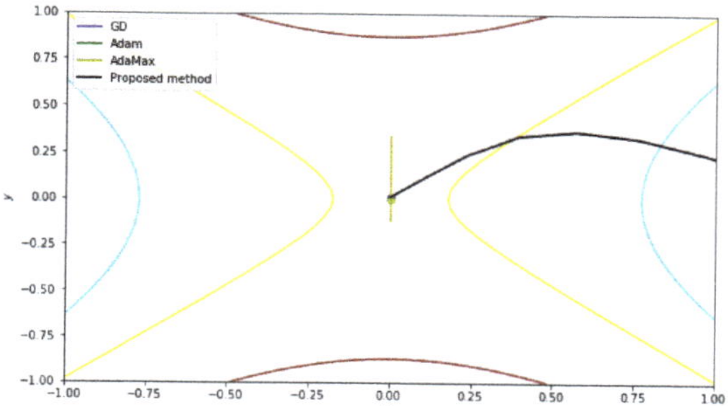

Figure 11. The result of the saddle point with an initial point of $(0, 0.01)$.

5.3. Two-Dimensional Region Segmentation Test

To see how one can divide the area in a more complicated situation, we introduce a problem of region separation in the form of the shape of a whirlwind. The value of 0 and 1 are given along the shape of the whirlwind, and the regions are divided by learning according to each method. Table 3b is the pseudo code for making the learning dataset for this experiment and Table 3a is a visualization of the dataset.

Table 3. Learning dataset.

(a) Image of *LS*	(b) Pseudo code of *LS*

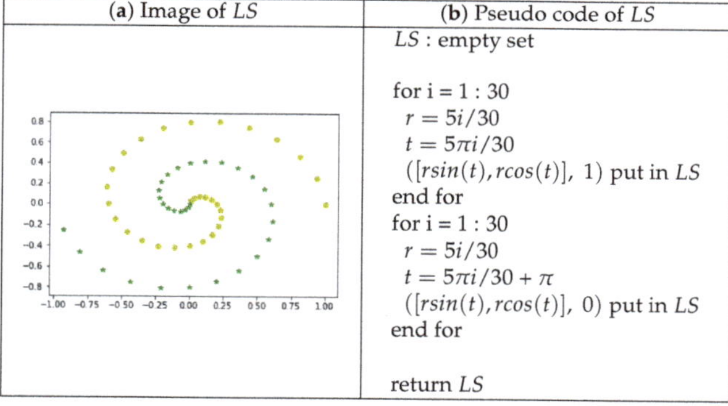	LS : empty set
	
	for i = 1 : 30
	  $r = 5i/30$
	  $t = 5\pi i/30$
	  $([rsin(t), rcos(t)],\ 1)$ put in LS
	end for
	for i = 1 : 30
	  $r = 5i/30$
	  $t = 5\pi i/30 + \pi$
	  $([rsin(t), rcos(t)],\ 0)$ put in LS
	end for
	
	return LS

To solve this problem, we use a 2-layer and 25-row neural network. First, the neural network is learned by each method. After that, $[-2, 2] \times [-2, 2]$ is divided into 60 equal sections, and each of the 3600 coordinates are checked to decide the parts where they belong.

The results are presented in Figure 12. The proposed method is better expressed in dividing the region according to the direction of the whirlwind, compared with other schemes.

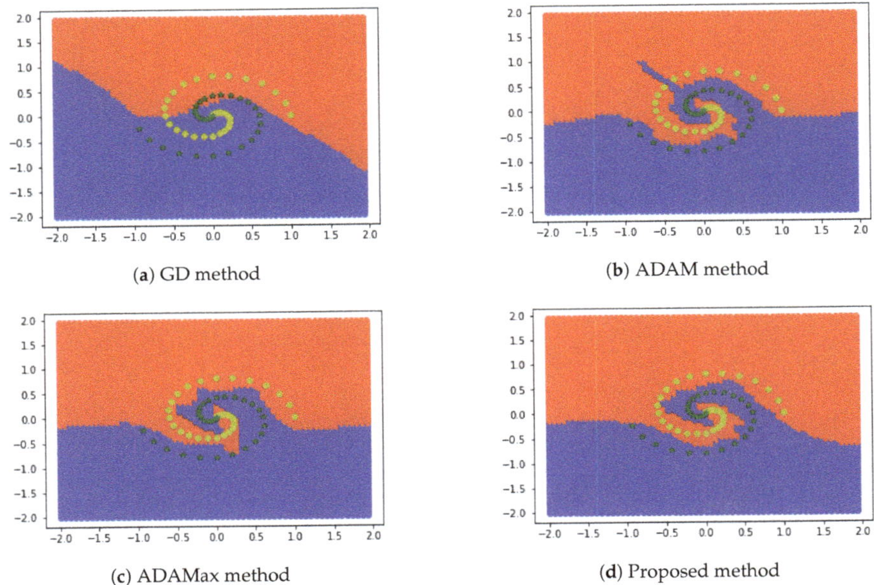

(**a**) GD method

(**b**) ADAM method

(**c**) ADAMax method

(**d**) Proposed method

Figure 12. comparing proposed scheme with other schemes.

In the same artificial neural network structure, different results were obtained for each method, which is related to the accuracy of location learning.

5.4. MNIST with CNN

In the previous sections, we have seen simple cases of classification problems using ANN. Here, we try a more practical experiment using a CNN (Convolutional Neural Network). As we all know, we use an ANN to increase the number of layers to solve more difficult and complex problems. However, as the number of layers increases, the number of parameters also increases, and the memory and operating time of the computer, likewise, also increase. For this reason, a CNN was introduced and the performance is shown at The ImageNet Large Scale Visual Recognition Challenge (ILSVRC) [11]. Therefore, in this experiment, we want to see how our method works with the CNN structure. In this experiment, TensorFlow was used, and for the fairness of the experiment, we used the MNIST problem using a CNN, which is one of the basic examples provided in TensorFlow (see https://github.com/ tensorflow/models/tree/master/tutorials/image/mnist). There were 8600 iterations, which output the Minibatch cost and validation error of each method as a result. The results are shown in Figure 13. As shown in Figure 13, all methods were found to reduce both cost and error.

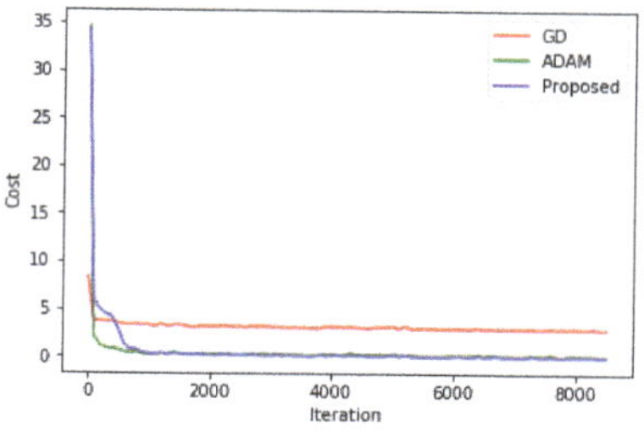

(**a**) Minibatch cost of each method

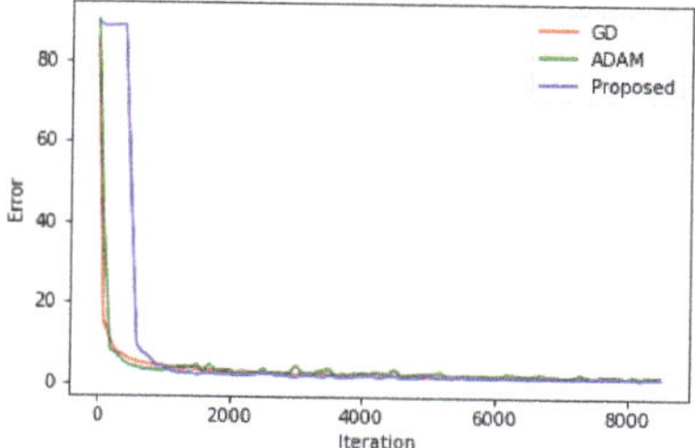

(**b**) Validation error of each method

Figure 13. Comparing the proposed scheme with other schemes.

For further investigation, we magnify the data from Figure 13 from the 6000th iteration, and show it in Figure 14.

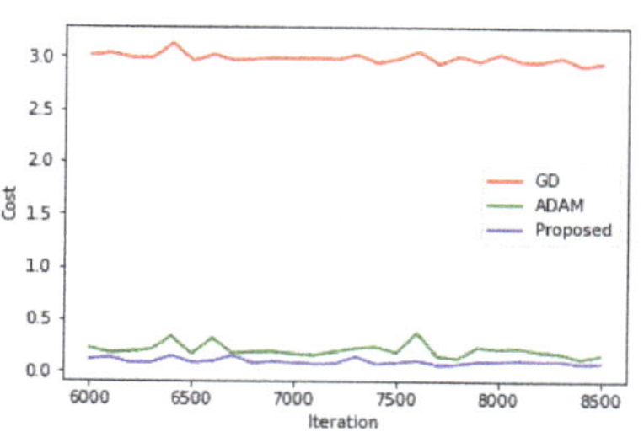

(**a**) Minibatch cost of each method

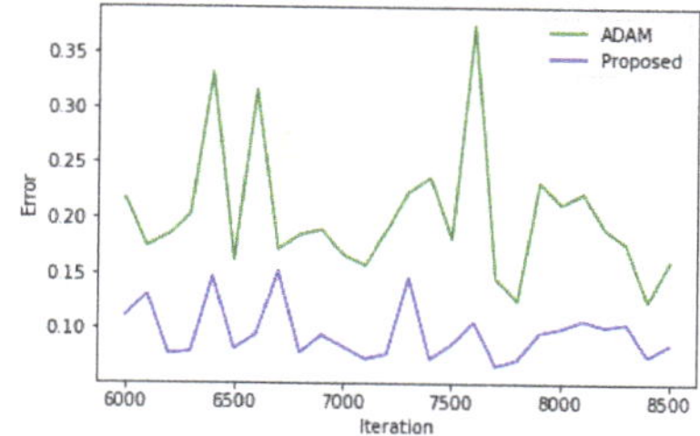

(**b**) Validation error of each method

Figure 14. Comparing the proposed scheme with other schemes with an iteration number over 6000.

Figure 14a shows that GD converges the slowest and the proposed method is more accurate than ADAM. Figure 14b shows the accuracies of the proposed method and the ADAM method, as measured on the test data. From this, we see that the proposed method works better for the classification of images than ADAM.

Figure 15 shows four sample digits from the MNIST test data that the proposed method classifies effectively but that ADAM failed to classify in this experiment.

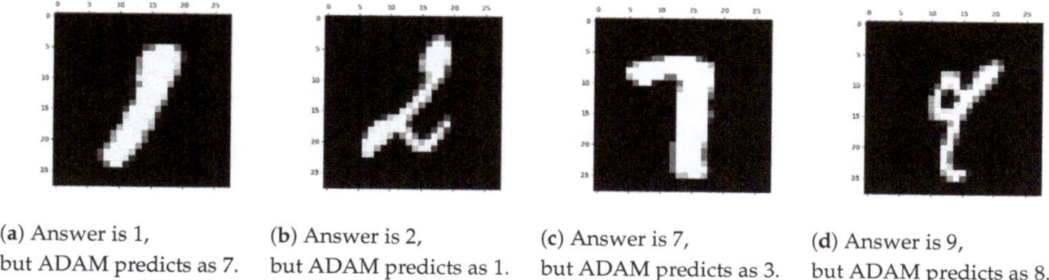

(**a**) Answer is 1,
but ADAM predicts as 7.

(**b**) Answer is 2,
but ADAM predicts as 1.

(**c**) Answer is 7,
but ADAM predicts as 3.

(**d**) Answer is 9,
but ADAM predicts as 8.

Figure 15. Examples that the proposed method predicted correctly, while the ADAM method does not.

5.5. CIFAR-10 with RESNET

We show in Section 5.4 that the proposed method is more effective than ADAM in the CNN structure using the MNIST dataset. Since the MNIST dataset is a grayscale image, the size of one image is $28 \times 28 \times 1$. In this section, we show that the proposed method is more effective than ADAM when using a dataset larger than the MNIST, such as the CIFAR10 (https://www.cs.toronto.edu/~kriz/cifar.html). CIFAR10 is a popular dataset for classifying 10 categories (airplane, automobile, bird, cat, deer, dog, frog, horse, ship, and truck) $32 \times 32 \times 3$ size RGB images. This dataset consists of 50,000 training data and 10,000 test data. In this experiment, as in Section 5.4, we used the example code (the same model and the same hyperparameter settings) provided by TensorFlow (see https://github.com/tensorflow/models/tree/master/tutorials/image/cifar10_estimator). In Section 5.4, we used a simple network using a CNN, but in this subsection, we use a more complex and high-performance network called RESNET (Residual network) based on CNN. RESNET is a well-known network that has good performance in ILSVRC, and we used RESNET 44 in this experiment.

Figure 16 shows the results of the training CIFAR10 dataset using the RESNET structure, and are compared with ADAM. In Figure 16a, the training cost of each method was calculated and plotted every 100th iteration. In Figure 16b, the validation cost of each method was calculated and plotted every 5000th iteration. This shows that the proposed method works effectively for more complex networks as well as simple networks.

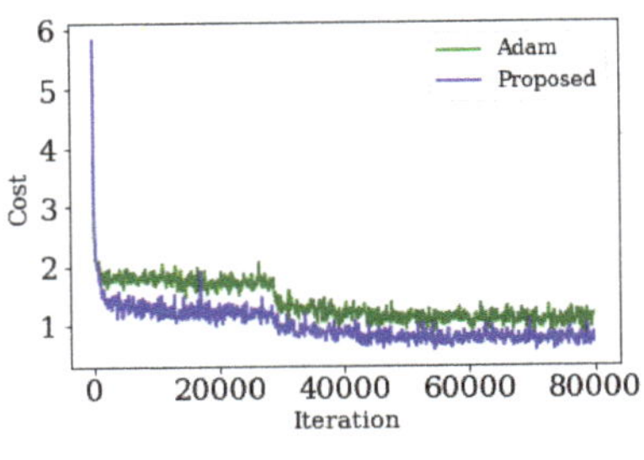

(a) Training cost of each method

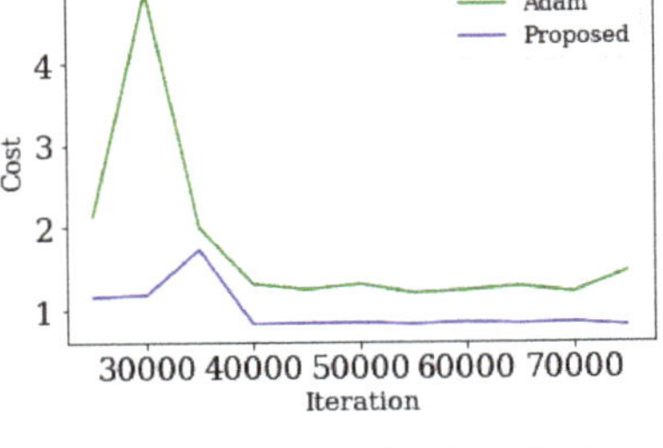

(b) Validation cost of each method

Figure 16. Result of CIFAR10 dataset using the RESNET 44 model. The following parameters are default values; iteration : 80000, batch size : 128, weight decay : 2×10^{-4}, learning rate : 0.1, batch norm decay : 0.997.

6. Conclusions

In this paper, we propose a new optimization method based on ADAM. Many of the existing optimization methods (including ADAM) may not work properly according to the initial point. However, the proposed method finds the global minimum better than other methods, even if there is a local minimum near the starting point, and has better overall performance. We tested our method only on models for image datasets such as MNIST and CIFAR10. Our future work is to test our method on various models such as RNN models for time series prediction, various models for natural language processing, etc.

Author Contributions: Conceptualization, D.Y.; Data curation, S.J.; Formal analysis, D.Y. and J.A.; Funding acquisition, D.Y.; Investigation, D.Y.; Methodology, D.Y. and J.A.; Project administration, D.Y.; Resources, S.J.; Software, S.J.; Supervision, D.Y. and J.A.; Validation, J.A. and S.J.; Visualization, S.J.; Writing—original draft, D.Y.; Writing—review & editing, S.J. All authors have read and agreed to the published version of the manuscript.

Funding: This work was supported by the basic science research program through the National Research Foundation of Korea (NRF) funded by the Ministry of Education, Science and Technology (grant number NRF-2017R1E1A1A03070311). Also, the second author Ahn was supported by research fund of Chungnam National University.

Acknowledgments: We sincerely thank anonymous reviewers whose suggestions helped improve and clarify this manuscript greatly.

Conflicts of Interest: The authors declare no conflict of interest.

References

1. Deng, L.; Li, J.; Huang, J.; Yao, K.; Yu, D.; Seide, F.; Seltzer, M.L.; Zweig, G.; He, X.; Williams, J.; et al. Recent advances in deep learning for speech research at microsoft. In Proceedings of the 2013 IEEE International Conference on Acoustics, Speech and Signal Processing—ICASSP 2013, Vancouver, BC, Canada, 26–31 May 2013.
2. Graves, A. Generating sequences with recurrent neural networks. *arXiv* **2013**, arXiv:1308.0850.
3. Graves, A.; Mohamed, A.; Hinton, G. Speech recognition with deep recurrent neural networks. In Proceedings of the 2013 International Conference on Acoustics, Speech and Signal Processing—ICASSP 2013, Vancouver, BC, Canada, 26–31 May 2013; pp. 6645–6649.
4. Hinton, G.E.; Srivastava, N.; Krizhevsky, A.; Sutskever, I.; Salakhutdinov, R.R. Improving neural networks by preventing co-adaptation of feature detectors. *arXiv* **2012**, arXiv:1207.0580.
5. Tieleman, T.; Hinton, G.E. *Lecture 6.5—RMSProp, COURSERA: Neural Networks for Machine Learning*; Technical Report; COURSERA: Mountain View, CA, USA, 2012.
6. Dean, J.; Corrado, G.; Monga, R.; Chen, K.; Devin, M.; Le, Q.; Mao, M.; Ranzato, M.; Senior, A.; Tucker, P.; et al. Large scale distributed deep networks. In Proceedings of the 25th International Conference on Neural Information Processing Systems—NIPS 2012, Lake Tahoe, NV, USA, 3 December 2012; pp. 1223–1231
7. Jaitly, N.; Nguyen, P.; Senior, A.; Vanhoucke, V. Application of pretrained deep neural networks to large vocabulary speech recognition. In Proceedings of the INTERSPEECH 2012, 13th Annual Conference of the International Speech Communication Association—ISCA 2012, Portland, OR, USA, 9–13 September 2012; pp. 2578–2581
8. Amari, S. Natural gradient works efficiently in learning. *Neural Comput.* **1998**, *10*, 251–276. [CrossRef]
9. Duchi, J.; Hazan, E.; Singer, Y. Adaptive subgradient methods for online learning and stochastic optimization. *J. Mach. Learn. Res.* **2011**, *12*, 2121–2159.
10. Hinton, G.; Deng, L.; Yu, D.; Dahl, G.E.; Mohamed, A.; Jaitly, N.; Senior, A.; Vanhoucke, V.; Nguyen, P.; Sainath, T.N. Deep neural networks for acoustic modeling in speech recognition: The shared views of four research groups. *Signal Process. Mag.* **2012**, *29*, 82–97. [CrossRef]
11. Krizhevsky, A.; Sutskever, I.; Hinton, G.E. Imagenet classification with deep convolutional neural networks. In Proceedings of the Neural Information Processing Systems–NIPS 2012, Lake Tahoe, NV, USA, 3 December 2012; pp. 1097–1105.
12. Pascanu, R.; Bengio, Y. Revisiting natural gradient for deep networks. *arXiv* **2013**, arXiv:1301.3584.
13. Sutskever, I.; Martens, J.; Dahl, G.; Hinton, G.E. On the importance of initialization and momentum in deep learning. In Proceedings of the 30th International Conference on Machine Learning—ICML 2013, Atlanta, GA, USA, 16–21 Jun 2013; PMLR 28(3), pp. 1139–1147.
14. Kochenderfer, M.; Wheeler, T. *Algorithms for Optimization*; The MIT Press Cambridge: London, UK, 2019.
15. Kingma, D.P.; Ba, J. ADAM: A method for stochastic optimization. In Proceedings of the 3rd International Conference for Learning Representations—ICLR 2015, San Diego, CA, USA, 7–9 May 2015.
16. Bottou, L.; Curtis, F.E.; Nocedal, J. Optimization Methods for Large-Scale Machine Learning. *SIAM Rev.* **2018**, *60*, 223–311. [CrossRef]
17. Roux, N.L.; Fitzgibbon, A.W. A fast natural newton method. In Proceedings of the 27th International Conference on Machine Learning—ICML 2010, Haifa, Israel, 21–24 June 2010; pp. 623–630.
18. Sohl-Dickstein, J.; Poole, B.; Ganguli, S. Fast large-scale optimization by unifying stochastic gradient and quasi-newton methods. In Proceedings of the 31st International Conference on Machine Learning—ICML 2014, Beijing, China, 21–24 June 2014; pp. 604–612.
19. Zeiler, M.D. Adadelta: An adaptive learning rate method. *arXiv* **2012**, arXiv:1212.5701.
20. Rumelhart, D.E.; Hinton, G.E.; Williams, R.J. Learning representations by back-propagating errors. *Nature* **1986**, *323*, 533–536. [CrossRef]
21. Becker, S.; LeCun, Y. *Improving the Convergence of Back-Propagation Learning with Second Order Methods*; Technical Report; Department of Computer Science, University of Toronto: Toronto, ON, Canada, 1988.

22. Kelley, C.T. Iterative methods for linear and nonlinear equations. In *Frontiers in Applied Mathematics*; SIAM: Philadelphia, PA, USA, 1995; Volume 16.

23. Reddi, S.J.; Kale, S.; Kumar, S. On the Convergence of ADAM and Beyond. *arXiv* **2019** arXiv:1904.09237.

24. Ruppert, D. *Efficient Estimations from a Slowly Convergent Robbins-Monro Process*; Technical Report; Cornell University Operations Research and Industrial Engineering: Ithaca, NY, USA, 1988.

25. Zinkevich, M. Online Convex Programming and Generalized Infinitesimal Gradient Ascent. In Proceedings of the Twentieth International Conference on Machine Learning—ICML-2003, Washington, DC, USA, 21–24 August 2003; pp. 928–935.

Article

Boundary Matching and Interior Connectivity-Based Cluster Validity Anlysis

Qi Li, Shihong Yue *, Yaru Wang, Mingliang Ding, Jia Li and Zeying Wang

School of Electrical and Information Engineering, Tianjin University, Tianjin 300072, China;
qili_2017@tju.edu.cn (Q.L.); yaruwang@tju.edu.cn (Y.W.); mlding@tju.edu.cn (M.D.); lijiajoyce@tju.edu.cn (J.L.);
wangzeying@tju.edu.cn (Z.W.)
* Correspondence: shyue1999@tju.edu.cn; Tel.: +86-22-2740-5477

Received: 11 January 2020; Accepted: 13 February 2020; Published: 16 February 2020

Abstract: The evaluation of clustering results plays an important role in clustering analysis. However, the existing validity indices are limited to a specific clustering algorithm, clustering parameter, and assumption in practice. In this paper, we propose a novel validity index to solve the above problems based on two complementary measures: boundary points matching and interior points connectivity. Firstly, when any clustering algorithm is performed on a dataset, we extract all boundary points for the dataset and its partitioned clusters using a nonparametric metric. The measure of boundary points matching is computed. Secondly, the interior points connectivity of both the dataset and all the partitioned clusters are measured. The proposed validity index can evaluate different clustering results on the dataset obtained from different clustering algorithms, which cannot be evaluated by the existing validity indices at all. Experimental results demonstrate that the proposed validity index can evaluate clustering results obtained by using an arbitrary clustering algorithm and find the optimal clustering parameters.

Keywords: clustering evaluation; clustering algorithm; cluster validity index; boundary point; interior point

---

1. Introduction

Clustering analysis is an unsupervised technique that can be used for finding the structure in a dataset [1–3]. The evaluation of clustering results plays a vital role in clustering analysis and usually is performed by a clustering validity index or several [4,5]. In the past decades, a large number of validity indices have been proposed to evaluate the clustering results and to determine the optimal number of clusters, which is an essential character of a dataset. These frequently used validity indices contain Davies-Bouldin measure [6], Tibshirani Gap statistics [7], Xie-Beni's separation measure [8], and etc. Moreover, the commonly used Bayesian Information Criterion (BIC) has been applied to estimate the number of clusters [9,10]. For example, [11] provided a closed-form BIC expression by imposing the multivariate Gaussian assumption on the distribution of the datasets. Then, a novel two-step cluster enumeration algorithm has been proposed by combining the cluster analysis problem. Thus, this new BIC method contains information about the dataset in both data-fidelity and penalty terms. Compared with the existing BIC-based cluster enumeration algorithms, the penalty term of the proposed criterion involves information about the actual number of clusters. Arbelaitz et al. [12] made comparisons among the existing validity indices. Recently, an unsupervised validity index [13] independent of any clustering algorithm was proposed for any dataset of spherical clusters. In addition, the idea has been proposed to deal with the clustering evaluation under the condition of big data [14,15], while very little work is available in the literature that discusses validity indices for big data. One of the few papers on this topic that we are aware of is [16], but the method does not consider the complex data structures in today's computing environment.

However, the existing validity indices are greatly restrained by the following three disadvantages at least.

Specific algorithm. Usually, the existing validity indices can only evaluate the clustering results obtained by a specified algorithm (e.g., C-means [17,18] or Fuzzy C-means [19,20]) rather than an arbitrary clustering algorithm. If the clustering algorithm chosen for the dataset is not suitable, the evaluating results will not be guaranteed.

Specific parameter. Different parameters in a clustering algorithm lead to different clustering results. Most cluster validity indices aim to select the best one among all clustering results, and thus they regard the clustering parameter as their variable. These existing indices can only regard the number of clusters as a variable rather than other clustering parameters, such as the density threshold in Density-Based Spatial Clustering of Applications with Noise (DBSCAN) algorithm [21] and the grid size in the CLIQUE algorithm [22]. Recently, a density peak-point-based clustering (DPC) algorithm [23] and its variants [24–26] have attracted considerable attention; but the number of peak points therein remains so uncertain that the correctness of clustering results is difficult to guarantee.

Untapped result. A cluster consists of a high-density center and a group of relatively low-density neighbors around the center in DPC, a group of core points and corresponding boundary points around these core points in DBSCAN, a center and a group of points that are assigned to the cluster by the nearest neighbor principle in C-means. Consequently, all points in a cluster are partitioned into two types. An accurate clustering partition must result from the correct identification of the two types of points. Although the existing algorithms may partition all points into the two types of points, and the partitioning results fail to be taken into a validity index to evaluate the clustering results. Especially when several clustering algorithms are performed in the same dataset, it is impossible to choose the best clustering result.

To solve the above problems, we propose a nonparametric measure to find all the boundary [27,28] and interior points in any dataset at first. Moreover, once a clustering algorithm is performed, all the boundary and interior points in any cluster can be found. After the measurement of boundary points matching and interior points connectivity between the entire dataset and all partitioned clusters, a novel validity index is proposed. Three typical clustering algorithms, i.e., C-means, DBSCAN, and DPC, are applied to evaluate the generality of the novel validity index. Two groups of artificial and CT datasets with different characteristics validate the correctness and generalization of the proposed validity index.

2. Related Work

In this section, we will firstly review three typical clustering algorithms, and then discuss a group of mostly used validity indices.

2.1. Typical Clustering Algorithms

Assume that $X = \{x_1, x_2, \dots, x_n\}$ is a dataset containing n data points, and $S_1, S_2, \dots, S_c$ are disjoint subsets of X. If the point x_j belongs to the i-th subset S_i, then we set u_{ij} equal to 1, or else 0. The binary membership function can be represented as follows

$$u_{ij} = \begin{cases} 1, & x_j \in S_i \\ 0, & x_j \notin S_i \end{cases}, \ i = 1,2,\dots,c; \ j = 1,2,\dots,n. \tag{1}$$

If each point belongs to one certain subset, then the partitioning of X is called a hard partitioning, satisfying

$$X = S_1 \cup S_2 \cup \dots \cup S_c, \ S_i \cap S_j = \phi, \ i \neq j, \ i,j = 1,\ 2,\ \dots,\ c. \tag{2}$$

There is a great volume of clustering algorithms, but the following three algorithms in them are representative.

2.1.1. C-Means Algorithm

C-means has been widely used in almost all fields owing to its simplicity and high efficiency. Its detailed steps are listed in Algorithm 1.

Algorithm 1. C-means Clustering Algorithm

Input: the number of clusters C and a dataset X containing n points.
Output: a set of C clusters that contains all n objects.
Steps:
1. Initialize cluster centers $v_1, v_2, \ldots, v_C$ by selecting C points arbitrarily.
2. Repeat;
3. Assign each point to one certain cluster according to the nearest neighborhood principle and a chosen measure;
4. Update cluster centers by $v_i = \sum_{j=1}^{n} u_{ij} x_j / \sum_{j=1}^{n} u_{ij}$ for $i = 1 - C$;
5. Stop if a convergence criterion is satisfied;
6. Otherwise, go to Step 2.

The key parameter in C-means is the number of clusters (C) which has to be determined in the clustering process.

2.1.2. DBSCAN Algorithm

In comparison to the C-means, DBSCAN has two parameters: a neighborhood radius of any point and the number of points within the neighborhood. However, in practice, the two parameters have to be turned to a density measure, which usually is their rate and is used to find the density differences and distributed characteristics of all points. In terms of a density threshold, DBSCAN distinguishes core points and border ones from all points. If the density of a point is higher than ε, then this point is called a core point; or else a border point.

DBSCAN starts with an arbitrary point p in X. If it is a core point, then assign all the points which are density-reachable from p with ε to the same cluster. If p is not a core point, then p is temporarily assigned to noise. Afterward, DBSCAN deals with the next point until all the points in X have been visited.

The density threshold ε in DBSCAN is a crucial parameter, and significantly affect the clustering results. As an example, a dataset with 176 points is evaluated (see Figure 1). Figure 1 shows the dataset is clustered into 5, 2, and 1 cluster with different values of ε, where points in different clusters are marked with different signs. It is clear that the number of clusters decreases as ε increases. Therefore, if ε is incorrectly chosen, then the number of clusters cannot be determined. Although Ordering Points to Identify the Clustering Structure (OPTICS) algorithm [29] can be applied to solve the value of ε, it neither provides clustering results explicitly nor is applied in high-dimensional data space.

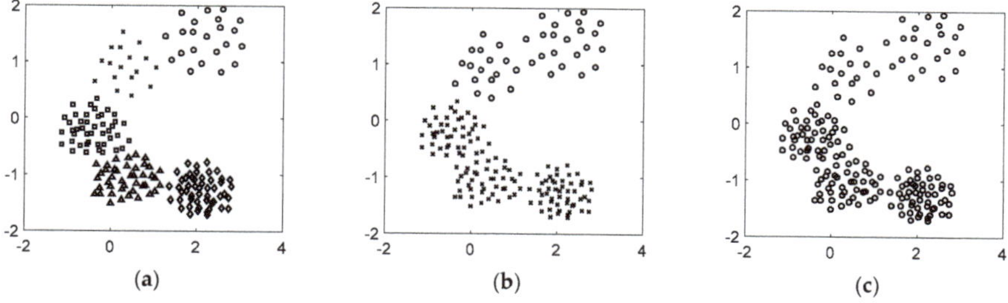

Figure 1. The number of clusters varies as ε increases: (**a**) five clusters; (**b**) two clusters; (**c**) one cluster.

2.1.3. DPC Algorithm

DPC algorithm is the latest clustering algorithm, which combines the advantages of C-means and DBSCAN algorithms. For each point x_i, DPC calculates its density ρ_i and its separated distance δ_i as follows.

$$\rho_i = \sum_j \chi(d(x_i, x_j) - d_c), \; s.t. \chi(x) = 1, \; if \; \chi(\bullet) < 0; \; else \; 0, \tag{3}$$

where d_c is a cutoff radius and $d(x_i, x_j) = \|x_i - x_j\|$. δ_i can be measured by finding the minimum distance between the data point x_i and other higher density points, i.e.,

$$\delta_i = \min_{j:\rho_j > \rho_i} d(x_i, x_j). \tag{4}$$

The cluster centers are these points that have relatively higher δ_i and ρ_i. After the calculation of $\gamma_i = \rho_i \delta_i$, $i = 1, 2, \ldots, n$, the points with higher γ_i are regarded as cluster centers. After determining the number of cluster centers, the clustering process can be used after scanning all points only once.

DPC, DBSCAN, and C-means have their own applicable ranges, respectively. Figure 2 shows three datasets with different characteristics: two nonspherical clusters in (a), three density-different clusters in (b), and three partially-overlapped clusters in (c), respectively. The clustering results by the above three algorithms are presented in these figures as well. Different numbers denote the real labels of the datasets, and different colors denote the clustering results by the corresponding clustering algorithms.

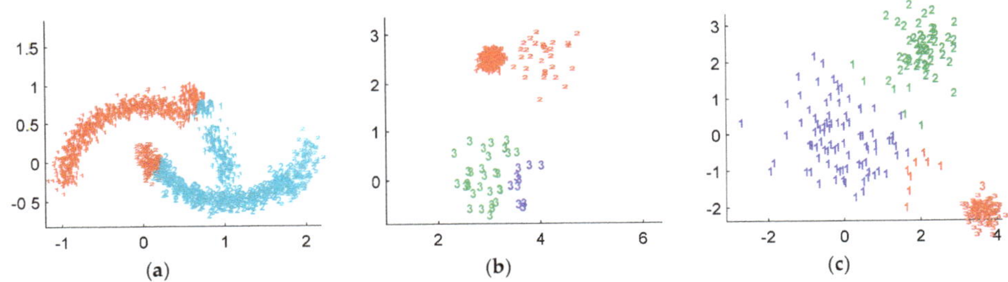

Figure 2. Three datasets with different characteristics and the clustering results by using density peak-point-based clustering (DPC), Density-Based Spatial Clustering of Applications with Noise (DBSCAN) and C-means: (**a**) two nonspherical clusters; (**b**) three density-different clusters; (**c**) three partially-overlapped clusters.

Figure 2a represents that DPC cannot work well when the investigated dataset contains a cluster with multiple density peaks. When one cluster has multiple peaks, DPC may regard these peaks as cluster centers, resulting in wrong clustering partition. Figure 2b shows the clustering results using the DBSCAN algorithm. In this case, there is none value of ε to make DBSCAN find a correct clustering result. Figure 2c shows that the two clusters in the left are partially overlapped and have a similar density. C-means cannot find boundary points correctly, which leads to wrong clustering results.

When dealing with datasets in high dimensional (HD) spaces, these three clustering algorithms show different characteristics. Different from C-means and DPC, DBSCAN cannot effectively cluster points in HD datasets. First, the computed densities in DBSCAN have little difference among various high-dimensional (HD) points and cause that core points are difficultly determined. Accordingly, the merging condition to the same cluster from one to another core points has high uncertainty. Second, to accelerate DBSCAN, an R* tree structure indexing all points have to be used to decrease computational complexity, but it is challenging to be built in an HD space. Inversely, C-means depends on distance computation rather than density computation, and the distance differences among points can be much larger than their density differences. Also, DPC uses not only density ρ but also separation measure δ to find all abnormal points as cluster centers, and can significantly avoid the HD problem in DBSCAN. Consequently, the HD problem plays little effect on C-means and DPC.

C-means algorithm needs users to provide cluster number as the input parameter, which has an essential effect on clustering results. DPC has the same problem as C-means, which needs to provide the number of clusters in advance. The clustering process of DBSCAN has relevance with point density. Different values of neighborhood radius will result in different numbers of clusters, which can affect the clustering results.

Table 1 shows the characteristics of typical clustering algorithms when dealing with different types of datasets. The sign "×" denotes that the algorithm cannot cluster the corresponding types of datasets effectively according to the accuracy and applicable range, while "√" has the opposite meaning. The sign "√/×" means that the algorithm can obtain correct clustering results sometimes, while in some cases will not.

Table 1. Applicable range of typical clustering algorithms.

Types/Algorithm	C-Means	DBSCAN	DPC
Arbitrary shape	×	√	√/×
Density-diversity	√/×	×	√/×
overlap	√/×	×	√
High-dimension	√	×	√/×
The number of clusters	×	×	×

2.2. Typical Cluster Validity Index

The validity index is a function, which regards the number of clusters (c) as its variable. This function can obtain its maximum or minimum value when c is the correct number of clusters. It can be formulated as follows

$$\text{Max(min) } z = f(c), c = 1, 2, \dots, C. \tag{5}$$

The intra-cluster distances denote the compactness of a cluster while inter-cluster distances estimate the separation among clusters [30,31]. The trial-and-error strategy can be used to find the optimum solution in (5), as shown in Figure 3.

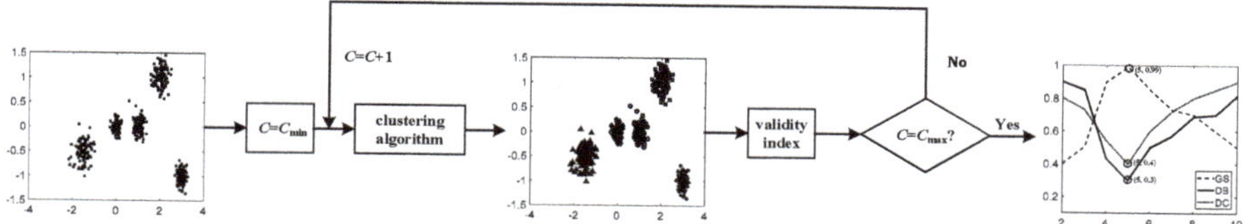

Figure 3. Evaluation processes of three existing validity indices.

Figure 2 and $c_{max} < \sqrt{n}$ [32] if there is no prior knowledge, where n is the number of points in dataset X. Afterwards, apply an applicable clustering algorithm to X with the value of c set from c_{min} to c_{max}. Calculate the corresponding value of (5). The maximum or minimum values of (5) indicates the optimal number of clusters. Note that different validity indices consist of different combinations of intra- and inter-cluster distances, and thus lead to different evaluation results.

In the following, three typical validity indices DB, GS, and, DC are illustrated. The evaluation process starts with $c_{min} = 2$, and ends with c_{max} that is large enough. The maximums or minimums of the validity indices denote the optimal number of clusters, as explained below.

2.2.1. Davies–Bouldin (DB) Index

Let Δ_i and z_i be the intra-cluster distance measure and cluster center of the i-th cluster, respectively. Let δ_{ij} denote the inter-cluster distance measure between clusters of C_i and C_j, and c can take values in $[c_{min}, c_{max}]$. The DB index [6] can be defined as

$$DB = \sum_{i=1}^{c} R_i / c, \ s.t., R_i = \max_{j, j \neq i}(\Delta_i + \Delta_j) / \delta_{ij}, \ \delta_{ij} = \|z_i - z_j\|, \ \Delta_i = \sum_{x \in C_i} \|x - z_i\| / |C_i|. \quad (6)$$

2.2.2. Dual-Center (DC) Index

For any clustering center v_i determined by a partitional clustering algorithm, assume $\dot{v}_i$ is the closest prototype to v_i, then the dual center is calculated as $\ddot{v}_i = (v_i + \dot{v}_i)/2$. Finally, a novel dual-center (DC) index [33] can be constructed, i.e.,

$$DC_c = \sum_{i=1}^{c} \Delta_i(c) / \sum_{i=1}^{c} \delta_i(c), \ s.t., \Delta_i = \sum_{j=1}^{n_i(c)} (x_j - v_i)^2, \ \delta_i = \sum_{j=1}^{\ddot{n}_i(c)} (x_j - \ddot{v}_i)^2, \quad (7)$$

where $n_i(c)$ and $\ddot{n}_i(c)$ are the number of points of the i-th cluster when the prototypes are regarded as v_i and $\ddot{v}_i$, respectively. Among the existing validity indices, DC has higher accuracy and robustness when dealing with both artificial datasets and real datasets in UCI [34].

2.2.3. Gap Statistic (GS) Index

The gap statistic (GS) index [7] firstly computes an intra-cluster measure as

$$W_c = \sum_{i=1}^{c} D_i / (2|C_i|), s.t., D_i = 2|C_i| \sum_{j \in C_i} \|x_j - \bar{x}\|, \bar{x} = \sum_{i=1}^{|C_i|} x_i / |C_i|. \quad (8)$$

Owing to the subjectivity of the detection of inflection point, GS can be formulated as

$$Gap_c = E^*[\log(W_c)] - \log(W_c), \ and \ W_c = \sum_{i=1}^{c} D_i / (2|C_i|), \quad (9)$$

where E^* denotes the expectation under a null reference distribution.

In sum, the above validity indices all take a trial-and-error way for a single specified clustering algorithm rather than general clustering algorithms. Moreover, these validity indices are the function of the number of clusters and are not designed for other possible clustering parameters. Therefore, an efficient and comprehensive method is necessary, which can evaluate clustering results for any clustering algorithm and arbitrary clustering parameters. In this paper, our proposed validity index presents an accurate solution to solve the above problem in a general way.

3. Materials and Methods

In a nonparametric way, we firstly partition all points into two groups, boundary and interior points, which are used to access the boundary matching degree and connectivity degree. By integrating these two quantities, a novel clustering evaluation index can be formed.

3.1. Boundary Matching and Connectivity

The density of any point in the existing clustering analysis is computed by counting the number of points in the point's neighborhood with a specified radius. However, the computed density only is a group of discrete integers such as 1, 2, ... , and thus many points have the same density which is indistinguishable. Moreover, the defined density may greatly be affected by the specified radius.

In this study, we first define a nonparametric density to find all boundary and interior points in any dataset. Assume $X = \{x_1, x_2, \ldots , x_n\}$ is a dataset in a D-dimensional space R^D. For any data point $x_k \in X$, its m nearest neighbors are denoted as, $x_{k,1}, x_{k,2}, \ldots , x_{k,m}$, with distances $d(x_k, x_{k,1})$, $d(x_k, x_{k,2})$, $\ldots , d(x_k, x_{k,m})$, where m is the integer part of $2D\pi$, $k = 1, 2, \ldots , n$. Here, $2D$ shows that one interval in any dimension in R^D can be measured by the two-interval endpoints, and π is a conversing coefficient when the m points are enclosed by a spherical neighborhood that is used in the existing density computation. Therefore, the density of any data point x_k in X is defined as

$$density(x_k) = \left\{ \sum\nolimits_{j=1}^{m} d(x_k, x_{k,j}) \right\}^{-1}, \ k = 1, 2, \ldots, n \tag{10}$$

Definition 1. *Boundary point and interior point. A point is called as a boundary or interior point if half of its m nearest neighbors have a higher or lower (equal) density than its density, respectively.*

The proposed notion of boundary and interior points have the following two characteristics.

(1) Certainty. Unlike the used density in other existing algorithms, the proposed density is fixed and unique for any point, which reduces the uncertainty in the clustering process. In fact, the clustering results in other algorithms may greatly be changed as the number of neighbors used for computing the density increases or decreases. Note that the effective estimation of the number of neighbors has been a difficult task, and so far, it remains unsolved [35].

(2) Locality. The classification of border or interior points is defined only by its m nearest neighbors, so the separating boundary or interior points presents local characteristics. Inversely, DBSCAN uses a global density to distinguish border or interior points in a density-skewed dataset, which even causes an entire cluster to be perfectly regarded as border points (see Figure 4). Therefore, the proposed density is a more reasonable local notion.

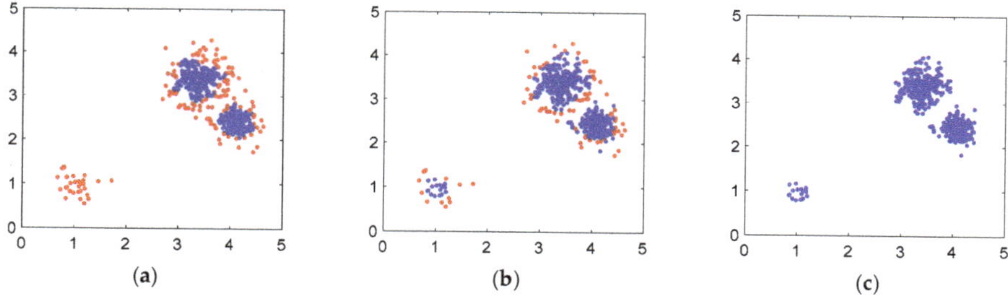

Figure 4. Comparison of boundary points determined by DBSCAN and (10): (**a**) global density; (**b**) local density; (**c**) interior points.

Figure 4a shows a density-skewed dataset with three clusters of large, medium, and small density, respectively. The red points in Figure 4a,b represent the boundary points computed by using DBSCAN and (10), respectively. DBSCAN can find no interior points in the cluster with the lowest density; in comparison, owing to the locality of (10), the red border points and blue interior points determined are distributed more reasonably. The interior points are located at the center of any cluster and surrounded by border points, while the border points construct the shape and structure of any clusters. Specifically, after the removal of border points in any dataset, the separation of clusters is greatly enhanced (see Figure 4c). Therefore, the real number of clusters can be determined more easily by any clustering algorithm.

In graph theory, a cluster is defined as a group of points that connect to each other [36,37]. In order to assess the connectivity of points in a dataset X, we calculate the density for all points in X and sort them in the order of increasing density. Assume x_{max} is the point with the highest density in X, and thereby, a connecting rule among points is defined as follows. For any point $x_k \in X$, the next point x_{k+1} is the point which is the nearest neighbor of x_k but has a higher density than x_k. Subsequantly, repeat the above steps until visiting the point x_{max}.

Definition 2. *Chain. A chain is a subset of X that starts with any data point x_i in X and stops at x_{max} based on the above connecting rule.*

There is a unique chain from any point in X since the nearest neighbor of each point is unique. The above steps are repeated until each point has been visited in X. Consequently, all n points in X respond to n chains, denoting them as $S_1, S_2, \ldots, S_n$. The largest distance between adjacent points in t-th chain S_t is denoted as $dis(S_t)$, $t = 1, 2, \ldots, n$. Figure 5 shows all chains in two datasets with various characteristics, where the arrow is the direction from low to high-density points. The green dotted circles denote the points with maximum densities.

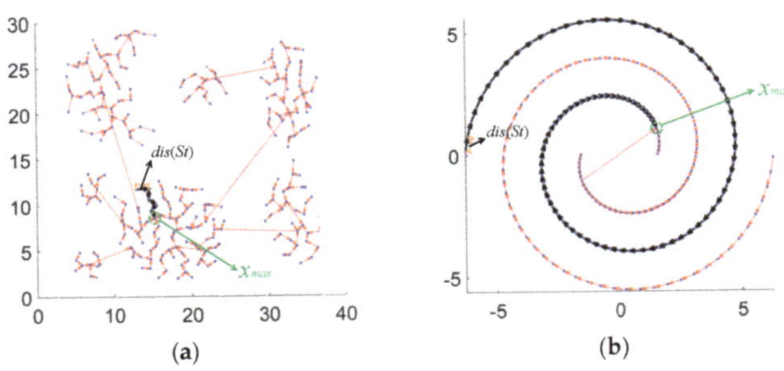

Figure 5. All chains in two datasets with different characteristics: (**a**) chains in a dataset of spherical clusters; (**b**) chains in a dataset of nonspherical clusters.

Figure 5 shows that the value of $dis(S_t)$ is small when a chain perfectly is contained in a cluster, but abnormally becomes large when a chain bridges a cluster and the other cluster. In views of the notation of the chain, we further define a notation of connectivity of X as follows.

Definition 3. *Connectivity. Let $S_1, S_2, \ldots, S_n$ be n chains in X, then the connectivity of all points in X is defined as*

$$con(X) = \left(\sum\nolimits_{t=1}^{n} dis(S_t) \right)^{-1}. \tag{11}$$

Along with the graph theory, the value of $con(X)$ indicates the degree of compactness of a dataset. It can reflect whether a chain is contained in a cluster, as explained and illustrated in the next section. In this paper, we use the notion of boundary matching degree and connectivity degree to access the clustering results obtained by any clustering algorithm.

3.2. Clustering Evaluation Based on Boundary and Interior Points

Once a clustering algorithm has partitioned a dataset X into c disjoint clusters, i.e., $X = C_1 \cup C_2 \cup \ldots \cup C_c$, we substitute X by $C_1, C_2, \ldots, C_c$, respectively, and find their boundary points using (10). The set of boundary points in C_k is denoted as BC_k, while the set of boundary points in X is BX. A boundary-point-matching index is defined as

$$bou(c) = \sum\nolimits_{k=1}^{c} \frac{|BX \cap BC_k|}{|BX \cup BC_k|}. \tag{12}$$

Equation (12) measures the matching degree of boundary points between the entire dataset X and the disjoint C clusters. In the mathematical meaning, it is clear that the values of bou (c) must fall in the interval $[0, 1]$. The following example can explain the cases that are smaller and equivalent to 1.

Figure 6a–c show the boundary and interior points determined by (10) when performing C-means algorithm at c is smaller, equal to or larger than 6, respectively, where interior points refer to the remaining points after removing boundary points.

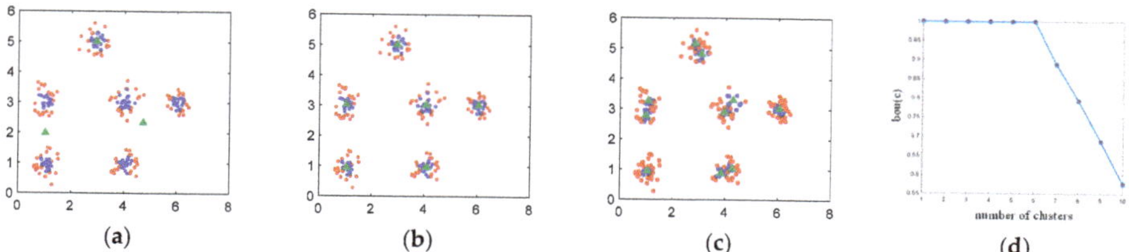

Figure 6. (**a**–**c**) shows the determined boundary points by using C-means and (10) when $c = 3, 6$, and 10, respectively. Blue and red points refer to the interior and boundary points, respectively; (**d**) shows the curve of *bou* (c) calculated according to (11); the sign Δ means the center of C-means.

The red boundary points at $c = 3$ and 6 are similar to these in X, but these boundary points at c equals 10 are different from those in X. Figure 6d shows that the values of *bou* (c) are nearly unchangeable when $c < 6$ but decrease fast when $c > 6$. When the number of clusters c is smaller than the actual one, and usually, any cluster does not be assigned two cluster centers. Therefore, the set of boundary points of all partitioned clusters are consistent with the entire set X, and *bou* $(c) = 1$. When c is larger than the actual number of clusters, there is at least one cluster the number of whose boundary points increase. Thus, *bou* $(c) < 1$. It can be seen that the values of *bou* (c) are helpful to find the real number of clusters for any clustering algorithm. Alternatively, we can regard any cluster in $C_1, C_2, \dots ,$ C_c as an independent dataset like X, and accordingly, x_{max} in X become these points with maximal density in $C_1, C_2, \dots ,$ and C_c, respectively. Thereby, we assess the connectivity among points according to (11) when $c = 1, 2, \dots , c_{max}$. The connectivity of X is reduced to the connectivity of each cluster. As c increases, the connectivity is enhanced since the number of maximal inter-cluster distances in these chains decreases.

Figure 7a–c show the connectivity calculated using (11) and C-means when c equals 3, 6, and 10, respectively. This value becomes smaller when $c < 6$, while it tends to be flat when $c > 6$, as shown in Figure 7d. Consequently, there is an inflection point on the curve in Figure 7d.

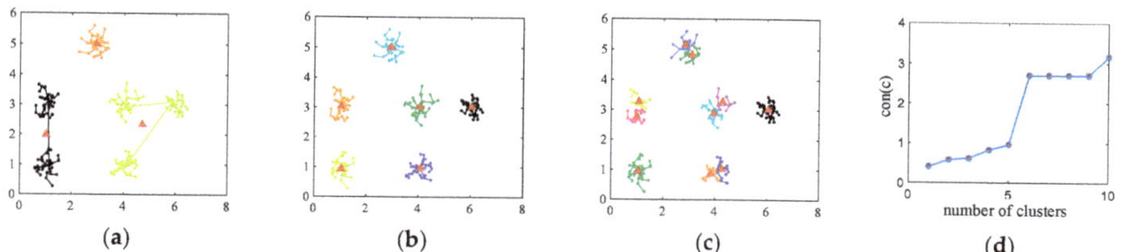

Figure 7. (**a**–**c**) The chains when c = 3, 6, and 10, respectively; (**d**) shows the curve of (11). Note that the points in the same color indicate that they are partitioned into the same cluster by using C-means algorithm and the sign Δ means the center of C-means.

As c increases, both the curves calculated according to (11) and (12) have inversely varying tendencies. It is expected that the real number of clusters is encountered at c^*, where the curve of (11) turns to be flat from fast-changing, and that of (12) becomes fast varying from slow-changing.

Considering the variances of *bou* (c) and *con*(c) can be calculated by curvature radius mathematically, we define a novel validity index according to *bou* (c) for boundary points and *con*(c) for interior points, respectively. By combining (11) and (12), we define a function as follows

$$F(c) = R_1(c) \bullet R_2(c)$$

$$s.t., \begin{cases} R_1(c) = \left|\Delta_1(c)\right|^2 / \left(1 + (\nabla_1(c))^2\right)^{3/2} \\ \Delta_1(c) = bou(c+1) + bou(c-1) - bou(c), \nabla_1(c) = bou(c+1) - bou(c-1) \\ R_2(c) = \left|\Delta_2(c)\right|^2 / \left(1 + (\nabla_2(c))^2\right)^{3/2} \\ \Delta_2(c) = con(c+1) + con(c-1) - con(c), \nabla_2(c) = con(c+1) - con(c-1) \end{cases} \qquad (13)$$

where the symbol Δ denotes a two-order difference operator of *bou* (c) and $con(c)$, aiming to locate the maximal inflection points on curves of *bou* (c) and $con(c)$, respectively. The optimal number of clusters c^* for any dataset is computed as

$$c^* = \mathrm{argmax}_c F(c). \qquad (14)$$

The proposed validity index has the following characteristics.

(1) Complementarity. The mathematical curvature and difference can reflect the varied tendency of a curve in (13), and thereby the real number of clusters c^* can be found. When $c < c^*$, $R_1(c)$ is nearly equivalent to $R_1(c^*)$ since the set of boundary points is approximately unchangeable, but $R_2(c) < R_2(c^*)$, since the number of center points successively increase. In sum, $F(c) < F(c^*)$. Inversely, when $c > c^*$, $R_1(c)$ successively decreases, and $R_2(c)$ tends to be flat; therefore, $F(c) < F(c^*)$. Consequently, (13) can attain a maximum when c^* appears.

(2) Monotonicity. Assume that c^* is the real number of clusters, and when c takes its two values c_1 and c_2 satisfying $c_1 < c_2 < c^*$, $F(c_1) < F(c_2) < F(c^*)$; otherwise, when $c^* < c_1 < c_2$, $F(c^*) > F(c_1) > F(c_2)$; Hence, $F(c)$ consists of two monotone functions at the two sides of c^*, respectively. Therefore, for arbitrary two values c_1 and c_2 satisfying $c_1 < c_2$, c_2 may refer to a more optimal clustering result. Usually, a larger value of $F(c^*)$ indicates a better clustering result.

(3) Generalization. Equation (13) can provide a wide entry for any clustering algorithm only if the clustering results are available and the corresponding numbers of clusters are taken as the variable of $F(c)$. Especially, a group of clustering results may result from different clustering algorithms and parameters, since any two clustering results are comparable according to the above monotonicity. For example, one clustering result with c_1 results from C-means, and others from DBSCAN, and so on. Equation (13) can evaluate the results of any clustering algorithm and parameter in a trial-and-error way. In comparison, the existing validity indices can mainly work for a specific algorithm and parameter of the number of clusters since the center in them has to be defined, especially for the C-means algorithm.

Hereafter, the cluster validity index of (13) based on boundary and interior points is called CVIBI. The evaluating process for any clustering results based on CVIBI is listed in Algorithm 2.

Algorithm 2. Evaluating Process Based on CVIBI

Input: a dataset $X \in R^D$ containing n points and clustering results from any clustering algorithm at $c = 1, 2, \ldots, c_{max}$.

Output: the suggested number of clusters.

Steps:

1. Calculate the density for each point in X according to (10);
2. Partition X into boundary and interior points;
3. Input clustering results at $c = 1, 2, \ldots, c_{max}$;
4. Partition each cluster into boundary and interior points;
5. Compute values of *bou* (c) or $con(c)$ at c equals $1, 2, \ldots, c_{max}$;
6. Solve the optimal value of (13);
7. Suggest an optimal number of clusters.
8. Stop.

---

4. Results and Discussion

We test the accuracy of CVIBI on two groups of typical datasets and compare it with three existing validity indices, i.e., DB, DC, and GS. In views of different characteristics of the investigated datasets, the clustering results are obtained using C-means, DPC, and DBSCAN algorithms, respectively, where the number of neighbors m in the experiments is fixed at the integer part of $2D\pi$.

4.1. Tests on Synthetic Datasets

Figure 8 shows five groups of synthetic datasets generated by the Matlab® toolbox, and each group consists of three datasets with different numbers of clusters. The determined boundary and interior points are in red and blue, respectively. The 15 datasets are denoted as sets 1–15, respectively. groups 1–4 contain the datasets with various densities, sizes, shapes, and distributions, respectively; and group 5 contains overlapped clusters.

The characteristics of these datasets are listed in Table 2. The first column of Table 2 denotes the names of these datasets. The second and fourth columns represent the numbers of clusters and objects of datasets, respectively. The third column denotes the dimensions of these datasets. The last column shows the number of objects of each cluster in datasets.

Table 2. Characteristics of 15 datasets in Figure 8.

Name	Clusters	Dimension	Number of Objects	Number of Each Cluster
Set 1	3	2	600	83/164/353
Set 2	4	2	350	30/60/120/240
Set 3	5	2	830	30/60/120/240/480
Set 4	3	2	420	60/120/240
Set 5	4	2	900	60/120/240/480
Set 6	5	2	1860	60/120/240/480/960
Set 7	3	2	1018	341/336/341
Set 8	4	2	404	134/90/90/90
Set 9	5	2	494	134/90/90/90/90
Set 10	3	2	360	120/120/120
Set 11	4	2	800	200/200/200/200
Set 12	5	2	1000	200/200/200/200/200
Set 13	3	2	600	200/200/200
Set 14	4	2	400	100/100/100/100
Set 15	5	2	1000	200/200/200/200/200

The number of neighbors m plays an essential role in CVIBI. Different values of m may lead to very different density and distance values, and thus to different evaluation results.

As an example, we solve the range of m in which each number can lead to the correct number of clusters when using DPC. Let $m = 1, 2, \ldots, n$, respectively. Along with these values of m, DPC is used to cluster all points in sets 1–15, respectively. Figure 9 shows the solved ranges in the 15 datasets that are represented by blue bars, where the red line in any bar denotes the value that is the integer part of $2D\pi$. All the experimental datasets used in Figure 8 are in the two-dimensional data space. Thus, $D = 2$ and the integer part of $2D\pi$ is just 12. Figure 9 shows that all the values of $2D\pi$ fall into the solved ranges corresponding to the correct number of clusters, which demonstrates the effectiveness and robustness of $2D\pi$. Consequently, we can fix the number of neighbors m at $2D\pi$ in experiments.

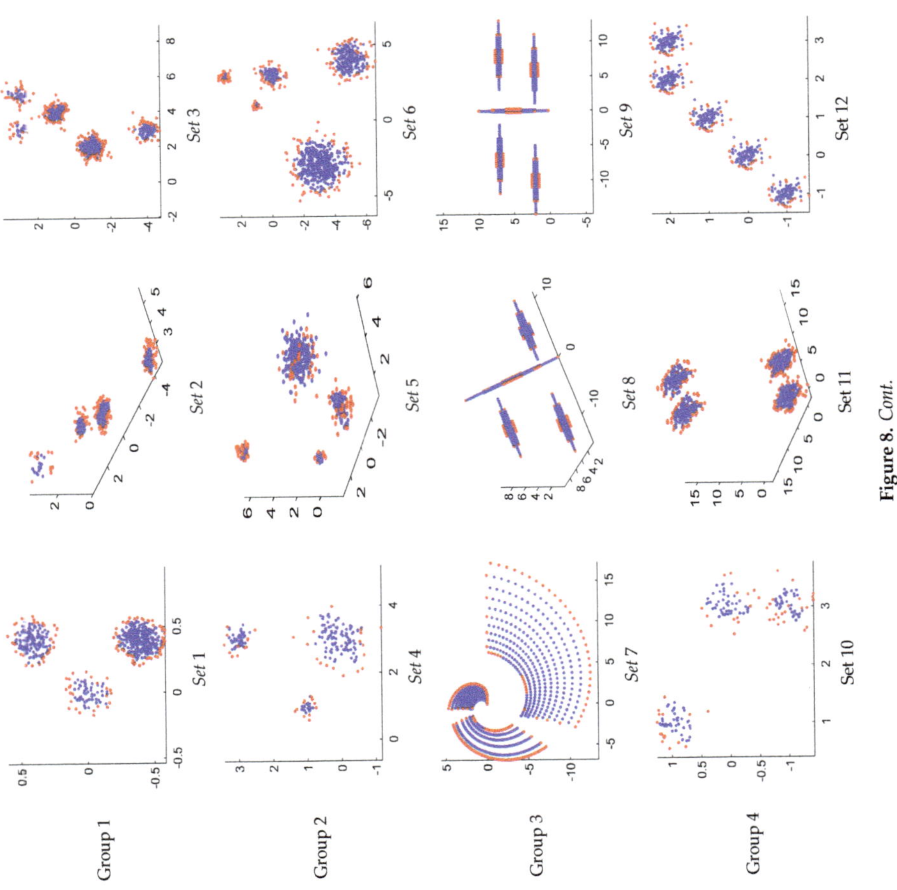

Figure 8. *Cont.*

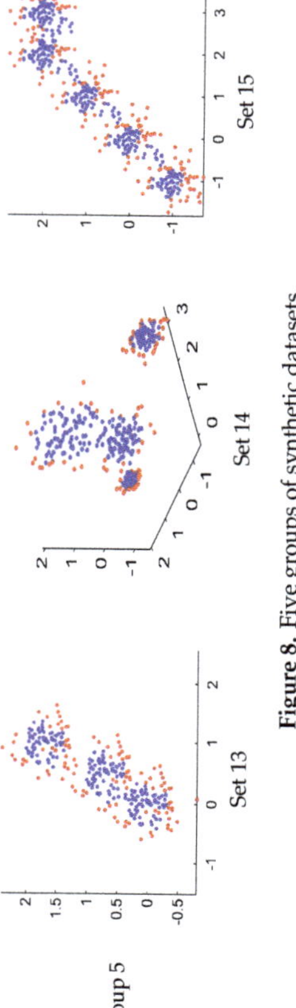

Figure 8. Five groups of synthetic datasets.

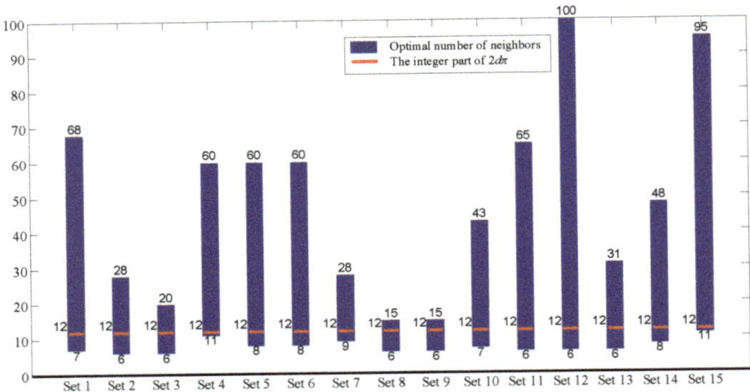

Figure 9. The solved range of the number of neighbors in 15 datasets.

4.1.1. Relationship between CVIBI and the Number of Clusters

The correctness of clustering evaluation depends on whether the correct number of clusters can be found by the validity index. In order to test the correctness of CVIBI, all points in these 15 datasets are partitioned into c clusters by DPC, C-means, and DBSCAN at $c = 1, 2, \ldots , c_{max}$, respectively, where the number of clusters in DBSCAN is obtained by taking various values of ε. Figure 10 shows the curves of CVIBI based on the three algorithms, respectively. The points marked by small circles in these curves are the suggested optimal values of (13).

Datasets in group 1 and group 2 contain clusters with different densities and sizes, respectively. CVIBI can point out these optimal clustering results on the above six datasets no matter what clustering algorithms are used. In terms of various shapes in *Group 3*, CVIBI works well based on DPC and DBSCAN, but C-means cannot. Because C-means originally is designed to partition spherical clusters rather than arbitrary shapes, and then incorrect partitions of boundary points and incorrect values of *con* (c) are caused. When the clusters in datasets contain different distributions (e.g., group 4), CVIBI based on DPC shows better performance than C-means and DBSCAN according to accuracy. Because C-means and DBSCAN cannot give correct clustering results, DPC can obtain relatively correct results. When the dataset has clusters that partially overlap with each other (e.g., group 5), DPC can obtain relatively correct clustering results than C-means and DBSCAN, as shown by the corresponding curves. The merit of CVIBI is to select the best one from any candidates of clustering results, no matter what clustering algorithm is used. However, if all available clustering results cannot contain the real number of clusters, CVIBI can find it as well.

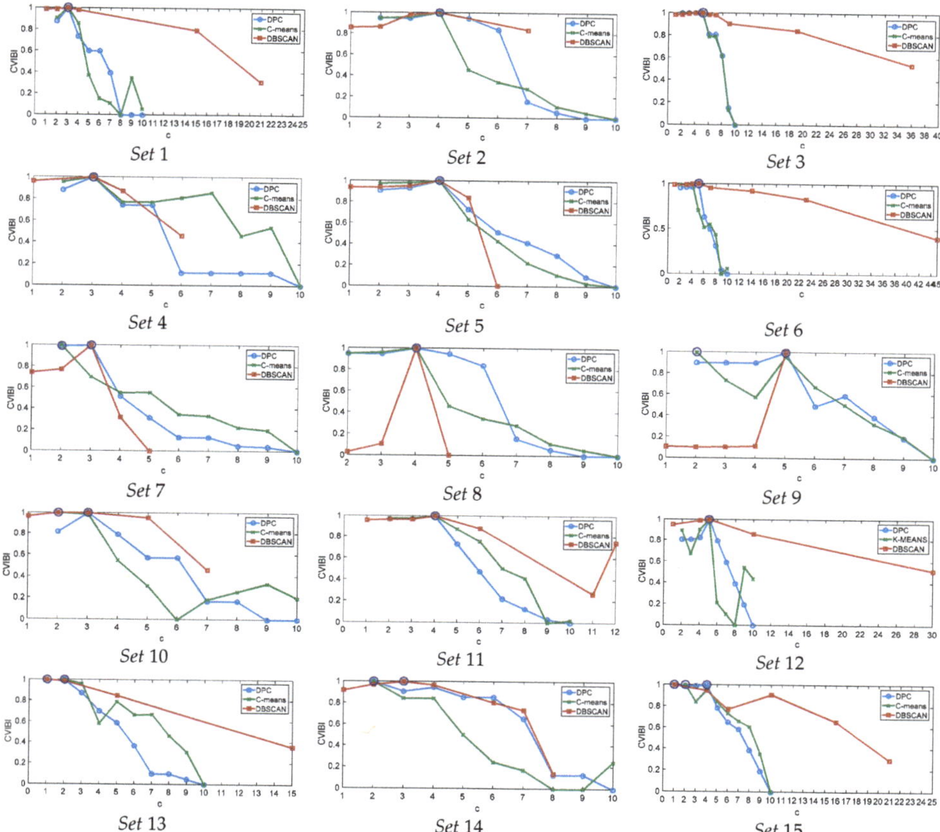

Figure 10. Evaluation results of CVIBI obtained using DPC, C-means, and DBSCAN.

4.1.2. Relationship between CVIBI and ε in DBSCAN

When using DBSCAN, any certain number of clusters responds to a continuous interval of *Eps* (i.e., ε). The clustering results of DBSCAN highly depend on the value of *Eps*. In order to solve the optimal values of *Eps*, we determine the possible range of *Eps* first. Then, a group of *Eps* values is evenly sampled to approximate the *Eps* variable itself. Then, we cluster the dataset using DBSCAN with the group of *Eps* values. Finally, the optimal values of *Eps* can be determined by CVIBI. Figure 11 shows the evaluation results of CVIBI with respect to *Eps*, where piecewise numbers chain in these CVIBI curves is the obtained number of clusters. The abscissa and vertical axes denote the value of *Eps* and CVIBI, respectively. The number along with each red point denotes the number of clusters computed by DBSCAN. If CVIBI can compute the real number of clusters, then a green box will be drawn and point out the optimal value of *Eps*. The maximum value of each curve in Figure 11 indicates the suggested optimal number of clusters. The corresponding values of *Eps* are the suggested optimal parameters. It also can be seen that the optimal *Eps* is not one certain number but a range of values. CVIBI with DBSCAN can point out the correct clustering results except *Set* 10 and group 5. Set 10 has three clusters and one cluster is separated from the other two clusters. The other two clusters are close to each other, which are easy to be treated as one cluster. CVIBI with DBSCAN algorithm regards these two clusters as one cluster, and the maximum value occurs when $c = 2$. The clusters in group 5 are significantly overlapped, so CVIBI cannot recognize the real numbers of clusters. Consequently, CVIBI can point out the optimal clustering results by finding the optimal parameter of *Eps*.

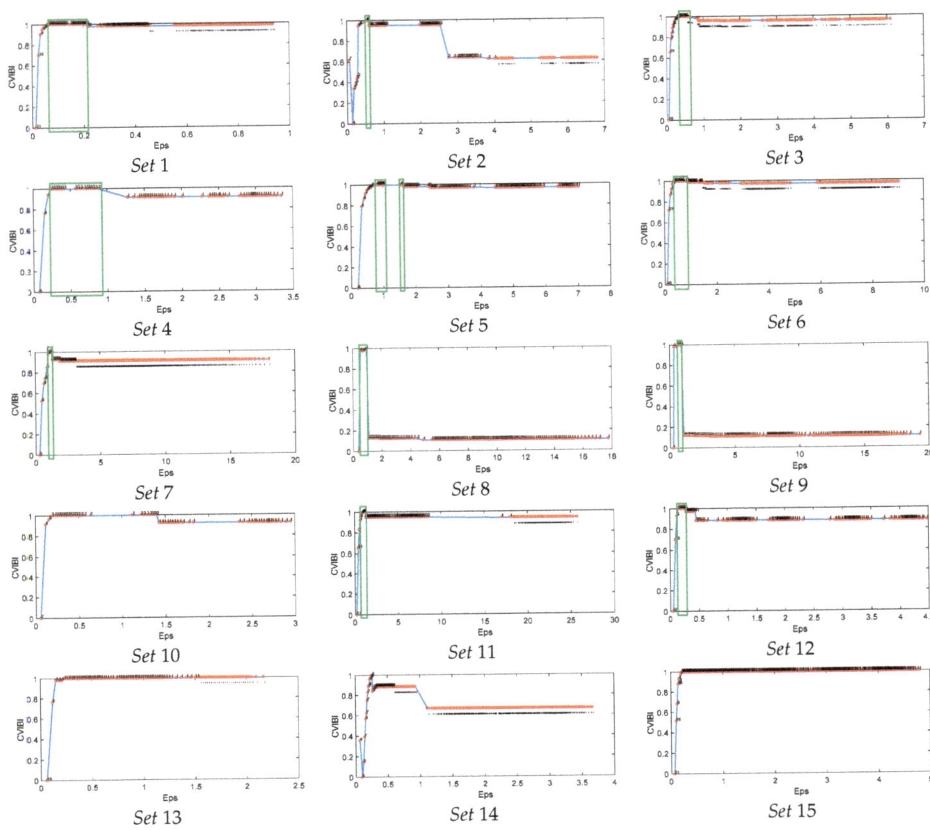

Figure 11. Evaluation results of CVIBI based on DBSCAN.

4.1.3. CVIBI Evaluation under Various Clustering Algorithms

Once a dataset is clustered by several clustering algorithms that lead to different clustering results, how one can select the optimal clustering result remain a challenging task for the existing validity indices. Nevertheless, CVIBI can realize this purpose. For example, CVIBI obtains three curves based on C-means, DBSCAN, and DPC algorithms. Each curve points out a suggested number of clusters, which may be different from the others. The optimal number of clusters can be selected by comparing the values of the boundary matching degree of *bou* (c).

Figure 12 shows the evaluating process of the multiple-peak dataset that is shown in Figure 2a. It shows that CVIBI based on DBSCAN suggests 2 as the optimal number of clusters, but CVIBI based on C-means and DPC both do 1. To assess their differences, we further analyze their values of *bou* (c). *bou* (c) = 1 in CVIBI based on C-means and DPC when c = 1, while *bou* (c) = 1 in CVIBI based on DBSCAN when c = 1 and 2. In views of the clustering process, we can conclude that any real cluster in all clusters is not separated when executing clustering algorithm if *bou* (c) = 1. Let Γ be the set of numbers of clusters whose *bou* (c) = 1. The maximum value in Γ indicates the optimal number of clusters in the multiple-peak dataset. So in the situation, both CVIBI and *bou* (c) can indicate that the real number of clusters is 2, and the clustering algorithm suitable for this dataset is DBSCAN.

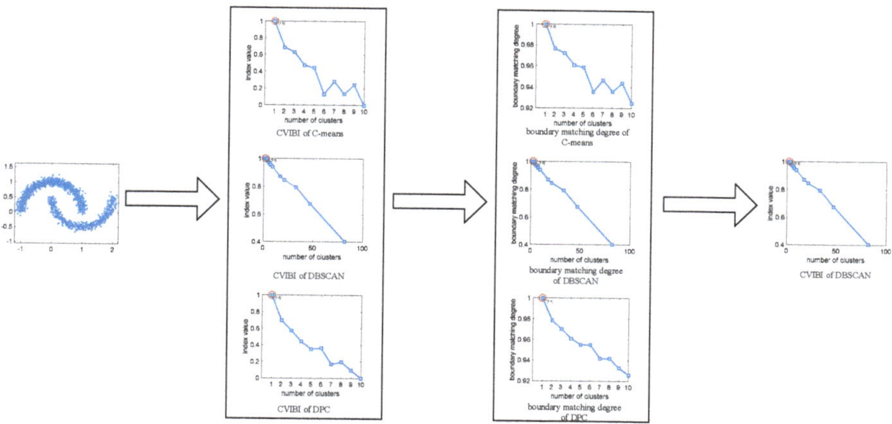

Figure 12. Evaluating the process of CVIBI on the multi-peak dataset.

Figure 13 shows the evaluating process of the density-diverse dataset in Figure 2b. We can conclude from Figure 13 that the suggested numbers of clusters are different. CVIBI with DBSCAN and DPC both suggest 2 as the optimal number of clusters while CVIBI based on C-means regards 3 as the optimal number of clusters. To assess which clustering result is best among the three clustering algorithms, the three curves of *bou* (*c*) are used to consist of Γ. Finally, the optimal number of clusters can be determined from Γ. It can be extracted that Γ set contains 1, 2, and 3 in total. So the suggested number of clusters is 3, which is the maximum value in Γ. The optimal algorithm for the density-diverse dataset is C-means algorithm.

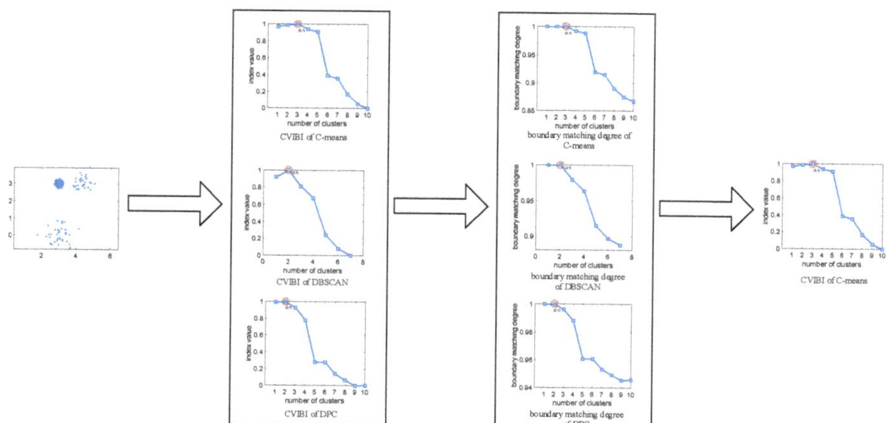

Figure 13. Evaluating process of CVIBI on density-diverse dataset.

Figure 14 shows the evaluating process of the dataset in Figure 2c. The results of the CVIBI index with three clustering algorithms are different. CVIBI based on C-means, DBSCAN, and DPC suggest 1, 2, and 3 as the real number of clusters, respectively. In order to obtain the optimal number of clusters, three curves of *bou* (*c*) are used. From the relationship between Γ and the real number of clusters, *bou* (*c*) of C-means, DBSCAN and DPC suggest 1, 2, and 3 as the optimal number of clusters, respectively. So the real number of clusters is 3, which is suggested by CVIBI with DPC. The suitable clustering algorithm of the investigate dataset is the DPC algorithm.

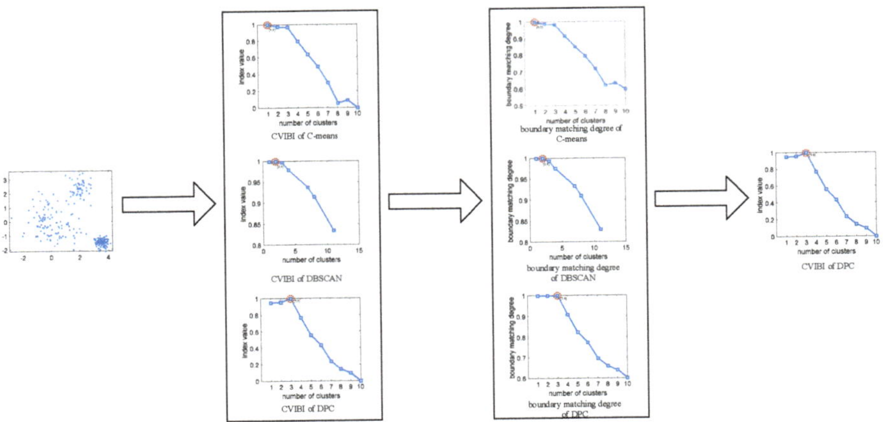

Figure 14. Evaluating the process of CVIBI on the overlapped dataset.

4.1.4. Comparison between CVIBI and Existing Validity Indices

Table 3 shows all evaluation results obtained by CVIBI, DB, DC, and GS when using the above three clustering algorithms. The four validity indices are analyzed as follows.

(1) Size and distribution. Groups 2 and 4 denote clusters with different sizes and distributions, and there are no overlapped clusters. The evaluation results show that except set 10, all validity indices are capable of determining the correct number of clusters no matter which clustering algorithm is used.

(2) Density and shape. Groups 1 and 3 contain clusters with different densities and shapes. When the densities among clusters are so skewed as in sets 1 and 3, the three validity indices cannot find the correct number of clusters. Compared with DC and DB, the cluster numbers calculated by CVIBI are closer to the real numbers of clusters.

(3) Overlap. Group 5 contains datasets with overlapped clusters. DPC is good at clustering such datasets, and the evaluation results of CVIBI are nearest to the real number of clusters. However, DC and GS with DPC give a relatively smaller number. Note the two indices are originally designed to evaluate the results for nonoverlapping clusters. Thus any two overlapped clusters may be incorrectly regarded as one cluster. For instance, the clusters in set 14 are most overlapped, and its evaluation results are the weakest. How to find the correct number of clusters in a dataset containing overlapped clusters is a difficult task for most existing validity indices, such as DB and GS; fortunately, CVIBI with DPC is effective for dealing with such a problem. In summary, DC shows performances better than CVIBI in the case of datasets containing spherical clusters.

Table 3. Evaluation of clustering results by CVIBI, Davies–Bouldin (DB), dual-center (DC) and gap statistic (GS) for 15 datasets.

Dataset and Algorithm	DB			DC			GS			CVIBI		
	C-Means	DBSCAN	DPC	C-Means	DBSCAN	DPC	C-Means	DBSCAN	DPC	C-Means	DBSCAN	DPC
Set 1	3√	2	3√	3√	2	3√	3√	2	3√	3√	3√	3√
Set 2	4√	3	4√	4√	3	4√	4√	2	4√	4√	4√	4√
Set 3	4	6	4	4	4	5√	3	3	5√	5√	5√	5√
Set 4	3√	3√	3√	3√	3√	3√	3√	3√	3√	3√	3√	3√
Set 5	4√	4√	4√	4√	4√	4√	4√	4√	4√	4√	4√	4√
Set 6	4	5√	4	5√	4	5√	4	5√	5√	5√	5√	5√
Set 7	2	2	2	2	2	2	2	2	2	2	3√	3√
Set 8	2	4√	2	4√	2	4√	2	2	4√	2	4√	4√
Set 9	2	5√	2	5√	2	5√	2	2	5√	2	5√	5√
Set 10	2	2	2	2	2	2	2	2	2	2	2	3√
Set 11	4√	4√	4√	4√	4√	4√	4√	4√	4√	4√	4√	4√
Set 12	5√	5√	5√	5√	5√	5√	5√	5√	5√	5√	5√	5√
Set 13	2	1	2	1	2	1	2	2	1	2	1	2
Set 14	2	2	2	2	2	2	2	2	2	2	3	2
Set 15	2	1	2	1	2	1	2	2	1	2	1	4

Note: for the sign 'Set x/y', x and y refer to the investigated dataset and the correct number of clusters, respectively.

4.2. Tests on CT Images

Digital Imaging and Communications in Medicine (DICOM) is a standard protocol for the management and transmission of medical images, which is widely used in healthcare facilities. Here, we choose CT images as test datasets, which are stored in accordance with the DICOM standard. Each image contains 512×512 pixels, and each pixel is identified by its CT value. Figures 15 and 16 show the clustering and evaluation results of CVIBI for a group of typical CT images. The first column represents the original CT images. Columns 2–4 represent the partition results, which are clustered by DPC and C-means, respectively. Here, pseudo colors denote different clusters. The fifth column is the curves of CVIBI.

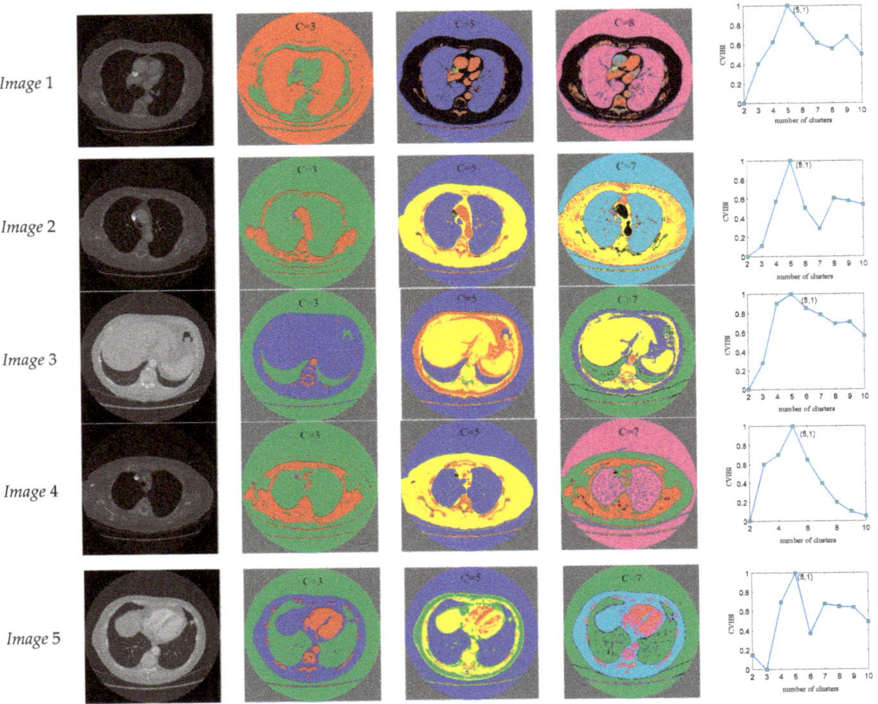

Figure 15. Tests of CVIBI on CT images with C-means.

Figures 15 and 16 show that we can obtain the optimal number of clusters when applying CVIBI with C-means and DPC to CT images, and the partitioned images show the shapes of various origins and tissues. Consequently, CVIBI with any clustering algorithms can take effect on automatic imaging segmentation in CT images, which can point out the categories of tissue in one particular CT layer.

Table 4 shows all evaluation results by using C-means and DPC with four indices, respectively, where x and y in sign image x/y refer to the investigated CT images and the correct number of clusters, respectively. The suggested numbers of clusters between the four validity indices are different; GS can identify the correct number of clusters no matter which clustering algorithm is applied; DC with C-means can identify the correct number of clusters, but DC with DPC fails for images 2 and 3; in terms of accuracy, CVIBI seems to be more efficient.

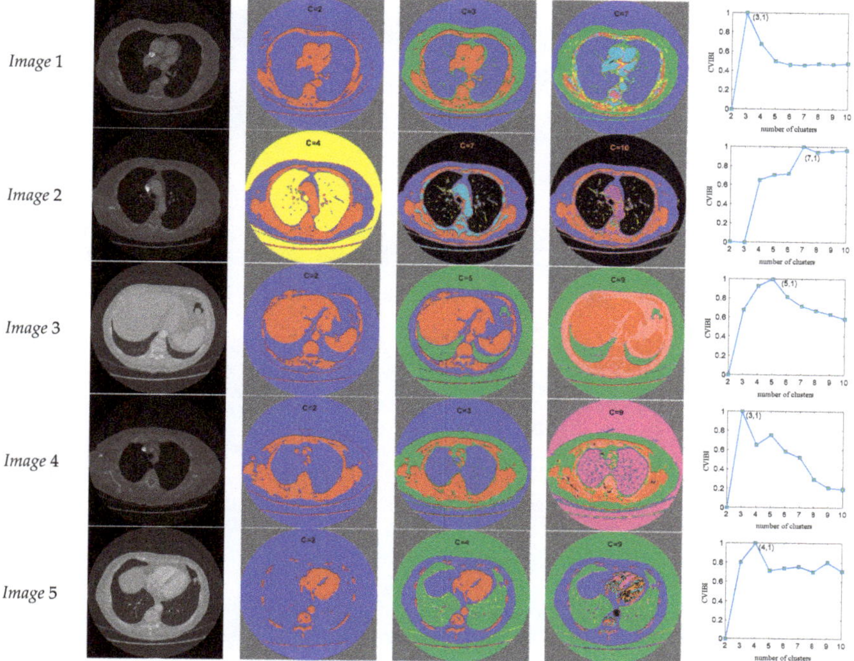

Figure 16. Tests of CVIBI on CT images with DPC.

Table 4. Evaluation of clustering results by CVIBI, DB, DC and GS for 15 datasets.

Dataset and Algorithm	DB		DC		GS		CVIBI	
	DPC	C-Means	DPC	C-Means	DPC	C-Means	DPC	C-Means
Set 1	4	5√	3√	5√	3√	5√	3√	5√
Set 2	2	3	7√	3	7√	5√	7√	5√
Set 3	2	4	5√	4	5√	5√	5√	5√
Set 4	2	5√	3√	5√	3√	5√	3√	5√
Set 5	4√	5√	4√	5√	4√	5√	4√	5√

5. Conclusions

The clustering evaluation is an essential but difficult task in clustering analysis. Currently, the existing validity evaluation has to depend on a specific clustering algorithm, a specific cluster parameter (or several), and specific assumptions, and has very limited applicable range. In this paper, we proposed a novel validity index, which can evaluate the clustering results obtained either by a single clustering algorithm or by several clustering algorithms. Especially, it can be applied to select any clustering parameters besides the typical number of clusters. To our knowledge, the kind of necessary applications cannot be realized by existing validity indices. This novel index outperforms the existing validity indices on some benchmark datasets in terms of accuracy and generality. Experimental results validate this index. The boundary matching degree and connectivity degree are important notions in graph theory. Our future work is to combine these notions with graph theory to reduce time complexity.

Author Contributions: Conceptualization, Q.L. and S.Y.; methodology, Q.L. and S.Y.; software, Q.L. and Y.W.; validation, Q.L., M.D. and J.L.; formal analysis, Z.W.; investigation, J.L. and Z.W.; resources, Z.W.; data curation,

Q.L. and M.D.; writing—original draft preparation, Q.L.; writing—review and editing, M.D. and J.L.; visualization, Y.W.; supervision, S.Y.; project administration, S.Y.; funding acquisition, S.Y. All authors have read and agreed to the published version of the manuscript.

Funding: This research was funded by the National Natural Science Foundation of China, grant number 61573251 and 61973232.

Conflicts of Interest: The authors declare no conflict of interest.

References

1. Wang, T.; Wang, X.; Zhao, Z.; He, Z.; Xia, T. Measurement data classification optimization based on a novel evolutionary kernel clustering algorithm for multi-target tracking. *IEEE Sens. J.* **2018**, *18*, 3722–3733. [CrossRef]
2. Nayak, P.; Vathasavai, B. Energy efficient clustering algorithm for multi-hop wireless sensor network using type-2 fuzzy logic. *IEEE Sens. J.* **2017**, *17*, 4492–4499. [CrossRef]
3. Dhanachandra, N.; Chanu, Y.J. A new image segmentation method using clustering and region merging techniques. In *Applications of Artificial Intelligence Techniques in Engineering*; Springer: Singapore, 2019; pp. 603–614.
4. Masud, M.A.; Huang, J.Z.; Wei, C.; Wang, J.; Khan, I. I-nice: A new approach for identifying the number of clusters and initial cluster centres. *Inf. Sci.* **2018**, *466*, 129–151. [CrossRef]
5. Wang, J.-S.; Chiang, J.-C. A cluster validity measure with a hybrid parameter search method for the support vector clustering algorithm. *Pattern Recognit.* **2008**, *41*, 506–520. [CrossRef]
6. Davies, D.L.; Bouldin, D.W. A cluster separation measure. *IEEE Trans. Pattern Anal. Mach. Intell.* **1979**, 224–227. [CrossRef]
7. Tibshirani, R.; Walther, G.; Hastie, T. Estimating the number of clusters in a data set via the gap statistic. *J. R. Stat. Soc. Ser. B (Stat. Methodol.)* **2001**, *63*, 411–423. [CrossRef]
8. Xie, X.L.; Beni, G. A validity measure for fuzzy clustering. *IEEE Trans. Pattern Anal. Mach. Intell.* **1991**, *13*, 841–847. [CrossRef]
9. Mehrjou, A.; Hosseini, R.; Araabi, B.N. Improved bayesian information criterion for mixture model selection. *Pattern Recognit. Lett.* **2016**, *69*, 22–27. [CrossRef]
10. Dasgupta, A.; Raftery, A.E. Detecting features in spatial point processes with clutter via model-based clustering. *J. Am. Stat. Assoc.* **1998**, *93*, 294–302. [CrossRef]
11. Teklehaymanot, F.K.; Muma, M.; Zoubir, A.M. Bayesian cluster enumeration criterion for unsupervised learning. *IEEE Trans. Signal Process.* **2018**, *66*, 5392–5406. [CrossRef]
12. Arbelaitz, O.; Gurrutxaga, I.; Muguerza, J.; PéRez, J.M.; Perona, I. An extensive comparative study of cluster validity indices. *Pattern Recognit.* **2013**, *46*, 243–256. [CrossRef]
13. Wang, Y.; Yue, S.; Hao, Z.; Ding, M.; Li, J. An unsupervised and robust validity index for clustering analysis. *Soft Comput.* **2019**, *23*, 10303–10319. [CrossRef]
14. Salloum, S.; Huang, J.Z.; He, Y.; Chen, X. An asymptotic ensemble learning framework for big data analysis. *IEEE Access* **2018**, *7*, 3675–3693. [CrossRef]
15. Chen, X.; Hong, W.; Nie, F.; He, D.; Yang, M. Spectral clustering of large-scale data by directly solving normalized cut. In Proceedings of the 24th ACM SIGKDD International Conference on Knowledge Discovery & Data Mining, London, UK, 19–23 August 2018.
16. Havens, T.C.; Bezdek, J.C.; Leckie, C.; Hall, L.O.; Palaniswami, M. Fuzzy c-means algorithms for very large data. *IEEE Trans. Fuzzy Syst.* **2012**, *20*, 1130–1146. [CrossRef]
17. Dhanachandra, N.; Manglem, K.; Chanu, Y.J. Image segmentation using K-means clustering algorithm and subtractive clustering algorithm. *Procedia Comput. Sci.* **2015**, *54*, 764–771. [CrossRef]
18. Bagirov, A.M.; Ugon, J.; Webb, D. Fast modified global k-means algorithm for incremental cluster construction. *Pattern Recognit.* **2011**, *44*, 866–876. [CrossRef]
19. Cai, W.; Chen, S.; Zhang, D. Fast and robust fuzzy c-means clustering algorithms incorporating local information for image segmentation. *Pattern Recognit.* **2007**, *40*, 825–838. [CrossRef]
20. Wu, K.-L.; Yang, M.-S. A cluster validity index for fuzzy clustering. *Pattern Recognit. Lett.* **2005**, *26*, 1275–1291. [CrossRef]

21. Ester, M.; Kriegel, H.-P.; Sander, J.; Xu, X. A density-based algorithm for discovering clusters in large spatial databases with noise. *KDD* **1996**, *96*, 226–231.
22. Ma, E.W.; Chow, T.W. A new shifting grid clustering algorithm. *Pattern Recognit.* **2004**, *37*, 503–514. [CrossRef]
23. Rodriguez, A.; Laio, A. Clustering by fast search and find of density peaks. *Science* **2014**, *344*, 1492–1496. [CrossRef] [PubMed]
24. Du, M.; Ding, S.; Xue, Y. A novel density peaks clustering algorithm for mixed data. *Pattern Recognit. Lett.* **2017**, *97*, 46–53. [CrossRef]
25. Ding, S.; Du, M.; Sun, T.; Xu, X.; Xue, Y. An entropy-based density peaks clustering algorithm for mixed type data employing fuzzy neighborhood. *Knowl. Based Syst.* **2017**, *133*, 294–313. [CrossRef]
26. Du, M.; Ding, S.; Jia, H. Study on density peaks clustering based on k-nearest neighbors and principal component analysis. *Knowl. Based Syst.* **2016**, *99*, 135–145. [CrossRef]
27. Krammer, P.; Habala, O.; Hluchý, L. Transformation regression technique for data mining. In Proceedings of the 2016 IEEE 20th Jubilee International Conference on Intelligent Engineering Systems (INES), Budapest, Hungary, 30 June–2 July 2016.
28. Yan, X.; Zhang, C.; Zhang, S. Toward databases mining: Pre-processing collected data. *Appl. Artif. Intell.* **2003**, *17*, 545–561. [CrossRef]
29. Ankerst, M.; Breunig, M.M.; Kriegel, H.-P.; Sander, J. OPTICS: Ordering points to identify the clustering structure. *ACM Sigmod Rec.* **1999**, *28*, 49–60. [CrossRef]
30. Hung, W.-L.; Yang, M.-S. Similarity measures of intuitionistic fuzzy sets based on Hausdorff distance. *Pattern Recognit. Lett.* **2004**, *25*, 1603–1611. [CrossRef]
31. Qian, R.; Wei, Y.; Shi, H.; Li, J.; Liu, J. Weakly supervised scene parsing with point-based distance metric learning. In Proceedings of the AAAI Conference on Artificial Intelligence, Honolulu, HI, USA, 27 January–1 February 2019.
32. Bandaru, S.; Ng, A.H.; Deb, K. Data mining methods for knowledge discovery in multi-objective optimization: Part A-Survey. *Expert Syst. Appl.* **2017**, *70*, 139–159. [CrossRef]
33. Yue, S.; Wang, J.; Wang, J.; Bao, X. A new validity index for evaluating the clustering results by partitional clustering algorithms. *Soft Comput.* **2016**, *20*, 1127–1138. [CrossRef]
34. Khan, M.M.R.; Arif, R.B.; Siddique, M.A.B.; Oishe, M.R. Study and observation of the variation of accuracies of KNN, SVM, LMNN, ENN algorithms on eleven different datasets from UCI machine learning repository. In Proceedings of the 2018 4th International Conference on Electrical Engineering and Information & Communication Technology (iCEEiCT), Dhaka, Bangladesh, 13–15 September 2018.
35. Nie, F.; Wang, X.; Huang, H. Clustering and projected clustering with adaptive neighbors. In Proceedings of the 20th ACM SIGKDD International Conference on Knowledge Discovery and Data Mining, New York, NY, USA, 24–27 August 2014.
36. Hubert, L.J. Some applications of graph theory to clustering. *Psychometrika* **1974**, *39*, 283–309. [CrossRef]
37. Yin, M.; Xie, S.; Wu, Z.; Zhang, Y.; Gao, J. Subspace clustering via learning an adaptive low-rank graph. *IEEE Trans. Image Process.* **2018**, *27*, 3716–3728. [CrossRef] [PubMed]

MDPI

St. Alban-Anlage 66

4052 Basel

Switzerland

Tel. +41 61 683 77 34

Fax +41 61 302 89 18

www.mdpi.com

Applied Sciences Editorial Office

E-mail: applsci@mdpi.com

www.mdpi.com/journal/applsci